Advances in Intelligent Systems and Computing

Volume 987

The series "Advances in Intelligent Systems and Computing" contains publications on theory, applications, and design methods of Intelligent Systems and Intelligent Computing. Virtually all disciplines such as engineering, natural sciences, computer and information science, ICT, economics, business, e-commerce, environment, healthcare, life science are covered. The list of topics spans all the areas of modern intelligent systems and computing such as: computational intelligence, soft computing including neural networks, fuzzy systems, evolutionary computing and the fusion of these paradigms, social intelligence, ambient intelligence, computational neuroscience, artificial life, virtual worlds and society, cognitive science and systems, Perception and Vision, DNA and immune based systems, self-organizing and adaptive systems, e-Learning and teaching, human-centered and human-centric computing, recommender systems, intelligent control, robotics and mechatronics including human-machine teaming, knowledge-based paradigms, learning paradigms, machine ethics, intelligent data analysis, knowledge management, intelligent agents, intelligent decision making and support, intelligent network security, trust management, interactive entertainment, Web intelligence and multimedia.

The publications within "Advances in Intelligent Systems and Computing" are primarily proceedings of important conferences, symposia and congresses. They cover significant recent developments in the field, both of a foundational and applicable character. An important characteristic feature of the series is the short publication time and world-wide distribution. This permits a rapid and broad dissemination of research results.

** **Indexing: The books of this series are submitted to ISI Proceedings, EI-Compendex, DBLP, SCOPUS, Google Scholar and Springerlink** **

More information about this series at http://www.springer.com/series/11156

Wojciech Zamojski · Jacek Mazurkiewicz ·
Jarosław Sugier · Tomasz Walkowiak ·
Janusz Kacprzyk

Editors

Engineering in Dependability of Computer Systems and Networks

Proceedings of the Fourteenth International Conference on Dependability of Computer Systems DepCoS-RELCOMEX, July 1–5, 2019, Brunów, Poland

 Springer

Editors
Wojciech Zamojski
Wrocław University of Science
and Technology
Wroclaw, Poland

Jacek Mazurkiewicz
Wrocław University of Science
and Technology
Wroclaw, Poland

Jarosław Sugier
Wrocław University of Science
and Technology
Wroclaw, Poland

Tomasz Walkowiak
Wrocław University of Science
and Technology
Wroclaw, Poland

Janusz Kacprzyk
Systems Research Institute
Polish Academy of Sciences
Warsaw, Poland

ISSN 2194-5357   ISSN 2194-5365   (electronic)
Advances in Intelligent Systems and Computing
ISBN 978-3-030-19500-7   ISBN 978-3-030-19501-4   (eBook)
https://doi.org/10.1007/978-3-030-19501-4

This Springer imprint is published by the registered company Springer Nature Switzerland AG
The registered company address is: Gewerbestrasse 11, 6330 Cham, Switzerland

Preface

We are pleased to present the proceedings of the Fourteenth International Conference on Dependability of Computer Systems *DepCoS-RELCOMEX* which took place in the Brunów Palace in Poland from 1 to 5 July 2019.

DepCoS-RELCOMEX is an annual conference series organized at Faculty of Electronics, Wrocław University of Science and Technology, and is devoted to problems of contemporary computer systems and networks, considered as complex systems, and their dependability, safety and security. The conference was started in 2006 by the Institute of Computer Engineering, Control and Robotics (CECR) and now is organized by the Department of Computer Engineering. Its idea came from the heritage of two other cycles of events: RELCOMEX (1977–1989) and Microcomputer School (1985–1995) which were organized by the Institute of Engineering Cybernetics (the previous name of CECR) under the leadership of Prof. Wojciech Zamojski, now also the DepCoS Chairman.

The proceedings of the previous DepCoS events were distributed (in historical order) by the IEEE Computer Society (2006–2009), Wrocław University of Technology Publishing House (2010–2012), and now by Springer in "Advances in Intelligent Systems and Computing" (AISC) volumes no. 97 (2011), 170 (2012), 224 (2013), 286 (2014), 365 (2015), 479 (2016), 582 (2017) and 761 (2018). Published by Springer Nature, one of the largest and most prestigious scientific publishers, the AISC series is one of the fastest growing book series in their programme. Its volumes are recognized in CORE—Computing Research and Education database, ISI Web of Science (now run by Clarivate Analytics), Scopus, Ei Compendex, DBLP, Google Scholar and SpringerLink, and many other indexing services around the world.

As the contents of these proceedings demonstrate, today's computer systems and networks are *really* complex and are applied in many different fields of contemporary life. Problems with their dependability are related to a broad variety of technologies or scientific disciplines which are seen in the diversity of paper topics. These kinds of systems cannot be interpreted only as (however complex and distributed) constructions built on the base of technical resources, but their analysis must take into account a unique blend of interacting people with their needs and behaviours, networks (with mobile activities, iCloud organization, Internet of Everything) and a large number of users dispersed geographically and producing an unimaginable number of applications interacting online. Dependability and performability came naturally as a contemporary answer to new challenges in the evaluation of reliability and efficiency in such environments. A growing number of research methods apply the latest results of artificial intelligence (AI) and computational intelligence (CI).

Dependability approach in theory and engineering of complex systems (not only computer systems and networks) is based on multidisciplinary approach to system theory, technology and maintenance. Dependability concentrates on efficient realization of tasks, services and jobs by a system which operates in real (and very often unfriendly) environment and is considered as a unity of technical, information and human assets—in contrast to "classic" reliability which is more restrained to analysis of technical resources (components and structures built from them). This difference has shaped the natural evolution of topics in subsequent DepCoS conferences which we have witnessed over the recent years and which can be seen also in the papers of this year's edition.

The Programme Committee of the conference, its organizers and the editors of this volume would like to gratefully acknowledge the participation of all reviewers who have evaluated conference submissions. We would like to thank, in alphabetic order, Andrzej Bialas, Ilona Bluemke, Eugene Brezhniev, DeJiu Chen, Manuel Gil Perez, Zbigniew Gomolka, Zbigniew Huzar, Igor Kabashkin, Vyacheslav Kharchenko, Wojciech Kordecki, Alexey Lastovetsky, Jan Magott, István Majzik, Jacek Mazurkiewicz, Marek Młyńczak, Yiannis Papadopoulos, Rafał Scherer, Mirosław Siergiejczyk, Robert Sobolewski, Janusz Sosnowski, Jaroslaw Sugier, Victor Toporkov, Tomasz Walkowiak, Bernd E. Wolfinger, Min Xie, Wojciech Zamojski and Wlodek Zuberek. Their cooperation has helped to refine the contents of this book and deserves our highest appreciation.

Last but not least, we would like to thank all the authors who decided to publish the results of their research with DepCoS conference. We hope that their works will contribute to new developments in design, implementation, maintenance and analysis of dependability aspects of computer systems and networks. Let the papers included in this volume create a valuable source material for scientists, researchers, practitioners and students who work in these areas.

The Editors

Organization

Fourteenth International Conference on Dependability of Computer Systems DepCoS-RELCOMEX

Brunów Palace, Poland, July 1–5, 2019

Programme Committee

Wojciech Zamojski (Chairman)	Wrocław University of Science and Technology, Poland
Ali Al-Dahoud	Al-Zaytoonah University of Jordan, Amman, Jordan
Andrzej Białas	Institute of Innovative Technologies EMAG, Katowice, Poland
Ilona Bluemke	Warsaw University of Technology, Poland
Wojciech Bożejko	Wrocław University of Science and Technology, Poland
Eugene Brezhniev	National Aerospace University "KhAI", Kharkiv, Ukraine
Dariusz Caban	Wrocław University of Science and Technology, Poland
De-Jiu Chen	KTH Royal Institute of Technology, Stockholm, Sweden
Frank Coolen	Durham University, UK
Mieczysław Drabowski	Cracow University of Technology, Poland
Francesco Flammini	University of Naples Federico II, Italy
Manuel Gill Perez	University of Murcia, Spain
Aleksander Grakowskis	Transport and Telecommunication Institute, Riga, Latvia
Zbigniew Huzar	Wrocław University of Science and Technology, Poland

Igor Kabashkin	Transport and Telecommunication Institute, Riga, Latvia
Janusz Kacprzyk	Polish Academy of Sciences, Warsaw, Poland
Vyacheslav S. Kharchenko	National Aerospace University "KhAI", Kharkiv, Ukraine
Mieczysław M. Kokar	Northeastern University, Boston, USA
Krzysztof Kołowrocki	Gdynia Maritime University, Poland
Wojciech Kordecki	The Witelon State University of Applied Sciences in Legnica, Poland
Leszek Kotulski	AGH University of Science and Technology, Krakow, Poland
Henryk Krawczyk	Gdansk University of Technology, Poland
Alexey Lastovetsky	University College Dublin, Ireland
Jan Magott	Wrocław University of Science and Technology, Poland
Istvan Majzik	Budapest University of Technology and Economics, Hungary
Henryk Maciejewski	Wrocław University of Science and Technology, Poland
Jacek Mazurkiewicz	Wrocław University of Science and Technology, Poland
Marek Młyńczak	Wroclaw University of Science and Technology, Poland
Yiannis Papadopoulos	University of Hull, UK
Ewaryst Rafajłowicz	Wrocław University of Science and Technology, Poland
Krzysztof Sacha	Warsaw University of Technology, Poland
Elena Savenkova	Peoples' Friendship University of Russia, Moscow, Russia
Rafał Scherer	Częstochowa University of Technology, Poland
Mirosław Siergiejczyk	Warsaw University of Technology, Poland
Czesław Smutnicki	Wrocław University of Science and Technology, Poland
Robert Sobolewski	Bialystok University of Technology, Poland
Janusz Sosnowski	Warsaw University of Technology, Poland
Jarosław Sugier	Wrocław University of Science and Technology, Poland
Victor Toporkov	Moscow Power Engineering Institute (Technical University), Russia
Tomasz Walkowiak	Wrocław University of Science and Technology, Poland
Max Walter	Siemens, Germany
Tadeusz Więckowski	Wrocław University of Science and Technology, Poland
Bernd E. Wolfinger	University of Hamburg, Germany

Min Xie   City University of Hong Kong, Hong Kong SAR,
                                       China
Irina Yatskiv   Transport and Telecommunication Institute, Riga,
                                       Latvia
Włodzimierz Zuberek   Memorial University of Newfoundland, St. John's,
                                       Canada

Organizing Committee

Chair

Wojciech Zamojski

Members

Jacek Mazurkiewicz
Jarosław Sugier
Tomasz Walkowiak
Mirosława Nurek

Contents

Water Quality Monitoring System Using WSN in Tanga Lake

Ali Al-Dahoud[1]([⊠]), Mohamed Fezari[2], and Hanene Mehamdia[2]

[1] Faculty of IT, Al-Zaytoonah University of Jordan, Amman, Jordan
aldahoud@zuj.edu.jo
[2] Laboratory of Automatic and Signals Annaba,
Badji Mokhtar Annaba University, BP 12, 23000 Annaba, Algeria
mouradfezari@yahoo.fr

Abstract. Continuous Lake water quality monitoring is an important tool to a different aspect which includes catchment management authorities, providing real-time data for environmental protection and tracking pollution sources in lakes around ANNABA reagent. In order to assist catchment managers to maintain the health of aquatic ecosystems taking into consideration a cost-effective water quality data collection a low-cost low power system is proposed. The proposed system is constituted by spreading the sensor nodes and the base station. The nodes and base stations are linked using WSN technology (e.g. Zigbee). Base stations will be connected to the Internet using the Ethernet shield. The results indicate that a reliable monitoring system can be designed with appropriate calibration of sensors; this will allow us to continue monitoring the quality of the water at higher spatial resolution.

1 Introduction

Lake Tonga is an area of unique international importance in the Mediterranean region. It has been listed on the Ramsar List since 1982 and is an integral part of the Biosphere Reserve. Located at 36° 51′N and 08° 30′E at the extreme north-east of the El Kala National Park (Wilaya of El Tarf) in Algeria.

Maintaining good water quality in rivers and streams benefits both humans and aquatic ecosystems. Currently, low-resolution water quality monitoring is conducted, and water samples are collected at regular periods for chemical analysis in the laboratory. The disadvantages of this approach are: (a) data collection is patchy in space and time, so sporadic pollution events can easily be missed; (b) it is time-consuming and expensive for personnel to collect water samples, return to laboratory to test and repeat the same procedure for different water resources; (c) there are certain biological and chemical processes such as oxidation-reduction potential that need to be measured on-site to ensure accuracy; (d) laboratory testing has a much slower turnaround time

This work is an extension of the project 3\14 – 2014 sponsored by Al-Zaytoonah University of Jordan.

W. Zamojski et al. (Eds.): DepCoS-RELCOMEX 2019, AISC 987, pp. 1–9, 2020.
https://doi.org/10.1007/978-3-030-19501-4_1

compared with on-site monitoring; (e) interpretation of data collected across different seasons is difficult, as the data is sparse both in space and time.

There are several systems that monitor water quality and report using telecommunication networks and different CPU. Rao et al. [5] demonstrated the use of WSN and GSM for tracking vehicles and mentioned the fact that it can be used for water quality measurement purposes. They used the Short Message Service (SMS) as a means to transfer data. Nasirudin et al. [6] used PIC16F876 as sensor node and GSM modems for updating the central database; measurements included temperature, pH, turbidity and DO. Wang et al. [7] used Atmel's AT91R40008 microprocessor as a sensor node and proposed Code Division Multiple Access (CDMA) based data transfer mechanism. Wang et al. in [7] measured DO, pH and conductivity for monitoring water quality and stores the data on integrated EEPROM. An example of the use of WiMAX is also found in [8], beside the node's communication, they implemented this long area network to provide data on the web in future; sensors included pH, conductivity, temperature, ORP, and DO as well.

Underwater acoustic communication networks, the cost and power consumption of underwater acoustic modem must come down. Commercial off the shelf (COTS) modems are not suitable for short range (100 m) underwater sensor nets since their power ranges and price points are designed for sparse, long range and expensive systems and this is an accepted fact that low power and low cost underwater acoustic modem is needed to monitor and surveying underwater ecological analyses [4].

The objective of our work is to develop a low-cost, wireless water quality monitoring system that aids in continuous measurements of water conditions. Each node contains five sensors and the nodes are placed in Plastic box overwater, some sensors have emerged into water such as pH, DO, Temperature, conductivity and Turbidity sensors. Information gathered from sensors is first processed by microcontroller and then the selected data is transmitted to the gateway via Xbee module. The base station is a personal computer with Graphic User Interface (GUI) developed using Matlab software. The communication of data between the gateway and base station is assured by a pair of Xbee-pro module. The GUI allows users to examine water quality data or alarm repeatedly when water quality detected when it is at below predetermined standards. The collected data is recorded on a hard disk; it can be analyzed using various software tools for future prediction and actions [1–3].

2 How Is Water Quality Measured?

In our sampling model, four dissolved parameters (Table 1) has been used; The pH values range from 7.10 to 7.43 and show that the water is slightly alkaline. As shown in Table 1. The water of the fourth station is relatively the most alkaline. The temperature varies from 20.9 to 21.4 °C and the third station has the highest value. On the other hand, the conductivity values range from 391.43 to 478.43 µS/cm; Station 3 shows the maximum value. However, dissolved oxygen varies from 3.66 to 5.13 mg/l; the third station is the richest.

Table 1. Five station parameters measurements from Lake Tonga done by biologists at Badji Mokhtar Annaba University.

Station parameters	S 1	S 2	S 3	S 4	S 5
pH	7.2 ± 0.5	7.10 ± 0.4	7.3 ± 0.5	7.5 ± 0.4	7.2 ± 0.5
Conductivity values (μS/cm)	391 ± 78.08	393.29 ± 86.81	478.4 ± 92	450 ± 7	411 ± 95
Temperature (°C)	20.9 ± 7	21.3 ± 7.8	21.4 ± 7.7	21 ± 7.5	21 ± 7
Oxygen concentration (mg/l)	4.47 ± 3	3.6 ± 2.71	5.2 ± 3.8	4.3 ± 3.2	4.6 ± 3.5

3   Wireless Sensor Network (WSN)

Water is a limited and essential resource for agriculture, industry and creature's existence on earth, including human beings. Monitoring of water quality is essential to control the biological characteristics of physical, chemical and water. For example, drinking water should not contain chemical substances that could be harmful to health; Water for agricultural irrigation should have low sodium content; Water for industrial uses should be low in some inorganic chemicals. In addition, monitoring of water quality can help in the detection of water pollution and releases of toxic substances.

A network of wireless water sensors (WWSN) prototype system developed for monitoring water quality in LVB is presented. Development was preceded by an environmental assessment, including the availability of cellular network coverage at the existing operating site. The system consists of micro-controller Arduino water quality sensors and a wireless network connection module. It detects water temperature, dissolved oxygen, pH and electrical conductivity in real time and disseminates information in graphical and tabular formats to stakeholders through a web portal and platforms, forms of mobile telephony.

Wireless Sensor Network (WSN) consists of a set of sensor nodes varying from a few tens of elements to several thousand, placed more or less randomly (for example by drop from a helicopter) in a geographical area called zone Catchment area, or area of interest, in order to monitor a physical phenomenon and to collect their data in an autonomous way. The sensor nodes use wireless communication to convey the captured data to a collector node called a sink node or base station. The well then transmits this data via the Internet or satellite to the "Task Manager" central computer to analyze these data and make decisions. Thus, the user can address requests to the other nodes of the network, specifying the type of data required, and then collecting the environmental data collected via the well node. Figure 1 shows the system design.

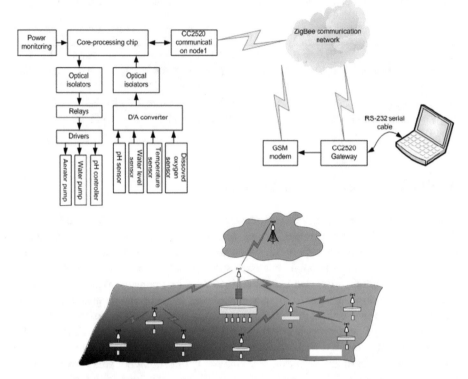

Fig. 1. System designed and synoptic of nodes distribution on work space.

This Synoptic focuses on the river and lake by monitoring water quality with low cost sensor nodes using the hierarchical communication structure. As a result, a large number of sensor nodes can be deployed to cover a large monitoring area with sufficient density.

The ultimate goal of this project is to implement a hierarchical water quality sensor network structure to reduce the cost of sensor arrays and increase the density of sensor node deployment. As the following figure shows, the array of sensors consists of a super-node and a number of small sensor nodes. Each small sensor node has a low-capacity panel and two low cost solar collectors (temperature sensor and dissolved oxygen sensor) connected and use a low ZigBee radio power for data transmission.

The super-node has a high-capacity solar panel and 4–5 connected commercial sensors connected and uses a powerful long-distance 802.11 Ethernet radio for data transmission. The whole network is divided into several groups depending on the intensity of the signal. Each group has a head node and the nodes of the cluster. The cluster head nodes send the collected data to the super-node (sink). All small sensor nodes use low-power Zigbee radios, and long-distance Ethernet radios are used between the super-node and the ground station.

3.1   Node Structure

Based on Fig. 2, our main sensor node contains the following sensors: DHT11 for temperature and humidity sensing, PH sensor, dissolution O_2 sensor, Conductivity sensors and a turbidity sensor. The main microcontroller used in the node is an Arduino "Arduino Mega" where the heart is an Atmega 2560 from Atmel. The transmission is provided by X-Bee Module using Zigbee protocol.

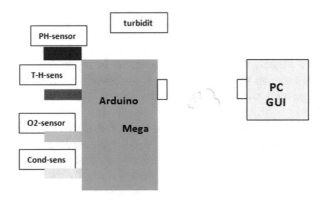

Fig. 2. Node Structure Design and conception with PC.

Used sensors: Analog pH meter Kit (SKU- SEN0161), DH Temperature and Humidity sensor, Turbidity Sensor, Conductivity sensor, O2 concentration sensor, Arduino Mega 2560 [9], Transmission Module.

Wireless Sensor Gateway Layer
The gateway is one of the most significant devices upon which the competence of the sensing action of a WSN depends. It collects useful data received from the nodes and makes this information obtainable usually via a wireless network. The conditions need a device intended approximately the subsequent constraints: low-power consumption, high storage capabilities, and pliable connectivity, low cost.

4   Experimental Results

In this part, a low-cost wireless water quality detection system is presented using Arduino Mega 2560. They measured the pH, light, turbidity, dissolved oxygen and temperature of the water. This work focuses on the use of several sensors to monitor and control the quality parameters of water of the lake in different areas in real time. The sensor nodes collect the water quality parameters and transmit them to the computer using Xbee modules. The software processing implemented on the host computer is monitored in real time and can analyze, calculate, display, print and process collected data from the implemented WSN. There are many platforms to the development of PC screen control such as C#, C++, and LabVIEW. In our study, MATLAB is chosen for the realization of software and the graphic user interface for human-machine interaction.

The monitoring center graphical interface allows the operator to monitor data obtained from the sensor nodes and to observe the behavior of the network in terms of the quality of the radio link of each sensor node state. And moreover, we used four nodes in the test, each installed in different zones within the lake. This system uses the following sensors:

- Temperature sensor DHT-11: to measure the temperature and humidity around the lake.
- 10 kΩ Potentiometer 10k: to simulate the variation of PH value. A designed Light sensor with photoresistance: to measure the turbidity of water.

In addition, sensor nodes have sensing elements to collect contextual information about the physical reality surrounding it. The nature of the information collected and the type of these elements strongly depend on the application in which the node is used. The designed sensor node can integrate more sensors of moisture, temperature and light.

Procedure: We carried out an experiment to emulate the turbidity of the water and the sensor used is a photoresist. For this purpose, we have carried out the assembly of Fig. 3, we have measured pure water, then gradually we added black color and we have noted the result of the variation of turbidity sensors in the following Table 2. We also took pictures illustrating our experimentation (see Fig. 3).

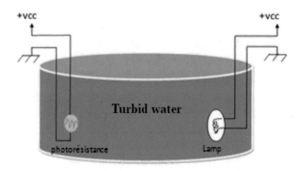

Fig. 3. Figure of designed turbidity sensor.

Table 2. Results of experiment on designed turbidity sensor

Number of dye strokes in 2L water	1	2	3	4	5	6	7	8
Measured resistance (kΩ)	6.320	9.90	12.38	26.27	61.66	87.38	142.30	194.20

The software part is also shared between the wireless sensor nodes and the server PC. The communication between the PC and the Arduino card is done via the USB port, by first installing an adapted driver (supplied by ARDUINO).

At the level of the wireless sensor node, two programs have been implemented allowing: sampling data from different sensors collect data from the sensor node that will be transferred to a computer source.

In the server PC, using the MATLAB software, an interface is used to control the nodes of wireless sensors and to receive the data sent by the sensor node via the wireless transmission module.

At the PC levels, using a Matlab program a graphical user interface was created for the water quality, water quality parameter that will be tracked in this project are temperature, environmental moisture, water turbidity, and pH. The GUI is designed with appropriate characteristics for monitoring purposes. The Graphical User Interface (GUI) has three main components which are a windowing system, an imaging model, and the application program interface. The application interface is the process where the user specifies how the windows and graphics appear on the screen. The Graphical user interface also has the ability to move data from one application to another i.e. store the data in Microsoft-Excel tables. A graphical user interface (GUI) has its own standard format to represent text and graphics. Technology and wireless Internet can make the system more reliable as users can access data no matter where they are.

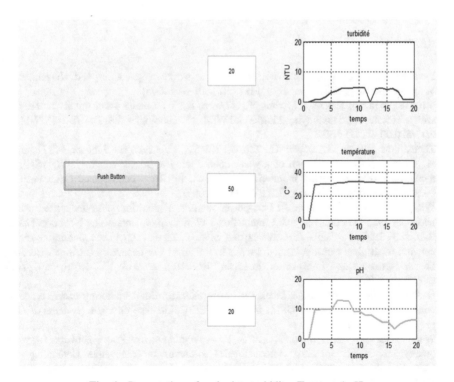

Fig. 4. Presentation of node data turbidity, Temp and pH.

The data acquisition of turbidity data, temperature and pH of the water can be represented on the graphical interface shown in Fig. 4. Furthermore, it shows the GUI of the real-time on simulation tests.

The acquisition of turbidity, temperature and water data was recorded in real-time. Figure 4 shows the recorded data during the testing process.

5 Conclusion

In this work, we proposed a river and lake water quality monitoring system based on the wireless sensor network that contributes in remote monitoring of the quality of water. We have studied and designed a system to monitor and analyze water lakes using adapted nodes in a wireless sensor network. These nodes can capture and detect quality parameters such as temperature, pH, turbidity, conductivity, etc.

We chose to work with a mega Arduino module for processing and acquisition. Using XBee modules for transmission and communication between the PC and the Arduino card. Furthermore, we developed the essential software for both the sensor node and the base station. The tests and experiments show the relative efficiency of our system despite the average price of these modules and the simplicity of the implementation. Taking into consideration, several constraints had to be met in order to achieve our objective.

References

1. Puccinelli, D., Haenggi, M.: Wireless sensor networks: applications and challenges of ubiquitous sensing. IEEE Circ. Syst. Mag. 5(3), 19–31 (2005)
2. Silva, S., Nguyen, H.N.G., Tiporlini, V., Alameh, K.: Web based water quality monitoring with sensor network: employing ZigBee and WiMax Technologies (2011). 978-1-4577-1169-5/11/$26.00 ©2011 IEEE
3. Zennaro, M., Floros, A., Dogan, G., Tao, S., Zhichao, C., Chen, H., Bahader, M., Ntareme, H., Bagula, A.: On the design of a water quality wireless sensor network (WQWSN): an application to water quality monitoring in Malawi. In: IEEE International Conference on Parallel Processing Workshops (ICPPW 2009) (2009)
4. Wills, J., Ye, W., Heidemann, J.: Low-power acoustic modem for dense underwater sensor networks. In: Proceedings of ACM International Workshop on Underwater Networks (2006)
5. Rao, A.S., Izadi, D., Tellis, R.F., Ekanayake, S.W., Pathirana, P.N.: Data monitoring sensor network for BigNet research testbed. In: 5th International Conference on Intelligent Sensors, Sensor Networks and Information Processing (ISSNIP), pp. 169–173. IEEE, New York (2009)
6. Nasirudin, M.A., Za'bah, U.N., Sidek, O.: Fresh water real-time monitoring system based on wireless sensor network and GSM. In: 2011 IEEE Conference on Open Systems (ICOS), pp. 354–357 (2011)
7. Wang, J., Li Ren, X., Li Shen, Y., Liu, S.-Y.: A remote wireless sensor networks for water quality monitoring. In: 2010 International Conference on Innovative Computing and Communication and 2010 Asia-Pacific Conference on Information Technology and Ocean Engineering (CICC-ITOE), pp. 7–12 (2010)

8. Silva, S., Nguyen, H.N., Tiporlini, V., Alameh, K.: Web based water quality monitoring with sensor network: employing ZigBee and WiMax technologies. In: IEEE High Capacity Optical Networks and Enabling Technologies (HONET), pp. 138–142 (2011)
9. Arduino: Arduno mega 2560 (2015). http://arduino.cc/en/Main/arduinoBoardMega2560,Mai

A Comparative Study of Statistical and Neural Network Models for PLC Network Traffic Anomaly Detection

Tomasz Andrysiak[⊠] and Łukasz Saganowski

Institute of Telecommunications and Computer Science,
Faculty of Telecommunication, Information Technology and Electrical
Engineering, UTP University of Science and Technology, Kaliskiego 7,
85-789 Bydgoszcz, Poland
{andrys, luksag}@utp.edu.pl

Abstract. Protection of systems and computer networks against novel, unknown attacks is currently an intensively examined and developed domain. One of possible solutions to the problem is detection and classification of abnormal behaviors reflected in the analyzed network traffic. In the presented article we attempt to resolve the problem by anomaly detection in the analyzed network traffic described with the use of three different models. We tested two class of models which differed in prediction. The first sorts was composed of ARFIMA and Holt-Winters models which are characterized by statistical dependences. The second sorts, on the other hand, included neural network auto-regression model which are characterized by single hidden layer and lagged inputs for forecasting univariate time series. In order to detect anomalies in the network traffic we used differences between real network traffic and its estimated model. The experiment results confirmed efficiency and effectiveness of the presented method.

Keywords: Anomaly detection · Statistical models ·
Neural network auto-regression model · Network traffic prediction

1 Introduction

For many years, there have been used safety systems based on formerly isolated and classified patterns of threats, named signatures. Anti-virus software, systems for detection and breaking-in counteraction and protection against information leaks are just examples from a long and diversified list of application of those techniques. Nevertheless, there is one aspect in common, namely, they are able to protect systems and computer networks from known attacks described by the mentioned patterns. However, does lack of traffic matching known signatures mean there is no threat?

A means to defend from novel, unknown attacks is a rather radical change in operation concept. Instead of searching for attacks' signatures in network traffic it is necessary to browse for abnormal behavior which is a deviation from the normal traffic characteristic. The strength of such an approach is visible in solutions which are not

© Springer Nature Switzerland AG 2020
W. Zamojski et al. (Eds.): DepCoS-RELCOMEX 2019, AISC 987, pp. 10–20, 2020.
https://doi.org/10.1007/978-3-030-19501-4_2

based on knowledge a priori of attacks' signatures but on what does not respond particular norms, profiles of the analyzed network traffic [1, 2].

The strength of such an approach is protection against so far unknown attacks, specially developed (targeted attacks), to attacks directed onto an information system or simply forming so called zero-day exploits. In such environments, anomaly detection systems may play a special role. Their task is then to detect (for the purposes of automatic reaction) not typical behaviors of the network traffic being symptoms of unauthorized activities directed onto the secured information resources [3, 4].

Currently intensively examined and developed methods for detection of intrusion/attacks are those utilizing phenomenon called anomaly in network traffic [5–7]. One of possible solutions is detection of abnormal behavior by means of statistical models, which describe the analyzed network traffic. The most often used are autoregressive models or exponential smoothing models [8, 9]. They allow for estimation of the characteristics of the analyzed network traffic. All the mentioned solutions require measure of similarity between the model of the network traffic and its real conduct. In literature, there can also be found methods included neural network for forecasting network traffic volatility [10].

In the present article we propose using estimation of statistical models (ARFIMA, Holt-Winters) and neural network auto-regression model for defined behavior profiles of a given smart lightning network traffic. The process of anomaly detection consisted in a comparison between the parameters of normal behavior (predicted on the basis of the tested models) and parameters of real network traffic.

This paper is organized as follows. After the introduction, in Sect. 2, the overview of security risks in smart lighting infrastructure is presented. In Sect. 3, there are generally discussed the statistical and neural network models used in methodology of anomaly detection system. Then, in Sect. 4, the experimental results of the proposed solution are shown. Conclusions are given thereafter.

2 Overview of Security Risks in Smart Lighting Infrastructure

In Smart Cities, security of critical infrastructures is essential for providing confidentiality, accessibility, integrity, and stability of the transmitted data. The use of advance digital technologies, which connect more and more complicated urban infrastructure, is risky because there may appear different types of abuses which may hamper or completely disenable proper functioning of a Smart City. Undoubtedly, one of the biggest frailty of a Smart City is the Smart Lighting system when taking into account the size of area where it functions, potentially big number of the system's devices, and the generated operational costs. Therefore, providing a proper level of security and protection becomes a crucial element of the Smart Lighting Communication Network (SLCN) solutions [11].

The task of a Smart Lighting system is not only lightning the streets. Depending on the kind of pavement, it must control the brightness of the lighting, its dimming, homogeneity, reflectivity, providing drivers and pedestrians with maximum safe visibility. Therefore, lighting installations with luminaires, in which they are used as light sources, must be easily controllable. Such controlling may include whole groups or even individual lamps, which may be turned on or off according to a specified schedule, dimmed up to any degree at specified times and the state of individual devices must be easy to control. In comparison to a traditional, autonomic lighting system, the Smart Lighting solutions are characterized by much bigger functionality and flexibility, however, due to their intelligent nature they may be liable to different type of abuses (attacks). Such actions may be realized by both the sole receiver of the service and intruders wanting to enforce a specific state of infrastructure [12, 13].

The receiver most often causes destructive actions to the SLCN, which interfere with the transmission of control signals (by active or passive influence) to achieve a change in period and/or intensity of the light. Increasing the intensity of lighting in front of the receiver's property allows for switching off the light on his land, which may result in significant economic benefits. However, a much bigger problem seems to be protection against intended attacks. There are numerous reasons for performing such attacks, the main of which is disturbing the controlling system in order to set a different value of lighting than the one established by the operator. Switching off the light or reduction of its intensity in some area may facilitate criminal proceedings. Another reason is malicious activity consisting in hindering the lives of neighbors or local authorities by forcing a change in the schedule of lighting (for instance, switching off the light at night, or turning it on during the day). However, a much more serious challenge seems to be protection against attacks realized for criminal purposes. Then, every potential smart lighting lamp may become a point by means of which an attack onto SLCN may be performed [14, 15].

Such actions, in particular in the area of the last mile, may have a conscious or unconscious nature. The unconscious interference most often happens when the low voltage (LV) network feeds both the streets and the users' households, where the included loads do not meet electromagnetic compatibility standards. The conscious form of interference in the communication system is related to deliberate activity that consists in switching into the SLCN infrastructure such elements as: capacitors, interfering generators or terminals emulating a hub. Loading such devices, even in LV networks dedicated only to the lighting, is not difficult, and using them by an intruder may remain unnoticed for a longer time [16].

Smart Lighting network security and protection from such attacks seems a harder task to solve then prevention from possible abuses (to achieve quantifiable but limited economic benefits) from the receivers' side.

3   The Proposed Solution - Predictive Anomaly Detection System

In this section we presented block scheme of proposed anomaly/attack detection method for Power Line Communication (PLC) smart lights network (Fig. 1). PLC traffic from smart lights network is acquired by means of traffic concentrator which plays also a role of gateway and translator to IP network. As a first step in our method we extracted PLC traffic features and then we calculate traffic features to achieve those presented in Table 1.

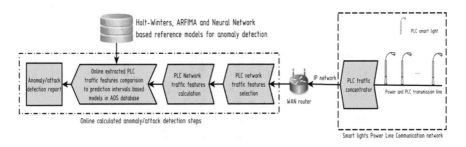

Fig. 1. Block scheme of proposed anomaly/attack detection method.

In the next step online extracted traffic features are compared to models based on prediction intervals stored in anomaly detection system (ADS) database models. ADS database models were calculated based on prediction intervals achieved for Holt-Winters, ARFIMA statistical models and Autoregressive Neural Network. PLC traffic for models calculation was gathered for 6 weeks with an assumption that there is no any anomalies in this period of time. Every PLC traffic features from Table 1 are compared to associated to this feature model. As an effect of comparisons anomaly/attack detection report is generated.

3.1 Holt-Winters Model

The Holt-Winters models belong to the most often used adaptive models based on exponential smoothing [17, 18]. They are utilized to predict variability of network traffic with trend and seasonal variations. There are two forms characterizing the seasonality of the model: additive and multiplicative.

The additive form of seasonality is as follows

$$l_t = \alpha(y_t - s_{t-m}) + (1 - \alpha)(l_{t-1} + b_{t-1}), \tag{1a}$$

$$b_t = \beta(l_t - l_{t-1}) + (1 - \beta)b_{t-1}, \tag{1b}$$

$$s_t = \gamma(y_t - l_t - b_{t-1}) + (1 - \gamma)s_{t-m}, \tag{1c}$$

$$\widehat{y}_{t+h|t} = l_t + b_t h + s_{t-m+h_m^+} \tag{1d}$$

and in the multiplicative form

$$l_t = \alpha\frac{y_t}{s_{t-m}} + (1 - \alpha)(l_{t-1} + b_{t-1}), \tag{2a}$$

$$b_t = \beta(l_t - l_{t-1}) + (1 - \beta)b_{t-1}, \tag{2b}$$

$$s_t = \gamma\frac{y_t}{l_{t-1} - b_{t-1}} + (1 - \gamma)s_{t-m}, \tag{2c}$$

$$\widehat{y}_{t+h|t} = (l_t + b_t h)s_{t-m+h_m^+}, \tag{2d}$$

where y_t i $\widehat{y}_t$ are, respectively, the value of the variable over time t and its prediction, $\alpha \in [0, 1]$ is a parameter called the smoothing coefficient, b_t is a weighted average of the increase in moment $t - 1$, s_t is the seasonal component, m is the period of seasonal variations, $h_m^+ = [(h-1) mod\ m] + 1$, a $\beta, \gamma \in [0, 1]$ is a parameter.

In exponential smoothing, trend is a combination of the level l, and an increase b. If in the above models we take $\alpha = 0$, then we obtain a level constant over time, $\beta = 0 -$ increase constant over time, and when $\gamma = 0$ – a constant over time seasonal component [19].

In the Holt-Winters model, we analyze time series with seasonal variations which may overlap the trend in an additive or multiplicative way. The multiplicative version is applicable when the assumption that the relative increments of the trend's variable value (beyond the time periods of change or trend breakdown) are approximately constant or change on regular basis [20].

3.2   ARFIMA Model

The Autoregressive Fractional Integrated Moving Average model called ARFIMA (p, d, q) is a combination of Fractional Differenced Noise and Auto Regressive Moving Average which are proposed by Grange, Joyeux and Hosking, in order to analyze the Long-Memory property [21, 22].

The ARFIMA (p, d, q) model for time series y_t is written as

$$\Phi(L)(1 - L)^d y_t = \Theta(L) \in_t, t = 1, 2, \ldots \Omega, \tag{3}$$

where y_t is the time series, $\in_t \sim (0, \sigma^2)$ is the white noise process with zero mean and variance σ^2, $\Phi(L) = 1 - \phi_1 L - \phi_2 L^2 - \cdots - \phi_p L^p$ is the autoregressive polynomial and $\Theta(L) = 1 + \theta_1 L + \theta_2 L^2 + \cdots + \theta_q L^q$ is the moving average polynomial, L is the backward shift operator and $(1 - L)^d$ is the fractional differencing operator given by the following binomial expansion

$$(1 - L)^d = \sum_{k=0}^{\infty} \binom{d}{k}(-1)^k L^k \tag{4}$$

and

$$\binom{d}{k}(-1)^k = \frac{\Gamma(d+1)(-1)^k}{\Gamma(d-k+1)\Gamma(k+1)} = \frac{\Gamma(-d+k)}{\Gamma(-d)\Gamma(k+1)}. \tag{5}$$

$\Gamma(*)$ denotes the gamma function and parameter d is the number of differences required to give o stationary series and $(1 - L)^d$ is the dth power of the differencing operator. When $d \in (-0,5; 0,5)$, the ARFIMA (p, d, q) process is stationary, and if $d \in (0, 0, 5)$ the process presents long-memory behaviour [21, 23].

Forecasting ARFIMA processes is usually carried out by using an infinite autoregressive representation of (3), written as $\Pi(L)y_t = \epsilon_t$, or

$$y_t = \sum_{i=1}^{\infty} \pi_i y_{t-i} + \epsilon_t, \tag{6}$$

where $\Pi(L) = 1 - \pi_1 L - \pi_2 L^2 - \ldots = \Phi(L)(1 - L)^d \Theta(L)^{-1}$.

In terms of practical implementation, this form needs truncation after k lags, but there is no obvious way of doing it. This truncation problem will also be related to the forecast horizon considered in predictions. From (6) it is clear that the forecasting rule will pick up the influence of distant lags, thus capturing their persistent influence. However, if a shift in the process occurs, this means that pre-shift lags will also have some weight on the prediction, which may cause some biases for post-shift horizons [21, 24].

3.3   Neural Network Auto-Regression Model

The nonlinear autoregressive model of order p, $NAR(p)$, defined as

$$y_t = h\big(y_{t-1}, \ldots, y_{t-p}\big) + \epsilon_t \tag{7}$$

is a direct generalization of linear AR model, where $h(\cdot)$ is a nonlinear known function [23]. It is assumed that $\{\epsilon_t\}$ is a sequence of random independent variables and identically distributed with zero mean and finite variance σ^2. The autoregressive neural network (NNARM), is a feedforward network constitutes a nonlinear approximation $h(\cdot)$, which is defined as

$$\widehat{y}_t = \widehat{h}\big(y_{t-1}, \ldots, y_{t-p}\big), \widehat{y}_t = \beta_0 + \sum_{i=1}^{I} \beta_i f\left(\alpha_i + \sum_{j=1}^{P} \omega_{ij} y_{t-j}\right), \tag{8}$$

where $f(\cdot)$ is the activation function and $\Theta = (\beta_0, \ldots, \beta_I, \alpha_1, \ldots, \alpha_I, \omega_{11}, \ldots, \omega_{IP})$ is the parameters vector, p denotes the number of neurons in the hidden layers [25].

The NNAR model is a parametric non-linear model of forecasting. Forecasting in this model is performed in two steps. In the first step, the order of auto-regression is determined for the given time series. The order of auto-regression indicates the number of previous values which the current value of the time series is dependent upon. In the second step, the NN is trained with a training set prepared by taking the order of auto-regression into consideration. The number of input nodes is determined from the order of auto-regression and the inputs to the NN are the previous, lagged observations in univariate time series forecasting. The predicted values are the output of the NN model. The number of hidden nodes is often decided by trial-and-error or through experimentation due to lack of any theoretical basis for selection [26]. The number of iteration should be selected properly in order to avoid the problem of over fitting.

4 Experimental Results

Experimental results were achieved based on PLC traffic taken from smart lights network. Test network consisted of 108 smart lamps using common transmission medium. For PLC common medium is a power line where concurrently power and transmission is provided to the smart lamps.

In PLC network there are two type of nodes: smart lamp node and PLC concentrator. Traffic is collected by means of traffic concentrator which is also a gateway between PLC network and IP network. We acquire PLC traffic from IP based interface of PLC concentrator. In the next step we extract PLC communication traffic features. We extracted traffic features presented in Table 1. We gathered traffic features mainly connected with PLC signal parameters such as PF_1 – RSSI (received signal strength indication for PLC node), PF_2 – SNR (signal-to-noise ratio), PF_3 – PER (packet error rate per time interval) or with parameters related to transmission protocol used in common communication medium for example: PF_5 – NNG (Number of neighbors for a given PLC node per time interval) or PF_6 – PR: Number of packet retransmissions for a given PLC node.

Table 1. PLC traffic feature description.

PLC feature	PLC traffic feature description
PF_1	RSSI: received signal strength indication for PLC node in [dBm]
PF_2	SNR: signal-to-noise ratio in [dBu]
PF_3	PER: packet error rate per time interval in [%]
PF_4	PPT: number of packets per time interval
PF_5	NNG: Number of neighbors for a given PLC node per time interval
PF_6	PR: Number of packet retransmissions by means of PLC communication (for a given PLC node)
PF_7	ACC: number of ACK/CANCEL copies received by PLC node (e.g. concentrator)

Traffic features from Table 1 are converted into a form of time series which are used for prediction in order to estimate variability of a given traffic feature. We have to choose model that is appropriate for our time series characteristic where for example there is no seasonality for time series representing PF_2 feature. Prediction intervals achieved for PF_2 feature by means of neural network auto-regression model is presented in Fig. 2.

We can see in Fig. 2 30 sample prediction period and two prediction intervals representing narrower 80% and wider 95% confidence level. These confidence levels are used to build models of traffic variability. For every traffic feature from Table 1 we calculate prediction intervals from PLC traffic. We assume that during model building there is no anomalies in captured PLC traffic. PLC traffic models are calculated based on traffic from 6 week period of time.

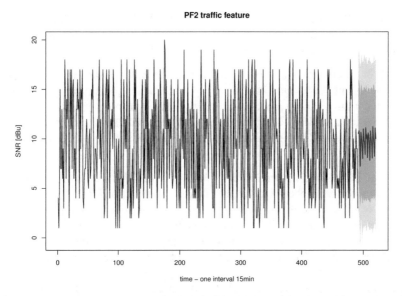

Fig. 2. Prediction intervals (narrower 80% and wider 95% confidence level interval) achieved for example PLC traffic feature SNR [dBu] by means of neural network auto-regression model

During normal work of proposed anomaly detection method we extract every traffic feature from Table 1 and check if online obtained values of traffic features are within prediction intervals set by model. If online extracted traffic features are outside prediction intervals set by models we indicate possible anomaly or attack.

Table 2. DR[%] and FP[%] results for three methods of anomaly or attacks generation.

PLC feature	DR[%]			PLC Feature	FP[%]		
	Holt-Winters	ARFIMA	NNARM		Holt-Winters	ARFIMA	NNARM
PF_1	92.23	94.52	97.38	PF_1	4.12	4.53	3.45
PF_2	93.12	95.22	98.14	PF_2	4.21	3.58	2.28
PF_3	93.36	94.55	96.72	PF_3	5.45	5.32	3.98
PF_4	89.14	93.75	96.45	PF_4	5.56	5.74	4.02
PF_5	88.81	92.57	97.44	PF_5	6.25	5.32	3.52
PF_6	87.46	91.38	95.62	PF_6	6.34	5.27	4.21
PF_7	90.74	94.56	96.67	PF_7	6.67	6.14	4.11

In order to evaluate efficiency of proposed method we simulated some attacks that have an impact on different traffic features from Table 1. We performed subsequent types of attacks or anomalies described below by Method 1 – Method 3:

Method 1: First method is realized by means of hardware modifications or impact on transmission quality by disturbing electrical parameters of PLC transmission. We connected for example in parallel additional capacitor (e.g. 4.7uF) near PLC modems in order to attenuate usable signal. Other method of disturbing transmission was injection of conducted disturbances into power line with frequencies from bandwidth used by PLC modem for transmission.

Method 2: Interference of communication process by installing additional PLC modem node, or concentrator that disturbs communication process between lamps, take a part in communication process and change PLC packets, generate random PLC packets. Such an activities has a big impact on communication process between PLC transmission nodes. Effect of this kind of attack can be seen in data link and network layer.

Method 3: Influence on communication process by adding to network new PLC node that replay received packets and transmit them to other transmission nodes or different segments of PLC network. Another type of performed attack was utilizing additional PLC nodes as a devices that take a part in communication process and transmit packets to other "fake" PLC node by transmission tunnel. This methodology of generating anomaly or attack will have an impact data link, network layer and also application layer.

Cumulative results of DR[%] and FP[%] that takes into consideration anomaly or attack detected by all extracted traffic features are presented in Table 2. We compare two statistical models Holt-Winter, ARFIMA and neural network auto-regression model. Based on traffic taken from our test network we achieved the best results by means of neural network auto-regression model where DR[%] changes from 95.62–98.14% while FP[%] 2.28–4.21%. Slightly worse results we achieved with ARFIMA model were DR[%] changes from 91.38–95.22% while FP[%] 3.58–6.14%.

We can also observe some coincidences between different methods of attack. For example hardware modifications and injected disturbances described in method 1 has an impact on communication process (routing protocol) between PLC nodes. In this case we can notice indirect impact on method 3 of attack, where we disturb routing mechanism by adding unauthorized PLC node. We can also notice similar effect for different PLC traffic features. When for instance signal-to-noise ratio PF_2 getting worse then packet error rate (PER) per time interval PF_3 and Number of packet retransmissions (PR) by means of PLC communication PF_6 also increases.

Achieved results in terms of DR and FP shows us that proposed models for detecting anomalies or attacks are promising (especially neural network auto-regression model) and can be used for anomaly detection purposes. FP values less than 5% are acceptable for anomaly detection class system and also for PLC based smart light network. In article we presented anomaly detection solution for PLC based smart lights network which is a subset of Internet of Things (IOT) applications. Smart light network are important for public safety and transport safety that's why such a solutions have to be implemented for existing and new installations.

5 Conclusion

Monitoring and protection of IT systems infrastructure against new, unknown attacks is currently an intensively examined and developed issue. An increasing number of new attacks, globalization of their range and growing complexity level enforces constant development of network protection systems. The most often implemented mechanism to ensure this safety are methods of detection and classification of abnormal behavior reflected in the analyzed network traffic. An advantage of such an approach is no necessity to prior defining and remembering patterns of such behaviors (abuse signatures). Thus, in the decision-making process it is only required to define what is and what is not an abnormal behavior in the network traffic in order to detect a potential unknown attack (an abuse).

In this article, we proposed solution of anomaly and attack detection algorithm for smart lights network based on PLC transmission. To detect the network traffic anomalies, there were used differences between the real network traffic and an estimated model of this traffic for the analyzed network parameters. We evaluated Holt-Winters, ARFIMA statistical models and neural network solution. Parameter estimations of the using models were realized on the basis of methodology described by Hyndman-Khandakar. Models of these three solutions were built taking into consideration prediction intervals achieved for PLC traffic without anomalies and attacks. We tested our solution with three different anomaly and attack scenarios. The best results were achieved for neural network solution where prediction intervals slightly better and faster approximate characteristic of PLC traffic. Efficiency of ARFIMA model was approximately 3% worse in case of detection rate and false positive. Worst results were achieved for Holt-Winters statistical model without long memory properties.

References

1. Esposito, M., Mazzariello, C., Oliviero, F., Romano, S.P., Sansone, C.: Evaluating pattern recognition techniques in intrusion detection systems. In: Proceedings of the 5th International Workshop on Pattern Recognition in Information Systems, PRIS 2005, In conjunction with ICEIS 2005, Miami, FL, USA, pp. 144–153, May 2005
2. Chondola, V., Banerjee, A., Kumar, V.: Anomaly Detection: a Survey. ACM Comput. Surv. **41**(3), 1–72 (2009)
3. Jackson, K.: Intrusion Detection Systems (IDS). Product Survey. Los Alamos National Library, LA-UR-99-3883 (1999)
4. Lim, S.Y., Jones, A.: Network anomaly detection system: the state of art of network behavior analysis. In: Proceedings of the 2008 International Conference on Convergence and Hybrid Information Technology, pp. 459–465 (2008)
5. Patcha, A., Park, J.M.: An overview of anomaly detection techniques: existing solutions and latest technological trends. Comput. Netw. **51**(12), 3448–3470 (2007)
6. Wei, L., Ghorbani, A.: Network anomaly detection based on wavelet analysis. In: EURASIP Journal on Advances in Signal Processing, vol. 2009 (2009)
7. Lakhina, A., Crovella, M., Diot, C.H.: Characterization of network-wide anomalies in traffic flows. In: Proceedings of the 4th ACM SIGCOMM Conference on Internet Measurement, pp. 201–206 (2004)

8. Yaacob, A., Tan, I., Chien, S., Tan, H.: Arima based network anomaly detection. In: Proceedings of the 2nd International Conference on Communication Software and Networks IEEE, pp. 205–209, (2010)

9. Zhou, Z.G., Tang, P.: Improving time series anomaly detection based on exponentially weighted moving average (EWMA) of season-trend model residuals. In: Proceedings of the 2016 IEEE International Geoscience and Remote Sensing Symposium (IGARSS), pp. 3414–3417 (2016)

10. Amini, M., Jalili, R., Shahriari, H.R.: RT-UNNID: a practical solution to real-time network-based intrusion detection using unsupervised neural networks. Comput. Secur. 25, 459–468 (2006)

11. Azkuna, I.: Smart Cities Study: International study on the situation of ICT, innovation and Knowledge in cities, The Committee of Digital and Knowledge-based Cities of UCLG, Bilbao (2012)

12. Mitchell, W.J.: Intelligent cities. Universitat Oberta de Catalunya (UOC) Papers: E-Journal on the Knowledge Society, no. 5. (2007). https://www.uoc.edu/uocpapers/5/dt/eng/mitchell.pdf

13. Wu, Y., Shi, Ch., Zhang, X., Yang, W.: Design of new intelligent street light control system. In: 8th IEEE International Conferences on Control and Automation (ICCA), pp. 1423–1427 (2010)

14. Abouzakhar, N.: Critical infrastructure cybersecurity: a review of recent threats and violations. In: 12th European Conference on Cyber Warfare and Security, pp. 1–10 (2013)

15. Elmaghraby, A.S., Losavio, M.M.: Cyber security challenges in smart cities: Safety, security and privacy. J. Adv. Res. 5(4), 491–497 (2014)

16. Kiedrowski, P.: Toward more efficient and more secure last mile smart metering and smart lighting communication systems with the use of PLC/RF hybrid technology. Int. J. Distrib. Sens. Netw. 2015, 1–9 (2015)

17. Holt, C.C.: Forecasting seasonals and trends by exponentially weighted moving averages, ONR Memorandum, vol. 52, Pittsburgh, PA: Carnegie Institute of Technology. Available from the Engineering Library, University of Texas at Austin (1957)

18. Winters, P.R.: Forecasting sales by exponentially weighted moving averages. Manage. Sci. 6, 324–342 (1960)

19. Archibald, B.C.: Parameter Space of the Holt-Winters' Model. Int. J. Forecast. 6, 199–209 (1990)

20. Gardner, E.S.: Exponential smoothing: the state of the art Part II. Int. J. Forecast. 22, 637–666 (2006)

21. Granger, C.W.J., Joyeux, R.: An introduction to long-memory time series models and fractional differencing. J. Time Ser. Anal. 1, 15–29 (1980)

22. Hosking, J.R.M.: Fractional differencing. Biometrika 68, 165–176 (1981)

23. Box, G., Jenkins, G., Reinsel, G.: Time series analysis. Holden-day San Francisco (1970)

24. Crato, N., Ray, B.K.: Model selection and forecasting for long-range dependent processes. J. Forecast. 15, 107–125 (1996)

25. Cogollo, M.R., Velasquez, J.D.: Are neural networks able to forecast nonlinear time series with moving average components? IEEE Lat. Am. Trans. 13(7), 2292–2300 (2015)

26. Zhang, G.P., Patuwo, B.E., Hu, M.Y.: A simulation study of artificial neural networks for nonlinear time series forecasting. Comput. Oper. Res. 28, 381–396 (2001)

Spreading Information in Distributed Systems Using Gossip Algorithm

Andrzej Barczak[1]($\boxtimes$) and Michał Barczak[2]($\boxtimes$)

[1] Faculty of Science, Institute of Computer Science, Siedlce University,
3 Maja 54, 08-110 Siedlce, Poland
andrzej.barczak@uph.edu.pl
[2] Mettler Toledo, Poleczki 21 d, 02-822 Warsaw, Poland
michal.barczak@mt.com

Abstract. In the following article problem of an information sharing in the distributed system is described. Ways of solving that problem with emphasize on Gossip protocol are presented. The main goal of the article is to examine and analyze the operation of Gossip algorithm, taking into consideration chosen mode and values of parameters. In the beginning, problem of exchanging information in the distributed system is presented and different algorithms solving it were shown. Afterward, Gossip algorithm is presented. The principle of its operation, its working modes and models in which Gossip algorithm can work are described.

As part of the article, an application in C# language was made, allowing to examine Gossip protocol in the laboratory environment. The laboratory environment contained minimum of six and a maximum of ten nodes. Number of iterations needed to achieve consistency for six, eight and ten connected computers was examined. Research was made for deferent working modes of Gossip protocol. To present operation and scalability of Gossip algorithm, the mathematical model of the algorithm and graph presenting percentage of infected nodes in individual iterations, is shown. In order to describe Gossip algorithm using a mathematical formula, a model known from epidemiology was used. At the end analysis of research's results was done. As well problems related to the Gossip algorithm and the method for solving them were described. The application of Gossip algorithm in commercial solutions was also presented.

Keywords: Gossip algorithm · Distributed systems ·
Information spreading algorithm

1 The Problem of Information Distribution in a Distributed Systems

1.1 Distributed Systems

Nowadays, more and more IT systems are distributed systems or systems that are using cloud solutions. We are calling a distributed system set of computers containing the cohesive software and connected with each other by the network. It contains few up to few thousands of connected computers called nodes. Applying distributed systems

© Springer Nature Switzerland AG 2020
W. Zamojski et al. (Eds.): DepCoS-RELCOMEX 2019, AISC 987, pp. 21–32, 2020.
https://doi.org/10.1007/978-3-030-19501-4_3

allows increasing speed of executing complex algorithms. It is taking place through the division of calculations into a lot of computational processes. These processes, in a concurrent way, perform calculations and exchange information between themselves or only synchronize the results.

Such a kind of systems allows to significantly increase computing power and capitalize the resources of each of the nodes in a more efficient way. One should not forget that applying distributed architecture increases systems reliability.

Distributed systems are characterized by several features. One of the most important of them is transparency. This feature guarantee that the user of a system is not aware of such a system's parameters like the geographical location of each of the nodes or system's size. The breakdown of nodes should as well be not perfectible by the user. Distributed system should guarantee as well the transparency of the methods of access to it. In that case end user has the impression that the system is consisting of a single node not many thousands of them. Another feature is the transparency of the transfer, that allows changing the position of some resources without knowledge of the end user. We can deal also with transparency in terms of redoubling. In this case, the user is not aware of redoubling certain resources.

Distributed system should be open to cooperation with other systems. It is pretty important that the functionality of distributed systems must be described using very precise interfaces. Another crucial feature of distributed systems is its' scalability taking into consideration both number of nodes and their geographical location. Another point is the concurrency of the systems that allows performing many tasks at one time. An easy way to configure and reconfigure is as well an important feature of distribution system. One of the most important features of distributed systems is their ability to share the same resources by all of the users. Their diversity in terms of the hardware is the next essential feature. Ensuring all features mentioned above is not an easy or trivial topic. Main feature that guarantees transparency is the ability to share and distribute information inside of a system. One of the most common solutions allowing to distribute information in a smooth way is Gossip algorithm (know as well as Gossip protocol).

1.2   One-to-One Algorithm

While working with huge distributed systems containing of hundred thousands of nodes, we need to deal with a problem of distributing information in an efficient and quick way. It is crucial that data, that was received by one of the nodes, will be sent to others in a fast way and without too much usage of network resources. There are several algorithms that help us to deal with that problem. The simplest, however also less efficient, algorithm of spreading information in distributed systems is algorithm one-to-one. Mentioned algorithm is just sending information node by node. In a first iteration node receiving a new piece of information and sends it to his neighbor. In the flowing iterations, information is distributed node by node. In case of one-to-one algorithm, the path of information flow is created. It leads to the decreasing of algorithms reliability. In case of a crash of network connection between two of the following nodes information is not farther sent. Due to the fact that one node sends information just to the one neighbor full procedure is very time consuming and not efficient. The function of information's propagation time according to the number of nodes is linear and can be described as

$$tp = ti * n$$

where tp is time of propagation, ti is time of one iteration and n is the number of nodes. Assuming that iteration time is 1 s and number of nodes is 1000 propagation time is 1000 s what is about 16 min. I would like to admit that one-to-one algorithm is not scalable, that means that the results will be even worse when the number of nodes increases.

1.3   All-to-All Algorithm

Another algorithm of spreading information in distributed systems is algorithm all-to-all. In this algorithm, each of the nodes, sends newly received information to all system's nodes, as soon as he gets it. Such a solution allows for increasing the reliability of the distributing of information. Compare to one-to-one algorithm crash of connection between two of the nodes does not have such negative effects. In case of problems with the connection between two of the nodes, information can be sent by a different path from another node. Problem with the propagation time of the information known from one-to-one algorithm has been solved as well. Full system receives the information almost immediately after its' appearance. Unfortunately, all-to-all algorithm is not free from flaws and limits. The major flaw is the amount of information sent in the system in each iteration which is N^2 where N is number of nodes. The limit of the nodes in the system derives directly from this flaw. In case of the systems containing tens of thousands of nodes, in each iteration, even thousands of millions of information can be sent. For both the computer network and a single node, it is impossible to handle such network traffic.

1.4   Gossip Algorithm

Xerox company at the end of the 1980 s struggled with a problem of replicating data on a few thousands of instances of the distributed database. Crucial aspects for Xerox were both time of spreading a big amount of information and the reliability of the full replication process. One of the propositions was Gossip algorithm called as well Epidemic algorithm. The rule of operation of this algorithm is very similar to spreading either gossip or the disease in some population [1]. Nodes of the system having information send it to the chosen neighbor in an iterative way. The way of selecting the neighbor might be random but it does not need to be such. Each of the nodes might calculate dynamic parameters of connection (like transmission time or quality parameters of transport medium). Based on those parameters for each of nodes there will be assigned a factor. Nodes with higher factor will be prioritized while selecting a neighbor for information sending. For nodes in a system, based on possessing of information, one of three following statuses (states) can be assigned [4]:

1. Infected by the information (I)
2. Susceptible (S). This status is assigned to the nodes not having information.
3. Removed (R). Nodes can be in state R when the information he has is old and is not distributed anymore.

There exist two main models of changing of statuses - Model SI and Model SIR. In Model SI there is a possibility only for changing state from S to state I, while in SIR Model change from I to R is also allowed. Gossip algorithm can work in three different modes: push, pull and hybrid push-pull. In the case of SI push implementation, nodes being in S state only are passively listening, waiting for new information. In the pull version, nodes with status S assigned, as well sends requests for a new piece of information. The hybrid model is a combination of Push and Pull. The pseudo code of the Gossip algorithm in the SI model is presented in listing 1.

```
1 public void Gossip()
2 {
3   while (true)
4   {
5       Thread.Sleep(timePeriod);
6       Node selectedNode = GetRandomNode();
7       if (this.push)
8       {
9           if (this.state == NodeState.Infected)
10          {
11              SendUpdate(selectedNode, info);
12          }
13
14      }
15      if (this.pull)
16      {
17          SendUpdateRequest(selectedNode);
18      }
19  }
20 }
21
22 private void OnGetUpdate(UpdateArgs args)
23 {
24      args.reciver.informations.Add(args.info);
25      args.reciver.state = NodeState.Infected;
26 }
27
28 private void OnGetUpdateRequest(UpdateArgs args)
29 {
30      if (this.state == NodeState.Infected)
31      {
32      SendUpdate(args.sender,args.info);
33      }
34 }
```

Listing 1. Gossip Algorithm in SI model

It is wealth mentioning that in SI model for both push and hybrid implementation data spreading will not be stopped even when all the nodes have received information. In case of pull implementation this problem do not occur, as long as all the nodes has information about amount of data that need to be redistributed. If all the nodes receives required number of information, they will stop sanding update requests. Supposing that nodes do not have knowledge about amount of information in a system, pull requests will be as well sent even when system is in consistent state. Constant sending requests and information may lead to overloading of a network. One of the solutions for described problem is SIR model. In SIR model for each piece of information the aging factor is added. After reaching this value information is set into state R and is not sent any more. Specifying aging factor is not an easy task. While doing that we need to take into consideration many aspects, like for example number of nodes and delays in computer network. Underestimation of the factor may lead to not sending information to some of the nodes. Overestimation however cause too big number of information and overloading the network. There are several methods of estimating the aging factor and most of them requires sending the return message after receiving already known information. In first implementation information can be set into status R after being sent redundantly defined number of times. Next one allows to stop sending information, with probability P, after sending every duplicated message. In last implementation one of the nodes having obsolete data is chosen to set the information in status R [3]. Pseudo code of Gossip protocol in a SIR model is presented in Listing 2. The differences in implementation between SI and SIR model were made in lines from 21 to 50.

```
1 public void Gossip()
2 {
3      while (true)
4      {
5          Thread.Sleep(timePeriod);
6          Node selectedNode = GetRandomNode();
7          if (this.push)
8              {
9              if (this.state == NodeState.Infected)
10                 {
11                     SendUpdate(selectedNode, info);
12                 }
13
14             }
15         if (this.pull)
16             {
17                 SendUpdateRequest(selectedNode);
18             }
19         }
20     }
21 private void OnGetUpdate(UpdateArgs args)
22 {
23     args.reciver.informations.Add(args.info);
24     args.reciver.state = NodeState.Infected;
25     SendResponse(args.sender, informationStatus);
26 }
27 private void OnGetResponse()
```

```
28 {
29     if (RemovalImplementation.Number == remImplementation && resendedMessages ==
obsolescence)
30     {
31         this.state = NodeState.Removed;
32     }
33     else if (remImplementation == RemovalImplementation.Probability)
34     {
35         if (GetProbablity())
36         {
37             this.state =NodeState.Removed;
38         }
39         else
40         {
41         resendedMessages++;
42         }
43     }
44 }
45 private bool GetProbablity()
46 {
47     var rand = new Random();
48     var p = rand.Next(1, obsolescence);
49     return p < obsolescence;
50 }
51 private void OnGetUpdateRequest(UpdateArgs args)
52 {
53     if (this.state == NodeState.Infected)
54     {
55     SendUpdate(args.sender, args.info);
56     }
57 }
```

Listing 2. Gossip Algorithm in SIR model

Other crucial problem, existing in Gossip algorithm, is gathering knowledge about system topology by each of a nodes. In case of huge systems it requires establishing very big list of connections on every node, what might have negative influence on scalability of the Gossip algorithm. Finding out newly joined nodes is as well problematic issue. It requires from nodes sharing information about the topology. Such a procedure generates additional big network traffic. Another solution states that one central managing node, that will have full knowledge about system structure, should be created. It is as well not perfect resolution cause the reliability of a system is much lower. In most efficient implementation every node has just partial information about a system. Assuming that combination of information from all the nodes allows to send information between every nodes it is best solution. More than that information about system topology can be as well sent with Gossip messages.

It is wealth mentioning that Gossip algorithm allows spreading more than one information in a system at the same time. Taking it into consideration it is possible that not all nodes will have the same information, opposite to all-to-all algorithm, full knowledge and coherence will be achieved after a few iterations. Gossip algorithm is an algorithm of final coherence, that not guarantee strong coherence of a system. That is why it cannot be used in systems requiring very strict coherence like transactional bank systems or internet stores. Some of the implementations of Gossip algorithm allows sending information to more the one neighbor in the same iteration. It for sure reduce the number of iterations and speeds up the process of information sharing, however, we need to remember that choosing to big number of neighbors to sent information will lead to creating almost all-to-all algorithm and overloading network.

1.5   The Description of a Application and Lab Environment

Algorithm Gossip has been implemented using .NET Remoting technology in C# programming language. The application allows to spread more than one information at one time and for changing the Gossip model (SI and SIR). As well it is possible to specify the number of neighbors for which information will be sent in one iteration and the aging factor. In the case of the SIR model, the user is able to select the method of choosing nodes that should be assigned to R status. Algorithm Mode can be parameterized as well. Application shows number of sent network packages corresponding with the Gossip protocol. As well the state of each node (number of information it has) can be shown. Base on the shown number of information in each iteration the efficiency of the Gossip algorithm is presented. The lab environment contains ten, connected with the network, physical servers (nodes), on which the test application is running. The nodes were connected in two different configurations. In the first one, the notes are connected into a complete graph while in the second one into a binary tree.

1.6   Analysis of Gossip Algorithm

The way Gossip algorithm in the SI model is working has been examined for six, eight and ten nodes. For six nodes binary tree topology presented in Fig. 1 was used.

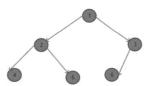

Fig. 1. Network topology for six nodes

One or two pieces of information were sent. They were propagated from nodes number one, six or both. When the information was sent from the first node, the most efficient occurred to be algorithm Gossip in the hybrid mode. System was coherent in the third iteration. Using push and pull modes, nodes needed five iterations to spread the information. While studying the distribution of information from node number six, we will also observe the highest effectiveness using the hybrid mode. All nodes received information in the fifth iteration. In the push mode information was distributed in the sixth iteration, while in pull mode not until eighth. We can observe similar results when the information's packages were sent from both node one and six at ones. Using hybrid mode algorithm was able to spread information in three iterations. For push and pull six iterations were needed to achieve such a result. For six nodes, as well as an examination of spreading information in the system of complete graph topology was done. One or two pieces of information were sent. In both cases, most efficient occurred to be hybrid mode. For both one and two pieces of information system get coherent in the second iteration. In pull mode, it took place in the third iteration for one information and forth for two pieces of information. In both cases, push mode needed four iterations to spread information over the system. Analogous tests were made for a system containing eight nodes. Binary tree topology of connections between servers is presented in Fig. 2.

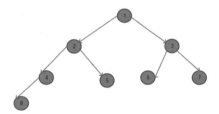

Fig. 2. Network topology for eight nodes

For hybrid and pull modes, when the infected node was that with number one, information was distributed in the second iteration. Algorithm Gossip in push mode needed six iterations to achieve it. For information distributed from node number eight, in hybrid mode, the system gets coherent in the sixth iteration. Using the pull method, the algorithm was able to spread information in seven iterations, while using the push in twelfth. Distributing two information's packages (one from node seven and one from node eight) lead to worse results. Gossip protocol in a push mode needed up to twenty-seven iteration to spread information. Pull method was able to redistribute information in eleven iterations and hybrid in nine. For hybrid mode sending three pieces of information was tested. They were distributed from nodes number eight, seven and six. Coherence was achieved in the eighth iteration. Next, measurements for nodes connected into a complete graph were done. Spreading one information took four iterations for the push, six for the pull and three for the hybrid mode. When the number of data packages was extended to two, the most efficient occurred to be hybrid and push modes. In both cases, the system was coherent in the fourth iteration. Pull mode needed one iteration more to achieve that. For hybrid mode, the number of selected neighbors, for information distribution in one iteration, was extended to two. It effected in spreading one information in two iterations. Similar experiments were made for a system containing ten nodes. The topology of network connections between servers is presented on Fig. 3.

When information was sent from node number one in push mode, the system achieved coherency in the fourteenth iteration. In the pull mode, this took place in the ninth iteration, while in the hybrid mode in the fifth. Considering the first node infected was node number ten, the Gossip algorithm, in both push and pull modes, spread the data in eleven iterations. In hybrid mode, each server received information after six

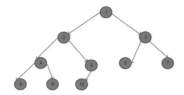

Fig. 3. Network topology for ten nodes

iterations. For two pieces of information sent from nodes nine and ten in push mode, the broadcast occurred in seven passes. For the pull mode this was done in the fifteenth iteration and for the hybrid mode in the fifth. After connecting all ten nodes in a complete graph topology, the results are much better. For one piece of information sent in the push mode, system was coherent after the fourth iteration. In the pull mode, it took three more iterations, while in the hybrid mode the message was sent within two iterations. In case of sending two messages in the push mode, the information was disseminated during six iterations, in the pull mode during seven, while in the hybrid mode during three.

In the SIR model, for five nodes connected in the topology of the complete graph, hybrid mode and one message, tests of various model configurations were made. In the SI model, every node within forty-five iterations sends ninety-two packages related to the exchange of information (requests to send a message and the messages themselves). With the larger number of nodes, it can be a significant load for the network. In order to minimize the number of requests and information sent, the SIR model was introduced, allowing to stop broadcasting information under certain conditions. The behavior of Gossip algorithm in three variants of the SIR mode was investigated. The first was to stop sending messages with a set probability P / *the probability of changing the status of information from the state I to R* / at the moment when the neighboring node sends information that it already received the particular message before. The tests began with a P value of 0.05. For this configuration, nodes sent from seventy-seven to one hundred and five packages during forty-five iterations. Considering that this value also contains the number of feedback messages and that the number of network packages is smaller than in the SI model, the effect is satisfying. Then the value of P was increased to 0.1. The result of the increase was visible and resulted in a reduction in the number of packages from individual nodes to a number between sixty and seventy. After another increase in the value of P to 0.2, the number of packages sent from one node was reduced to a value between fifty and fifty-seven. A further increase in the value of P did not bring such large changes. For P equal to 0.3, nodes send out fifty to fifty-five packages, while for a P of 0.5, the result is between forty-five and fifty packages. For the higher probability values, it was already possible to observe the lack of final system coherency, cause the information was set in R status on all nodes before it was sent out over the entire system. Another implementation of the Gossip algorithm in the SIR model assumes switching to the R state after sending n redundant messages. Assuming n equal three, each node sent between fifty and fifty-three packages. The results for n equal to five are identical. At n of six, this number increased to fifty-four - fifty-seven packages. Assuming n equal to eight, we can observe another increase in the number of packages to a value between fifty-five and sixty. The next implementation allowed sending by a predetermined management node to a randomly selected neighbor containing a message, request to change the information's state to R. In this implementation, each node sent within seventy messages. The summary of the test results for the number of packages sent is presented in Table 1.

Table 1. Number of sent packages in Gossip algorithm. Source: own work

	Minimal number of packages	Maximal number of packages
SI		
	92	92
SIR		
With Probability *P*		
0.05	77	105
0.1	60	70
0.2	50	57
0.3	50	55
0.5	45	50
After *n* redundant messages		
3	50	53
5	50	53
6	54	57
8	55	60
Random	68	73

The researches were made on a relatively small number of nodes, so the question arises about the efficiency of the algorithm for much larger systems. In the case of a full graph topology, it can be assumed that the probability p of choosing each of the n neighbours is equal to $1 / n$. The Gossip algorithm of the SI model and in the push mode can be described using a mathematical model known from epidemiology, showing the number of infected people (in our case nodes) depending on the time. The formula for the number of infected nodes is shown below [1]:

$$U(t) = \frac{n+1}{1 + n * e^{-(n+1)*p*t}} \tag{1}$$

where: n is the number of nodes, and t is the number of iterations. Assuming that the number of nodes is one hundred thousand, we need only twenty-four iterations to achieve coherency. For a million nodes, the number of iterations only increases to twenty-eight. For the pull and hybrid models, the results should be even better. Figure 4 shows the percentage share of infected nodes in iterations for ten, one hundred, one thousand, ten thousand one hundred thousand and million nodes.

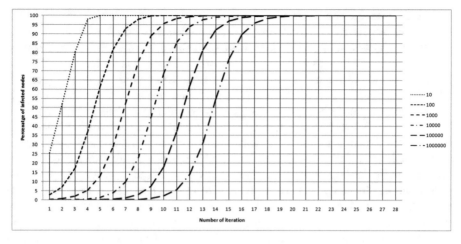

Fig. 4. Percentage of infected nodes in each of the iterations

The chart shows that after the seventeenth iteration system coherence for the number of nodes less than a million is more than ninety percent, and from the nineteenth iteration more than ninety-nine percent.

2 Summary

Based on the results of the tests carried out and the calculations made, it can not be argued that the Gossip algorithm is perfectly suited for the spreading of information in distributed systems. However, it should be taken into account that the effectiveness of the algorithm depends on many factors. The topology of connections between nodes has a fundamental influence on the operation of Gossip algorithm. In hierarchical structures, which is examined tree topology, information needs more iteration to reach all the nodes than in topology of the complete graph. In the tree structure, node from which the information will be sent is not without significance. The deeper the infected server is in the network structure, the more iterations are needed to bring about system integrity. The choice of algorithm's operating modes also affects the efficiency of its work. The fastest mode is the hybrid mode. Slightly worse results can be observed for the pull mode. The weakest effects give the use of the push mode. The hybrid mode, unfortunately, is not free of drawbacks. The main one consists of sending a large number of network packages. Each of the nodes operating in this mode, during each iteration, sends out one to two packages. The first package is responsible for sending a request message. The second is sent if the table of the information is not empty. The amount of sent packages can be significantly reduced using the SIR model of Gossip algorithm. It allows not to broadcast information that has already been sent to many nodes. There are several implementations that allow setting R status for the information, causing the information not to be distributed. The most effective implementation was the transition to the status of R with a constant probability P, in the case of obtaining a message about redundant information. For the probability P from 0.2–0.5, the number of packages sent has been halved. Implementation allowing to determine the status R after sending redundant message n times gives slightly worse results. Random selection of nodes to set the R status for information is the least effective and seems to be the riskiest of all described approaches. In specific situations, incorrect selection of the node may cause the system will not achieve the final coherence. This happens when the node being drawn is the only one that has a connection to a part of the system and will not be able to send information there. I think that the interesting aspect of the Gossip algorithm is its scalability. In the hybrid mode and the complete graph topology, I did not notice a significant difference in the number of needed iterations while sending one message for six nodes and sending two messages to ten nodes. During the study of Gossip protocol in the push mode, in particular in the hierarchical network topology, one could notice a large dependence of the protocol's effectiveness on the selection of neighboring nodes when sending a message. In the case of two neighbors, the probability of drawing each of them is equal to 0.5. It may happen that in the course of several iterations, the same neighbor will be drawn. This will result in the information not reaching the other of the adjacent nodes and extending the information transfer process. The solution to this problem is to create a local dictionary to determine for which nodes the information was sent. This would eliminate from the drawing the nodes that received the message. The described problem does not occur when we have many adjacent nodes, because the probability of drawing each of them becomes smaller as the number of connected nodes increases. It is obvious that during the implementation of the Gossip algorithm special attention should be paid to the problems of mutual distributed exclusion. Exclusion can occur when multiple nodes at the same time try to send a message to the same node. The solution may be to use an algorithm that solves this problem, such as the Ricard Agrawal's algorithm based on requests. It should be mentioned that the Gossip algorithm is highly scalable, as shown by mathematical methods. This favors the popularity of solutions based on this algorithm. It is used, among others, in the Apache Cassandra software, which is a free tool for managing no-SQL distributed databases. Another system using the epidemiological algorithm is SERF [2], used to manage server farms. The Gossip protocol has also been applied to Amazon Web Services [5].

References

1. Demers, A., Greene, D., Hauser, C., Irish, W., Larson, J., Shenker, S., Sturgis, H., Swinehart, D., Terry, D.: Epidemic algorithms for replicated database maintenance
2. HashiCorp: Gossip protocol
3. Jelasity, M.: Gossip-based protocols for large-scale distributed systems
4. Lopez, F.: Introduction to gossip
5. The Amazon S3 Team: Amazon s3 availability event, 20 July 2008

Structurization of the Common Criteria Vulnerability Assessment Process

Andrzej Bialas[(✉)] [iD]

Institute of Innovative Technologies EMAG, Leopolda 31,
40-189 Katowice, Poland
Andrzej.Bialas@ibemag.pl

Abstract. The paper deals with the Common Criteria Evaluation Methodology
(CEM), especially with its part related to the vulnerability assessment. The aim
of the paper is better structurization of the vulnerability assessment process,
allowing its future automatization. The ontological approach will be applied to
develop the models of processes and data. The elementary evaluation processes
are defined on the basis of the analysis of the CEM vulnerability assessment.
The process activities, input and output information, are identified and specified
in a pseudocode. The process verification against CEM is performed. The
conclusions summarize the verification and propose future works to build the
ontology, knowledge base and the vulnerability assessment tool.

Keywords: Security assurance common criteria · Vulnerability assessment ·
Knowledge management · IT security evaluation

1 Introduction

The paper concerns the security assurance methodology specified in the ISO/IEC
15408 Common Criteria (CC) standard [1, 2]. The CC methodology provides confidence that countermeasures applied in an IT product (hardware, software, firmware,
system) are adequate to ensure that the IT product is secure in its operational environment. The assurance is measurable using EALs (Evaluation Assurance Levels) in
the range from EAL1 to EAL7. The CC methodology comprises three basic processes:

- IT security development process of the IT product called TOE (Target of Evaluation); as a result of different security analyses the Security Target (ST) is worked
  out, containing the security problem definition (SPD), security objectives, security
  requirements and functions;
- TOE development process (according to EAL); as a result the TOE and its documentation (evidences) are developed;
- security evaluation process of the TOE is carried out in an independent, accredited
  laboratory (called ITSEF – IT security Evaluation Facility) according to the
  Common Criteria evaluation methodology (CEM, ISO/IEC 18045) [3].

© Springer Nature Switzerland AG 2020
W. Zamojski et al. (Eds.): DepCoS-RELCOMEX 2019, AISC 987, pp. 33–45, 2020.
https://doi.org/10.1007/978-3-030-19501-4_4

The Common Criteria methodology is described in publications worldwide [4, 5] and the author's publications [6, 7]. The paper is the continuation of research presented in the author's papers concerning CEM: [8] (attack potential method), [9] (general CEM ontology).

Security functions (behaviors) are expressed by security functional requirements (SFR) components – grouped by families and families grouped by classes. The assurance is expressed similarly by security assurance requirements (SAR) components – grouped similarly. The given EAL embraces the defined subset of SARs.

Each SAR component (in CEM called "sub-activity"), has the D (evidence ought to be delivered for the evaluator), C (required content and presentation of evidence) and E (how it will be evaluated) elements. For example AVA_VAN.3 concerning the "Focused vulnerability analysis", has elements: AVA_VAN.3.1D, AVA_VAN.3.1C, AVA_VAN.3.1E to AVA_VAN.3.4E. For each E element the evaluation verdict (Pass/Fail/Inconclusive) and its justification are issued.

CEM supplements CC to ensure common understanding of evaluation activities. For each E-element a certain number of work units (D-, C-based) are defined to describe more precisely the evaluation sub-activity ensuring more repeatable evaluations made by different evaluation labs. The key point of this methodology is to determine the existence of flaws or weaknesses (vulnerabilities) in the IT product, exploitable in the operational environment. The evaluation according to CEM [3] embraces components (sub-activities) of different families of assurance classes:

- examination of the provided documentation (evaluation evidences), including the security target assurance class (ASE), design documents (ALC-development, ADV-design and implementation, ATE-testing) and user guidance (AGD),
- vulnerability analysis (AVA_VAN family) and independent testing (ATE_IND).

The paper is focused on the key point of CEM, i.e. on the Common Criteria Vulnerability Assessment (CCVA). The vulnerability assessment is very complex, iterative, interrelated process considering many different factors, specific for the evaluated IT product. It is embedded into the entire IT security evaluation process performed according to CEM and it depends on the EAL, the TOE character and range. Therefore it is difficult to plan this process for a given IT product. The paper motivation is to better organize and facilitate the CCVA process.

The general aim of the research presented in the paper is to identify the CCVA-relevant processes and data, allowing to create the ontology, knowledge base and, in the future, the evaluation supporting tool. Better structurization of the vulnerability assessment should lead to its better organization and preciseness as well as its easier planning and running. The paper contribution is:

- to help the evaluators to plan the evaluation process,
- to elaborate process and data models for the software tool supporting the CCVA process.

Section 2 presents the current state of research in the paper domain. Section 3 presents the elaborated elementary evaluation processes embracing the general CCVA process. The verification of processes against the CEM is summarized in Sect. 4. The conclusions are focused on the further research aiming at the development of the knowledge base and supporting tool.

2 Current State of Research

The CC standard defines the AVA_VAN assurance components [1], while CEM [3] deals with their evaluation. The Japanese guide [10] supplements and refines the CCVA methodology. The Common Criteria community exchanges knowledge and experience about the CCVA implementation, mainly during the International Common Criteria Conferences (ICCC). These presentations usually deal with the specific CCVA issues raised in ITSEFes during evaluation.

The publication [11] provides a critical discussion of the vulnerability assessment process defined in CEM. To minimize basic drawbacks of the CCVA process, like a very generic vulnerability classification and poorly defined methodology, the author proposes to strengthen positive elements of CCVA, i.e. attack patterns and repeatability. The attack patterns can use existing solutions like CAPEC [12], and the method can be more aligned to SAR classes and families represented by evaluation evidences (AGD – misuse, ALC – delivered vulnerabilities, ATE – malfunction, ADV_ARC/TDS, ASE_SPD – attack path). This approach is fully compatible with the processes presented in the paper.

The publication [13] concerns test automation. Currently the automated test platforms, scripted- and classic tools are used. Best practices, advantages and limitations are discussed. This issue is similar to the automation of vulnerability assessment.

The presentation [14] concerns a new approach to the patch management in the context of vulnerability and risk management. It concerns activities after the IT product certification.

Data models, including ontology-based models are important for the CCVA processes modeling and to build supporting tools. The paper [15] embraces a very extensive literature survey focused on "the security assessment ontologies", and concludes that:

- "Most of works on security ontologies aim to describe the Information Security domain (more generic), or other specific subdomains of security, but not specifically the Security Assessment domain";
- there is "a lack of works that address the research issues: Reusing Knowledge; Automating Processes; Increasing Coverage of Assessment; Secure Sharing of Information; Defining Security Standards; Identifying Vulnerabilities; Measuring Security; Protecting Assets; Assessing, Verifying or Testing the Security".

The work [16] presents an architecture and development methodology of Cyber ontology, integrating existing ontologies. The [17] proposes a reference ontology to be used in global exchange of operational information related to cybersecurity. The results of this research can be used to integrate the presented CCVA processes with the existing vulnerability-related data sources.

The report [18] specifies methods and tools for vulnerability analyses, which can be used in penetration tests in CCVA.

The CC vulnerability assessment is still under research. There is no holistic approach to the CCVA automation. To build supporting tools, first models of processes and data should be elaborated, and this is what this paper is focused on.

3   Model of the Common Criteria Vulnerability Assessment Process

The paper deals with the AVA_VAN.1-4 components. The elementary evaluation processes EEP and related to them input/output data are introduced (see Fig. 1) as a result of the AVA_VAN components/sub-activities analysis [1]/pp. 311–346, [3]/pp. 182–188, Annex B. Each of the AVA_VAN.x components, where x = 2, 3, 4, has four E elements, except AVA_VAN.1, which has only three. This exception influences the EEP definitions.

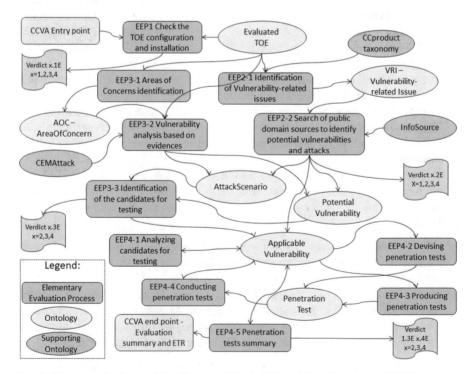

Fig. 1. The general scheme of the Common Criteria Vulnerability Assessment (CCVA) process – elementary evaluation processes and related data.

Each E element has at least one EEP assigned. Please note that the element AVA_VAN.x.1E requires only one process (EEP1) to issue a verdict, while AVA_-VANx.2 – two of them (EEP2-1, EEP2-2). Processes are shown as green rectangles with rounded corners. Input/output information (ontologies) are shown as pink ellipses. Supporting ontologies designed for many projects are marked orange.

Elementary processes are specified in a simple way using a pseudocode. Each of them has information about the concerned E element, input/output information and activities.

CCVA embraces all elementary processes from EEP1 to EEP4-5. The Evaluated TOE represents the TOE and all delivered evaluation evidences to be used in all EEPs. In EEP1 the evaluator verifies correctness of the TOE configuration and installation. As a result, the x.1E verdict is issued.

```
PROCESS EEP1 Checking the TOE configuration and
installation
COMMENT It concerns elements AVA_VAN.x.1E, x= 1,2,3 or 4
INPUT The TOE, the evidences: ASE_INT, ALC_CMC, AGD_PRE
ACTIVITIES
1. Examination whether the TOE test configuration is
consistent with the configuration under evaluation as
specified in the Security Target (ST)
2. Examination whether the TOE has been installed
properly and is in a known state
3. Assigning and justifying the verdict
OUTPUT Evaluation verdict for AVA_VAN.x.1E
(PASS/FAIL/INCONCLUSIVE) and its justification
ENDPROCESS
```

In the next CCVA step, the evaluator shall examine the sources of publicly available information to identify potential vulnerabilities in the TOE. Please note that the IT products may have diversified hardware-, software-specific vulnerabilities, and the information sources are numerous and diversified too. For this reason this task is partitioned to EEP2-1 (search planning) and EEP2-2 (searching).

EEP2-1 embraces the analysis of evidences and the identification of the Vulnerability related issues (searched keywords, short statements of different categories, etc.) expressed by ontology classes. The CC product taxonomy used in [2] is helpful.

```
PROCESS EEP2-1 Identification of Vulnerability-related
issues
COMMENT It concerns elements AVA_VAN.x.2E, x= 1,2,3 or 4
INPUT The TOE evidences, mainly ASE_INT, ADV
INPUT The CC IT product taxonomy
ACTIVITIES
1. Review of the TOE characteristics (TOE type, TOE
description, TOE overview) and the general CC products
taxonomy to orientate the search of potential
vulnerabilities in the public information sources
2. Identification of the issues, e.g. keywords, statement,
relevant for the potential vulnerabilities search
3. Placement of these issues into the
Vulnerability related Issue (VRI) knowledge base
OUTPUT Knowledge in the Vulnerability-related Issue
orientating the potential vulnerabilities search
ENDPROCESS
```

The supplementary **Information source** knowledge base is used for EEP2-2. It embraces any kind of sources: services, repositories, mailing lists, blogs, data feeds, search engines, scientific publications, etc. Particular sources are represented by individuals of the ontology classes. The sources are searched using contents included in the **Vulnerability related issues**. The aim is to identify **Potential vulnerabilities** and related **Attack Scenarios** concerning the TOE. These knowledge bases contain full information sets about two kinds of identified vulnerabilities (based on the public information search and evaluation evidences search – EEP3-2) and attacks, including description, provenance, attack theater and methods, tools, context, score, links to external bases, etc.

Additionally, for AVA_VAN.1, the assessment is done whether these vulnerabilities are applicable in the TOE operational environment. All applicable vulnerabilities are placed in the **Applicable Vulnerability** knowledge base. This knowledge base embraces extra information related to the attack potential, severity, exploitability, rationale for penetration testing, test report, etc. When the process ends, the verdict is issued.

PROCESS EEP2-2 Search of public domain sources to identify potential vulnerabilities and attacks
COMMENT It concerns elements AVA_VAN.x.2E, x= 1,2,3 or 4, but AVA_VAN.1.2E is slightly specific
INPUT The TOE evidences, mainly ASE_INT (TOE type, TOE description, TOE overview), ADV
INPUT The Vulnerability-related Issue knowledge base, which expresses "what to search"
INPUT The Information source knowledge base, which expresses "where to search"
ACTIVITIES
1. Searching of the publicly available sources of information to identify possible attack scenarios on the evaluated TOE or similar products, and the vulnerabilities exploited by these attacks
2. Potential vulnerabilities are registered in the Potential Vulnerability knowledge base (the first group of vulnerabilities – based on public domain search). The related scenarios are placed into the Attack Scenario knowledge base
3. Assigning and justifying the verdict
COMMENT For AVA_VAN.1 there is no further vulnerability identification based on the evidences. For this reason the vulnerabilities-candidates for further testing should be selected now
IF AVA_VAN.1
4. Considering the operational environment description (ASE_SPD and ASE_OBJ), analyze if the given potential vulnerability is applicable to the TOE in its operational environment
5. If yes, mark it as the candidate for testing, place it additionally into the Applicable Vulnerability knowledge base and report in the ETR (Evaluation Technical Report)
6. Consider results in the AVA_VAN.1.2E verdict
ENDIF AVA_VAN.1
OUTPUT Update of the Potential Vulnerab. knowledge base
OUTPUT Update of the Attack scenario knowledge base
IF AVA_VAN.1
OUTPUT Update of the Applicable Vulnerab. knowledge base
OUTPUT Update of the ETR document
ENDIF AVA_VAN.1
OUTPUT Evaluation verdict for AVA_VAN.x.2E.
ENDPROCESS

The next step concerns the identification and analysis of vulnerabilities encountered during the detailed evidences review (for AVA_VAN.1 omitted).

EEP3-1 is a preparative stage analogous to EEP2-1. It is focused on the identification of **Areas of concerns** – suspected vulnerability-prone areas of design.

```
PROCESS EEP3-1 Areas of Concerns identification
COMMENT It concerns elements AVA_VAN.x.3E, x= 2,3 or 4
INPUT The TOE evidences, mainly ASE, ADV, AGD
ACTIVITIES
1. Reviewing the TOE evaluation evidences, especially the
suspicious areas of the TOE (e.g. too complex), which may
include identified vulnerabilities
2. Placement of these issues into the Area of concern
(AOC) knowledge base
OUTPUT Knowledge in the Area of concern orientating the
potential vulnerabilities search and analysis
ENDPROCESS
```

EEP3-2 is focused on the identification and analysis of **Potential vulnerabilities** (and related **Attack scenarios**) existing in the TOE, but this analysis uses the TOE evaluation evidences. Different methods depending on the AVA_VAN.x rigour and the **CEM-Attack taxonomy** ([3]/Annex B) are applied, which considers the bypassing, tampering, misuse, monitoring and direct attack subcategories [8].

```
PROCESS EEP3-2 Vulnerability analysis based on evidences
COMMENT It concerns elements AVA_VAN.x.3E, x= 2,3 or 4
INPUT The TOE evidences, mainly ASE, ADV, AGD
INPUT Knowledge in the Area of concern orientating the
potential vulnerabilities analysis
INPUT The CEM-Attack taxonomy and knowledge base
ACTIVITIES
1. Identification of the potential vulnerabilities and
the related attack scenarios by analyzing the TOE
evidences, focusing on the areas of concerns. This is the
second group of vulnerabilities - evidence-based
2. The identified potential vulnerabilities are added to
the Potential Vulnerability knowledge base. The related
scenarios are added to the Attack Scenario knowledge base
COMMENT The unstructured, focused or methodical
approaches are possible, depending on the AVA_VAN.x.3
rigor. The flaw hypothesis methodology can be used and the
CEM attack taxonomy may be helpful
OUTPUT Update of the Potential Vulnerab. knowledge base
OUTPUT Update of the Attack scenario knowledge base
ENDPROCESS
```

EEP3-3 is focused on the analysis of potential vulnerabilities in the context of the TOE operational environment. The vulnerabilities recognized as applicable are placed in the **Applicable vulnerabilities** as the candidates for further analysis and penetration testing.

```
PROCESS EEP3-3 Identification of the candidates for
testing
COMMENT It concerns elements AVA_VAN.x.3E, x= 2,3 or 4
INPUT The TOE evidences, mainly ASE, ADV, AGD
INPUT Knowledge in the Potential Vulnerability knowledge
base
INPUT Knowledge in the Attack scenario knowledge base
COMMENT Check all vulnerabilities - based on the public
domain search and on the evaluation evidences
ACTIVITIES
1. Considering the operational environment description
(ASE_SPD and ASE_OBJ), analyze if the given potential
vulnerab. is applicable the TOE operational environment
2. If yes, mark it as the candidate for testing, place it
in the Applicable Vulnerability knowledge base and report
in the ETR
OUTPUT Update of the Applicable Vulnerab. knowledge base.
OUTPUT Update of the ETR document.
OUTPUT Evaluation verdict for AVA_VAN.x.3E.
ENDPROCESS
```

The last step concerns penetration testing of applicable vulnerabilities. EEP4-1 preselects candidates for testing using the attack potential method, discussed in the [8].

```
PROCESS EEP4-1 Analyzing candidates for testing
COMMENT It concerns elements AVA_VAN.x.4E, x= 1,2,3 or 4
INPUT The TOE evidences, mainly ASE, ADV, AGD
INPUT Knowledge in the Applicable Vulnerability base
INPUT Knowledge in the Information source base, which may
contain important third party information about
vulnerabilities (from national schemes, SOG-IS, etc.)
ACTIVITIES
1. Calculation of the attack potential
2. Prioritizing candidates and preparing them for testing
3. Update of the Applicable Vulnerability knowledge base
by this information
OUTPUT Supplemented Applicable Vulnerab. knowledge base
ENDPROCESS
```

On this basis the penetration tests are devised (EEP4-2), produced (EEP4-3) and conducted (EEP4-4). They are stored in the **Penetration Test** base, containing test description, method, tool, prerequisites, expected results and other documentation.

PROCESS EEP4-2 Devising penetration tests
COMMENT It concerns elements AVA_VAN.x.4E, x= 1,2,3 or 4
INPUT The TOE evidences, mainly ASE, ADV, AGD
INPUT Knowledge in the Applicable Vulnerability base
INPUT Knowledge from the Potential Vulnerability base
INPUT Knowledge from the Attack scenario base
ACTIVITIES
1. Work out the concepts of penetration tests
2. Place them into the Penetration Test knowledge base
OUTPUT Tests concepts in the Penetration Test base
ENDPROCESS

PROCESS EEP4-3 Producing penetration tests
COMMENT It concerns elements AVA_VAN.x.4E, x= 1,2,3 or 4
INPUT The TOE evidences, mainly ASE, ADV, AGD
INPUT Knowledge in the Applicable Vulnerability base
INPUT Knowledge from the Potential Vulnerability base
INPUT Knowledge from the Attack scenario base
INPUT Tests concepts in the Penetration Test base
ACTIVITIES
1. Project and implementation of the penetration tests and preparing their documentation
2. Place them into the Penetration Test knowledge base
OUTPUT Documented tests in the Penetration Test base
ENDPROCESS

PROCESS EEP4-4 Conducting penetration tests
COMMENT It concerns elements AVA_VAN.x.4E, x= 1,2,3 or 4
INPUT The TOE evidences, mainly ASE, ADV, AGD
INPUT Knowledge in the Applicable Vulnerability base
INPUT Knowledge from the Potential Vulnerability base
INPUT Knowledge from the Attack scenario base
INPUT Documented tests in the Penetration Test base
ACTIVITIES
1. Performing the penetration tests
2. Place results into the Penetration Test base
COMMENT Different tools, like debuggers, scanners, disassemblers, forensics analysis tools, etc. may be applied
OUTPUT Testing results in the Penetration Test base
ENDPROCESS

EEP4-5 summarizes CCVA, especially penetration test results, and gives the 1.3E/x.4E verdict. The TOE resistance to an attacker of basic/enhanced basic/moderate/high attack potential is determined and all exploitable and residual vulnerabilities are recorded in the ETR.

```
PROCESS EEP4-5 Penetration tests summary
COMMENT It concerns elements AVA_VAN.x.4E, x= 1,2,3 or 4
INPUT Testing results in the Penetration Test base
ACTIVITIES
1. Examination of all tests results - to determine that
the TOE in its operational environment is resistant to an
attacker possessing required attack potential
2. Summary of the results - preparing input about the
exploitable and residual vulnerabilities for the ETR
3. Updating the Penetration Test base by conclusions
OUTPUT Conclusions in the Penetration Test base
OUTPUT Update of the ETR document
OUTPUT Evaluation verdict for AVA_VAN.x.4E
ENDPROCESS
```

4 Verification

The main aim of the verification is to show that elementary evaluation processes cover all work units defined in CEM, which is shown in Table 1. AVA_VAN.1 does not contain a vulnerability analysis based on evidences. It causes an irregularity in numbering work units.

Please note that EEP2-1 and EEP3-1 have a preparatory character for the search and are related to automation.

Table 1. Elementary evaluation processes vs. work units.

Process	Work units for component AVA_VAN.1	Work units for components AVA_VAN.x, x = 2, 3, 4
EEP1 TOE Checking the TOE configuration and installation	AVA_VAN.1-1	AVA_VAN.x-1
	AVA_VAN.1-2	AVA_VAN.x-2
EEP2-1 Identification of Vulnerability-related issues		
EEP2-2 Search of public domain sources to identify potential vulnerabilities and attacks	AVA_VAN.1-3	AVA_VAN.x-3
EEP3-1 Areas of Concerns identification		
EEP3-2 Vulnerability analysis based on evidences		AVA_VAN.x-4

(continued)

Table 1. (*continued*)

Process	Work units for component AVA_VAN.1	Work units for components AVA_VAN.x, x = 2, 3, 4
EEP3-3 Identification of the candidates for testing	AVA_VAN.1-4	AVA_VAN.x-5
EEP4-1 Analyzing candidates for testing		
EEP4-2 Devising penetration tests	AVA_VAN.1-5	AVA_VAN.x-6
EEP4-3 Producing penetration tests	AVA_VAN.1-6	AVA_VAN.x-7
EEP4-4 Conducting penetration tests	AVA_VAN.1-7	AVA_VAN.x-8
	AVA_VAN.1-8	AVA_VAN.x-9
EEP4-5 Penetration tests summary	AVA_VAN.1-9	AVA_VAN.x-10
	AVA_VAN.1-10	AVA_VAN.x-11
	AVA_VAN.1-11	AVA_VAN.x-12

5 Conclusions

The paper concerns research on the vulnerability assessment methodology specified in CEM to define a set of elementary evaluation processes fully covering this methodology and allowing the CCVA automation in the future. The paper is not focused on data modelling. The research on this issue is performed concurrently – the ontological data models are elaborated and the related knowledge bases are worked out using the concepts and relations identified during the research presented here.

The specification of the introduced elementary evaluation processes with the use of a pseudocode allows for better structurization of CCVA. A clear view of vulnerability assessment is obtained allowing to better plan this process and continue works on the supporting tool.

Acknowledgement. The paper deals with the KSO3C (National scheme of the Common Criteria evaluation and certification) project, financed by the Polish National Centre for Research and Development as part of the second CyberSecIdent – Cybersecurity and e-Identity competition (CYBERSECIDENT/381282/II/NCBR/2018).

References

1. Common Criteria for IT Security Evaluation. Part 1-3, version 3.1 rev. 5 (2017)
2. CC Portal. https://www.commoncriteriaportal.org/. Accessed 7 Jan 2019
3. Common Methodology for IT Security Evaluation. version 3.1 rev. 5 (2017)
4. Hermann, D.S.: Using the Common Criteria for IT Security Evaluation. CRC Press, Boca Raton (2003)
5. Higaki, W.H.: Successful Common Criteria Evaluation. A Practical Guide for Vendors, Copyright 2010, Lexington, KY (2011)
6. Bialas, A.: Common criteria related security design patterns for intelligent sensors—knowledge engineering-based implementation. Sensors **11**, 8085–8114 (2011)

7. Bialas, A.: Computer-aided sensor development focused on security issues. Sensors **16**, 759 (2016)
8. Bialas, A.: Software support of the common criteria vulnerability assessment. In: Zamojski, W., et al. (eds.) Advances in Intelligent Systems and Computing, vol. 582, pp. 26–38. Springer, Cham (2017)
9. Bialas, A.: Common criteria IT security evaluation methodology – an ontological approach. In: Zamojski, W., et al. (eds.) Advances in Intelligent Systems and Computing, vol. 761, pp. 23–34. Springer, Cham (2019)
10. Vulnerability assessment guide for developers. IPA (2013)
11. Tallon Guerri, J.: Vulnerability analysis taxonomy achieving completeness in a systematic way. In: International Common Criteria Conference, Tromso (2009)
12. CAPEC – Common Attack Pattern Enumeration and Classification. https://capec.mitre.org/. Accessed 7 Jan 2019
13. Turner, L.: Test Automation for CC. Best Practices (CCUF Test Automation WG). In: International Common Criteria Conference, Amsterdam (2018)
14. Guerin, F.: Return from study period in ISO SC27 WG3 on patch management evaluation for common criteria. In: International Common Criteria Conference, Amsterdam (2018)
15. de Franco Rosa, F., Jino, M.: A survey of security assessment ontologies. In: Rocha, Á., et al. (eds.) Recent Advances in Information Systems and Technologies. WorldCIST 2017. AISC, vol. 569. Springer, Cham (2017)
16. Obrst, L., Chase, P., Markeloff, R.: Developing an Ontology of the Cyber Security Domain, The MITRE Corporation (2012)
17. Takahashi, T., Kadobayashi, Y.: Reference Ontology for Cybersecurity Operational Information, The British Computer Society (2014). (open access article)
18. Goertzel, K.M., Winograd, T. (contributor): Information Assurance Tools Report – Vulnerability Assessment, 6th edn. Information Assurance Technology Analysis Center (IATAC), USA (2011)

Anomaly Detection in Network Traffic Security Assurance

Andrzej Bialas$^{(\boxtimes)}$, Marcin Michalak , and Barbara Flisiuk

Institute of Innovative Technologies EMAG, Leopolda 31,
40-189 Katowice, Poland
{andrzej.bialas,marcin.michalak,
barbara.flisiuk}@ibemag.pl

Abstract. The paper focuses on a selected element of network security assurance, which is anomaly detection in network traffic monitoring. The anomaly detection component is developed as part of Regional Security Operation Center (developed in the RegSOC project) – a local instance of the Security Operational Center (SOC) – to detect incidents or their symptoms in terms of outlier observations in data. The objective of the research is to assess and select for implementation methods and tools satisfying the requirements of the performed RegSOC project. The paper discusses the role and placement of such tools in the general SOC architecture and requirements to be satisfied by these tools in a view of the specific RegSOC project needs. Next, a review of available methods and tools is performed to select the most useful ones. Using the selected tool, a general concept of security analysis component is presented and assessed against the project requirements.

Keywords: Incident management · Security operation center ·
Security analysis methods and tools · Anomaly detection · Outlier analysis

1 Introduction

The security of the network traffic is one of the most important features and expectations of the modern world. Irrespective of a private opinion of John Snowden's network activity, it is worth to consider his words about privacy in the Internet: "Technology can actually increase privacy. The question is why are our private details that are transmitted online or why are private details that are stored on our personal devices any different than the details and private records of our lives that are stored in our private journals." [1]. That sentence brings us to the question: How can we increase the Internet network data security, traffic safety and reliability?

One of the possible solutions of the problem is to establish specified network traffic monitoring institutions whose main goal would be to observe the network traffic and prevent the unwanted network usage. This unwanted network usage may be considered a kind of anomaly in the general network activities.

This paper is a study of potential improvement in the area of network traffic anomaly monitoring and detection. The presented approach is strongly connected with the notions of SOC (Security Operating Center) and its regional network adaptation.

© Springer Nature Switzerland AG 2020
W. Zamojski et al. (Eds.): DepCoS-RELCOMEX 2019, AISC 987, pp. 46–56, 2020.
https://doi.org/10.1007/978-3-030-19501-4_5

The paper features some researches that are preliminary activities of the RegSOC (Regional Security Operation Center) project [2]. The RegSOC project aims at the development of a methodology and tools to create regional security operation centers in Poland for public and business organizations. The implementation of the specialized SOC embracing a huge number of public and business entities and covering a significant part of the country, requires the research aimed at:

- the possibility to use existing security analysis methods and tools in the project,
- the development of new methods to solve RegSOC specific problems and implement them in the project.

In the paper a proposal of a dedicated network traffic anomaly detection component is presented. The component is planned to extend the functionality of RegSOC components of the network traffic analysis.

The paper is organized as follows: it starts with a discussion of security analysis tools in the general SOC architecture, afterwards a short description of the RegSOC project security analysis requirements is presented, then a short introduction to anomaly detection is provided, the next section contains a discussion about the possible data and algorithms for the purpose of network traffic anomaly detection, finally a general concept of anomaly detection component in the RegSOC structure is discussed; the paper ends with some final conclusions and perspectives of further works.

2 The Security Analysis Tools in the General SOC Architecture

SOC is a mixed, organizational and technical solution, which encompasses:

- people, their organization – a management structure, knowledge and skills needed for the SOC operation, training and awareness;
- processes, focusing on security monitoring, security incident management, threat identification, digital forensics and risk management, vulnerability management, security analysis, etc.;
- technology (different methods or tools used by people to support processes), solutions for the security monitoring, network infrastructure readiness, events collections, correlation and analysis (security analysis), security control, log management, vulnerability tracking and assessment, communication, threat intelligence, etc.

These three elements are important for all SOCs, but particular SOCs differ with respect to the organization, maturity, used technology and its advancement.

The key SOC process is information security incident management. Generally, the SOC team monitors indicators of compromise of the protected assets, detects security events, classifying some of them as incidents, and ensures the right reaction to incidents. The security incident management is performed in the similar way in SOCs and is well specified in standards, e.g. ISO/IEC 27035 [3]. The organization implementing SOC should be well prepared to manage incidents – processes should be established, people should be trained and equipped with supporting tools. Incident management is based on the incident live cycle.

Different actors (employees, contractors, partners, internet service providers, third parties, etc.) can report security events using different channels (e-mails, phone calls, web forms, etc.). With respect to the SOC character, an additional, specific actor, i.e. a SOC analytics tool, is considered. The tool provides the event-related information to the security analyst, which can be considered here a specific information channel to report events. All security-related events are placed into a centralized data collection, sometimes called a point of contact (POC).

The next activity is called the event triage and encompasses the event verification, classification and the assignment of a responsible person to handle the event. In addition, the severity level is assessed using a predefined scale and with respect to the possible event impact. A security incident is a security event that causes damages, like data loss.

Further activities embrace: incident investigation related to the data collection and analysis, resolution research and its implementation, incident recovery, and "lessons learnt".

Collecting and analyzing the relevant data are the key activities of SOC. Data collection is related to the monitoring of IT infrastructure, assets and processes belonging to the organization. SOC collects data originating from different sources and having different formats. The collecting process should be supported by time synchronization. The quantity of collected data should be necessary and sufficient to infer incidents or their symptoms during the further data analysis.

The following data sources are usually considered:

- logging messages from:
    - security-related equipment, like firewalls, intrusion detection/prevention systems (IDS/IPS), web proxies, malware detection systems;
    - network infrastructure components, e.g. routers, switches, access points, gateways;
    - operating systems, virtualization platforms, databases, network applications;
    - physical security components and others;
        - net flow parameters, network packets;
        - files, especially configuration files, hash values, HTML files, etc.

The logic messages collection is based on the syslog protocol. The NetFlow technology introduced by Cisco plays the key role in the network traffic monitoring, collecting information about the source and destination of traffic, the class of service, and the causes of traffic congestion. The unsampled (all packets) or sampled (each nth packet of their certain number) information flow collection is possible. For the forensics purposes, the entire packets should be collected.

The collected data should be ordered and prepared for further security analyses, i.e. should be parsed and normalized. The parser, e.g. logstash [4], transforms a raw input string, e.g. logging message, to predefined fields (e.g. to JSON format), allowing the event specification in a unified way. Parsers can use regular expressions (regex patterns). Thanks to the normalization, similar extracted events coming from multiple sources can be treated in the same way.

During the security analysis the sampled and ordered data are researched to reveal potential known and unknown threats. The following groups of security analysis methods, usually implemented in SIEM tools (Security Information and Event Management) or collaborating with them, can be considered [5]:

- rule-based correlation,
- anomaly-based correlation,
- risk-based correlation,
- mixed-mode-combinations of the above.

In reference to the presented groups, the RegSOC anomaly detection component will belong to the second mentioned group. The anomaly-based correlation is a two-step methodology. In the first step the baseline is established with the use of statistical profiling. The baseline expresses a reference behavior of the monitored IT environment.

The second step concerns the detection of activity patterns which deviate from activities embraced by the baseline. The baselines express different aspects of behavior, e.g. traffic rates, port or protocol usage in time period, CPU or memory usage, number of login attempts, etc.

3 RegSOC Project Requirements Related to the Security Analysis Methodology and Tools

The main result of the RegSOC project will be a cybersecurity monitoring platform supported by hardware and software appliances and supplemented by a procedural and organizational model of operation.

The RegSOC research is multidisciplinary and embraces:

- technical aspects of IT systems, networks, their environments and cybersecurity,
- information processing, including statistics, data analysis, modeling, artificial intelligence, machine learning,
- users' behavior, including social engineering and forensics aspects.

Implemented methods should consider:

- variability of IT systems and their environment in time,
- current and real data about security breaches (not only the archival ones),
- relationships between subjects on the low level (e.g. network devices) and on the high level (user-user),
- non-technical aspects of systems use,
- adapting the anomaly detection system to the characteristics of the observable object,
- performance parameters.

4   Anomaly Detection

The notion of anomaly is also known in the literature as outlier and the issue of anomaly detection is known as outlier detection and outlier analysis. It is not easy to provide the formal definition of an outlier.

One of the definitions [6] claims that "an outlying observation is one that appears to deviate markedly from other members of the sample in which it occurs". Further in this work two possible causes of this occurrence are pointed out: an extreme manifestation of the random variability inherent in the data or the result of the gross deviation from a prescribed experimental procedure or an error in data acquisition.

Another definition of an outlier and its origin was proposed by Hawkins [7] and it states that an outlier is an observation which deviates so much from the other observations as to arouse suspicions that it was generated by a different mechanism.

Barnett and Lewis [8] define an outlier as the observation inconsistent with the remaining data, while Weisberg [9] brings a more detailed explanation: an outlier is the observation that does not follow the same model as the rest of the data.

5   Review of Available Methods and Tools

For the purpose of the anomaly detection component development, it is necessary to define the goal of its analysis. The goal definition is strongly connected with the available data to be analyzed and, finally, the data availability is the logical consequence of the network devices functionality. This part of the paper presents a more detailed description and analysis of each of three requirements of anomaly detection.

5.1   Data Sources

The network traffic can be monitored with many different types of devices and systems. Their very wide classification can be found in [5] where data sources are divided into the following groups:

- security elements,
- network elements,
- operating systems,
- virtualization platforms,
- applications,
- databases,
- physical security elements,
- SCADA systems or Distributed Control Systems.

Security elements are some software solutions such as firewalls, antivirus programs, intrusion detection and prevention systems or malware analysis tools. The network elements may be represented by hardware elements, such as switches, routers or wireless access points. The other natural data sources are operating systems and their inner logs. The mentioned group of applications consists of web servers, DNS servers, e-mail gateways and many others.

From the RegSOC project point of view not all of the mentioned elements may be considered the potential data sources.

5.2   Collected Data

The network traffic may be described with data that come originally from the mentioned data sources. But the same network traffic may be also described with data called derived variables. Original variables are signals measured by the data sources while derived variables are a result of the original variables transformation.

The natural original variables that should be considered for analysis from the outlier detection point of view are as follows:

- IP address of the package sender,
- IP address of the package receiver,
- Port number from which the package was sent,
- Port number to which the package was sent,
- Package size,
- Type of the protocol used,
- Date and time of the package detection.

On the basis of the mentioned original variables, the below proposal of the derived variables is presented:

- Date and time derived variables:
  - Binary variable that shows whether the day was a working day or a day-off,
  - Day of the week,
  - Week of the year,
  - Month,
  - Shift number – in the case of monitoring the network traffic in the company with the three-shift working order,
- IP address derived variables:
  - Binary variables that show whether the IP address of the sender belongs to a specified group of IPs.

Apart from the mentioned variables (original and derived) it is also worth to consider an analysis of the aggregated data. The variables from both mentioned groups may be aggregated into one-minute, one-quarter, one-hour (or any other time interval) variables. The type and the meaning of the aggregate may be different for different input variable types:

- numerical variables: in the case of numerical variables (such as a package size) the aggregated value may be a minimal, maximal, average value or standard deviation;
- binary and categorical variables: in such a case each possible value of the considered categorical variable can be transformed to a single aggregated variable, i.e. the percentage of observed packages for which the categorical variable took a specified value (for the binary variables only one new variable must be defined).

It is also worth to consider the context-based aggregation of all types of data (binary, categorical and real – e.g. package size). The context may be the a specific day or time (one of the shifts, day-offs or working days) or the package source/destination IP. Then, the percentage (for binary or categorical variables) or some statistics (min, max, mean, etc. for the package size) of packages sent/received from the specific IP or the pool of IPs will be derived and aggregated in the presented way.

All above presented variables – original and derived and their aggregations – may become a target of outlier detection. But the best-to-have variable for better models of outlier detection development should point out packages or time intervals in which the nature of the network traffic was suspicious or even obviously unwanted. This variable should be given by a security network expert.

5.3   Methods of Anomaly Detection and Analysis

The methods of anomaly detection can be divided into two groups: statistical and based on spatial proximity.

In the case of one-dimensional methods with normal distribution, a sample criterion or Grubb's test can be applied [8]. In the case of multi-dimensional data spatial proximity-based methods one has to take into consideration the Euclidean or Mahalanobis distance measure. In the paper [10] the observation p is marked as an outlier if no more than k points in the data set are at a distance of d or less than p. Similarly, in the paper [11] if p of the k nearest neighbors of the data point (p < k) is closer than the specified threshold D, then the point is a normal observation – otherwise it is an outlier. The simplification of this approach is presented in [12]: only the distance to the mth closest neighbor is compared with the threshold D. One of the state-of-the-art methods of density-based clustering – DBSCAN [13] – can also be used for the outlier detection: observations that did not become members of any created clusters may be interpreted as outliers. A modification of one well-known classifier – SVM (Support Vector Machines [14]) – called one-class SVM [15] – tries to find the optimal hyperplane which separates all typical points of the data from the remaining ones which are interpreted as outliers. For the purpose of a non-numerical data outlier analysis other ways of (dis)similarity observations should be applied – an interesting review of different measures can be found in [16].

The selection of the anomaly detection algorithm will be considered in the context of real network traffic data from the local network provider. This will be available in the further RegSOC project development.

6   General Concept of the Security Analysis Component

6.1   RegSOC Structure Overview

The general scheme of RegSOC is presented in Fig. 1.

RegSOC monitors the whole netflow in the internet that consists of multiple providers (local or regional). The providers support the access to the internet (the cloud in the lower part of the scheme) for their clients. It may be assumed that there exist

NIDSes (Network Intrusion Detection Systems) between the providers and their inner networks of clients. NIDSes monitor the network traffic, report traffic statistics and accidents to RegSOC. Apart from that, additional data between network operators and RegSOC are exchanged. It is assumed that different RegSOCs may exchange data such as network traffic reports and statistics.

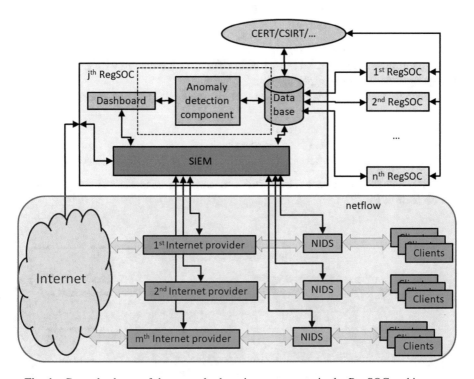

Fig. 1. General scheme of the anomaly detection component in the RegSOC architecture.

Looking into the inner structure of RegSOC, two most important elements must be mentioned: the SIEM module (Security Information and Event Management) and the database. In this scheme SIEM is considered in a bit wider way: it may be a classical SIEM module but also it may contain any of popular network traffic monitoring systems such as Snort, Suricata or many others. The data from SIEM and from other RegSOCs are stored in the database. RegSOC has also a connection with the external CERT/CSIRT and similar systems.

The other important part of RegSOC is the Dashboard – the visualization of the present network traffic statistics and the interpretation of unwanted events in the network traffic. And finally, the anomaly detection component is placed between the database and the dashboard.

6.2   Anomaly Detection Component

The anomaly detection component is presented as a single block in the diagram (Fig. 1) but its range extends partially into the database and the RegSOC dashboard, which is marked with the dashed line.

In general, the presented component analyzes the data of network traffic and generates cyclical reports and current alarms. From a more detailed point of view, its features can be listed as follows:

- observing the incoming data,
- deriving new features from the variables stored in the database by SIEM,
- aggregating raw values into aggregated variables,
- analyzing new data (incoming, derived or aggregated) in the context of outlier detection,
- reporting the results of analyses:
  - cyclically as a report (stored in the database),
  - on the dashboard when an anomaly in the network traffic is detected.

The presented requirements lead to the following structure of the component that can be seen in Fig. 2.

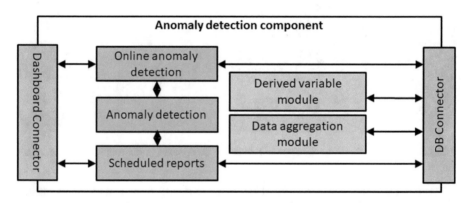

Fig. 2. Scheme of the anomaly detection component.

All of the mentioned tasks are strongly related to the database. In this case the database is understood more widely than just a relational database or a data warehouse (rather a meta-database). SIEMs and similar tools report the network events into text files or JSONs. The anomaly detection methods can use just the raw results produced by these tools and the files may be considered a part of the meta-database. On the other hand, the raw results can be stored in the relational way, together with the aggregated and derived variables. This leads to the necessity to develop a proper relational database or data warehouse. Their structures and technology will become a goal of further research. The access to the database will be performed with the DB Connector module.

Two independent modules are developed for the raw variables processing: Data Aggregator and Derived Variable Module. The former is responsible for cyclical data aggregation into the new variables, while the latter derives new variables from the original ones. The results of the analysis provided by these modules are stored back in the database.

The heart of the component is the Anomaly Detection module. It will contain the algorithms of the outlier analysis that are not accessible in SIEM. These algorithms will be used in two different ways: on-line and cyclical. The on-line analysis will be performed by the On-line Anomaly Detection module. Its tasks will be: to observe the newly incoming data in the context of anomaly detection, to save the most important results into the database and to provide the most critical alerts to the RegSOC operator's dashboard. The second module – Scheduled Reports – will have a similar functionality: it will generate periodic reports which will be saved in the database and presented on the dashboard in one of two cases – on the operator's demand or when the content of the report points at potentially unwanted anomalies.

7   Conclusions and Further Works

The project called RegSOC [2] is a dedicated project of the regional Security Operation Center. For this reason, the paper deals with the issue of network security assurance. This comprehensive issue is considered in a special aspect: how the detection of network traffic anomalies should improve the network dependability.

As part of this project, a component of network traffic anomaly detection was intended. Such a point of view led us to propose a component that could extend or complement the anomaly analysis functionality of the Security Information and Event Management module – also a part of the real RegSOC instance.

The paper concerns research aimed at the security analysis methodology focused on the security events/incidents detection. The existing methods and tools will be assessed with respect to the basic project requirements. Some methods can be considered candidates for implementation. In addition, some gaps and new R&D fields related to these methods will be identified.

As the project is in the initial phase of development, many open issues are planned to be solved in the near future by the project consortium: the choice of the SIEM module, development of the network traffic data model, points of the RegSOC application. The results of the above issues will imply the further development of the anomaly detection model, especially in the area of technology, used algorithms and data analysis results presentation.

Acknowledgements. RegSOC – Regional Center for Cybersecurity. The project is financed by the Polish National Centre for Research and Development as part of the second CyberSecIdent – Cybersecurity and e-Identity competition (agreement number: CYBERSECIDENT/381690 /II/NCBR/2018).

References

1. Jarre, J.M., Snowden, E.: Exit, Electronica 2: The Heart of Noise, Columbia Records (2016)
2. RegSOC: http://www.ibemag.pl/en/news/item/324-the-regsoc-project-has-started. Accessed 18 Jan 2019
3. ISO/IEC 27035 Information Technology – Security techniques – Information security incident management. ISO, Geneva (2011)
4. Logstash: https://www.elastic.co/products/logstash. Accessed 18 Jan 2019
5. Muniz, J., McIntyre, G., AlFardan, N.: Security Operations Center: Building, Operating, and Maintaining Your SOC. Cisco Press, Indianapolis (2016)
6. Grubbs, F.E.: Procedures for detecting outlying observations in samples. Technometrics **11** (1), 1–21 (1969)
7. Hawkins, D.M.: Identification of Outliers. Monographs on Applied Probability and Statistics. Springer, Dordrecht (1980)
8. Barnett, V., Lewis, T.: Outliers in Statistical Data, 3rd edn. Wiley, Chichester (1994)
9. Weisberg, S.: Applied Linear Regression. Wiley Series in Probability and Statistics, 3rd edn. Wiley, Hoboken (2005)
10. Ramaswamy, S., Rastogi, R., Shim, K.: Efficient algorithms for mining outliers from large data sets. SIGMOD Rec. **29**(2), 427–438 (2000)
11. Knorr, E.M., Ng, R.T.: Algorithms for mining distance-based outliers in large datasets. In: Proceedings of the 24th International Conference on Very Large Data Bases, pp. 392–403 (1998)
12. Byers, S., Raftery, A.E.: Nearest-neighbor clutter removal for estimating features in spatial point processes. J. Am. Stat. Assoc. **93**(442), 577–584 (1998)
13. Ester, M., Kriegel, H.P., Sander, J., Xu, X.: A density-based algorithm for discovering clusters in large spatial databases with noise. In: Proceedings of the Second International Conference on Knowledge Discovery and Data Mining, pp. 226–231 (1996)
14. Boser, B.E., Guyon, I.M., Vapnik, V.N.: A training algorithm for optimal margin classifiers. In: Proceedings of the Fifth Annual Workshop on Computational Learning Theory, pp. 144–152 (1992)
15. Schoelkopf, B., Williamson, R.C., Smola, A.J., Shawe-Taylor, J., Platt, J.C.: Support vector method for novelty detection. In: Solla, S., Leen, T., Mueller, K. (eds.) Advances in Neural Information Processing Systems, vol. 12, pp. 582–588. MIT Press, Cambridge (2000)
16. Boriah, S., Chandola, V., Kumar, V.: Similarity measures for categorical data: a comparative evaluation. In: Proceedings of the SIAM International Conference on Data Mining, pp. 243–254 (2008)

Development of the Multi-platform Human-Computer Interaction for Mobile and Wearable Devices

Agnieszka Bier and Zdzisław Sroczyński[✉]

Institute of Mathematics, Silesian University of Technology,
Kaszubska 23, 44-100 Gliwice, Poland
{agnieszka.bier,zdzislaw.sroczynski}@polsl.pl

Abstract. The paper concerns the problem of dependability of multi-platform mobile applications developed with FMX framework. Intended for simultaneous development of applications dedicated to multiple mobile platforms, FMX enables faster testing at desktop environment, convenient porting the project to another target platform, as well as internationalization and localization of the software. In this paper we discuss the details of several case-studies illustrating the functionalities listed above and provide comprehensive experimental examples with indication of encountered issues, alongside with the general solutions for the proper user experience design. Our observations show possible weaknesses and advantages of multi-platform applications developed with FMX and therefore can be a valuable reference for designers of multi-platform human-computer interaction and software engineers dealing with the mobile and wearable development and testing.

1 Introduction

Mobile devices became ubiquitous in contemporary modern world, especially when it comes to wearables. This is a kind of very intimate equipment, which includes smartglasses (like Android-based Google Glass) [1], hand-worn smart-watches (running Android Wear, Apple watchOS or the other dedicated embedded OS), smartbands, body-dressed items and even shoes [2] providing the monitoring of game performance for athletes. The wearable devices can extend the set of sensors of the mobile smartphone working as a peripheral device which shares the results of measurements of human physical activity: heartbeat, gait characteristics, steps and distance, or even electrocardiography. On the other hand, there are categories of wearable devices that can operate independently, as for example smartwatches with Android Wear 2.0, able to run every application compatible with relatively small screen and limited input methods [3].

In general, the development of human-computer interaction for mobile devices differs from the conventional desktop or web page projects in the significant way. Besides smaller screen, the designer should also take into account the touch-based and motion gestures, voice commands, continuous operating of

© Springer Nature Switzerland AG 2020
W. Zamojski et al. (Eds.): DepCoS-RELCOMEX 2019, AISC 987, pp. 57–68, 2020.
https://doi.org/10.1007/978-3-030-19501-4_6

the applications and necessary cooperation with the use of notifications [4]. The energy efficiency should be addressed as well, because of the limited capacity of batteries. These factors constitute specific environment with special requirements, which have to be fulfilled in order to acquire the acceptable level of user experience. On the other hand, the software development tools needed for mobile development are essentially different one from another, making the programming and user interface design harder and more time consuming with every mobile platform added to the project schedule.

For these reasons, there is a noticeable interest in simultaneous multi-platform software development as a solution reducing the necessary efforts and setbacks related to differences between particular destination platforms. This goal is often realized by a common API established for this purpose while the integration with underlying mobile operating system is provided by the virtual machine, runtime module or a set of native plugins. A key issue in such systems are the quality and dependability of so-developed applications on different platforms and this became the motivation for the presented research.

In the following, we address the problem of the quality and dependability of software developed simultaneously for different mobile platforms. Our goal is to examine and assure the most convenient, minimal-effort software development method, maintaining the possibly closest interface to native methods of the mobile operating system. We have reviewed potential solutions taking into account the following features:

1. the wide set of target platforms, including the most popular iOS, Android and Android Wear,
2. fast executables, natively compiled, preferably from classic computer languages, shortening learning path,
3. the strong integration of the development environment and ease of the target platform exchange,
4. GUI designer with automatic theming according to the given mobile operating system, providing the appropriate user experience level,
5. the common, platform independent API for run-time library and device's peripherals,
6. the convenient testing module without the need of the deployment to the actual mobile device,
7. the proper internationalization, allowing the fluent translations of the user interface according to the mobile system settings.

Therefore, for further examination we have chosen the FMX framework and toolset, considering they meet the majority of the requirements stated above. We are aware that some of the features, considered useful in above list, are the matter of continuous discussions and will possibly never become absolutely objective. The representative example here may be the preference for natively compiled or managed languages (i.e. C++ vs C#) [5]. Additionally, we have compared some essential features of the FMX with a few other general-purpose solutions available in the market, discussing the differences between the development platforms within the following sections.

On the other hand, we have intentionally omitted some specialized solutions, especially game and multimedia frameworks as Unity 3D, which usually base on a hand-made user interface and do not provide extended support for the design of HCI compliant with mobile manufacturers' guidelines.

2 FMX Multi-platform Software Development Framework

The FMX Framework – RAD Multi-Device Application Platform (called Fire-Monkey or FM in previous editions) developed by Embarcadero is fully object-oriented software library/framework designed to develop multi-platform applications with multimedia and stunning visual effects [6]. An object-oriented architecture allows to embed controls one into another, easily building the innovative solutions of human-computer interaction (HCI) [7]. Furthermore, there is a style engine incorporated into the platform, which provides an easy way to fit the look-and-feel of the application to the particular operating system.

The possible target operating systems are desktop Windows and OS X workstations, Linux servers, Apple's iOS (in iPhones, iPads, iPods Touch) as well as Android [8]. Although there are some dedicated frameworks with similar level of portability, as for example Unity 3D (specialized for game development), FMX appears to be the most sophisticated general usage tool for multi-platform development [9]. GUI client applications built with FMX platform use efficient 2D and 3D vector graphics [10] to build the user experience, the IDE allows RAD (Rapid Application Development) [11] approach with the use of visual designer, enriched by animation effects and transformations of bitmapped graphics. In case of mobile systems, the resulting application is compiled into executable by the native cross-compilers for the ARM architecture. Moreover, the FMX applications take full advantage of capabilities and sensors of the mobile OS and hardware, network connection, geolocation and maps, camera interface, as well as direct connection to platform APIs with all their details [12].

RAD Studio – the programming IDE used for FMX development as well as all the cross-compilers, require a Windows computer to operate. The deployment of the executable to the actual target operating system is done with the use of special software agent called Platform Assistant Server (PAServer). PASever helps also to sign the code for the iOS with the Apple's Xcode code-signing tools. Moreover, the programmer can test applications directly on Windows with the mobile preview option or with the use of pure Windows visual theme. This approach is the major simplification of the development for mobile targets, and makes it similar to the common desktop application design, where the result of the compilation is almost instant.

2.1 Software Localization Issues in FMX

The availability of national versions of the software is very important for users. Contemporary applications have to communicate with the user in his language and automatically switch the language settings according to the settings in the mobile operating system. Software environment of FMX platform makes it easy to setup a correct language in the source code of the app, using special TLang component and multiplatform service called IFMXLocaleService. The exemplary procedure for acquiring two-letter national language code (for example "pl" for Polish, "en" for English etc.) is given in Listing 1.1.

Listing 1.1. The function returning two-letter code for the current device's locale

```
function multiLangID : String ;
var
  LocaleSvc : IFMXLocaleService ;
begin
  LocaleSvc := TPlatformServices . Current . GetPlatformService (←
      IFMXLocaleService )                    as IFMXLocaleService ;
  Result := LocaleSvc . GetCurrentLangID ;
end ;
```

The proper localization support in FMX requires some refactorings in the source code of the library. First, the original `GetCurrentLangID` method sets always only the English locale in iOS operating system. This should be replaced with call to native method from iOS Foundation framework: `NSLocale.preferredLanguages`.

Similarly, for Android the method `GetCurrentLangID` should be corrected as well with the use of `Locale.getLanguage` method.

The three screenshots of the test multi-platform application, developed with FMX framework are presented in Fig. 1. The refactorings mentioned above were necessary to develop an efficient user experience matching the language settings in the mobile operating system.

(a) Android 5/English   (b) Android 9/Polish   (c) iOS 12/Polish

Fig. 1. The test application (puzzle game) developed with the use of FMX framework, localized for different national languages and themed for different operating systems.

2.2   Testing Mobile Applications in MS Windows Environment

The MobilePreview feature of FMX framework offers a possibility to test the app in the MS Windows environment with the GUI drawn in a mobile way (the system uses special graphical style somewhat similar to classic iOS skeuomorphic one). This can be done by adding `FMX.MobilePreview` unit to the `uses` clause in the source code of the project.

Although not all features are available in this mode (mostly because of the lack of some peripherals, as accelerometer in common desktops or laptops), the graphic effects, network connection, all the visual controls can be thoroughly tested on Windows system (see Fig. 2) with much faster compilation and without necessity to pack, sign, and deploy executables to mobile device [13]. As the deployment process to the actual mobile device can take rather long, sometimes minutes, it significantly delays the work of a manual tester. There are the automated testing features available for FMX as well, so these kind of tests can be performed faster and extensively in this mode to provide more stable and reliable applications.

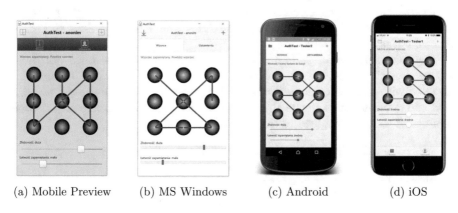

(a) Mobile Preview   (b) MS Windows   (c) Android   (d) iOS

Fig. 2. An example of the mobile application developed with FMX framework: tested on desktop Windows, themed with MobilePreview (a), run at desktop Windows (b), run at Android (c), run at iOS (d).

On the other hand, there are many functionalities that cannot be completely tested without the use of the proper mobile device like smartphone or tablet. The above mentioned accelerometer, photo camera, compass and GPS sensors – they all need real testing in real world. This in particular concerns the wearable devices, which put a challenge by the details of the design, especially by the extremely tiny screen size. In this case the manual testing is superior than automated one due to unspecified objective parameters – the user experience for this category of the mobile equipment has not been thoroughly examined yet. It is worth to note, that automated tests of the user experience/user interface are the one the most significant challenges in the software engineering area, where

Test-Driven Development is successfully used for the business logic algorithms and the methods for the evaluation of the tests are developed as well [14].

Certainly, the simulation cannot assure all real conditions which the mobile application will be facing during usage. Nevertheless, MobilePreview does not depend on desired target of the application (i.e. iOS, Android, Windows) and basic applications working well in this mode are also supposed to work at final mobile operating systems without any alterations, which is undoubtedly an advantage of this solution.

3   Testing Multi-platform Development with FMX Platform

The multi-platform capabilities of the FMX library were tested by porting one OS version to another version of the same project. The portability was verified in two directions. Firstly, the iOS version of the a puzzle game was ported into Android environment. secondly, the reverse procedure was tested with the application designed for pattern lock-code ratings. Additionally, some rather basic applications, displaying calendar and weather forecast from the public REST service, were tested on both mobile platforms and Google Wear OS (former Android Wear).

The complete source code of the FMX project is common for every target software platform, as well as definition of the UI. The definition of the UI is encoded with the use of domain-specific language, which ensures easy serialization and deserialization of the objects representing visual controls and the other non-visual components (see Listing 1.2) [15].

Listing 1.2. The part of the description of some controls in FMX project

```
object TabItemSettings: TTabItem
  CustomIcon = <
    item
    end>
  IsSelected = False
  Size.Width = 177.000
  Size.Height = 49.000
  Size.PlatformDefault = False
  StyleLookup = 'tabitemcontacts'
  TabOrder = 0
  Text = 'Ustawienia'
  ExplicitSize.cx = 50.000
  ExplicitSize.cy = 49.000
  object MemoLog: TMemo
    Touch.InteractiveGestures = [Pan, LongTap, DoubleTap]
    DataDetectorTypes = []
    StyledSettings = [Family, Style, FontColor]
    Align = Client
  ...
```

This approach is different than the other multi-platform tools. For example Xamarin, Cordova/PhoneGap or Adobe Air require a separate design of the user interface for each platform or force manual checks for operating system version to apply the appropriate theme. React Native provides the wrapper to native controls, but uses JavaScript engine for code execution.

In contrast to these solutions, FMX application is not a hybrid one, the user interface is rendered by the application itself using automatic theme, consistent with the mobile operating system, and the code is compiled directly to ARM7 opcodes, which ensures efficiency and utilization of all computational power of the mobile device. On the other hand, this approach requires a new compiler for every new platform or even platform version. This happened for example when Apple introduced a requirement for 64 bit image for the applications distributed by AppStore (from the beginning of 2015). RAD Studio implements LLVM architecture for mobile compilers, so the needed alterations are not extremely complicated, but they take a lot of time and great effort for testing.

The result of the setup of the new target is a new form design in the FireUI designer module of RAD Studio. The icons, colors and placement of some specific controls are fit to the requirements of particular mobile operating system. Comparing Figs. 2(c) and (d) it is clear, that the look and feel of the app changes according to the regulations and guidelines of the target operating system. The most evident example in this case is tabs placement, which is top for Android and bottom for iOS. There are also native dialog boxes available automatically without the need of re-coding. One more factor worth noting is the ability to adapt to different screen sizes. Taking picture from the built-in camera, accessing photo library or sharing a picture also involve standard dialogs of target operating system.

It is possible to arrange some special behaviour of the application using the conditional compilation thanks to directives: {$IFDEF IOS}, {$IFDEF ANDROID} and {$IFDEF WIN32} [16]. This way the developer can design a proper human-computer interaction, taking into account user's habits and some design patterns common at given software platform. For example, the application can respond to resizing of the main window at Windows, while such an action is impossible on mobile platforms, where the app occupies the whole screen (except for special static areas as status bar on the top). In particular, conditional compilation (see Listing 1.3) was necessary in our experiments to determine the exact model of Android device to setup screen presentation according to the shape of the smartwatch (round or rectangular).

Listing 1.3. Conditionally compiled code to determine the model of Android mobile device

```
...
var   WatchModel: String;
begin
  {$IFDEF ANDROID}
  WatchModel:=JStringToString(TJBuild.JavaClass.MODEL);
  if (WatchModel='G Watch R') or (WatchModel='HUAWEI WATCH') then
  begin
  ...
```

The same approach was used to deal with power manager issues at Wear OS smartwatch connecting with remote REST server in weather forecast test application. Due to relatively long initialization time of the connection we encountered frequent timeouts because of the application going automatically into the background. It was necessary to turn off the energy saving options for this particular application and conditional compilation was convenient for this task, because it was completely platform-dependant (Fig. 3).

(a) Round smartwatch   (b) Rectangular smartwatch

Fig. 3. Calendar FMX application tested on different Wear OS devices: round (a) and rectangular (b) one.

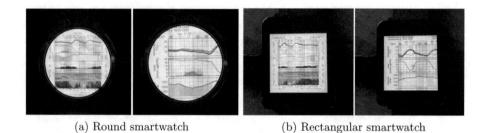

(a) Round smartwatch   (b) Rectangular smartwatch

Fig. 4. Weather forecast FMX application tested on different Wear OS devices: round (a) and rectangular (b) one.

For the purpose of our experiments we developed two test applications designed particularly with smartwatches in mind: weather forecast and calendar. It is worth noting that even properly designed user interface, automatically fitting into different screen sizes, ratios and resolutions, needs some extra tuning. The main requirement is the navigation based solely on gestures.

The weather forecast is a kind of application which encounters a typical problem for small smart devices: the visualization of the detailed graphics is hardly readable due to scale. Certainly it needs to be enlarged and moved on user's request. Therefore, we introduced a set of gestures to ensure the following tasks: swipe up, down and left moving the weather chart, chevron gesture to reset the horizontal position (swipe right is restricted for the operating system and terminates the application) and long tap to enlarge/reset the size of the chart (see Fig. 4).

For the calendar application we have implemented only up and down gestures, avoiding the conflict with the operating system gesture. There are also two active touch fields to the left and to the right from the calendar control, enabled only at rounded watches. With this application we tested the standard list to choose the value (for months and years). The lists were operational in general, but rounded watch made it harder to use, because only the items in the middle of the screen were fully visible. This inconvenience is present in the standard applications of Wear OS, for example in the settings, so that is not specific for the FMX framework.

Generally speaking, we did not encounter any issues during the portability tests. As long as the programmer keeps up with the FMX framework, which supports the extensive set of features of the mobile devices, the simultaneous development for many operating systems is fluent. Certainly, going into some very advanced features requires the direct access to the API of the particular mobile operating system. Nevertheless, majority of educational, business, data visualization application and logical games can be a potential target for FMX development.

3.1   Efficiency of the Applications Based on FMX Framework

During our experiments we have tested the application with Huawei Watch smartwatch, equipped with 1.2 GHz quad-core Qualcomm Snapdragon 400 processor and running Google Wear OS 2.2/Android 7.1). We observed an average delay of 2 to 5 s (depending on the particular app) at launch of the different FMX-developed applications in comparison to the typical apps from the operating system. Further operations, reactions on users activities and screen refreshing were immediate and fluent. The FMX application executable for Android is stored in the dynamically linked native library, which is relatively large, because it contains the whole framework, run-time modules and a part of the necessary assets. The result size of the unpacked empty FMX app is about 20 MB. This size does not grow significantly with the further development of the app, although it takes time to load it for the first time after launch and it takes more space than traditional Android Java-based apps. Therefore, the probable cause of the recorded launch delay at tested wearable device may be the time necessary to load the large compiled module of the app and the memory manager operations on the limited operational memory (512 MB), which can be considered too small for a typical Android device. Despite these inconveniences, we consider the overall user experience for FMX-based application at wearable devices as acceptable.

The main test applications: pattern unlock simulator and puzzle game were developed with the use of "test at Windows first" approach. Pattern unlock simulator is an essential part of the system for collection and rating the strength of pattern passwords on the mobile platform. It was tested by a few hundreds of users with different devices and no operational issues were reported [17].

The iOS app version was tested using iPhone 4s, iPhone SE, iPhone 7 and iPhone 8 with the up to date version of iOS system. The application launch took about 2–3 s for the slowest device and less than second for the fastest one, which is similar to the other applications, including built-in system services, as phone book or telephone module. The overall efficiency, taking into account responsiveness for user actions was also at the standard level.

The same application, compiled for Android target, was tested using Samsung Galaxy S4 mini i9195 (2013 medium-level smartphone with dual-core Qualcomm Snapdragon 400 1.7 GHz processor, running Android 4.2.2 Jelly Bean), SONY Experia Z1 Compact with Android 5.1 and SONY Experia XZ1 with Android 9. The application run smoothly not only at the high-end smartphones, but at medium-level as well.

4   Discussion and Conclusions

We have presented the evaluation of the multi-platform FMX development framework in terms of efficiency, maturity and stability. We considered the overall efficiency of the examined solution as the performance of the developed applications and the convenience of the development process, IDE features and the comprehensiveness of the framework API. We assumed that performance similar to the standard, built-in application from the mobile operating system should be sufficient for the production development. We have also chosen some different types of applications (graphical, with remote back-end, communicating with device's sensors, running at wearables) to conclude the efficiency of the development environment, which in the subjective rating also indicates the suitability for the production use.

The maturity of the development framework can be generally defined as the lack of critical errors ("show stoppers") and additionally the effective support from the community and manufacturer, as well as comprehensive documentation. Although we did not refer to the support and documentation before, this factors are at adequate level in case of FMX, allowing the comfortable development of different categories of mobile apps, that were created during our experiments. The FMX platform is without a doubt far above the proof of concept stage.

The stable development platform gives the assurance of uninterrupted operations at the most of the available hardware regardless the operating system version. In fact, certain test applications developed during our experiments were used by hundreds of users equipped with many different devices. They had very small screens and big tablet ones, relatively old and contemporary high-end ones – all run the software developed with the use of FMX framework without locks or unexpected lagging.

The results of experiments with Android and iOS versions of the test applications conducted during our case-study indicate that FMX platform can be considered as a mature, stable and efficient solution for multi-platform software development.

We have investigated the issues in the field of internationalization and localization of the mobile software. It can be concluded that FMX and the other programming libraries should be tested in the real environment, using appropriate national settings, as vendors have limited possibilities to check their work against all the languages available in the mobile operating system.

The core feature examined during our case-study was the portability of the applications, i.e. the ability to re-compile and run them at different mobile operating systems without the need of repeating the GUI design or even source code fine-tuning. This part of experiments was successful as we obtained the multi-platform applications whose performance and user interactions (the "look and feel") were at the standard level of native apps. Moreover, we have formulated some general recommendations for the human-computer interaction design for wearable devices.

The presented results of testing multi-platform development of mobile applications with FMX platform provide new insights, valuable for programmers dealing with the mobile development targeting iOS and Android operating systems. Several observations, remarks, requirements and solutions prepared during the described case-studies can concern the general aspects of software engineering in terms of the design of human-computer interaction for mobile and wearable devices.

The mobile HCI development with the use of the multi-platform frameworks including FMX is certainly worth further investigations on user experience, portability and energy efficiency in more complex software projects.

References

1. Bulling, A., Kunze, K.: Eyewear computers for human-computer interaction. Interactions **23**(3), 70–73 (2016)
2. Mewara, D., Purohit, P., Rathore, B.P.S.: Wearable devices applications & its future. In: Science [ETEBMS-2016], vol. 5, no. 6 (2016)
3. Sroczyński, Z.: Internet of things location services with multi-platform mobile applications. In: Proceedings of the Computational Methods in Systems and Software, pp. 347–357. Springer (2017)
4. Zhang, H., Rountev, A.: Analysis and testing of notifications in android wear applications. In: Proceedings of the 39th International Conference on Software Engineering, pp. 347–357. IEEE Press (2017)
5. Bluemke, I., Gawkowski, P., Grabski, W., Grochowski, K.: On the performance of some C# constructions. In: Advances in Dependability Engineering of Complex Systems, pp. 39–48. Springer (2017)
6. Sroczyński, Z.: Human-computer interaction on mobile devices with the FM application platform. In: Rostański, M., Pikiewicz, P. (eds.) Internet in the Information Society. Insights on the Information Systems, Structures and Applications, pp. 93–106. University of Dabrowa Gornicza Press (2014)
7. Harrison, R., Flood, D., Duce, D.: Usability of mobile applications: literature review and rationale for a new usability model. J. Interact. Sci. **1**(1), 1 (2013)
8. Sroczyński, Z.: Designing human-computer interaction for mobile devices with the FMX application platform. Theor. Appl. Inf. **26**(1–2), 87–104 (2014)
9. Arsjentiev, D.A., Pashkov, P.S.: The Environment for Multi-device Application Development Delphi XE5. Bulletin of Moscow State University of Printing Arts, Moscow (2013). (Russian edn.)
10. Kovačević, Ž.: RAD studio: developing mobile applications. In: 16. CARNetova korisnička konferencija-CUC 2014 (2014)

11. Kralev, V.S., Kraleva, R.S.: Methods and tools for rapid application development. In: International Scientific and Practical Conference World science, vol. 1, no. 4, pp. 21–24. ROST (2017)
12. Teti, D.: Delphi Cookbook. Packt Publishing Ltd., Birmingham (2016)
13. Sroczyński, Z.: Actiontracking for multi-platform mobile applications. In: Computer Science On-line Conference, pp. 339–348. Springer (2017)
14. Derezińska, A., Trzpil, P.: Mutation testing process combined with test-driven development in. NET environment. In: Theory and Engineering of Complex Systems and Dependability, pp. 131–140. Springer (2015)
15. Chandler, G.: FireMonkey Development for iOS and OS X with Delphi XE2. Coogara Consulting, Melbourne (2012)
16. Glowacki, P.: Expert Delphi. Packt Publishing Ltd., Birmingham (2017)
17. Bier, A., Kapczyński, A., Sroczyński, Z.: Pattern lock evaluation framework for mobile devices: human perception of the pattern strength measure. In: International Conference on Man–Machine Interactions, pp. 33–42. Springer (2017)

Tool for Assessment of Testing Effort

Ilona Bluemke[(⊠)] and Agnieszka Malanowska

Institute of Computer Science, Warsaw University of Technology,
Warsaw, Poland
I.Bluemke@ii.pw.edu.pl, a.h.malanowska@gmail.com

Abstract. Testing, being one of crucial factors in providing high quality software, needs a significant amount of resources of the whole project. Estimation of amount of expenditures required to test the software, called testing effort, would considerably facilitate the project management process. Early estimation of the test effort, e.g. during design process, can significantly facilitate project management and resources distribution optimization. We made an attempt to prepare such an estimate. The tool estimating time necessary to test the system, on the basis of its UML model, called IoTEAM, was designed and developed. The tool is based on two methods: mapping UML class and sequence diagrams into the results of Function Point Analysis and Test Point Analysis. The choice of those methods was preceded by a comprehensive study of the current state of art. Some results of this study are presented in this paper. The usage of IoTEAM is also described.

Keywords: Testing effort · FPA · Function Point Analysis ·
Test Point Analysis · TPA · UML

1 Introduction

Software testing is a key part of the software development process, especially for quality assurance, but it requires a lot of time and resources. It is estimated that testing activities consume more than a half of the cost of the whole software development process [1, 2]. Consequently, an estimation of the effort necessary for software testing at early stage of the system development would be very useful for project managers. Unfortunately, studies dedicated to the assessment of testing effort are rare. We made a systematic literature review and we were looking for methods, which can be applied in early stages of software development and can be semi-automatic. The results of our literature review can be found in [3] and in Sect. 2.1 we briefly recall some of those results.

We found only one method which can be implemented i.e. Test Point Analysis, TPA [4]. This method is based on function points [5]. As currently many systems are being developed using object-oriented approach, we decided to implement automated calculation of function points on the basis of UML [6] diagrams. This transformation is based on Uemura, Kusumoto and Inoue algorithm [7, 8].

A tool named IoTEAM (Implementation of Testing Effort Assessment Method) has been designed and is being implemented at the Institute of Computer Science Warsaw

© Springer Nature Switzerland AG 2020
W. Zamojski et al. (Eds.): DepCoS-RELCOMEX 2019, AISC 987, pp. 69–79, 2020.
https://doi.org/10.1007/978-3-030-19501-4_7

University of Technology. This tool works as an extension (a plug-in) of Microsoft Visual Studio Enterprise 2015 [9]. IoTEAM estimates the testing effort on the basis of class and sequence diagrams and some information given by a user. The estimation of testing effort is expressed in the number of hours required to test the system. To our best knowledge, it is the first tool calculating test effort from UML diagrams in semi-automatic way. Unfortunately we were not able to verify this tool on real-world projects. Software companies are not willing to provide data about testing effort, neither to provide the project (UML diagrams). Although there are several public repositories of UML models or diagrams available on the Internet e.g. [10–12], they do not contain any significant projects that could be used to test IoTEAM. Even if we test our tool on UML models from such repositories, we would not be able to verify the results of the estimation, since no information about testing time is provided. We made an exemplary project (shown in Sect. 4) to present how IoTEAM tool can be used.

The paper is organized as follows. Section 2.1 contains related work, Sect. 2.2 identifies key features of transformation from UML diagrams into function points, while in Sect. 2.3 calculation of testing effort is presented. In Sect. 3 the architecture of IoTEAM is briefly described. In Sect. 4 main steps of the test effort estimation are presented for an exemplary system. Finally, Sect. 5 concludes the paper, highlighting some issues and indicates future research directions.

2   Theoretical Background

There are a lot of publications dedicated to the reduction of testing effort. Related work on test effort estimation is briefly presented in Sect. 2.1.

The goal of IoTEAM is the transformation from UML diagrams into test hours estimated for the project. In the first phase function points are estimated form UML diagrams using method proposed by Uemura, Kusumoto and Inoue [7, 8]. This process is briefly described in Sect. 2.2. Next, Test Point Analysis (TPA) [4] is applied. Main features of this analysis are presented in Sect. 2.3.

2.1   Related Work

Different approaches are described in review articles [2] and [1]. Bareja and Singhal [2] concentrate on the existing machine learning methods to classify software modules as fault-prone or non-fault-prone. The main idea of those distinction between modules is that non-fault prone modules require less attention while being tested. Elberzhager et al. [1] conducted a systematic mapping study of methods of reduction of testing effort. They distinguished five categories of methods: prediction of fault-prone elements of the system, automation of testing, reduction of the size of test input, usage of quality assurance methods in earlier stages of the development process and different test strategies [1]. All the described approaches could possibly help reducing test effort, however, our main concern is to estimate the size of an effort, not to reduce it. None of those methods calculates estimation, so they cannot be applied in our case.

Sharma and Kushwaha [13] propose the prediction of effort on the basis of the improved requirement based complexity (IRBC) measure. They attempt to predict the effort of the whole software development process. As a source for IRBC calculation, they use Software Requirement Specification (SRS) document. Estimation of the whole software development effort is also the main concern of Idri et al. [14]. They use ensembles of classical or fuzzy analogy methods to predict the effort. Both recalled articles are examples of different approaches for effort estimation. The first one can be called analytical and is based on some measures or metrics. The second one uses artificial intelligence methods, especially machine learning. Unlike authors of [13] and [14], we focus on the testing effort only, so theirs propositions are unacceptable for us.

There are also some approaches to optimally allocate resources for testing. Fiondella and Gokhale [15] assume that more efforts should be allocated to those parts of the system, which require more reliability. They identified two key characteristics that have an impact on system reliability: system architecture and relationship between effort and reliability of the particular component [15]. Lo [16] considers using sensitivity analysis in order to allocate resources so that number of remained faults or testing effort are minimized.

Only a few researches consider testing effort estimation as a separate problem. De Almeida et al. [17] proposed an analytical method of such estimation on the basis of the use case model. Badri et al. [18] study the impact of the use case model on the size of test suites. Aranha and Borba [19] and Nguyen et al. [20] developed methods of effort estimation on the basis of the test specifications. The former use test specification written in controlled natural language, the latter is based on the test cases. None of these approaches were implemented as an automatic or semi-automatic tool.

2.2   Transition from UML Diagrams into Function Points

Function points (FP) were proposed many years ago by Albrecht [21]. Nowadays, there are several versions of Function Point Analysis (FPA) method, among which IFPUG [5] version is one of the most common. IFPUG version of Function Point Analysis consists of seven steps, as described in [7] (details can be found in [7] or [3]):

1. Determination of the type of function points count – one of the following: development project, enhancement project or application.
2. Determination of the system boundary i.e. which functions belong to the scope of the system.
3. Determination of data functions and their complexity. There are two types of data functions: internal logical files (ILFs) and external interface files (EIFs). Their complexity is counted on the basis of DET (data element type) and RET (record element type) factors.
4. Determination of transactional functions and their complexity. There are three types of transactional functions: external inputs (EI), external outputs (EO) and external inquiries (EQ). Their complexity is counted on the basis of DET and FTR (file type referenced) factors.
5. Determination of the unadjusted function points, which is the sum of FPs assigned to all data functions and transactional functions.

6. Determination of the value adjustment factor. This factor allows to include the impact of general system characteristics on the size of the system expressed in the function points.
7. Determination of the adjusted function points – on the basis of the unadjusted function points, value adjustment factor and the type of function points count.

Uemura, Kusumoto and Inoue proposed the method to conduct the first five steps of the IFPUG version of FPA on the basis of the UML class and sequence diagrams. These steps are recalled below [7]:

1. Only two types of function points count are taken into consideration in this approach, that is, development project or enhancement project. Applicable UML diagrams for the selected type of count must be available.
2. There is a distinction between objects presented in the sequence diagrams. There are two categories of objects: actor and non-actor objects. Uemura, Kusumoto and Inoue [7] assume that actor objects are outside the system, what implies that only non-actor objects belongs to the system.
3. Only non-actor objects can be considered as data functions, for only those objects are in scope of the system. The candidate for data function is every non-actor object (from the sequence diagram) which has at least one attribute and exchanges data with non-actor objects [7]. Data exchange is understood as an exchange of message with at least one parameter in the sequence diagram. Then, it is necessary to define the type and complexity of each candidate. Uemura, Kusumoto and Inoue propose the following approach. ILFs are those candidates that 'have attributes changed by the operations of other objects', while the rest of candidates are defined as EIFs. Determination of the distinction between ILFs and EIFs requires information about the transactional functions identified in the system. If the candidate for data function is connected with at least one EI transactional function, it is treated as an ILF [7]. Otherwise, it is assumed that the data function is an EIF. The value of DET factor for the data function is the number of attributes of the object, including attributes inherited from super-classes. The RET factor has the constant value equal to 1 [7].
4. Transactional functions are formed from single messages or sequences of messages from the sequence diagrams. Those messages have to provide data exchange, as defined above. Only sequences of messages initiated by the actor objects are taken into consideration in this step. To determine the type and complexity of each transactional function (i.e. sequence of messages), there are five patterns of sequences defined in [7]. In each pattern, DET is equal to the number of parameters of a message and FTR is the number of data functions objects that appeared in the sequence. These patterns were proposed for UML 1.0. Because of the UML modifications, we **have proposed two additional modifications of the original patterns**, described in [3]. The tool described in this paper uses both original and modified patterns to recognize data functions.
5. As in the original FPA method, the unadjusted function points is calculated as a sum of points assigned to all data functions and transactional functions.

2.3   Test Point Analysis

Test Point Analysis (TPA), introduced by van Veenendaal and Dekkers [4], can be used to estimate the time of system or acceptance tests, which means that it is designed to estimate the time of black box testing. One of its input data are Function Point Analysis result. As it is stated in [4], an estimation of white box testing time is contained in the function points count. The conclusion is, that combination of FPA and TPA allows to calculate estimated time of the whole testing process – from unit tests to acceptance tests. Estimating black box testing three elements must be considered [4]:

1. size of the system being tested,
2. test strategy (tested components, quality attributes, coverage) and
3. productivity.

Information about system size and test strategy is used to calculate test points, which can be defined as a kind of measure of testing effort independent of any specific testing organization. The size of the system can be given by the number of function points assigned to it (Sect. 2.2). Productivity factor is the number of hours necessary to perform given testing work [4] (i.e. to perform one test point). Combination of these three elements enables us to express testing effort estimation in hours.

User has to choose appropriate values defined in TPA method (detailed values given in [4]). Unlike in FPA, the main unit of system in TPA is a function, instead of data functions and transactional functions. Some transformation from FPA terms into TPA terms has to be conducted. We have proposed such a transformation in [3] and it is implemented in IoTEAM.

TPA method starts with calculation of dynamic test points which depends on two factors: function-dependent and dynamic quality-dependent. In function-dependent factor five parameters of a function (i.e. user-importance, usage-intensity, interfacing, complexity and uniformity) are taken into consideration. Quality-dependent factor is calculated on the basis of the information about the importance of dynamically explicitly tested quality attributes and the number of dynamically implicitly tested quality attributes. Dynamic test points are calculated separately for each function on the basis of the two above factors and the number of function points assigned to the function. The number of static test points is calculated on the basis of the number of quality attributes tested statically. Total number of test points is calculated using the number of dynamic and static test points, with the formula presented in [4]. The number of primary test hours is influenced by total number of test points and values of productivity and environmental factor. Environmental factor value is determined on the basis of six parameters which describe the manner of system development and testing i.e. test tools, development testing, test basis, development environment, test environment and testware. The value of primary test hours represents work involved in preparation, specification, execution and completion of tests [4]. Next, management activities like planning and control of testing, must be included. The size of allowance for these activities reflects the size of the test team and the level of automation of management process. Combination of primary test hours and management allowance gives the total number of test hours i.e. testing effort estimation.

An example of TPA usage is presented in Sect. 4. The detailed values for several parameters used in TPA method are given in [4]. Every parameter has some nominal value defined, which can be used if the proper value is unknown. It is not clear on what basis these values were chosen.

3   IoTEAM Tool

Our tool IoTEAM (Implementation of Testing Effort Assessment Method) works as an extension (a plug-in) of Microsoft Visual Studio Enterprise 2015 [9]. It can be accessed through graphical user interface. The Windows Presentation Foundation [22] – WPF – technology was used for presentation. To separate logic and presentation layers we used MVVM (Model – View – ViewModel) design pattern [23]. The usage of MVVM resulted in IoTEAM architecture shown in Fig. 1.

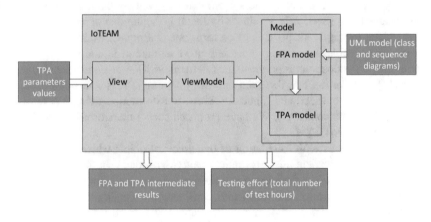

Fig. 1. Structure of IoTEAM

Structure of IoTEAM consists of three modules: `View`, `ViewModel` and `Model`. `View` implements graphical user interface, through it the user inserts some parameters used in the Test Point Analysis (TPA parameters). XAML (Extensible Application Markup Language) and C# languages were used for its implementation. C# code of the `View` module, according to the MVVM idea, is not implementing any logic, neither any services. `ViewModel` is a mediator between presentation and logic layers of an application. Module `Model`   is responsible for all analysis and contains two parts. The first one (`FPA model`) transforms UML class and sequence diagrams into results of Function Point Analysis (data functions and transactional functions) and is fully automatic. The second part (`TPA model`) transforms data functions and transactional functions into TPA functions and calculates testing effort (in hours). The result strongly depends on parameters entered by user (TPA parameters). IoTEAM produces total number of test hours. All intermediate results of the analysis (coefficients used in TPA method) are also presented to the user (examples are shown in next Section). More details about the IoTEAM implementation can be found in [3].

4 Example

To present the usage of IoTEAM a special simple system was designed. This system is supporting functioning of a small guesthouse with three- and four-person rooms. The potential guest is able to make a reservation of appropriate room and for desirable period, cancel the reservation and pay, even in few parts. The employee of the guesthouse is also able to make reservations, and has some additional functions like check in/check out guest. This simple system stores information about reservations and available rooms. For this exemplary system class and sequence diagrams were prepared.

The first step of testing effort calculations is the function point evaluation, which is fully automated. The detailed information about the results of Function Point Analysis is presented to the user, as in an example in Fig. 2.

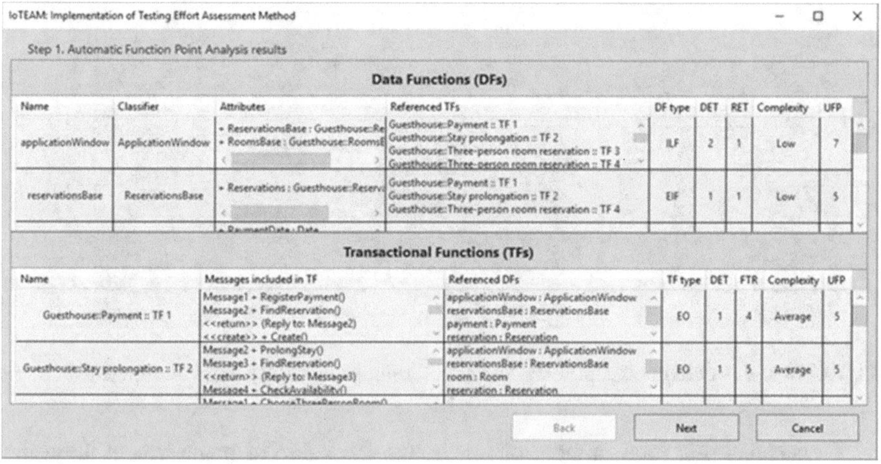

Fig. 2. Step 1 – results of function points calculations

In step 2 the results of function points calculations (data functions and transactional functions) are automatically transformed into data for TPA analysis (functions in TPA meaning). For each function values of five parameters have to be defined (user importance, complexity, etc.). Each parameter has some nominal value, as proposed in

[4], but these values can be modified by user in the third step. In step 4 user has to determine the importance of dynamically, explicitly measured quality attributes: suitability, usability, security, efficiency. Those attributes are chosen by van Veenendaal and Dekkers [4] from ISO 9126 standard. Each attribute can be defined as: unimportant for test purposes, relatively unimportant but requires testing, normal importance, very and extremely important, as proposed in [4]. In next step (5) user chooses dynamically, implicitly measured quality attributes. In this case, all sub-characteristics of ISO 9126 quality characteristics (functionality, usability, reliability, efficiency, maintainability, portability) are equally treated as quality attributes.

The function- and quality-dependent factors influence the number of dynamic test points assigned to a function. The number of tested quality attributes (defined as above i.e. sub-characteristics of six ISO 9126 quality characteristics) also influences the number of static test points (step 6). User has to determine which of them will be tested by static tests. In next step (7) IoTEAM calculates primary test hours using productivity and environmental factors - Fig. 3. Environmental factor value depends on six parameters described in Sect. 2.3.

Fig. 3. Step 7 – determining primary test hours using productivity and environmental factors

To calculate the total number of test hours, the value of management allowance (planning and control allowance [4]) has to be counted. This value is determined concerning size of team and used management tools in step 8. Finally (step 9), the results of the whole analysis are presented as shown in Fig. 4. The last row of the table contains the final result of the analysis i.e. testing effort expressed in test hours.

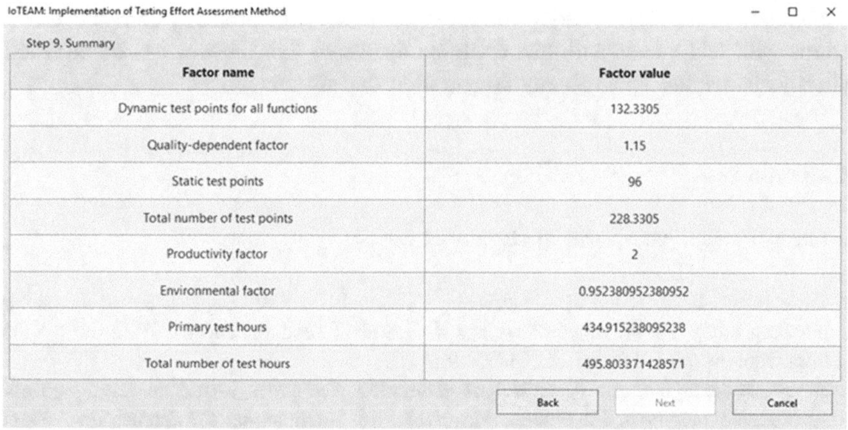

Factor name	Factor value
Dynamic test points for all functions	132.3305
Quality-dependent factor	1.15
Static test points	96
Total number of test points	228.3305
Productivity factor	2
Environmental factor	0.952380952380952
Primary test hours	434.915238095238
Total number of test hours	495.803371428571

Fig. 4. Step 9 – results summary

5 Conclusions

It is obvious that knowledge about the time required for testing activities could support information project management. Such estimation, especially in an early stage of the project, would be useful not only to managers, but also to the other members of the team, e.g. developers or testers. Still it is very difficult to find a method or a tool supporting such estimation. Our tool – IoTEAM – fills this gap. It calculates the number of hours necessary to perform all testing activities on the basis of the system class and sequence diagrams and information about the development and testing process. The tool is based on the transformation of UML diagrams into function points and the Test Point Analysis. Such combination of FPA and TPA allows to calculate estimated time of the whole testing process – from unit tests to acceptance tests.

Our contribution is not only the implementation of the methods proposed in literature, but also adaptations of these methods and preparing transformation between the output of the first one and the input of the second one. The combination of those methods is unique and, to our best knowledge, IoTEAM is the first tool that allows to estimate testing time using UML model of the system. The conducted literature review indicates that there are not any automatic or semi-automatic tools for testing effort assessment which can be applied in early stages of the development process at all.

IoTEAM needs some improvement. Both methods used in analysis (Sect. 2) were published in 1999 and are based on outdated versions of standards. In the current version of IoTEAM, the new elements of UML 2.x and the new software quality standard, ISO 25010, are not supported. Further work includes modifications of the methods used, in order to comply with the new standards. Particularly, new elements of sequence diagrams, such as combined fragments or interaction uses, are planned to be included in the analysis in near future.

It would be useful to verify IoTEAM on real-world projects, compare the estimations with real values and then calibrate results so that the tool can be adjusted to real projects, we are open on any cooperation on this subject.

References

All Internet pages were valid in December 2018

1. Elberzhager, F., Rosbach, A., Eschbach, R., Münch, J.: Reducing test effort: a systematic mapping study on existing approaches. Inf. Softw. Technol. **54**(10), 1092–1106 (2012). https://doi.org/10.1016/j.infsof.2012.04.007
2. Bareja, K., Singhal, A.: A review of estimation techniques to reduce testing efforts in software development. In: Proceedings 2015 Fifth International Conference on Advanced Computing & Communication Technologies, Haryana, IEEE, pp. 541–546 (2015). https://doi.org/10.1109/acct.2015.110
3. Malanowska, A.: Testing effort assessment, BSc thesis, Institute of Computer Science, Warsaw University of Technology (2017). (in Polish)
4. van Veenendaal, E.P.W.M., Dekkers, T.: Test point analysis: a method for test estimation. In: Kusters, R., et al. (ed.), Project Control for Software Quality. Shaker Publishing, Maastricht (1999). http://www.erikvanveenendaal.nl/NL/files/Testpointanalysis%20a%20method%20for%20test%20estimation.pdf
5. ISO/IEC 20926:2009: Software and systems engineering—Software measurement IFPUG functional size measurement method, ISO/IEC (2009). https://www.iso.org/obp/ui/#iso:std:iso-iec:20926:ed-2:v1:en
6. OMG Unified Modeling Language: Version 2.5 (2015). http://www.omg.org/spec/UML/2.5/PDF/
7. Uemura, T., Kusumoto, S., Inoue, K.: Function-point analysis using design specifications based on the Unified Modelling Language. J. Softw. Maint. Evol. Res. Pract. **13**(4), 223–243 (2001). https://doi.org/10.1002/smr.231
8. Uemura, T., Kusumoto, S., Inoue, K.: Function point measurement tool for UML design specification. In: Proceedings: Sixth International Software Metrics Symposium, pp. 62–69. IEEE, Boca Raton (1999). https://doi.org/10.1109/metric.1999.809727
9. Welcome to Visual Studio 2015, in: Microsoft Developer Network, Microsoft (2017). https://msdn.microsoft.com/en-us/library/dd831853.aspx
10. The Open Model Initiative. http://openmodels.org/
11. Metamodel Zoos. http://web.emn.fr/x-info/atlanmod/index.php?title=Zoos
12. GenMyModel. https://repository.genmymodel.com/public/0
13. Sharma, A., Kushwaha, D.S.: Applying requirement based complexity for the estimation of software development and testing effort. ACM SIGSOFT Soft. Eng. Notes **37**(1), 1–11 (2012). https://doi.org/10.1145/2088883.2088898
14. Idri, A., Hosni, M., Abran, A.: Improved estimation of software development effort using Classical and Fuzzy Analogy ensembles. Appl. Soft Comput. **49**, 990–1019 (2016). https://doi.org/10.1016/j.asoc.2016.08.012
15. Fiondella, L., Gokhale, S.S.: Optimal allocation of testing effort considering software architecture. IEEE Trans. Reliab. **61**(2), 580–589 (2012). https://doi.org/10.1109/tr.2012.2192016

16. Lo, J.H.: An algorithm to allocate the testing-effort expenditures based on sensitive analysis method for software module systems. In: TENCON 2005 - 2005 IEEE Region 10 Conference, pp. 1–6. IEEE, Melbourne (2005). https://doi.org/10.1109/tencon.2005.301151

17. de Almeida, É.R.C., de Abreu, B.T., Moraes, R.: An alternative approach to test effort estimation based on use cases. In: 2009 International Conference on Soft. Testing Verification and Validation Workshops, Denver, pp. 279–288 (2009). https://doi.org/10.1109/icst.2009.31

18. Badri, M., Badri, L., Flageol, W.: On the relationship between use cases and test suites size: an exploratory study. ACM SIGSOFT Soft. Eng. Notes **38**(4), 1–5 (2013). https://doi.org/10.1145/2492248.2492261

19. Aranha, E., Borba, P.: Test effort estimation models based on test specifications. In: Proceedings of the Testing: Academic and Industrial Conference, Practice and Research Techniques, pp. 67–71. IEEE, Windsor (2007). https://doi.org/10.1109/taic.part.2007.29

20. Nguyen, V., Pham, V., Lam, V.: qEstimation: a process for estimating size and effort of software testing. In: Proceedings of the 2013 International Conference on Software and System Process, pp. 20–28. ACM, New York (2013). https://doi.org/10.1145/2486046.2486052

21. Albrecht, A.J.: Function point analysis. Encyclopedia of Software Engineering, vol. 1, pp. 518–524. Wiley, Chichester (1994)

22. Windows Presentation Foundation, in: Microsoft, Microsoft (2017). https://docs.microsoft.com/en-us/dotnet/framework/wpf/index

23. The MVVM Pattern, in: Microsoft Developer Network, Microsoft (2012). https://msdn.microsoft.com/en-us/library/hh848246.aspx

Flow Shop Problem with Machine Time Couplings

Wojciech Bożejko$^{(\boxtimes)}$, Radosław Idzikowski, and Mieczysław Wodecki

Faculty of Electronics, Wroclaw University of Science and Technology,
Janiszewskiego Street 11/17, 50-372 Wrocław, Poland
{wojeciech.bozejko,radoslaw.idzikowski,
mieczyslaw.wodecki}@pwr.edu.pl

Abstract. The work presents a multi-machine flow shop problem with minimization of the completion time of all tasks. Different temporal couplings concerning machine operation have been proposed, resulting from the analysis of practical cases. Their mathematical models were presented, an exact algorithm and two approximate algorithms were implemented.

Keywords: Task scheduling · Time constraints · Limited idle · Limited waiting

1 Introduction

In the process of managing the production line or planning construction works, there is a problem of scheduling tasks, in which there are many limitations due to the technologies available. Such a process is identified in the literature as a classic flow shop problem [2], belonging to the class of strongly NP-hard problems, which organic use of algorithms accurate only for small problem instance sizes. For larger sizes, metaheuristic algorithms are usually used [5].

In the work we present various time couplings (also see: [1]) dependent on the machine. In the literature one can meet the limit of continuous work of machines (called textitno-idle constraint), then we propose two other constraints, where there will be acceptable breaks in the schedule.

2 Problem Formulation with Additional Constraints

In a flow problem, there is a set of n tasks:

$$J = \{J_1, J_2, \ldots, J_n\}, \tag{1}$$

which should be executed on m machines from the set:

$$M = \{M_1, M_2, \ldots, M_n\}. \tag{2}$$

© Springer Nature Switzerland AG 2020
W. Zamojski et al. (Eds.): DepCoS-RELCOMEX 2019, AISC 987, pp. 80–89, 2020.
https://doi.org/10.1007/978-3-030-19501-4_8

Each task consists of m operations $O_{i,j}$ which should be executed in $p_{i,j}$ time on a dedicated machine M_j:

$$J_i = \{O_{i,1}, O_{i,2}, \ldots, O_{i,m}\}. \tag{3}$$

Tasks are executed in the same order (permutation) π on each machine. The problem comes down to determining the permutation of tasks so that the end time of all tasks (makespan) is minimal.

2.1   Mathematical Model

236/5000 To create a schedule by $S_{\pi(i),j}$ will mark the start time of the $\pi(i)$ task on the M_j machine and the $C_{\pi(i),j}$ end moment of this task tasks on the same machine. Tasks perform continuously, so:

$$C_{\pi(i),j} = S_{\pi(i),j} + p_{\pi(i),j}. \tag{4}$$

Only one task can be performed on each machine at a time:

$$C_{\pi(i+1),j} \geq C_{\pi(i),j} + p_{\pi(i+1),j}. \tag{5}$$

As part of the tasks, the technological order must be kept:

$$C_{\pi(i),j+1} \geq C_{\pi(i),j} + p_{\pi(i),j+1}. \tag{6}$$

Assuming that the starting moment of the first permutation task has the machine M_1 is 0: $S_{\pi(1),1} = 0$. Based on the formulas (4)–(6), we can conclusively state that the end time of all tasks is the moment of ending the last task (due to the permutation) π on the machine M_m:

$$C_{max} = C_{\pi(n),m}. \tag{7}$$

Table 1. The times of operations of tasks

Task	1	2	3	4	5	6
Machine 1	2	3	3	6	5	6
Machine 2	3	1	1	2	5	4
Machine 3	3	4	4	3	3	5

Table 1 shows an example of input data with all operation times $p_{i,j}$ for 5 tasks ($n = 5$) and 3 machines ($m = 3$). We will consider ordering tasks of the classic flow problem for natural permutation of $\pi = (1, 2, 3, 4, 5)$. The exact schedule is shown in the Gantt diagram in Fig. 2, where it is seen that the end time for all tasks is $C_{\max} = 34$.

For the natural permutation $\pi = (1, 2, 3, \cdots, n-1, n)$ we design a directed graph

$$G = (V, E) \tag{8}$$

with weighted vertices and edges (arcs), where (Fig. 1):

- set of vertices $V = \bigcup_{j=1}^{m} \bigcup_{i=1}^{n} \{(i,j)\}$,
- set of edges (arcs) $E = E_1 \cup E_2$, where:
  - horizontal arcs $E_1 = \bigcup_{j=1}^{m} \bigcup_{i=1}^{n-1} \{(i, i+1)\}$,
  - vertical arcs $E_2 = \bigcup_{j=1}^{m-1} \bigcup_{i=1}^{n} \{(j, j+1)\}$,
- vertices weights: $W : V \to \mathbb{R}$, $W(i,j) = p_{ij}$.

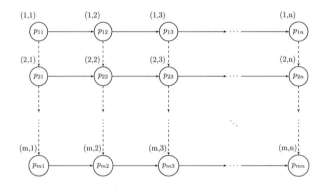

Fig. 1. Classic graph for the flow shop problem.

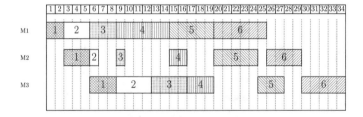

Fig. 2. Example Gantt chart for classic flow shop problem.

2.2 Inserted Idle Time

If we assume, that a task should wait by *at least* $\hat{r}_j$ time on the machine j, than the following additional constraints appears:

$$S_{\pi(i),j+1} \geq S_{\pi(i),j} + p_{\pi(i),j}, \tag{9}$$

$$S_{\pi(i+1),j} \geq S_{\pi(i),j} + p_{\pi(i),j} + \hat{r}_j. \tag{10}$$

Table 2. The times of operations and the shortest waiting

Task	1	2	3	4	5	6	$\hat{r}_j$
Machine 1	2	3	3	6	5	6	1
Machine 2	3	1	1	2	5	4	2
Machine 3	3	4	4	3	3	5	0

For a natural permutation $\pi = (1, 2, 3, \cdots, n-1, n)$ we design a following directed graph

$$G = (V, E) \tag{11}$$

with weighted arcs and edges, where (Fig. 3):

- vertices set $V = \bigcup\limits_{j=1}^{m} \bigcup\limits_{i=1}^{n} \{(i, j)\}$,
- edges set $E = E_1 \cup E_2 \cup E_3$, where:
  - horizontal arcs $E_1 = \bigcup\limits_{j=1}^{m} \bigcup\limits_{i=1}^{n-1} \{(i, i+1)\}$,
  - vertical arcs $E_2 = \bigcup\limits_{j=1}^{m-1} \bigcup\limits_{i=1}^{n} \{(j, j+1)\}$,
- vertices weights: $W : V \to \mathbb{R}, W(i, j) = p_{ij}$,
- arcs weights: $L : E \to \mathbb{R}, L(i+1, i) = \begin{cases} \hat{r}_j & \text{if } (i+1, i) \in E_1, \\ 0 & \text{if } (i+1, i) \in E \backslash E_1. \end{cases}$

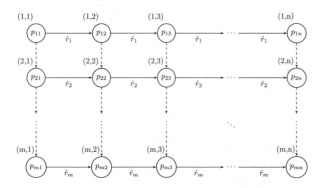

Fig. 3. Graph model for inserted idle time case.

2.3 Limited Idle

When analyzing practical examples for a flow shop problem with 'no idle' constraint, there was a need to develop this limitation, as it often turned out that there was no need to force continuous machines on all machines. By $\hat{d}_j$ we will

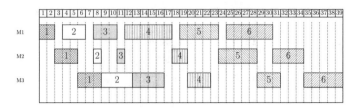

Fig. 4. Example Gantt chart of inserted idle time case.

mark the longest possible idle time after completing the operation on the machine M_j. Such a join will be called limited idle. Equation (5) is modified again:

$$C_{\pi(i+1),j} \leq C_{\pi(i),j} + p_{\pi(i+1),j} + \hat{d}_j. \tag{12}$$

A special case is when all $\hat{d}_j$ are equal to 0, then we are talking about the no idle constraint. When, for example, from Table 1 and natural permutation we accept $\hat{d}_1 = 2$, $\hat{d}_2 = 4$ and $\hat{d}_3 = 0$ we will get the schedule shown in Fig. 6. In this case, we again have $C_{\max} = 34$, despite the fact that on the machine M_3 we forced continuous work.

Table 3. The times of operations and the longest waiting

Task	1	2	3	4	5	6	$\hat{d}_j$
Machine 1	2	3	3	6	5	6	2
Machine 2	3	1	1	2	5	4	4
Machine 3	3	4	4	3	3	5	0

As for previous cases, for a natural permutation $\pi = (1, 2, 3, \cdots, n-1, n)$ we design a following directed graph

$$G = (V, E) \tag{13}$$

with weighted arcs and edges, where (Fig. 5):

– vertices set $V = \bigcup\limits_{j=1}^{m} \bigcup\limits_{i=1}^{n} \{(i, j)\}$,

– edges (arcs) set $E = E_1 \cup E_2 \cup E_3$, where:

  • horizontal arcs $E_1 = \bigcup\limits_{j=1}^{m} \bigcup\limits_{i=1}^{n-1} \{(i, i+1)\}$,

  • vertical arcs $E_2 = \bigcup\limits_{j=1}^{m-1} \bigcup\limits_{i=1}^{n} \{(j, j+1)\}$,

  • return arcs $E_3 = \bigcup\limits_{j=1}^{m-1} \bigcup\limits_{i=2}^{n-1} \{(i+1, i)\}$,

- vertices weights: $W : V \to \mathbb{R}$, $W(i,j) = p_{ij}$,
- arcs weights: $L : E \to \mathbb{R}$, $L(i+1,i) = \begin{cases} -p_{ij} - \hat{d}_j & \text{if } (i+1,i) \in E_3, \\ 0 & \text{if } (i+1,i) \in E \backslash E_3. \end{cases}$

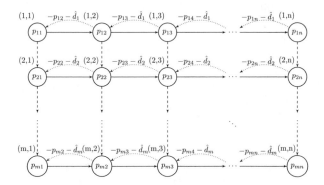

Fig. 5. Graph model for limited idle constraint.

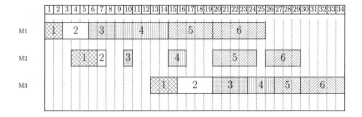

Fig. 6. Example Gantt chart for limited idle constraint.

3   Case Study

The flow problem with time couplings is applicable in the case of foundations for high-voltage power grid poles. A particular example is the piling foundations made using continuous flight auger piles, which are often used in wetlands. The process of forming a pile can be divided into 4 stages:

1. Drilling to the desired depth.
2. Pulling the bit while pumping the concrete under pressure with a bit.
3. Removal of spoil removed by the drill.
4. Embedding reinforcement in concrete.

Each of the stages should be performed immediately after completion of the previous stage, which is why we will treat the whole as one process, i.e. as one M_2 machine in the flow problem. The process requires the use of specialized multi-tone equipment (crane), which generates large costs, therefore it is recommended to perform the piling process at all positions without downtime. Prior to piling, the site should be prepared by making a shallow excavation, clearing the passage or, in special cases, arranging a temporary road from specialized boards. Of course, the position should also be adequately protected from third parties. We will treat the whole preparatory process as a M_1 machine, where you do not have to force downtime.

The next stage will be the preparation of the pile heads, which we will mark as the M_3 machine. First, the excavation should be deepened to the level of the concrete underlay. Then make a layer of pre-concrete. Disassemble the pile heads 5 cm above the base concrete layer and set the reinforcing bars according to the design. In order to proceed to the stage performed on the M_3 machine, you must first wait a minimum of 6–7 days to achieve concrete strength. However, you can not wait for more than 9–10 days, because then the concrete will reach too high strength and can cause extensive cracks when forging. That's why you have to do it in the time window, preferably between 6 and 8 days.

The fourth process, M_4, will be preparing the foundation slab. First, reinforce the slab by joining it with the reinforcement of piles. Another cork is the execution of the formwork and flooding the whole with a suitable mixture of concrete. After this stage, wait about 21 days before proceeding to the next stage. This time there is no limited waiting.

The last stage is the dismantling of the foundation and leveling the area around the site. If a temporary road was arranged, it should be dismantled. This process will mean the machine M_5.

4   Algorithms

For all the time couplings discussed, one exact algorithm based on the Branch and Bound approach [6] and two approximate algorithms: Simulated Annealing [8] and Tabu Search [3].

4.1   Branch and Bound

The Branch and Bound method is based on browsing the state tree, which represents all possible paths that can be found in all solutions stored in the leaves. Reviewing all paths and leaves requires too much computing effort, so the algorithm for each node determines the lower [4] (meaning *Lower Bound*) of the goal function's value in the subtree. In the example we used the best first strategies (*Best First*), therefore nodes with a partial solution are placed on the priority queue according to their lower limit. Sub-trees for which the lower limit is worse than some upper bound (*Upper Bound*) are discarded and treated as indirectly viewed. By the upper bound we mean the best solution's goal function value found so far.

4.2   Simulated Annealing

The simulated annealing algorithm is a probabilistic technique of searching the neighborhood inspired by the metallurgy annealing process. We start with the solution of a random and high initial temperature, when the material, in this case the solution, is subject to change and accept a worse solution with a certain probability (better we always accept). The work of the algorithm ends when the temperature of the material reaches the final temperature, which falls according to the given coefficient. In order to improve the efficiency of this method, we can repeat the search operation a set number of times for each temperature value.

4.3   Tabu Search

The Tabu Search algorithm is a metaheuristics that chooses the best solution from the neighborhood (here, 'swap' type). In order to avoid getting the algorithm stuck in a local minimum, a list of prohibitions (called *tabu list*) is used, where we place the attributes of movements that led to the current solution. Most often taboo list is implemented in the form of a matrix or array.

5   Computational Experiments

In order to examine the effectiveness of the heuristic algorithm and check the impact of time couplings on its quality. A series of tests was carried out and the results were compared with the exact method. The research was carried out on a PC-class computer equipped with an Intel i7-6700K processor clocked at 4.00 GHz, 16 GB RAM and an SSD operating under the Microsoft Windows 10 operating system.

The algorithms were tested on the examples generated due to limitations of the partitioning method and constraints and the lack of sample data for limitations with limited downtime.

The percentage relative deviation (PRD) was used as a measure of solutions quality.

$$PRD(\pi) = 100\%(C_{max}(\pi) - C_{max}(\pi^{ref}))/C_{max}(\pi^{ref}) \qquad (14)$$

where

- π – solution found,
- π^{ref} – reference solution,
- $C_{\max}(\pi)$ – value of the cost function for a solution π.

As the reference algorithm, the Branch and Bound method was taken. Then, the algorithm of simulated annealing and the search with prohibitions with the parameter 5,000 was tested. number of iterations. For both algorithms to have an approximate duration of time in simulated annealing, the neighbor search for $\frac{n^2}{3}$ was repeated for each temperature. In order to eliminate errors, approximate algorithms were included 50 times for all examples, checking all proposed time couplings, then the results were averaged and collected in Tables 4, 5 and 6, where:

- t_{BB} – Branch and Bound algorithm runtime,
- t_{SA} – Simulated Annealing algorithm runtime,
- t_{TS} – Tabu Search algorithm runtime,
- PRD_{SA} – percentage relative deviation of Simulated Annealing results,
- PRD_{TS} – percentage relative deviation of Tabu Search results.

Table 4 presents the results for a classic flow problem for the generated 80 sets of test data, 10 items for a given problem size. In total, 8 different instance sizes were tested, changing the number of tasks and machines. The files have been saved in the format proposed by Taillard [7]. Due to the complexity of the Branch and Bound method and the function calculating boundaries, measurements were made for a maximum of 16 tasks on 10 machines.

Table 4. Results for flow shop problem

Instance	Dimension			Measures		
	$n \times m$	$t_{BB}[ms]$	$t_{SA}[ms]$	$t_{TS}[ms]$	$PRD_{SA}[\%]$	$PRD_{TS}[\%]$
ri001–ri010	10 × 5	16.20	33.02	24.49	0.31	0.00
ri011–ri020	10 × 10	38.50	54.53	49.07	1.57	0.00
ri021–ri030	12 × 5	291.50	52.85	42.43	0.60	0.01
ri031–ri040	12 × 10	364.30	89.99	86.77	1.52	0.01
ri041–ri050	14 × 5	1736.90	77.37	66.47	0.58	0.01
ri051–ri060	14 × 10	5948.30	136.20	139.60	1.50	0.16
ri061–ri070	16 × 5	4529.20	108.20	97.97	0.34	0.05
ri071–ri080	16 × 10	10853.30	194.36	208.30	1.51	0.31

In the second stage of testing, the same input files were used as for the classical flow problem, then the time feedback was tested without idle times (no idle). The results are presented in the Table 5. Due to the scheduling and the more complex function of the target, instances of 12 tasks on 10 machines were examined to the maximum.

Table 5. Results for no idle flow shop problem

Instance	Dimension			Measures		
	$n \times m$	$t_{BB}[ms]$	$t_{SA}[ms]$	$t_{TS}[ms]$	$PRD_{SA}[\%]$	$PRD_{TS}[\%]$
ri001–ri010	10 × 5	289.50	80.30	86.06	0.89	0.00
ri011–ri020	10 × 10	649.00	138.22	163.41	2.55	0.00
ri021–ri030	12 × 5	4010.60	131.59	143.38	0.90	0.00
ri031–ri040	12 × 10	69680.80	352.49	412.74	2.15	0.00

In the next stage of research, the previously used input data had to generate the longest possible downtime on the machines ($\hat{d}_j$ in $[0.9]$) to test the flow problem with limited idle. Six sizes of the problem were tested and the results are presented in the Table 6.

Table 6. Results for limited idle flow shop problem

Instance	Dimension			Measures		
	$n \times m$	$t_{BB}[ms]$	$t_{SA}[ms]$	$t_{TS}[ms]$	$PRD_{SA}[\%]$	$PRD_{TS}[\%]$
ri001–ri010	10 × 5	101.70	89.53	98.74	1.44	0.00
ri011–ri020	10 × 10	567.70	182.81	211.84	2.74	0.00
ri021–ri030	12 × 5	464.40	144.57	164.38	1.27	0.00
ri031–ri040	12 × 10	31774.30	320.63	382.19	2.41	0.00
ri041–ri050	14 × 5	89053.90	234.09	273.88	1.20	0.04
ri051–ri060	14 × 10	114069.13	495.22	606.86	3.90	0.04

6 Conclusions

The article presents the problem of scheduling from the construction practice with the limitation in the form of machine couplings (no idle constraint) and with limited machine idle times (limited idle). The operation time of all presented algorithms increased compared to the unlimited version, which was influenced by the fact that the purpose functions required a larger amount of calculations and there was no possibility to use the so-called accelerators used for classic flow shop problems. Despite the extension of the implementation time of the proposed algorithms, the quality of the results obtained has not worsened despite the addition of additional restrictions.

Acknowledgement. The paper was partially supported by the Wrocław University of Science and Technology internal grant PWR-RUDN no. 45WB/00001/17 and National Science Centre of Poland, grant OPUS no. DEC 2017/25/B/ST7/02181.

References

1. Adiri, I., Pohoryles, D.: Flowshop/no-idle or no-wait scheduling to minimize the sum of completion times. Nav. Res. Logist. Q. **29**(3), 495–504 (1982)
2. Bożejko, W., Hejducki, Z., Wodecki, M.: Applying metaheuristic strategies in construction projects management. J. Civ. Eng. Manag. **18**(5), 621–630 (2012)
3. Glover, F.: Tabu search. Part I. ORSA J. Comput. **1**, 190–206 (1989). Second edition (1992)
4. Gowrishankar, K., Rajendran, C., Srinivasan, G.: Flow shop scheduling algorithms for minimizing the completion time variance and the sum of squares of completion time deviations from a common due date. Eur. J. Oper. Res. **132**(3), 643–665 (2001)
5. Ruiz, R., Maroto, C.: A comprehensive review and evaluation of permutation flow-shop heuristics. Eur. J. Oper. Res. **165**(2), 479–494 (2005)
6. Smutnicki, C.: Algorytmy szeregowania zadań. Politechnika Wrocławska (2012)
7. Taillard, E.: Benchmarks for basic scheduling problems. Eur. J. Oper. Res. **64**(2), 278–285 (1993)
8. van Laarhoven, P.J.M., Aarts, E.H.L.: Simulated Annealing. Springer, Netherlands (1987)

Correspondent Sensitive Encryption Standard (CSES) Algorithm in Insecure Communication Channel

Rafał Chałupnik, Michał Kędziora$^{(\boxtimes)}$, Piotr Jóźwiak,
and Ireneusz Jóźwiak

Faculty of Computer Science and Management,
Wroclaw University of Science and Technology, Wroclaw, Poland
michal.kedziora@pwr.edu.pl

Abstract. The purpose of the research was to implement the correspondent sensitive encryption algorithm for encryption of files and sending them through insecure communication channel. Currently, in practical applications correspondents are using either one known to them pair of keys (what has negative influence on security), use Public Key Infrastructure or as the simplest solution they are exchanging key through another channel. Additional benefits of our proposed solution are integrity control and ability to limit decryption capability to known recipients. The solution we propose allows correspondents to pair their devices fast and easily, so they could securely and without passwords transfer file through insecure channel.

Keywords: Cryptography · Cyber security ·
Correspondent Sensitive Encryption Algorithm

1 Introduction

There are several approaches to the solution of secure communication through unsecure communication channel using correspondent aware systems. Most popular approach is to use public key infrastructure [9]. Public Key Infrastructure consist of a certificate authority that store, issue and sign digital certificates, a registration authority which verifies the identity, central directory, a certificate management system and certificate policy stating requirements and procedures. Unfortunately, PKI has some disadvantages first it has large complexity which supports attacks example is compromising CA organizations or theft of issued certificates. Another threat vector for PKI are Denial of Service Attacks. There are ideas to design and implement an attack resilient public key infrastructure as alternative, but solution is still proof of concept [10]. Part of PKI weaknesses is connected with TLS protocol [14, 15], therefore there are works towards enhancements, but solution is still proof of concept [11]. Other solutions like Certificate-less authenticated encryption adds authentication to ID-based encryption but also needs trusted center that handles authenticating process [12, 13]. Our goal was to propose simple correspondent sensitive algorithm which doesn't need complex infrastructure to work properly.

© Springer Nature Switzerland AG 2020
W. Zamojski et al. (Eds.): DepCoS-RELCOMEX 2019, AISC 987, pp. 90–98, 2020.
https://doi.org/10.1007/978-3-030-19501-4_9

2 Current Solutions

The idea of an asymmetric algorithm is first response to the described problem. RSA was designed by Ron Rivest, Adi Shamir, and Leonard Adleman. It is best known asymmetric key cryptosystems for key exchange, digital signatures and encryption. RSA uses a variable size of key and encryption and is an asymmetric (public key) cryptosystem based on number theory. RSA is using two prime numbers to generate public and private keys. These two keys are then used for encryption and decryption process. At first sender encrypts the message using receivers public key and when he transmits message to receiver. The receiver can decrypt it using his own private key [2, 3]. RSA has many disadvantages in its design and therefore is not preferred for the commercial use for pure encryption and decryption of large portions of data. When small values of p and q are selected for the key creation then the encryption process becomes too weak and unsecure [5]. If large p and q lengths are selected, then it consumes more time and the performance gets degraded in comparison with symmetric encryption algorithms. Also, the algorithm requires of similar lengths for p and q, which is not practical while encryption large amount of data. Padding techniques are required [1]. Key Generation Procedure [4] is as follows:

1. Choose two distinct large random prime numbers p and q such that $p \neq q$.
2. Compute $n = p \times q$.
3. Calculate: $phi(n) = (p - 1)(q - 1)$.
4. Choose an integer e such that $1 < e < phi(n)$
5. Computed to satisfy the congruence relation $d \times e = 1 \bmod phi(n)$; d is kept as private key exponent.
6. The public key is (n, e), and private key is (n, d).
7. Keep all the values d, p, q and phi secret.

Encryption process requires plaintext $P < n$ and ciphertext: $C = P^c \bmod n$, while decryption of ciphertext C is $P = C^d \bmod n$. Practical use of asymmetric algorithm usually requires use of symmetric algorithm for encryption of data. Most popular symmetric encryption algorithm is Advanced Encryption Standard (AES) which was encryption standard recommended by NIST to replace DES in 2001 [16]. AES algorithm can support key length of 128, 192, and 256 bits. The algorithm is referred to as AES-128, AES-192, or AES-256, depending on the key length. During encryption and decryption process, AES algorithm goes through 10 rounds for 128-bit keys, 12 rounds for 192-bit keys, and 14 rounds for 256-bit keys in order to deliver final ciphertext or to retrieve the original plaintext [6]. AES divides a 128-bit data length into four basic operational blocks treated as array of bytes and organized as a matrix of the order of 4×4 that is called the state.

For both encryption and decryption, the cipher begins with an Add Round Key stage. However, before reaching the final round, this output goes through nine main rounds, during each of those rounds four transformations are performed; (1) Sub-bytes, (2) Shift-rows, (3) Mix-columns, (4) Add round Key. In the final of (10th) round, there is no Mix-column transformation [7]. Decryption is the reverse process and uses inverse functions such as Inverse Substitute Bytes, Inverse Shift Rows and Inverse Mix

Columns [17]. There are five confidentiality modes of operation for use with an underlying symmetric key block cipher algorithm defined: Electronic Codebook (ECB), Cipher Block Chaining (CBC), Cipher Feedback (CFB), Output Feedback (OFB), and Counter (CTR). In purpose of our solution we will use mostly CBC (Fig. 1) in which blocks of plaintext is XORed which previous ciphertext block before encrypted. This results that each ciphertext block depends on all plaintext blocks which where processed up to that point. Which can be presented as:

$$C_i = E_K(P_i \oplus C_{i-1}) \tag{1}$$

$$C_0 = IV$$

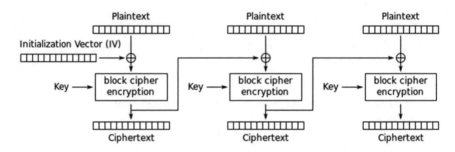

Fig. 1. Cipher Block Chaining (CBC) mode encryption

To construct Correspondent Sensitive Encryption Algorithm, use of Secure Hash Algorithm will be necessary. For our purpose use of SHA-1 will be sufficient despite of known attacks and collisions. SHA-1 completes a message digest based on same principles as in MD5 message digest algorithm but is able to generate a larger hash value (160 bits vs 128 bits in MD5).

3   Correspondent Sensitive Encryption

In order to be able to implement the proposed solution, the CSES (Correspondent-Sensitive Encryption Standard) algorithm was designed. As the name suggests, the application of the algorithm is used to encrypt the content, but with the knowledge of correspondents (senders and defined recipients). This chapter presents the design and operation of the algorithm.

For the algorithm to work, correspondents must be known. In the context of the entire algorithm (both the encryption and decryption process), the correspondent can be present as a tuple:

$$\{identifier, \; public\,key, \; private\,key\} \tag{2}$$

The correspondent identifier must be unique on a global scale. The algorithm for generating such identifiers does not matter for the CSES itself, but in the implementation was based on the GUID, due to its popularity and very low probability of collision (the total number of combinations is 2^{128}). Thanks to this, such identifiers can be generated offline, without the participation of any central server, practically without the risk of accidentally generating two identical entities. Each of the correspondents must also have their asymmetric algorithm key pair. According to its assumption, during the process of both encryption and decryption, the correspondent must use his private key and the public key of his sender/recipient.

3.1   Encryption

The key functionality of the algorithm is file encryption. To be able to have file content securely transmitted by any means of communication, it must have form of cryptogram. The CSES algorithm uses AES in a 256-bit configuration, using two of its modes. First is Electronic Codebook (ECB) and second one is Cipher Block Chaining (CBC) - the addictive cryptogram of a given block from the block previous.

The implementation of the CSES algorithm uses both of these modes. Due to absence of the need to store information about the initialization vector, ECB mode was used to encrypt the metadata. To encrypt the contents of a file where it is needed the highest security level, CBC mode was used. It is worth mentioning that storing the initialization vector is not a problem, because it is encrypted using it ECB mode and attached to file metadata.

3.2   Digital Signature

To be able to verify the sender, which is one of the main benefits of CSES algorithm, and at the same time facilitate the verification of owned keys, it was decided to attach signature to metadata. The process of its obtaining is as follows: the sender of the file signs the symmetric key shortcut of the encryption algorithm with its private key. As CSES assumes the use of a single encryption key, one and only once, binding it inextricably with the contents of the file, a symmetric encryption key can be unambiguously associated with the file. The sender, signing this key, forces the recipient its decryption with the sender's public key, which immediately confirms that the metadata was not infringed by a third party. The process on the recipient's side looks a bit different: after decrypting his pair obtained from the metadata {keyfile, initialization vector} The recipient of the file may decrypt the signature with the sender's public key. If the key shortcuts obtained in this way are identical, the recipient can assume the correctness of the key and the initialization vector (which reduces the chances of finding inaccuracies already during the decrypting process of the file).

3.3   Metadata

One of the key issues allowing the interpretation of the file content is its extension. Thanks to extension, it is known without looking inside that the file is for example photo. To encrypt the file, original extension should change otherwise it could suggest to the user that nothing happened with the file, which results in a surprise during opening attempts (most likely it would be a misleading "file is damaged" message). The second, more important reason is security issue. Such extension could be used as information leak, indicating what expect after decrypting the file content. To sum up these problems and draw conclusions, it was decided to extract the extension from the file name, to encrypt it (using the previously mentioned ECB mode) and attached to the file metadata. The encrypted file gets new extension that uniquely connects it to the encryption application. Metadata header may look as follows:

```
<CqrFile>
  <Correspondents>
    <Sender Id="70967ccb•02e1•4fa6•9e67•446cd7cbae8d"/>
    <Recipient Id="9a00838b•5bdc•4d7d•8468•cd2cf85d4b7"/>
    <Recipient Id="5917ba9c•0742•43ac•ab0d•d38afedf7f1"/>
  </Correspondents>
  <CryptoData>
    <Pair
      Key="471a6aa777b5cc0ab519ee76c1c25014b9660619db
      57e24714c8a1585c3d53"
      IV="71c22997a6ab440c764c92127fc10a17"/>
    <Pair
      Key="3ad3346b425912a10f53b68c37f749458c913fabfbc8
      93b7e26fb45bbc25d776"
      IV="e00544fb300aa0fde1ebd4f28dee2807"/>
  </CryptoData>
  <Extension Value="95506321349d7df4e8843e86f9b478bc"/>
  <Signature Value="cc0676fea97423efa1b47d21a60cbb4c"/>
</CqrFile>
```

Analyzing the header from above: the first encountered value is the unique identifier of the sender of the file. Thanks to it the recipient can verify who sent the file. Below is a list of unique recipient IDs. The recipient can decrypt the file if it is in this list. He learns which pair {key, vector} belongs to him. The CryptoData section stores these pairs of cryptographic values. The recipient, knowing which pair belongs to him, decrypts them with his private key. Thanks to such a mechanism, you can attach many recipients to the file, and each of them can decrypt the necessary data only with a known key. Next in the header there is an encrypted file extension. The recipient, already knowing the cryptographic data they need, can decrypt it and, as a result, find out how interpret the decrypted file. At the end of the header there is a signature. As it was mentioned, if decrypted by the sender's public key, the signature equals the shortcut of the key obtained during encryption, metadata is correct.

4   CSES Implementation

The steps that must be followed to apply correctly are outlined below on Fig. 2.

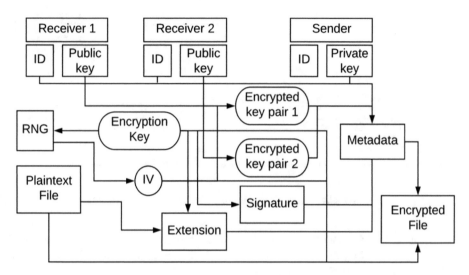

Fig. 2. Visualization of encrypting the file with the CSES algorithm

Content Sensitive Encryption Standard algorithm to encrypt the file need to perform the following steps. At first, we need to generate a random encryption key and initialization vector. Then we must encrypt encryption key and initialization vector for each recipient. Next, we encrypt the file extension with a symmetric algorithm in ECB mode. Afterward we generate the signature by calculating the encryption key abbreviation and encrypting it with its asymmetric algorithm using the private key. As a next step we must place our ID in the header (as the sender). Then we put the recipient IDs in the header. Next, we place the {key, vector} pair for the recipients in the header, taking into account their order as for the identifiers. Then we place the encrypted file extension in the header. Next, we place the file's signature in the header. Afterward we encrypt the file contents with a symmetric algorithm in CBC mode using generated encryption key and initialization vector. Last step is to Place the generated header at the beginning of the encrypted file (Fig. 3).

There are ten steps that must be followed to make Content Sensitive Encryption Standard algorithm to decrypt the file. Visualization of process is also presented in Fig. 2. First step is to read the metadata from the beginning of the encrypted file. As second step we must read the sender of the file - check if it is known to us. If known, download its public key from its resources. If it's not, stop decrypting and notify the user of an unknown sender. Third step is to read the recipients of the file - check if we

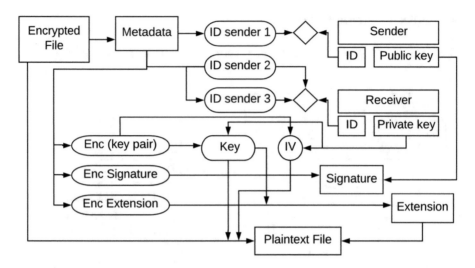

Fig. 3. Visualization of decrypting the file with the CSES algorithm.

are on the list. If so, continue. Remember which one you are on the list. If not, stop decrypting and notify the user that the message is not intended for him. Next, we read the pair {key, vector} from the index on which your ID was. Then we decrypt the pair {key, vector} with an asymmetric algorithm using private key. Next, we decrypt the signature by an asymmetric algorithm using the sender's public key. Afterward we generate the shortcut of the previously obtained encryption key and compare it with the signature. If they are compatible, continue. If not, notify of a potential violation of the metadata by a third party. Next step is to decrypt the file extension with a symmetric algorithm in ECB mode using the previously obtained encryption key. Then we decrypt the contents of the file with a symmetric algorithm in CBC mode with use of the previously obtained encryption key and the initialization vector. Last step is to change the output file extension to previously decrypted.

The use of the CSES algorithm, in addition to enabling the solution of the problem of secure file transfer, with the knowledge of correspondents (senders and defined recipients) also introduced several advantages. Below are the benefits of using this algorithm. First the content of the file is secure (thanks to the symmetric AES algorithm). The sender is sure that only the recipient can decrypt the file. The sender can send the file to several recipients at once. The recipient is sure that the sender has certainly encrypted and sent the file. The recipient is sure that the file has arrived untouched. For each communication a different encryption key and initialization vector are generated. Analyzing the above features, a problem arises: correspondents must know each other's public keys. The solution proposed assumes an exchange these keys at the beginning of using the application, and then each time to generate keys it did not require sending any information separately.

5 Conclusion and Future Work

The purpose of the research was to implement the correspondent sensitive encryption algorithm to use in purpose to encrypt and decrypt files and sent them through insecure communication channel. In summary, the security of the CSES algorithm is based directly on the security of the algorithms it uses. They should be chosen so as to guarantee how the highest standard of user's privacy as a whole. Therefore, in implementation the following configuration was used: As asymmetric algorithm: RSA (with key length: 2048 bits), for symmetric encryption there was used AES (key length: 256 bits, block length: 128 bits) and as hashing algorithm: SHA (with length of the shortcut: 256 bits). Summing up the description of the technical side of the algorithm, it can be concluded that with a little more work than using a standard encryption algorithm, we can achieve much secure correspondent aware solution. In addition, the encryption key is every time randomly generated, making it difficult for a potential cryptographic attack compared to a scenario where the key is always the same. Presently, in real-world solutions correspondents are using either Public Key Infrastructure, or they are exchanging key through another channel, which generates issues with complexity and need of third party involved. Additional benefits of our proposed solution are integrity control and ability to limit decryption capability to known recipients. The solution we propose allows correspondents to pair their devices fast and easily, so they could securely and without passwords transfer file through insecure channel. Our future work is focused on full practical implementation of CSES algorithm in mobile application used to encrypt and decrypt data and sent it through insecure network.

References

1. Kakkar, A., Singh, M.L., Bansal, P.K.: Comparison of various encryption algorithms and techniques for secured data communication in multimode network. Int. J. Eng. Technol. **2**(1), 87–92 (2012)
2. Kumar, A., Jakhar, S., Makkar, S.: Comparative analysis between DES and RSA algorithm's. Int. J. Adv. Res. Comput. Sci. Softw. Eng. **2**(7), 386–391 (2012)
3. Zhou, X., Tang, X.: Research and implementation of RSA algorithm for encryption and decryption. In: The 6th International Forum on Strategic Technology, pp. 1118–1121 (2011)
4. Somani, U., Lakhani, K., Mundra, M.: Implementing digital signatures with RSA encryption algorithm to enhance the data security of cloud in cloud computing. In: 1st International Conference on Parallel, Distributed and Grid Computing (PDGC), pp. 211–216 (2010)
5. Singh, G., Supriya, S.: A study of encryption algorithms (RSA, DES, 3DES and AES) for information security. Int. J. Comput. Appl. (0975-8887) **67**(19) (2013)
6. Singh, G., Singla, A., Sandha, K.S.: Cryptography algorithm comparison for security enhancement in wireless intrusion detection system. Int. J. Multi. Res. **1**(4), 143–151 (2011)
7. Chowdhury, J.Z., Pishva, D., Nishantha, G.G.D.: AES and confidentiality from the inside out. In: The 12th International Conference on Advanced Communication Technology (ICACT), pp. 1587–1591 (2010)
8. Eastlake, D., Jones, P.: US Secure Hash Algorithm 1 (SHA1). RFC 3174 (2001)

9. Lozupone, V.: Analyze encryption and public key infrastructure (PKI). Int. J. Inf. Manage. **38**(1), 42–44 (2018)
10. Basin, D., Cremers, C., Kim, T.H., Perrig, A., Sasse, R., Szalachowski, P.: Design, analysis, and implementation of ARPKI: an attack-resilient public-key infrastructure. IEEE Trans. Dependable Secure Comput. **15**(3), 393–408 (2018)
11. Lee, T., Pappas, C., Szalachowski, P., Perrig, A.: Towards sustainable evolution for the TLS public-key infrastructure. In: Proceedings of the 2018 on Asia Conference on Computer and Communications Security (ASIACCS 2018). ACM, New York (2018)
12. Mandt, T.K., Tan, C.H.: Certificateless authenticated two-party key agreement protocols. In: Okada M., Satoh I. (eds.) Advances in Computer Science - ASIAN 2006. Secure Software and Related Issues. ASIAN 2006. Lecture Notes in Computer Science, vol 4435. Springer, Heidelberg (2007)
13. Al-Riyami, S.S., Paterson, K.G.: Certificateless public key cryptography. In: Laih, C.S. (eds.) Advances in Cryptology - ASIACRYPT 2003. Lecture Notes in Computer Science, vol. 2894. Springer, Heidelberg (2003)
14. Rescorla, E.: The Transport Layer Security (TLS) Protocol Version 1.3, RFC 8446, August 2018. https://doi.org/10.17487/rfc8446
15. Sheffer, Y., Holz, R., Saint-Andre, P.: Summarizing known attacks on transport layer security (TLS) and datagram TLS (DTLS), RFC 7457, February 2015. https://doi.org/10.17487/rfc7457
16. National Institute of Standards and Technology, and National Institute of Standards and Technology. Advanced Encryption Standard (AES). US Department of Commerce, National Institute of Standards and Technology (2017)
17. Bhanot, R., Hans, R.: A review and comparative analysis of various encryption algorithms. Int. J. Secur. Its Appl. **9**(4), 289–306 (2015)

Numerical Analysis of the Building Materials Electrical Properties Influence on the Electric Field Intensity

Agnieszka Choroszucho(ID) and Adam Steckiewicz$^{(\boxtimes)}$(ID)

Bialystok University of Technology, Wiejska 45D, 15-351 Bialystok, Poland
{a.choroszucho,a.steckiewicz}@pb.edu.pl

Abstract. The paper discusses the numerical analysis of the electromagnetic waves propagation effects within an area containing a non-ideal, non-homogenous and absorbing dielectric. The relation between homogeneous and heterogeneous structures of building materials and electric conductivity are examined. Different ranges of the conductivity were analyzed at frequency 2.4 GHz with respect to commonly used materials like the concrete, aerated concrete, solid brick and brick with holes. The analysis was performed using the FDTD method. The obtained values of the electric field intensity provide the determination of an attenuation coefficient for the different walls constructions. The detailed analysis of the different buildings constructions will make possible better understanding of the wave phenomena as well as counteract a local signal fading at planning of the wireless networks systems.

Keywords: Wireless networks · Indoor propagation model ·
WLAN signal attenuation · Building materials ·
Finite-Difference Time-Domain method (FDTD)

1 Introduction

The most frequently analyzed issues related to wireless networks operation and distribution of the electromagnetic (EM) field are: losses created on the way between the base station and the receiver which results in signal fades, attenuation and reflection coefficient, EM wave incident angle. In order to determine these factors the following methods are used: empirical, numerical (e.g. FDTD, FDFD, ray tracing) as well as theoretical methods using wave optics [3].

The increasing number of the mobile WiFi devices have made the indoor localization an attractive research area, especially for the Internet of things applications [10]. Since Global Positing System (GPS) does not perform well in the indoor environment, several indoor localization approaches have been proposed in the past few years [10]. In modern world, the advent of fully interactive environments within Smart Cities and Smart Regions requires the usage of multiple wireless systems. In the case of user-device interaction, which have multiple applications (e.g. Intelligent Transportation Systems or Smart Grids) the large amount of transceivers are employed in order to achieve anytime, anyplace and any device connectivity. The resulting combination of

© Springer Nature Switzerland AG 2020
W. Zamojski et al. (Eds.): DepCoS-RELCOMEX 2019, AISC 987, pp. 99–109, 2020.
https://doi.org/10.1007/978-3-030-19501-4_10

many wireless networks has fundamental limitations derived from coverage and capacity relations, as a function of required quality of service parameters, required bit rate and energy restrictions. In this way, inherent transceiver density poses challenges in overall system operation, given by multiple node operation that increases overall interference levels. In [9] a deterministic based analysis, presented as a function of topological node distribution, is applied to variable density wireless sensor network operation within complex indoor scenarios. However, only the extensive analysis provides sufficient interference characterization for conventional transceivers.

The detailed analysis of the influence of building materials on the electromagnetic (EM) field distribution has been discussed in a number of studies [1–4]. The wave propagation analysis in the high frequency regime is associated with taking into account the construction itself and the building materials with different electrical properties, i.e. the electric permittivity ($\varepsilon = \varepsilon_0\varepsilon_r$) and conductivity ($\sigma$) [3, 4]. Both the periodic structures as well as special elements of the material (reinforcement), subject the wireless transmission channel to the interference, signal delays and signal loss. The construction of resistible communication networks requires taking into account several factors affecting the EM field distribution at the stage of a system design (for example building geometry or complex material structures between transmitter and receiver). These problems are especially visible in low range wireless networks (i.e. Wi-Fi) inside the buildings.

In this paper the numerical method (FDTD) is presented. The aim is to determine the maximum field strength and, for example, estimate the damping factor based on the formulas [3]. The electric field intensity behind the wall made of homogeneous (concrete, brick, aerated concrete) and heterogeneous building material were analyzed. Three kinds of brick were considered: solid brick (without the air holes) and two types of clinker brick (with 18 or 30 vertical holes). The influence of the brick's conductivity on the electric field values was discussed. The main purpose of the analysis was investigation of the different building materials (brick characterized by specific electrical parameters) other than the concrete or aerated concrete discussed by several authors [1, 5]. The presented results may serve as a knowledge source for the EM field distribution analysis and dependability problems of wireless computer systems in an indoor area containing non-ideal, absorbing dielectrics.

2 Mathematical Model

To determine the distribution of EM field, the electric field intensity is analyzed in a model solved by the Finite-Difference Time-Domain method (FDTD) [6–8]. The FDTD method is based on Maxwell's equations in time and space [4, 6–8]:

$$\nabla \times \mathbf{E} = -\frac{\partial \mathbf{B}}{\partial t}, \tag{1}$$

$$\nabla \times \mathbf{H} = \mathbf{J}_P + \mathbf{J}_D + \mathbf{J}_I, \tag{2}$$

$$\nabla \cdot \mathbf{D} = \rho, \tag{3}$$

$$\nabla \cdot \mathbf{B} = 0, \tag{4}$$

$$\nabla \cdot \mathbf{J} = -\frac{\partial \rho}{\partial t}. \tag{5}$$

where: $\mathbf{E}$ is an electric field (vector) in [V/m], $\mathbf{H}$ is a magnetic field (vector) in [A/m]. The sources of both fields are the electric charges and currents, which can be expressed as local densities, namely charge density ρ and current density $\mathbf{J}$. Whereas $\mathbf{J}_P$ is the density of a conduction current, $\mathbf{J}_D$ is the density of displacement current and $\mathbf{J}_I$ is the vector of current density which forces the field.

The field formulation using Eqs. (1)–(5) allows to find the distribution of EM field in a stationary and transient state. After the transformation of the Maxwell's Eqs. (1)–(5), the dynamics of electromagnetic phenomena inside continuous area is expressed using the wave equation

$$\nabla \times \left(\frac{1}{\mu_0 \mu_r} \nabla \times \mathbf{E} \right) + \sigma \frac{\partial \mathbf{E}}{\partial t} + \varepsilon_0 \varepsilon_r \frac{\partial^2 \mathbf{E}}{\partial t^2} = -\frac{\partial \mathbf{J}_I}{\partial t}. \tag{6}$$

Equations (1)–(5) determine the field distribution in the transient state of the time-varying field excitations [6, 8]. After the decomposition Eqs. (1)–(2) are in the form of six conjugate first order differential equations, that describes the electric and magnetic field components [6–8]

$$\frac{\partial E_x}{\partial t} = \frac{1}{\varepsilon} \left(\frac{\partial H_z}{\partial y} - \frac{\partial H_y}{\partial z} - \sigma E_x \right), \tag{7}$$

$$\frac{\partial E_y}{\partial t} = \frac{1}{\varepsilon} \left(\frac{\partial H_x}{\partial z} - \frac{\partial H_z}{\partial x} - \sigma E_y \right), \tag{8}$$

$$\frac{\partial E_z}{\partial t} = \frac{1}{\varepsilon} \left(\frac{\partial H_y}{\partial x} - \frac{\partial H_x}{\partial y} - \sigma E_z \right), \tag{9}$$

$$\frac{\partial H_x}{\partial t} = \frac{1}{\mu} \left(\frac{\partial E_y}{\partial z} - \frac{\partial E_z}{\partial y} \right), \tag{10}$$

$$\frac{\partial H_y}{\partial t} = \frac{1}{\mu} \left(\frac{\partial E_z}{\partial x} - \frac{\partial E_x}{\partial z} \right), \tag{11}$$

$$\frac{\partial H_z}{\partial t} = \frac{1}{\mu} \left(\frac{\partial E_x}{\partial y} - \frac{\partial E_y}{\partial x} \right). \tag{12}$$

2.1   FDTD Method and the Area Discretization

In three-dimensional scheme of the FDTD method the Yee cell is used [4, 6]. The FDTD method is based on the division of the analyzed area into an appropriate number of cells. The propagation of the EM field inside the structures of a single building system model can be simplified to two-dimensional variant. In this case the impact of external structural elements and changes of the field distribution in the vertical direction are omitted. In the 2D model, the development of the scheme is based on a rectangular grid with staggered components of the field (Fig. 1).

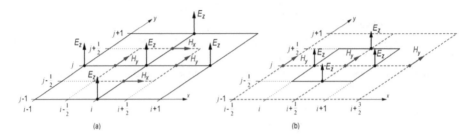

Fig. 1. Discretization of the Maxwell's equations, two-dimensional area (the TM$_z$ version).

In FDTD method, the integration of the Maxwell's equations in the time-domain is based on the two-step approach. In a selected time where the distribution of the EM field is determined, the component vector values of the magnetic field are shifted in time by $\Delta_t/2$. The determination of component intensity vectors of the electric field E_x, E_y, E_z is possible with the prior appointment of the component intensity vectors of the magnetic field H_x, H_x, H_x in the previous time step.

Applying the Euler central difference scheme [4, 6–8] for the approximation of partial derivatives in the area and time, the differential Eq. (9) takes the form [3, 6]:

$$\frac{{}^{n+1}_{i,j,k}E_z - {}^{n}_{i,j,k}E_z}{\Delta_t} = \frac{1}{\varepsilon_{i,j,k}} \left(\frac{{}^{n+\frac{1}{2}}_{i+\frac{1}{2},j,k}H_y - {}^{n+\frac{1}{2}}_{i-\frac{1}{2},j,k}H_y}{\Delta_x} - \frac{{}^{n+\frac{1}{2}}_{i,j+\frac{1}{2},k}H_x - {}^{n+\frac{1}{2}}_{i,j-\frac{1}{2},k}H_x}{\Delta_y} - \sigma_{i,j,k} {}^{n+\frac{1}{2}}_{i,j,k}E_z \right)$$

(13)

The values ${}^{n+\frac{1}{2}}_{i,j,k}E$ in the Eq. (13) are approximated by the arithmetic average of solutions in following steps n and $n + 1$ (semi-implicit approximation) [8, 10]

$$ {}^{n+\frac{1}{2}}_{i,j,k}E_z = \frac{{}^{n+1}_{i,j,k}E_z + {}^{n}_{i,j,k}E_z}{2} $$

(14)

3   The Geometry of the Analyzed Models

The aim of the analysis was to identify the influence of building materials electrical properties on the electric field intensity. Numerical models with the walls made of a homogeneous (solid brick, concrete, aerated concrete) and heterogeneous material (brick with holes) were considered. The electrical properties of the analyzed material are presented in Table 1. The height (h), width (b) and length (l) of all bricks were $0.065 \times 0.12 \times 0.25$ m, respectively (Fig. 2). Three variants of bricks were analyzed:

- solid brick (without the air holes),
- brick with 18 vertical holes (model *B18*) (Fig. 2a),
- brick with 30 vertical holes (model *B30*) (Fig. 2b).

Table 1. Commonly accepted in the literature electrical properties for the description of the popular building materials (for $f = 2.4$ GHz).

Material	Relative electrical permittivity (ε'_r) [-]	Conductivity (σ) [S/m]	Literature
Brick	4.44	0.01	[4]
Concrete	6	0.00195	[3]
Aerated concrete	2.25	0.01	[2, 3]

The typical dimensions of the bricks and also the width of the holes within them (s_h) are presented in Fig. 2. Used dimensions were given on average, because the forming, drying and firing processes of ceramic materials may cause the variations in dimensions [3, 4]. The numerical analysis also take into account bricks with the various whole width (s_h) along length of the brick ($l = 0.25$ m). After all processes (drying and hardening) some parts of the brick might have an untypical size. In this reason the influence of a different relative volume of the clay mass on the values of electric field was analyzed. This factor was modified (Table 2) by changing the size of air holes (s_h). The percentage of the total brick volume inside the brick with different size of the holes was calculated using the ratio

$$V_{\%mc} = \frac{V_c - V_d}{V_c} \cdot 100\%, \tag{15}$$

where V_d is the relative volume of all holes inside the brick, $V_c = h \cdot b \cdot l$. The typical average size of air holes in the brick *C18* is 0.011 m and in *C30* it is $s_h = 0.15$ m.

The dimensions of the bricks and their electrical image, where $\lambda_b = 0.0593$ m is the length of the wave in a brick material and $\lambda_a = 0.125$ m (in an air) were shown (Fig. 2). The dimensions and numerical conditions of the analyzed area in one of the variants of the walls have been presented in Fig. 2c.

Table 2. Percentage of the ceramic volume inside the brick connected with the variable size of the holes inside two kinds of bricks.

Geometric size (s_h) [m]	Geometric size (s_h) [m]	Relative volume of the clay mass in the brick $C18$ $(V\%_{mc})$	Relative volume of the clay mass in the brick $C30$ $(V\%_{mc})$
0.005	0.40 λ_a	90.40	87.50
0.007	0.56 λ_a	86.56	82.50
0.009	0.72 λ_a	82.72	77.50
0.011	0.88 λ_a	78.88	72.50
0.013	0.104 λ_a	75.04	67.50
0.015	0.120 λ_a	71.20	62.50
0.017	0.136 λ_a	67.36	57.50
0.019	0.152 λ_a	63.52	52.50

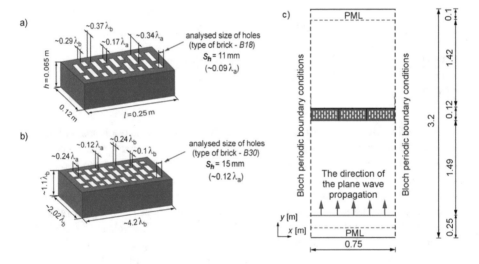

Fig. 2. Geometry of analyzed model: (a) brick $B18$, (b) brick $B30$ and (c) boundary conditions.

The field was induced by a plane sinusoidal wave at frequency $f = 2.4$ GHz. The absorption of the EM wave was obtained using PML (*Perfectly Matched Layer*) boundary conditions [2]. The EM field in the system was a harmonic plane wave polarized linearly, propagating in the direction parallel to axis Oy ($\mathbf{k} = \mathbf{1}_y$)

$$\mathbf{E}(x, y, t) = E_z\mathbf{1}_z = \sin(\omega t) \cdot 1(t) \cdot \mathbf{1}_z. \tag{16}$$

The model length was reduced to three bricks, whereas in the edges perpendicular to the Ox axis Bloch's periodic boundary conditions were assumed [3, 7].

Due to large differences of the materials electrical properties in the literature, the variability of the conductivity within the range $\sigma = 0 \div 0.2$ S/m was also taken into account. The stability of the time marching explicit scheme requires satisfying the Courant-Friedrichs-Lewy (CFL) condition [7, 10], which determines the relationship between time step in the leap-frog scheme of FDTD method and the maximum size of Yee cell [3, 6–8]. The area was composed of the Yee cells equal to 0.001 m.

4   The Results of Numerical Analysis

The FDTD method made possible to obtain instantaneous field images. For this reason it was necessary to develop additional algorithm. The target of it was to prepare the map describing the maximum values of electric field in the processing sequence of instantaneous field distribution calculated at regular intervals. An algorithm was formulated and a software tool developed to determine the maximum values of the analyzed field component (Figs. 3, 4, 5 and 6).

The presented results are shown when the steady state of the EM field distribution had been achieved. At the beginning of the calculations, the correctness of the adopted assumptions was checked. The results of the numerical analysis were compared with the results obtained by analytical method [3]. The usage of the analytical method is possible only for models containing a wall made of a homogeneous material such as solid brick, concrete or aerated concrete. On the basis of the results it can be seen that the correct size of the Yee cell has been accepted ($\Delta_x = \Delta_y = 0.001$ m) (Fig. 3). Due to this, the length of the wave in the air (λ_a) corresponded to 75 Yee cells.

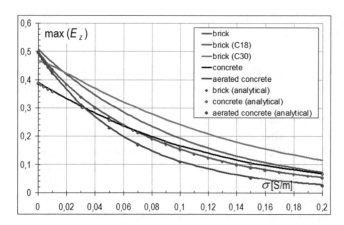

Fig. 3. The dependence between the conductivity and the relative maximum values of E_z component behind the wall made of various building materials (analytical and numerical).

The analysis shows that for low conductivities, the lowest relative maximum values of E_z component were obtained for a wall made of the concrete. While above conductivity equal to 0.01 in two models made of bricks with holes (*B18*, *B30*), the highest values of the max(E_z) were obtained (Fig. 3). A porous material made of air holes and little percentage of the ceramic volume inside the brick cause this result.

In the Fig. 4 the distribution of max(E_z) values was shown for the models made of hollow bricks (*B18, B30*) with the parameters described in Table 1. During the passing over the different areas of bricks, the local changes of the wave velocity cause to appear the instantaneous images of the field. Indicated phenomenon is related to interactions of EM wave on the boundary medium inside the clay hollow bricks, which cause multiple reflections from edges of the holes. It is the reason of the local increase and decrease of the electric field values behind the wall. The field behind the wall made of brick *B30* has higher both minimum and maximum values.

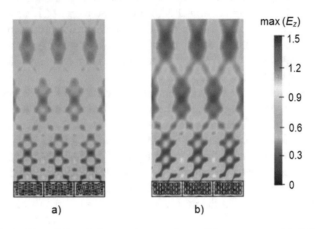

Fig. 4. The distribution of the relative maximum values of E_z component: (a) *B18*, $f = 2.4$ GHz, (b) bricks *B30*, $f = 2.4$ GHz.

Figures 5 and 6 show the relative maximum values of the E_z component and their dependencies on the size of holes (s_h) and the conductivity. The analysis has shown that regardless of the conductivity (where $\sigma > 0.04$ S/m) the decrease of brick's $V_{\%mc}$ results with higher values of electric field. When the conductivity is equal to 0.035 S/m, all of the characteristics have a very similar value (max)E_z, except model with wall made of bricks without air holes. Considering the typical size of the brick hole in model *B18* ($s_h= 0.011$ m) with conductivity $\sigma = 0.1$ S/m, it can be seen that the value of (max)E_z is 10% higher than in the model with a solid brick.

The model with bricks *B30* at $\sigma = 0.035$ S/m showed similar values of the relative maximum values of the E_z component ((max)$E_z = 0.38$) in models with $s_h\{0.005,$ 0.007, 0.015 m} (Fig. 6). For characteristics connected with $s_h\{0.017, 0.019$ m}, an increase value of (max)E_z can be seen. Indicated tendency is a result of the multiple reflections occurring within the material containing almost 50% relative volume of the clay in the brick ($s_h = 0.017$ m - $V_{\%mc} = 57.5\%$ and $s_h = 0.019$ m - $V_{\%mc} = 52.5\%$).

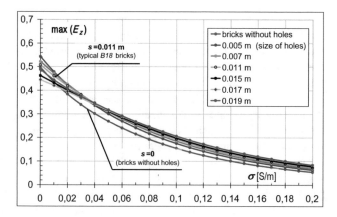

Fig. 5. The dependence between the conductivity and the relative maximum values of E_z component behind the wall made of bricks *B18*.

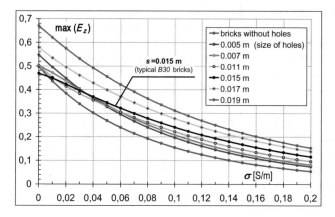

Fig. 6. The dependence between the conductivity and the relative maximum values of E_z component behind the wall made of *B30* bricks.

In Fig. 6 for the characteristics with the size of brick's holes s $\in$ {0.017, 0.019 m}, an increase in the relative maximum values of the E_z component is observed. This trend is a result of the multiple reflections occurring within the material containing almost 50% of the clay in the brick ($V_{\%mc} = 57.5\%$ for $s_h = 0.017$ m and $V_{\%mc} = 52.5\%$ for $s_h = 0.019$ m). Due to the norms and standards such brick structures are not acceptable in mass production, therefore these characteristics should not be analysed in detail.

5 Conclusion

FDTD method enables the estimation of the electric field intensity and received power levels with increased precision. It provides information related with interference levels, which can be given by intra-system or inter-system interference in the case of networks coexisting within the analyzed model. The paper propose the usage of FDTD method to electric field intensity analysis and estimation of attenuation coefficient. This numerical method and considered case indicates that complex building structure analysis allows precise evaluation of the electric field intensity distribution.

Presented methodology may be helpful in predicting signal outages that contributes to the loss of network security. The analysis of the presented models shows the effectiveness and flexibility of the FDTD method. Due to numerical approach, it is possible to investigate EM wave behavior and predict decay or signal amplification. Some reflections, diffractions and absorption in the wall, containing dielectric and air inside the structure, changes the final distribution of EM field in the front and behind the wall. If percentage of the brick's volume decreases, the distribution of EM field becomes irregular and the local value change of a wall's EM near-field also appears.

However, the non-monotonous distribution such as the electric field damping may be explained by the local interference. The mentioned variability possesses a group of characteristics for conductivity $\sigma < 0.01$ S/m. During the EM field propagation through a porous environment the wave amplitude is multiplied by the refracted and reflected waves within a structure with a low damping factor.

A localized speed variation of the wave passing through the various brick areas is seen in the electric field distribution and the presence of interference. Divergence in the field values is greater for the bricks *B30*. A higher percentage of the wall's ceramic material causes the lower wavefront deformations in the area behind the wall.

In the case of a macroscopic analysis of the EM field inside the buildings the homogenization is necessary to simplify the numerical model construction. This approach is imposed by the limited calculation efficiency of the computers when the complex structures are numerically analyzed. The obtained values of electric field allow determining the attenuation coefficient for the different walls variants.

Acknowledgment. This work was prepared under scientific work S/WE/2/2018 and supported by the Polish Ministry of Science and Higher Education.

References

1. Tan, S.Y., Tan, Y., Tan, H.S.: Multipath delay measurements and modeling for interfloor wireless communications. IEEE Trans. Veh. Technol. **49**(4), 1334–13341 (2000)
2. Stavrou, S., Saunders, S.R.: Review of constitutive parameters of building material. IEEE Trans. Antennas Propag. **1**, 211–215 (2003)
3. Choroszucho A.: An analysis of the electromagnetic waves propagation in construction elements with a complex structure in the range of wireless communication. PhD dissertation, Bialystok (2014). (in Polish)
4. Choroszucho, A., Butryło, B.: The numerical analysis of the influence the incidence angle of the plane wave on the values of the electric field intensity inside the models with the complex construction of a wall. Przegląd Elektrotechniczny **90**(12), 21–24 (2014)
5. Pinhasi Y., Yahalom, A., Petnev, S.: Propagation of ultra wide-band signals in lossy dispersive media. In: IEEE International Conference on Microwaves, Communications, Antennas and Electronic Systems, COMCAS, pp. 1–10 (2008)

6. Taflove, A., Hagness, S.C.: Computational Electrodynamics, the Finite-Difference Time-Domain Method, 3rd edn. Artech House, Norwood (2005)
7. Oskooi, A.F., Roundyb, D., Ibanescua, M., Bermelc, P., Joannopoulosa, J.D., Johnson, S.G.: MEEP: A flexible free-software package for electromagnetic simulations by the FDTD method. Comput. Phys. Commun. **181**, 687–702 (2010)
8. Elsherbeni, A.Z., Demir, V.: The Finite-Difference Time-Domain Method for Electromagnetics with MATLAB Simulations, 2nd edn. SciTech Publishing, Raleigh (2015)
9. Lopez-Iturri, P., Aguirre, E., Azpilicueta, L., Astrain, J.J., Villandangos, J., Falcone, F.: Challenges in wireless system integration as enablers for indoor context aware environments. Sensors **17**, 1616 (2017)
10. Shen, X., Xu, K., Sun, X., Wu, J., Lin, J.: Optimized indoor wireless propagation model in WiFi-RoF network architecture for RSS-based localization in the internet of things. In: 2011 International Topical Meeting on Microwave Photonics Jointly Held with the 2011 Asia-Pacific Microwave Photonics Conference, pp. 274–277 (2011)

Framework to Verify Distributed IoT Solutions for Traffic Analysis in ATN Stations

Bogdan Czejdo[1] and Wiktor B. Daszczuk[2]

[1] Department of Mathematics and Computer Science,
Fayetteville State University, Fayetteville, NC 28301, USA
bczejdo@uncfsu.edu

[2] Institute of Computer Science, Warsaw University of Technology,
Nowowiejska Street 15/19, 00-665 Warsaw, Poland
wbd@ii.pw.edu.pl

Abstract. IoT networks continuously evolve and require new theoretical and practical studies. Complex cooperation between IoT devices, based on inter-action with their internal states especially needs to be based on new significant scientific solutions. To pursue this goal we propose a dual formalism for a distributed systems being IoT networks. We refer to it as the Integrated Model of Distributed Systems (IMDS), implemented in the Dedan framework. In this dual but integrated framework, the two views of a distributed system are available: the server view of cooperating modules or the agent view of migrating threads. The Dedan framework automatically finds deadlocks and checks distributed termination in a modeled system, observed in servers communication or in sharing resources by agents. Partial deadlocks/termination are also identified, i.e., some activities may be performed in a system that is partially deadlocked/terminated. Automated verification supports the rapid development of IoT protocols. In this paper, we also discuss the problem of how the exhaustive search in the process of deadlock detection can be improved by probabilistic search using machine learning.

Keywords: IoT negotiation protocols · Autonomous vehicles ·
Automated Transit Networks · ATN stations · Automated verification

1 Introduction

The Internet of Things, or IoT, is probably the most challenging "frontier" [1] now in terms of new devices and applications. There is a variety of IoT devices in production. Some of them are relatively simple like a thermostat, and some are really "smart" like autonomous cars. All of them, however, represent only "the tip of the iceberg" [1] of this new "frontier". The Gartner definition of IoT "The network of physical objects that contain embedded technology to communicate and sense or interact with their internal states or the external environment" [2] emphasizes the need to account for both simple sensing and possibly complex interactions with internal states of IoT devices.

A more intense, complex task-oriented cooperation between IoT devices, based on interaction with their internal states will require new significant scientific and practical

© Springer Nature Switzerland AG 2020
W. Zamojski et al. (Eds.): DepCoS-RELCOMEX 2019, AISC 987, pp. 110–122, 2020.
https://doi.org/10.1007/978-3-030-19501-4_11

advances [3]. Before it can take place still a significant effort needs to be extended to solve fundamental challenges of data synchronization issues such as efficiency, reliability and privacy in IoT devices, taking into consideration intermittent connectivity, efficient energy use, devices that may have minimal resources in terms of compute and storage power "making them more dependent on servers to process and store data" [1] drifting towards client-server architectures. Unfortunately, such architectures can be only partially adopted to massive networks composed "largely of many tiny, low-power devices" [4]. Probably the most important modification is for the server and client software so that all control for data acquiring and pushing is done on the server.

Nevertheless it is claimed in most of the literature that the IoT network configuration is poised to look very different in the future [1, 3, 4]. The goal of this paper is to develop a framework for a distributed system that take into consideration the current progress of design of IoT networks cooperation framework that would address the future needs. "IoT is a wildly diverse ecosystem" [1], and a range of different need for solutions can emerge out of it as it continues to evolve [4]. To be specific we assume that there will be the need for a more intense, complex task oriented cooperation between IoT devices based on interaction between their internal states. To achieve this goal we propose a dual formalism for a distributed system for IoT networks. We refer to it as Integrated Model of Distributed Systems (IMDS [5–7]) and implemented it as the Dedan framework [8]. In this dual but integrated framework, the two views of a distributed system are available: the server view of cooperating modules or the agent view of migrating threads. The Dedan framework automatically finds deadlocks and checks distributed termination in modeled system, observed in server's communication or in sharing resources by agents. Partial deadlocks/termination are identified, i.e., some activities may be performed in a system that is partially deadlocked/terminated.

The Dedan framework can be especially applied for IoT applications that include *human-to-intelligent_device* and *intelligent_device-to-intelligent_device* sophisticated interactions for several reasons. First, the server view can support the networks with an "intelligent" server and multiple simple devices. Second, the agent view can support a complex devices and their interactions. Most of the other frameworks described in the literature support either server view (which can be compared to a Client-Server model [9]) or agent view (which corresponds to the Remote Procedure Call paradigm [9]) and require a costly and error prone translation. Third, our dual formalism allows for migration of agents meaning that IoT devices can communicate with multiple servers changing associations dynamically. Fourth, we can use our framework to specify a multi-level hierarchy of tasks to be accomplished by a cooperating devices. As a result our dual framework supports a successful deployment of a system with high reliability and assurances about task progress/accomplishment in a timely manner.

The most current model checking techniques [3] explore the entire behavior of a system in systematic manner but by checking all the possibilities. We are introducing some improvements such as using multilevel tasks/diagrams [10] but the explosion of states is still a problem. We are currently addressing this challenge using our dual framework for probabilistic verification as is described in the Future Work section. We take an advantage of our framework that we can explicitly generate a graph of cooperation starting from any moment and supporting look-ahead analysis of future states and possibly real-time solutions.

As an example of application we chose new automation techniques in urban transport related to autonomous vehicles moving in urban traffic [11–13]. We concentrated on Automated Transit Networks (ATN, also called Personal Rapid Transit - PRT [14]). The movement takes place on a separate track, typically raised above the ground level. This approach liberates the designers from most problems of recognition of the environment. Main targets are keeping up the track [15], routing and empty vehicles management [16].

Modern ATN systems are decentralized, where decisions are made using simple communication protocols between autonomous modules. For example, a delivery of empty vehicles may follow this principle [16, 17]. The distributed cooperation of autonomous modules on Internet of Things paradigm (IoT), using simple negotiation protocols [18] may be used in ATN station maneuvering, where track segment controllers guide vehicles between charging lot, parking lots and boarding/alighting lots [19–22]. Such a maneuvering may cause traffic conflicts, if more than one vehicle takes part in a change of places. An example of an automatic vehicle guidance system based on cooperating track segment controllers is described in [23].

In the paper, our dual framework is used to specify three different mini-tasks related to simple "changing places" of ATN vehicles. More specifically the three tasks are presented: using turnout lot, alternative tracks and backtracking. The verification methodology is presented that allows automated checking the correctness of the cooperation solution in a "push the button" style, i.e., to prepare the specification and just run the verification procedure. The procedure finds total and partial deadlocks in both communication and resource sharing, and checks for inevitability of each agent termination. The termination of a task means successful reaching its target position.

Section 2 presents an overview of the IMDS formalism. In Sect. 3, distributed algorithm for "change places" mini-tasks, with a possibility of deadlock, is introduced. Then, in Sect. 4 three solutions are given. The conclusions and future work are presented in Sect. 5.

2   Integrated Model of Distributed Systems

Our framework for a dual IoT distributed system has been implemented on the base of the IMDS formalism [6, 7]. The formalism is based on two sets: the *states* of servers (distributed modules), and the *messages* used by agents (representing the distributed computations). The agents migrate between servers by means of messages. A system *configuration*[1] is a set of servers' states and agents' messages.

A system changes its configuration by executing a service called by an agent with its message. As the effect, the server changes its state and a new message is issued in the context of the agent. It is the *action* relation (*message, state*) Λ (*new_message, new_state*). Interleaving model of concurrency is assumed [24], i.e., one action at a

---

[1] We do not use the term 'state' for the system to avoid ambiguity, as it is attributed to a server.

time is executed. A special type of action is an agent-termination, in which no new message is generated, only new state is on action's output: (*message, state*) Λ (*new_state*). A system starts with its *initial configuration*, containing *initial states* of all servers and *initial messages* of all agents. All actions of a server may be grouped to a server process, while alternatively all actions of an agent may be grouped to an agent process. The two projections of a system onto its processes are the views: the server view and the agent view.

The semantics of IMDS system is defined as a Labeled Transition System (LTS, [25]), where the set of *nodes*[2] contains the configurations, the *initial node* is the initial configuration, and the *transitions* are actions.

For the programming purposes, The servers' states are defined as pairs (*server, value*). A message is an invocation of a service of a server in a context of an agent, therefore it is a triple (*agent, server, service*).

A temporal logic is defined over IMDS, and general temporal formulas (not related to the structure of a verified system) are constructed for deadlock/termination identification. The generality of formulas means that they do not depend of individual system's structure and they may be checked by a user without any knowledge of temporal logic.

3   A "Change Places" Example

An example of ATN station structure is presented in Fig. 1. It consists of two lots: boarding lot on the right and alighting lot on the left. We assume that the two lots offer additional features: the left-hand lot allows to charge the vehicle's batteries while the right-hand one is a parking lot for an empty vehicle (if no passengers wait for a vehicle). If a vehicle in a parking lot needs some charging, and a vehicle in the left-hand lot is fully charged, they should switch their positions. This may be managed centrally by a station controller, but we assume that autonomous track segment controllers may operate the vehicles. It is illustrated schematically in Fig. 2a, showing the three distributed controllers: for edge lot controllers *markerE* (charging one and parking one), and for a middle segment controller *markerM* in between. The vehicles are called *AMP*s – Autonomous Moving Platforms.

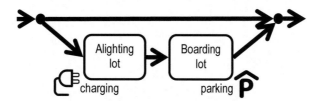

Fig. 1. A general scheme of an ATN station

---

[2] 'Nodes' are used instead of 'states' for the same reason.

The source code of IMDS specification contains a definition of server types (lines 1,12), then a declaration of server and agent instances (l. 21,22), and finally an initialization part assigning actual parameters for the formal ones (l. 24,25) and defining initial states of the servers (l. 24,25) and initial messages of the agents (l. 26). Each server type is equipped with sets of services (l. 2,13), states (l. 3,14) and actions (l. 5-10,16-19). An action (*message*=(*a,s,r*), *state*=(*s,v*)) Λ (*new_message*=(*a,s',r'*), *new_state*=(*s,v'*)) has the form {a.s.r, s.v} -> {a.s'.r', s.v'}.

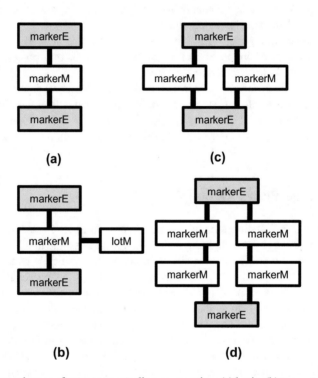

Fig. 2. Various schemes of segment controllers cooperation: (a) basic, (b) turnout, (c) two ways, (d) retreat

```
 1. server:   markerE(agents AMP[N];servers markerM),
 2. services  {start,tryM[2],okM[2],takeM},
 3. states    {free,resM,resL,occ,end},
 4. actions   {
 5. <i=1..N>      {AMP[i].markerE.start, markerE.occ} ->
                  {AMP[i].markerM.tryE[i], markerE.occ},
 6. <i=1..N><j=1..2>{AMP[i].markerE.okM[j], markerE.occ} ->
                  {AMP[i].markerM.takeE[j], markerE.free},

 7. <i=1..N><j=1..2>{AMP[i].markerE.tryM[j], markerE.free} ->
                  {AMP[i].markerM.okE[j], markerE.resM},
 8. <i=1..N><j=1..2>{AMP[i].markerE.tryM[j], markerE.resL} ->
                  {AMP[i].markerM.notE[j], markerE.resL},
 9. <i=1..N><j=1..2>{AMP[i].markerE.tryM[j], markerE.occ} ->
                  {AMP[i].markerM.notE[j], markerE.occ},
10. <i=1..N>          {AMP[i].markerE.takeM, markerE.resM} ->
                  {markerE.end},
11. }
12. server:   markerM(agents AMP[N];servers markerE[2]),
13. services  {tryE[2],okE[2],notE[2],okL[2],takeE[2],switch[2]},
14. states    {free,resE[2],occ},
15. actions   {
16. <i=1..N><j=1..2>{AMP[i].markerM.tryE[j], markerM.free} ->
                  {AMP[i].markerE[j].okM[j], markerM.resE[j]},
17. <i=1..N><j=1..2>{AMP[i].markerM.takeE[j], markerM.resE[j]}->
                  {AMP[i].markerM.switch[3-j], markerM.occ},
18. <i=1..N><j=1..2>{AMP[i].markerM.switch[j], markerM.occ} ->
                  {AMP[i].markerE[j].tryM[j], markerM.occ},
19. <i=1..N><j=1..2>{AMP[i].markerM.okE[j], markerM.occ} ->
                  {AMP[i].markerE[j].takeM, markerM.free},
20. }

21. servers   markerE[2],markerM;
22. agents    AMP[N];

23. init   -> {
24. <j=1..2>  markerE[j](AMP[1..N],markerM).occ,
25.              markerM(AMP[1..N],markerE[1,2]).free,

26. <j=1..2>  AMP[j].markerE[j].start,
27. }.
```

The track segments may be in states (l. 3,14): occ – occupied, res – reserved, $free$ – not occupied, end – occupied, in final position. This may concern edge lot – E, or middle segment – M.

The protocol is based on three messages: a controller of occupied segment sends try (l. 5,18), and when acquires a response ok (l. 7,16, target segment becomes reserved), issues a $take$ message (l. 6,19) which frees the actual segment and catches the target segment. The protocol is illustrated as a message sequence in Fig. 3a. The case of rejecting the order because of the segment occupation is presented in Fig. 3b. The $switch$ message sent by $markerM$ controller to itself causes the change of a number of $markerE$ with which it communicates (switch 1 => 2 or 2 => 1).

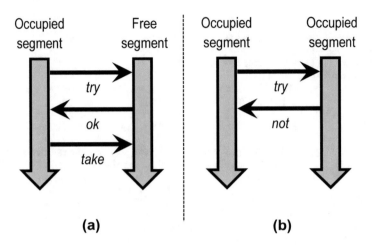

Fig. 3. The inter-controller negotiation protocol: (a) granting sequence, (b) rejecting sequence

As the vehicles must meet somewhere in between the lots, the conflict is obvious and therefore a deadlock in the system is identified in Dedan. It is illustrated in the server view in Fig. 4, as a sequence diagram leading from the initial configuration to the deadlock configuration.

4   Solutions of the Problem

4.1   Solution 1 - Turnout

The problem may be solved by introducing an additional lot *lotM* for turnout (Fig. 2b). One of the vehicles may diverge for this additional lot and let the other vehicle to bypass. A controller of occupied track segment may answer *not* for a request, which directs a vehicle to *lotM*. This is realized by the new actions of *markerM*:

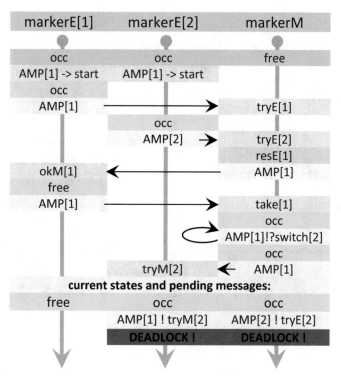

Fig. 4. The sequence diagram of the basic version of controllers communication (the server view), leading to a deadlock

```
1. <i=1..N><j=1..2>{AMP[i].markerM.notE[j], markerM.occ} ->
              {AMP[i].lotM.try[j], markerM.occ},
2. <i=1..N><j=1..2>{AMP[i].markerM.okL[j], markerM.occ} ->
              {AMP[i].lotM.take[j], markerM.wait},
```

The solution is successful: no deadlock is reported and common termination of both agents is given as the sequence diagram in Fig. 5. It is the agent view, where servers' history is presented on left, and agents' history is shown on the right. Dedan finds only inevitable termination, which is absent if a deadlock occurs or when an agent falls into infinite loop without escape.

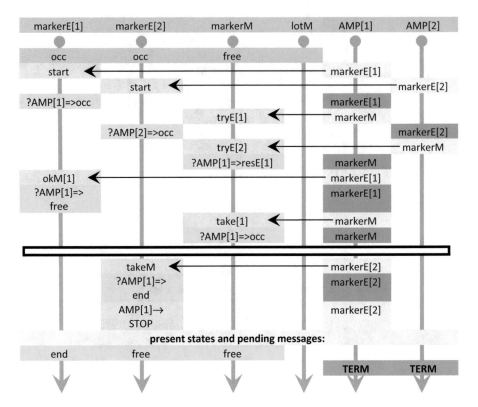

Fig. 5. The sequence diagram of solution 1 (*Turnout*) in the agent view – successful termination. A large part of the sequence is suppressed between the two black lines

4.2   Solution 2 – Two Ways

The second solution is based on using two alternative tracks for bypassing (Fig. 2c). If a vehicle attempts to take the occupied track – it switches to the free one. The alternative tracks are modeled by two middle segments *markerM[2]*. If *markerE* controller gets *not* response (l. 3), it switches to the opposite track (1 => 2, 2 => 1). The full set of actions of *markerE* controller is:

```
1.  <i=1..N><j=1..2>{AMP[i].markerE.start, markerE.ini} ->
                    {AMP[i].markerM[1].try[i], markerE.occ[1]},
2.  <i=1..N><j=1..2>{AMP[i].markerE.ok, markerE.occ[j]} ->
                    {AMP[i].markerM[j].take[i], markerE.free},
3.  <i=1..N><j=1..2>{AMP[i].markerE.not, markerE.occ[3-j]} ->
                    {AMP[i].markerM[j].try[i], markerE.occ[j]},

4.  <i=1..N><k=1..2>{AMP[i].markerE.try[k], markerE.free} ->
                    {AMP[i].markerM[k].ok, markerE.res},
5.  <i=1..N>        {AMP[i].markerE.take, markerE.res} ->
                    {markerE.end},
```

This solution is successful as well: no deadlock is reported and common termination of both agents is shown just as in the *Turnout* solution.

4.3   Solution 3 – Retreat

The third solution is based on a retreating possibility – a vehicle that discovers a conflict retreats to its starting position and attempts to use alternative track (Fig. 2d). To let the vehicles retreat – additional track segments are used, one on every alternative track. This leads to a set of four track segment controllers *markerM[4]*. In the symmetric solution, a deadlock or a thrashing occurs between cooperating *markerM* servers. Therefore, a priority is assigned to one *markerM* in every pair. The retreating protocol messages have the *_b* suffix, for example in the *markerE* controller (l. 6,7):

```
1. <i=1..N><j=1..2>{AMP[i].markerE.start, markerE.ini} ->
                {AMP[i].markerM[j].try, markerE.occ[j]},
2. <i=1..N><j=1..2>{AMP[i].markerE.ok, markerE.occ[j]} ->
                {AMP[i].markerM[j].take, markerE.free},
3. <i=1..N><j=1..2>{AMP[i].markerE.not, markerE.occ[3-j]} ->
                {AMP[i].markerM[j].try, markerE.occ[j]},

4. <i=1..N><k=1..2>{AMP[i].markerE.try[k], markerE.free} ->
                {AMP[i].markerM[k].ok, markerE.res},
5. <i=1..N>       {AMP[i].markerE.take, markerE.res} ->
                {markerE.end},

6. <i=1..N><k=1..2>{AMP[i].markerE.try_b[k], markerE.free} ->
                {AMP[i].markerM[k].ok_b, markerE.res_b},
7. <i=1..N><k=1..2>{AMP[i].markerE.take_b[3-k],markerE.res_b}->
                {AMP[i].markerM[k].try, markerE.occ[k]},
```

The third solution has no deadlock as the previous solutions. The common termination of both agents is reported.

5   Conclusions and Future Work

Distributed systems, including current and future Internet of Things (IoT) networks might require modeling that reflects the cooperation of devices/applications. Automated verification of deadlocks and distributed termination can be accomplished in our framework. Our framework supports the communication duality that allows the integration of remote procedure call and client-server paradigm into a single, uniform model. The paper presents an original formalism for the modeling and verification of distributed systems. The Integrated Model of Distributed Systems (IMDS) defines a distributed system as two sets: states and messages, and the relationship of the "actions" between these sets. General temporal formulas over IMDS, independent of the structure of the verified system, allow automated verification. These formulas distinguish between deadlocks and distributed termination, and between communication deadlocks and resource deadlocks. Partial deadlocks and partial termination can also be checked. The implemented dual framework of IMDS formalism (Integrated Model of Distributed Systems) incorporates model checking using general temporal formulas, and supports all the listed requirements. Automated verification allows for rapid development of solutions, generated even on-line during distributed system operation.

The current version of the framework allows the user to explore the entire behavior of a system in a systematic manner. We can explicitly generate a graph of cooperation starting from any point of interest and supporting look-ahead analysis. This feature of the framework has created a good foundation for novel non-deterministic model checking based on duality of the framework. More specifically the server in a distributed system, being in a given state, accepts the message and executes the action. This action creates a new state of the server. These two states constitute a 2-node path that can be easily generalized to n-node path. All n-node paths from the given state can be recorded and the table of paths can be generated. This table can be enhanced by an additional column(s) (label) describing if there is any problem with the last state in the path e.g. deadlock, partial deadlock or lack of termination, etc. We have used such extended table as a training set for machine learning algorithms to optimize graph traversal in our model checking. The crucial importance of the probabilistic method is that it can allow us to avoid checking all graph paths but help us to identify with high probability those paths that can cause problems. Specifically, in our preliminary investigation we have applied Support Vector Machine algorithm to learn what paths lead to a problematic state.

Similarly we used the input message, together with the output one, as a fragment of the distributed computation called agent. The distributed system is described as a relation of "actions" over finite sets of states and messages, composed in pairs (*message, state*). Several messages connected by actions can be recorded and the table of various paths can be created. These table can be enhanced by a label describing if there is any problem with the last message in the path and used as a training set for machine learning. The advantage of our method is that we are learning rules with respect to servers and separately on agents, giving two probabilistic views of a distributed systems. Actually, we are also in the process of comparing these view with an integrated view based on integration of the training sets. One of many research questions is to determine the best path length to optimize the local and global prediction.

References

1. Tozzi, C.: IoT and Data Syncing: Tips for Developers (2016). https://sweetcode.io/iot-data-syncing-tips-developers/
2. IT Glossary. Internet of Things. https://www.gartner.com/it-glossary/internet-of-things
3. Ratana, H., Mohamad, S.M.S.: Towards model checking of network applications for IoT system development. J. Telecommun. Electron. Comput. Eng. **9**(3–4), 143–149 (2017). http://journal.utem.edu.my/index.php/jtec/article/view/2934
4. Burns, P.: 5 Reasons Why Synchronization is Critical to IoT. The clocks on your IoT devices are way more important than you think (2017). https://www.iotforall.com/iot-synchronization/

5. Daszczuk, W.B.: Communication and resource deadlock analysis using IMDS formalism and model checking. Comput. J. **60**(5), 729–750 (2017). https://doi.org/10.1093/comjnl/bxw099

6. Daszczuk, W.B.: Specification and verification in integrated model of distributed systems (IMDS). MDPI Comput. **7**(4), 1–26 (2018). https://doi.org/10.3390/computers7040065

7. Daszczuk, W.B.: Integrated Model of Distributed Systems. SCI, vol. 817, 238 p. Springer, Cham (2020). https://doi.org/10.1007/978-3-030-12835-7

8. Dedan. http://staff.ii.pw.edu.pl/dedan/files/DedAn.zip

9. Jia, W., Zhou, W.: Distributed network systems. From Concepts to Implementations. In: NETA, vol. 15, 513 p. Springer, New York (2005). https://doi.org/10.1007/b102545

10. Czejdo, B., Bhattacharya, S., Baszun, M.: Use of multi-level state diagrams for robot cooperation in an indoor environment. In: ICDIPC 2011: Digital Information Processing and Communications, Ostrava, Czech Republic, 7–9 July 2011, Part II. CCIS, vol. 189, pp. 411–425. Springer, Heidelberg (2011). https://doi.org/10.1007/978-3-642-22410-2_36

11. Parkinson, S., Ward, P., Wilson, K., Miller, J.: Cyber threats facing autonomous and connected vehicles: future challenges. IEEE Trans. Intell. Transp. Syst. **PP**(99), 1–18 (2017). https://doi.org/10.1109/tits.2017.2665968

12. Kim, T.U., Lee, J.W., Yang, S.: Study on development of autonomous vehicle using embedded control board. In: 2016 11th International Forum on Strategic Technology (IFOST), Novosibirsk, Russia, 1–3 June 2016, pp. 599–603. IEEE (2016). https://doi.org/10.1109/ifost.2016.7884331

13. Zhang, S., Yen, I.-L., Bastani, F., Moeini, H., Moore, D.: A semantic model for information sharing in autonomous vehicle systems. In: 2017 IEEE 11th International Conference on Semantic Computing (ICSC), San Diego, CA, 30 January–1 February 2017, pp. 32–39. IEEE (2017). https://doi.org/10.1109/icsc.2017.93

14. McDonald, S.S.: Personal Rapid Transit (PRT) system and its development. In: Encyclopedia of Sustainability Science and Technology, pp. 7777–7797. Springer, New York (2012). https://doi.org/10.1007/978-1-4419-0851-3_671

15. Kozłowski, M., Choromański, W., Kowara, J.: Parametric sensitivity analysis of ATN-PRT vehicle (Automated transit network – personal rapid transit). J. VibroEng. **17**(3), 1436–1451 (2015). https://www.jvejournals.com/article/16019

16. Daszczuk, W.B., Mieścicki, J., Grabski, W.: Distributed algorithm for empty vehicles management in personal rapid transit (PRT) network. J. Adv. Transp. **50**(4), 608–629 (2016). https://doi.org/10.1002/atr.1365

17. Chebbi, O., Chaouachi, J.: A decentralized management approach for on-demand transit transportation system. In: Abraham, A., Wegrzyn-Wolska, K., Hassanien, A., Snasel, V., and Alimi, A. (eds.) Second International Afro-European Conference for Industrial Advancement, AECIA 2015, Paris, Villejuif, France, 9–11 September 2015. AISC, vol. 427, pp. 175–184. Springer, Cham (2015). https://doi.org/10.1007/978-3-319-29504-6_18

18. Garofalaki, Z., Kallergis, D., Katsikogiannis, G., Ellinas, I., Douligeris, C.: Transport services within the IoT ecosystem using localisation parameters. In: 2016 IEEE International Symposium on Signal Processing and Information Technology (ISSPIT), Limassol, Cyprus, 12–14 December 2016, pp. 87–92. IEEE (2016). https://doi.org/10.1109/isspit.2016.7886014

19. Won, J.-M., Choe, H., Karray, F.: Optimal design of personal rapid transit. In: 2006 IEEE Intelligent Transportation Systems Conference, Toronto, Canada, 17–20 September 2006, pp. 1489–1494. IEEE (2006). https://doi.org/10.1109/itsc.2006.1707434

20. Muller, P.J.: Open-guideway personal rapid transit station options. In: Twelfth International Conference, Automated People Movers, Atlanta, Georgia, 31 May–3 June 2009, pp. 350–360. American Society of Civil Engineers, Reston (2009). https://trid.trb.org/view/920208

21. Lowson, M., Hammersley, J.: Maximization of PRT station capacity. In: 90th Annual Meeting Transportation Research Board, Washington, DC 1317 (2011). https://www.researchgate.net/publication/258344039_Maximization_of_PRT_Station_Capacity
22. Fatnassi, E., Chaouachi, J.: Discrete event simulation of loading unloading operations in a specific intermodal transportation context. In: Silhavy, R., Senkerik, R., Oplatkova, Z.K., Silhavy, P., and Prokopova, Z. (eds.) 5th Computer Science On-line Conference 2016 (CSOC 2016), Software Engineering Perspectives and Application in Intelligent Systems, vol. 2, Prague, Czech Republic, 27–30 April 2016. AISC, vol. 465, pp. 435–444. Springer, Cham (2016). https://doi.org/10.1007/978-3-319-33622-0_39
23. Czejdo, B., Bhattacharya, S., Baszun, M., Daszczuk, W.B.: Improving resilience of autonomous moving platforms by real-time analysis of their cooperation. Autobusy-TEST 17(6), 1294–1301 (2016). arXiv:1705.04263
24. Godefroid, P., Wolper, P.: Using partial orders for the efficient verification of deadlock freedom and safety properties. In: 3rd International Workshop, CAV 1991, Aalborg, Denmark, 1–4 July 1991. LNCS, vol. 575, pp. 332–342. Springer, Heidelberg (1992). https://doi.org/10.1007/3-540-55179-4_32
25. Reniers, M.A., Willemse, T.A.C.: Folk theorems on the correspondence between state-based and event-based systems. In: 37th Conference on Current Trends in Theory and Practice of Computer Science, Nový Smokovec, Slovakia, 22–28 January 2011, LNCS, vol. 6543, pp. 494–505. Springer, Heidelberg (2011). https://doi.org/10.1007/978-3-642-18381-2_41

The Picking Process Model in e-Commerce Industry

Alicja Dąbrowska⬤, Robert Giel$^{(\boxtimes)}$ ⬤, and Marcin Plewa⬤

Wroclaw University of Science and Technology, Wroclaw, Poland
robert.giel@pwr.edu.pl

Abstract. The subject of this article is related to the dynamic development of the Polish e-commerce market. The e-commerce market generates the need for professional handling and order processing in logistics system. Particularly in the preparation of goods for shipment to the customer.

This article describes the model of picking process carried out in accordance with the person-to-goods principle. The person-to-goods principle is currently most commonly used in warehouse systems serving the e-commerce market. According to this principle, the person responsible for collecting the goods moves between the storage racks. Currently, many solutions are being implemented to improve this process, e.g. by showing to the person responsible for collecting the goods the order of the collection points in the warehouse.

The aim of the article is to present the model of picking process in e-commerce industry. This model is used to evaluate the route selection algorithms used by the automatic queuing system. Currently, one criterion - the time criterion - is used to assess such systems. This article presents a evaluation based on more than one criterion, including the number of collisions occurring on the routes and the average distance between picking points.

Keywords: Storage · Modelling · Path planning

1 Introduction

Popularization of Internet access initiated the possibility of conducting e-commerce. Over the years increasing number of e-commerce enterprises located in Poland can be observed [1]. This dynamic development of e-commerce market caused the need for changes in logistics processes [2, 3]. Enterprises are continuously searching and implementing new solutions, which allow them to achieve advantage over competitors. They are trying to reduce the time of executing customers' orders, increase customer service level and improve efficiency of logistics processes [4, 5].

Taking into consideration basic parts of warehouse process: receiving, storage, picking and shipment, picking, which is collecting products according to picking lists created based on customers' orders is the component, which has the biggest influence on delivering orders in designated time [6]. In classification based on way of moving, there are three principles of picking: person-to-goods, goods-to-person and combination of these two. Modern warehouses are increasingly using goods-to-person principle, however there are still many enterprises, which are using person-to-goods principle.

© Springer Nature Switzerland AG 2020
W. Zamojski et al. (Eds.): DepCoS-RELCOMEX 2019, AISC 987, pp. 123–131, 2020.
https://doi.org/10.1007/978-3-030-19501-4_12

This principle depends on sending the picker to the locations from picking list, where products ordered by customers are stored. Picking process based on person-to-goods principle has the following subprocess: moving to the given location, searching for the right product, quality control and collecting product. Moving to the given location is the subprocess, which takes the most time among all picking subprocesses [7, 8].

Not using pathfinding algorithms in picking process can lead to low efficiency of pickers, long distances of paths and increased number of unwanted collisions between pickers. Occurrence frequency of mentioned events can be reduced by finding and adjusting to owned picking area way of finding the best path, which is understood as a path that allows to execute given picking list in required time and with the smallest number of collisions. The aim of this article is to create simulation model reflecting way of moving in picking process. Simulation model was used to examine impact of pathfinding algorithms on values of chosen evaluation parameters.

2 Model Description

Picking process reflected in model is a part of warehouse process, which takes place in e-commerce enterprise located in Wroclaw. This enterprise offers a wide range of products such as cloths, books, electronics, household appliances, cosmetics and food. Delivery of orders to customers has limited time of execution dependent on chosen delivery option and customer location.

Picking process is performed in storage area, along which roller-belt conveyor is located. On both sides of the conveyor are racks set in rows, which come in two different dimensions called short and long racks. Between racks are alleys and per-pendicular to shelves are main corridors. Each rack is located in one of four blocks separated by main corridors. Block A with short racks and block B with long racks is settled on the left side of the conveyor and block C and D with respectively long and short racks is settled on the right side of the conveyor.

Picking lists are assigned to pickers and they contain specified number of locations to visit. Picker's route in model has the character of an undirected graph, in which he can move between specific vertices. Every vertex is a point marked $P_{x,y}$, where $x \in \{0, 1, 2, \ldots, 65\}$ is a point number and $y \in \{0, 1, 2, \ldots, 25\}$ is an alley number. Scheme of storage area is presented on Fig. 1.

On Fig. 1 the starting point is marked as S letter and it is the beginning and the ending of picking list execution according to given algorithm. Starting point is an entry to the system and an exit from the system.

Created model is characterized by parameters showed in Table 1. These are con-stant parameters from the actual system.

In actual system picker equipped in scanner and trolley with place for two con-tainers goes to the given locations from picking list, which are displayed on the scanner screen, scans the container, location's barcode, finds the right product, makes quality control, scans product's barcode and puts the product into the container. Time needed to execute these activities is not included in the model, because only time needed to move between locations from given picking list was taken into consideration. Picking list is a list of products to be taken from inventory with an indication of their locations

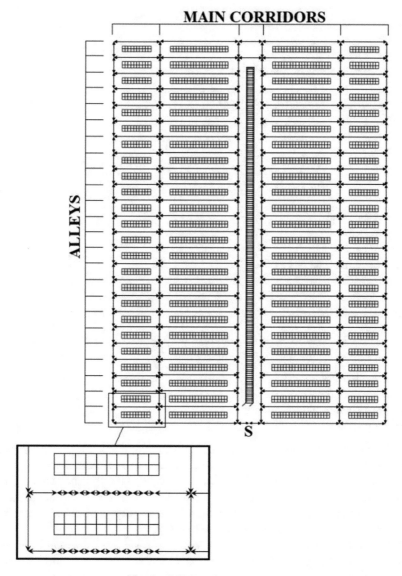

Fig. 1. Scheme of storage area

in the storage area. Visiting all locations from given picking list is understood as execution of the order. In the model was assumed that locations on the picking list can not be repeated and picker can not stop at any location. Searching and collecting processes are ignored.

The model allows to execute simulation of moving between locations during picking process with the use of 5 pickers. Every picker executes one picking list

Table 1. Model's constant parameters

Parameter	Value	Unit
Picking area dimensions	41000 × 30550	mm
Distance between locations with products to collect within the rack	300	mm
Distance between points on main corridors	1600	mm
Number of alleys	26	unit
Number of short racks	50	unit
Number of long racks	50	unit
Short rack dimensions	3000 × 600 × 1800	mm
Long rack dimensions	6000 × 600 × 1800	mm
Alley width	1000	mm
Main corridor width	2000	mm
Picker's velocity	2	m/s
Number of pickers	5	unit
Number of locations on picking list	100	unit

generated on the beginning of the simulation and moves between given location according to four pathfinding algorithms, which are:

- random algorithm,
- S-shape algorithm,
- nearest neighbor algorithm,
- BFS algorithm.

Every picker starts execution of his order in starting point, where starting time of used algorithm is saved. At the beginning picker has the route from the random algorithm, after finishing order he is coming back to the starting point and according to the next algorithms he is executing the same order. Total time of executing picking list and number of collisions are saved after finishing every algorithm. General scheme of the simulation model's algorithm is showed on Fig. 2.

After generating table with all possible locations to visit, 100 different locations are taken from it according to uniform distribution. In this way the picking list is created. It is executed in subsequent iterations according to given algorithm. Getting to the starting point is a finish of the algorithm and a start of simulation with another algorithm. In random algorithm locations are visited in the order in which they were given on picking list. Created order in the picking list is not modified.

For creating the route according to S-shape algorithm [9–11], four blocks (A, B, C and D) designated in the picking area were used. The S-shape algorithm assumes moving on the route in shape of the letter S. Thanks to the information stored in every point of the path about alley number and point number, which shows considered block, it is possible to change the order of locations on picking list compatibly with S-shape algorithm. Due to that it was possible to program moving from the beginning to the end

Fig. 2. General scheme of the simulation model's algorithm

in blocks A and C and moving from the end to the beginning in blocks B and D and maintaining assumed shape of route. Picker starts his work from block A and he goes through block B, C and at the end D.

Another used algorithm is the nearest neighbor algorithm [12]. This algorithm depends on choosing unvisited point located in the nearest distance to the current position. It is a greedy algorithm selecting the best part solution at the considered moment. For this algorithm distances between all possible combinations of pairs of locations from the picking list was designated and all locations was marked to give information about state of visit. The procedure of searching the nearest point to the present one was repeated until all locations were marked as visited.

The last used algorithm is breadth-first search algorithm, in short BFS [13, 14]. This algorithm helps to find the shortest route, which is understood as a route with the smallest number of connections between locations. Distances between locations are not taken into consideration.

3   Results of Simulation

One hundred simulations were executed. During every simulation the picking list with one hundred locations was generated, which order was changed three times in the next stages of simulation according to the S-shape, nearest neighbor and BFS algorithms to examine the influence of pathfinding algorithm on chosen evaluation parameters. Used evaluation parameters are average time of executing single order with standard deviation, minimum and maximum time of executing single order, average distance between visited locations and number of collisions. Table 2 shows average total, minimum and maximum time of executing single order and average distances between locations for analyzed algorithms.

Table 2. Average total, minimum and maximum time of executing single order and average distances between locations for analyzed algorithm

Algorithm name	Average time of executing single order [s]	Standard deviation	Minimum time of executing single order [s]	Maximum time of executing single order [s]	Average distance between two locations [m]
Random	970	81	789	1158	19, 4
S-shape	289	8	270	308	15, 8
Nearest neighbor	374	215	245	1492	7, 5
BFS	342	17	301	385	6, 8

Presented results connected with time of executing single task was showed also in the form of a graph on Fig. 3 for easier comparison.

Taking into consideration time of executing single task, the best average result, which is 289 s was given in S-shape algorithm and the worst result in random algorithm, in which this time was over three times longer (970 s). It is also worth noting that the average result of the order execution time in S-shape algorithm had the lowest standard deviation among all considered. The lowest possible time of executing single task was given in the nearest neighbor algorithm, however average time of executing single task in this algorithm had the highest standard deviation. It shows that collection of locations on the picking list and their arrangement in the storage area have the significant influence on given results. There are such collections of locations for which the best algorithm can be the nearest neighbor algorithm and such collections of locations for which better solution is S-shape. In case of average distance between locations there is the same dependence as in the case of average time of executing single order because of constant velocity of picker.

Another examined parameter was the number of collisions, which is number of meetings of one picker with the rest. Given values of this parameter are presented in Table 3.

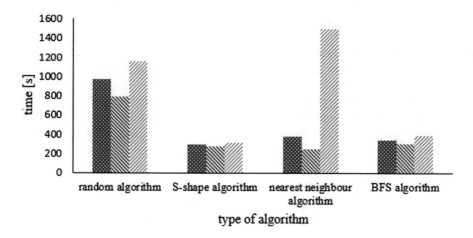

Fig. 3. Diagram of average, minimum and maximum times of executing single order for different algorithms

Table 3. Average, minimum and maximum numbers of collisions for analyzed algorithms

Algorithm name	Average number of collisions	Minimum number of collisions	Maximum number of collisions
Random	225	154	409
S-shape	11	0	42
Nearest neighbor	11	1	30
BFS	12	1	28

The highest number of collisions, which is 225 was observed in random algorithm and the lowest number of collisions (11 collisions) in nearest neighbor algorithm and S-shape algorithm. Collision occurrence is a reason for extending the order execution time and creating the danger of an accident due to the small widths of the alleys. Because of this it is desirable to lower the number of collisions. Given results of minimum number of collisions show that it is possible to almost completely eliminate them.

To define picker's efficiency the enterprise is using indicator E_w described by the formula (1), in which maximum allowed time to execute single order t_{max} and time of executing single order, which is time of route transition R_t are taken into consideration.

$$E_w = \frac{t_{max}}{R_t} \qquad (1)$$

Value of efficiency indicator above 1 or equal 1 means that the order was executed in lower time than maximum allowed time to do this or was equal to maximum allowed time. Value below 1 informs that the maximum allowed time to execute single order was exceeded. Maximum allowed time to execute single order is understood as a time given to move between locations from the picking list with excluding processes of searching and collecting products and in the enterprise the value of it is 400 s. Based on simulation results and known value of allowed time to execute single order it was possible to calculate average efficiency indicator of pickers. Results of calculations are showed in Table 4.

Table 4. Efficiency indicator of analyzed algorithms

Algorithm name	Efficiency indicator
Random	0, 4
S-shape	1, 4
Nearest neighbor	1, 1
BFS	1, 2

All algorithms except the random algorithm allows the pickers to achieve efficiency indicator below 1. The best average efficiency expressed by indicator was given in S-shape algorithm, where indictor was 1, 4. Using random algorithm caused exceeding the allowed time, which is unacceptable and leads to delays in execution of orders.

4   Summary and Conclusions

Results of simulations show that thanks to using appropriate pathfinding algorithm, it is possible to reduce time of order execution, average distance between locations, number of collisions and to achieve the required efficiency by pickers. Achieving these results may not only improve the timeliness of deliveries to customers, but also create safer work conditions due to reduced number of collisions.

Taking into consideration average time of executing single order with standard deviation, average distance between locations and number of occurred collisions S-shape algorithm is best suited for the current storage area. However, minimum achieved time of executing single task shows that big influence on results has given collection of locations in the picking list. There are such sets of locations for which the algorithm of the nearest neighbor turns out to be a better algorithm than S-shape algorithm. Maximum achieved times of executing single task gives very similar conclusion. There are such sets of locations for which the algorithm of the nearest neighbor can be worse than random algorithm. After implementing S-shape algorithm there are no results significantly different from the rest, the difference between the lowest and the highest result is only 38 s. S-shape algorithm assumes moving along four blocks in strictly specified order. There are no coming back to the same alleys within the block, so additional distances are eliminated and moving from one side of the conveyer to another is limited to minimum. The nearest neighbor algorithm is a greedy algorithm, so it is based on choosing the best solution at the considered moment, which for some sets of locations can cause returning to the same alleys and frequent moving from one side of the conveyer to another. It can be the reason for occurrence of extremely high results in relation to the rest of them. However, there are also some sets of locations, for which the nearest neighbor algorithm can give better results than S-shape algorithm.

By examining for what kind of collections of locations can give better results using the nearest neighbor algorithm instead S-shape algorithm and which collections give the worst results in the nearest neighbor algorithm it is possible to find a method of selecting the best pathfinding algorithm by arrangement of locations from the picking list in storage area. Using pathfinding algorithms can help to achieve better results in picking process thanks to adjusting available algorithms to received set of locations in orders.

References

1. Rokicki, T.: E-commerce market in Poland. Inf. Syst. Manage. **5**, 563–572 (2016)
2. Żuchowski, W.: The impact of e-commerce on warehouse operations. LogForum **12**, 95–101 (2016)
3. Żurek, J.: E-commerce influence on changes in logistics processes. LogForum **11**, 129–138 (2015)
4. Kawa, A.: Fulfillment service in e-commerce logistics. LogForum **13**, 429–438 (2017)
5. Płaczek, E.: New challenges for logistics providers in the e-business era. LogForum, 39–48 (2010)
6. Park, B.C.: Order picking: issues, systems and models. In: Manzini, R. (ed.) Warehousing in the Global Supply Chain. Springer, London (2012)
7. Bartholdi, J.J., Hackman, S.T.: Warehouse and Distribution Science. Georgia Institute of Technology, Atlanta (2012)
8. Tompkins, J.A., White, Y.A., Bozer, E.H., Tanchoco, J.M.A.: Facilities Planning, 4th edn. Wiley, Hoboken (2010)
9. Jacyna, M., Bobiński, A., Lewczuk, K.: Modelowanie i symulacja obiektów magazynowych 3D, 1st edn. PWN, Warsaw (2017)
10. Le-Duc, T.: Design and control of efficient order picking processes. Ph. D. thesis, RSM Erasmus University, Rotterdam (2005)
11. Gałązka, M., Jakubiak, M.: Simulation as a method of choosing the order picking concept. Logist. Transp. **11**, 81–88 (2010)
12. Johnson, D.S., McGeoch, L.A.: The traveling salesman problem, a case study in local optimization. In: Aarts, E.H.L., Lenstra, J.K. (eds.) Local Search in Combinatorial Optimization. Wiley, New York (1997)
13. Even, S.: Graph Algorithms, 2nd edn, pp. 11–14. Cambridge University Press, Cambridge (2011)
14. Gross, J.L., Yellen, J.: Handbook of Graph Theory. CRC Press, Boca Raton (2004)

Evaluation of Design Pattern Utilization and Software Metrics in C# Programs

Anna Derezińska$^{(\boxtimes)}$ [ID] and Mateusz Byczkowski

Institute of Computer Science, Warsaw University of Technology,
Nowowiejska 15/19, 00-665 Warsaw, Poland
A.Derezinska@ii.pw.edu.pl

Abstract. Utilization of design patterns is supposed to have a considerable impact on software quality and to correlate with different software metrics. Much experimental research has considered these issues but almost none related to C# programs. This paper examines utilization of Gang-of-Four design patterns combined with results of software metrics calculated on a set of C# programs. The design patterns have been automatically detected in source code. Analyzed applications with design patterns evaluated to be more complex but in the same time better maintainable than applications without design patterns. Usage of design patterns contributed to a growth in class encapsulation. Classes that implemented design patterns were more complex that other classes used in both types of applications with and without design patterns. The outcomes could be of importance in software maintenance, reverse engineering and program refactoring.

Keywords: Design Patterns · Software metrics · Design Pattern detection · Gang of Four · C#

1 Introduction

Design Patterns (DP in short) presented by Gang of Four [1] are programming artefacts commonly used in object-oriented software development that could influence software quality, especially its ability to be tested, refactored, maintained, etc. However, the detailed impact of design pattern usage on software quality is still an open question, as ambiguous conclusions have been drawn from many experiments [2–4].

Utilization of DP has been analyzed in numerous programming environments [4, 5]. Different variants of DP implementation in C# have been presented by Metsker [6] and Sarcar [7], and applied in practice. However, to the best of our knowledge, no research, except of the works published by Gatrell et al. [8, 9], has been performed on usage of DP in C# programs. Furthermore, despite a lot of studies considering utilization of DP in relation to various software metrics have been performed [2, 3, 5], there is a gap in such analysis in the area of C# programs.

We have tried to fill in this gap. Therefore, we have refined detection rules of DP in C# programs [10]. While using an enhanced DP detector integrated with software metric tools we have conducted experiments on 20 C# programs. In this paper we have presented results of software analysis at the class level, comparing classes participating

W. Zamojski et al. (Eds.): DepCoS-RELCOMEX 2019, AISC 987, pp. 132–142, 2020.
https://doi.org/10.1007/978-3-030-19501-4_13

in DP instances (DP classes) with the remaining classes. We have also related this distinction of classes to the outcomes of selected software quality metrics [11].

Our experiments have showed that applications with design patterns seemed to be more complex but in the same time better maintainable than applications without design patterns. Membership of DP was associated with a high encapsulation of classes. Classes that implemented design patterns were more complex than other classes used in both types of applications with and without design patterns.

The reminder of this paper is organized as follows. The next Section describes some related work. Selected software metrics are surveyed in Sect. 3. A brief overview of C# Analyzer is given in Sect. 4. Results of the experiments are presented and discussed in Sect. 5. Finally, Sect. 6 concludes the paper.

2   Related Work

Utilization of GoF design patterns and their impact on the software quality have been examined in various studies [2, 5]. Evaluation of DP has been performed through two main activities: (1) computing code metrics, (2) using expert opinion (e.g. surveys). In this paper we refer to the first approach.

Gatrell et al. have reported on investigation of development of a commercial C# system [8, 9]. Different software features in DP-based and other classes were recognized. The authors focused on DP introduced intentionally, mentioned in documentation, and manually detected after code inspection. Classes participating in DP were found to be changed more frequently than other classes [8]. Developers could have been more familiar with these classes that made them easy to adopt and change. As the highest change prone patterns were observed *Adapter*, *Method*, *Proxy*, *Singleton*, and *State*. Pattern-based classes were also more fault-prone than non-pattern classes, both in terms of the number of changes made to a class, and the size of the changes required to fix a fault [9]. *Adapter*, *Method* and *Singleton* patterns were said to be the most fault prone patterns. Coupling of classes had a significant relationship with the fault proneness of classes in the system. Those results are not directly comparable to our work as the authors did not use any automatic detection of DP and focused on a system evolution while comparing many versions of a program under development.

The majority of other experimental studies on DP usage, combined often with software metric measurement, considered C++ and Java programs, or referred to a class model level, and thus could be considered language independent.

Hussain et al. observed a significant correlation between a structural complexity of Java programs and usage of *Template*, *Adapter*, *Command*, *Singleton*, and *Factory Method* design patterns [12].

Gravino and Risi investigated 10 Java tools and founded that DP classes had a greater number of LOC and a greater number of comments, better LCOM (Lack of Cohesion in Methods) in comparison to other classes, and were more complex in terms of WMC (Weighted Method per Class) [13].

An influence of DP on stability has been studied on a sizeable set of Java classes [14]. Classes that play exactly one role in a GoF design pattern were found to be more stable than classes that play zero or more than one role. However, the statistical results varied across projects from different application domains and different types of patterns.

Coupling and cohesion of classes have also been investigated in a set of Java applications in [15]. Assessment of CBO (Coupling Between Objects) and LCOM metrics showed that DP related classes are more coupled and less cohesive than DP non-participant classes.

3 Measurement of Software Metrics for C# Programs

Different software metrics [11] have been considered in the context of programs with DP, while the most often there are subsets of the CK (Chidamber and Kemerer) and QMOOD (Quality Model of Object Oriented Design) sets of software metrics [12, 13, 15–17]. In the study discussed in this paper we have used metrics supported by four static analysis tools of C# code [18–21]. As many of these metrics are similar, we have limited our discussion to the following subset of metrics:

1. LOC - Lines Of Code. Code lines are counted as program statements possible to be executed. Instructions of type "for", "if", "while" that create a simple block enclosed by "{", "}" are counted as one line, "if-else" statements as two lines.
2. DIT - Depth of Inheritance. It returns a depth level in the inheritance tree. All classes inherit from *System.Object* and have at least DIT = 1. Interfaces have DIT = 0.
3. CBO - Coupling Between Objects (Class Coupling) [11]. It counts how many elements of other classes (methods, attributes) are used by a considered class.
4. CC - Cyclomatic Complexity (McCabe metric) [11]. It gives a structural complexity of program control dependencies (for a program, class or method).
5. MI - Maintainability Index [22]. It estimates a relative effort to maintain a program and takes values from 0 to 100. In Visual Studio it is calculated as:

$$MI = max(0, (171 - 5.2 * ln(HalsteadVOL) - 0.23 * CC - 16.2 * ln(LOC)) * 100/171) \qquad (1)$$

Where Halstead Volume [11] is defined as:

$$HalsteadVOL = (OP + OD) * log2(UOP + UOD) \qquad (2)$$

OP – the total number of operators, OD – the total number of operands,
UOP – the number of distinct operators, UOD – the number of distinct operands.

6. lLOC - Logical statements Lines of Code – The number of statements ended with semicolon ";".
7. IC - Interface Complexity. The sum of the number of input parameters and the number of return points.
8. NM - Number of Methods (public, protected, private, total).
9. NA - Number of Attributes (public, protected, private, total).

4   C# Analyzer

Automatic evaluation of DP utilization in C# programs and calculation of software metrics have been the main goals of a prepared tool. The C# Analyzer tool has been implemented as an extension of Visual Studio. It integrates some VS facilities with a program developed for DP detection and external tools for software metric assessment. Analyzer consists of three main modules: a general manager, a software metric manager, and a DP detector (Fig. 1). The general manager collaborates directly with VS and supervises a system configuration, a user interface, as well as other components. It collects results from other modules and supports their evaluation.

The second module detects DP according to a set of structural rules. It is an enhanced version of the tool implemented by Nagy and Kovari [24]. The main enhancements have referred to recognizing of five additional design patterns and refinement of criteria of pattern detection [10]. The criteria have been adjusted to specific features of the C# programing language. In the current version, the following patterns can be recognized: *Singleton, Factory Method, Proxy, Decorator, Composite, Adapter, Mediator, Chain of Responsibility, Observer,* and *Visitor.*

In general, the improvements have resulted in higher number of recognized pattern instances than using the previous tool [10]. It was mainly due to the additional types of

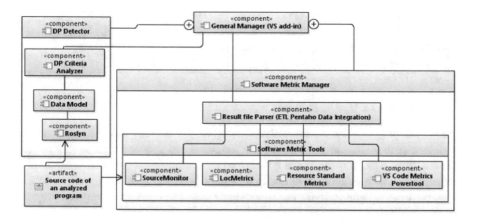

Fig. 1. Architecture of the C# analyzer.

pattern detected. However, the more refined criteria have eliminated some of pattern instances accepted formerly, as they have now been counted as false positive.

The third module is devoted to software metric calculation. It calls the tools assigned to the system. The current version cooperates with Visual Studio Code Metrics Powertool [18], which is a built-in VS static analysis facility, and three external tools: SourceMonitor [19], Resource Standard Metrics [20], and LocMetrics [21]. The result data of metrics are further processed by an ETL (Extract, Transform, and Load) tool. The ETL submodule is realized with ETL-Pentaho Data Integration [23]. Given workflows control transformations corresponding to specific software metric tools.

5   Experiments on Design Patterns and Software Metrics in C#

In this section, we describe the experimental analysis performed on a set of C# programs with support of C# Analyzer, and discuss the results.

5.1   Experiment Subjects

The experiments have been conducted on 20 real programs (Table 1). The programs were prepared by various developers in different purposes, not associated with the project under concern. Within this set, 13 programs: (1), (2), and (10–20) were libraries or other open source applications originated from GitHub resources. Remaining 7 programs (3–9) were developed by students from our University (WUT) in the context of courses they had attended before or due to other their activities.

Table 1. Programs analyzed in experiments.

Id	Program	Id	Program	Id	Program	Id	Program
(1)	TDDEvaluation	(6)	LuaLinter	(11)	Figures	(16)	Mario Objects
(2)	Log4net	(7)	WPFCalculator	(12)	PacManDuel	(17)	RedDog
(3)	CentralOffice	(8)	SPGenerator	(13)	Cat.Net	(18)	DogeSharp
(4)	AngularCalculator	(9)	BinaryStructure	(14)	Cars	(19)	Play
(5)	HuffmanCoder	(10)	FactoryPattern	(15)	PlainElastic.Net	(20)	Catnap

5.2   Detection of Design Patterns

The subject programs have been analyzed using the DP detector module. The numbers of design pattern instances found in the programs are shown in Table 2. The first column lists the patterns that could be detected by the tool. Next columns correspond to programs in which patterns were detected. The last column gives a sum of the pattern instances in all programs. In the last row, all occurrences of patterns in a given program are summarized.

Entirely, 61 occurrences of design patterns were observed. Only in 8 programs some DP have been detected. In the remaining 12 programs no patterns from the considered DP list were found. The most pattern instances referred to *Factory Method* (40). In several cases *Singleton* and *Proxy* were recognized. Only few code extracts were identified as *Decorator*, *Visitor*, or *Composite*. The remaining four patterns were not detected in any of programs from this set.

Among 8 programs with DP detected, three programs have been prepared by WUT students. Therefore, we could verify the results of automatic detection with intentions of the developers. The authors of programs (6), (8), and (9) endorsed that intentionally developed such design patterns in their programs in the locations indicated by the tool. In this way, we have additional confirmation of correct detection of *Singleton*, *Factory Method*, *Decorator* and *Visitor*, and approved DP occurrence in 23 cases.

Table 2. Numbers of design pattern instances detected in programs.

Design Pattern	Programs								Sum
	(2)	(6)	(8)	(9)	(10)	(13)	(16)	(17)	
Singleton	5		1						6
Factory method	13	13	6		2	1		5	40
Proxy	9								9
Decorator	1			2					3
Composite							1		1
Adapter									0
Mediator									0
Chain of respon.									0
Observer									0
Visitor				1			1		2
Sum	28	13	7	3	2	1	1	6	61

5.3   Utilization of Classes with Design Patterns

In comparison to the former tool [24], the current DP detector returns not only information about an occurrence of a given pattern with its main class but also identifies all classes that are members of the detected pattern [10].

In all programs, 2392 classes have been analyzed. It was recognized that 112 classes are members of DP, which corresponded to 4.7% of all classes. While taking into account only projects with DP, classes that are DP members cover 13.5% of all classes. Though, this feature differs noticeably among different projects. Characteristics of programs with DP are shown in Table 3. Some programs consist of several projects, and in those cases the results are given separately for member projects denoted by (A), (B), (C). The numbers of all classes in a project (column *#classes*) are compared with the numbers of classes covered by DP. The next column gives a percentage of those classes. Except of the program (10), which is small and specially devoted to design pattern usage, classes used in DP covered from a few to twenty-several percent of all project classes. For three projects it was about ¼ of the entire code, which indicates that it is an important factor in further code evaluation and maintenance.

For comparison, DP-based classes in the C# project examined in [8, 9] represented 8.85% of all classes. However, in that case study the most popular patterns turned out to be Singleton and Strategy, while Factory Method was rarely used. According to [13], DP classes covered from 6 to 34% of all classes in commonly used Java tools.

The last three columns of Table 3 give information about depth of inheritance tree, coupling between objects, and maintainability index.

In projects with DP, we have compared classes participated in DP with the remaining classes. The data are given in two rows for each project, denoted as DP *yes* or *no* (Table 4). The results of the following metrics are shown: Logical Statement Lines of Code - lLOC, Interface Complexity - IC, Cyclomatic Complexity - CC, as well as the number of *public*, *protected*, and *private* methods and attributes, accordingly.

Table 3. Selected metrics of projects with design patterns

Program	Project	LOC	#classes	# DP classes	% of DP cl.	DIT	CBO	MI
(2)	(A) Log4net	7834	294	32	10.9%	5	423	85
	(B) Log4net.Tests	2715	67	1	1.5%	5	243	78
(6)	LuaLinter	1347	125	28	22.4%	4	187	86
(8)	(A) SPGenerator. Generator	179	38	1	2.6%	3	73	90
	(B) SPGenerator.Model	55	11	2	18.2%	3	6	95
	(C) SPGenerator. SharePoint	739	29	7	24.1%	3	155	84
(9)	BinaryStructureLib	653	55	7	12.7%	3	77	85
(10)	FactoryPattern	10	7	5	71.4%	1	5	98
(13)	Cat.Net	1034	39	10	25.6%	2	100	85
(16)	MarioObjects	3381	72	3	4.2%	7	186	91
(17)	(A) RedDog. ServiceBus	349	32	4	12.5%	3	88	85
	(B) RedDog.Messanger	602	64	12	18.8%	3	136	89
Sum		18898	833	112	13.5%			

Summary of mean metrics calculated over three sets of classes are given in Table 5. The selected metrics are the same as in Table 4. Two upper rows refer to projects with DP. They include mean values for classes that cover DP (*Yes* in the DP column) and other classes from the same projects (*No*). The last row contains values for classes not covering DP from all kinds of projects considered in the experiments (*No-all*).

Table 4. Mean metrics of classes with and without DP in projects using DP (DP classes > 10%)

Project	DP	ILOC	IC	CC	Number of methods			Number of attributes		
					public	prot.	priv.	public	prot.	priv.
(2A)	Yes	28.3	11.0	13.7	3.6	0.8	1.0	0.3	0.1	3.5
	No	20.1	10.2	9.5	2.8	0.7	0.9	0.3	0.1	2.0
(6)	Yes	6.3	4.6	1.7	1.4	0.0	0.0	0.0	0.0	1.2
	No	4.5	4.2	2.7	1.3	0.0	0.3	0.0	0.1	1.1
(8B)	Yes	6.0	0.0	0.0	0.0	0.0	0.0	1.0	0.0	0.0
	No	4.1	1.1	0.4	0.4	0.0	0.0	0.4	0.0	0.1
(8C)	Yes	25.3	16.0	13.3	2.4	2.1	0.7	0.7	0.7	1.7
	No	18.3	10.0	6.4	1.6	0.6	1.1	0.3	0.0	3.4
(9)	Yes	8.57	5.4	3.7	2.0	0.0	0.9	0.4	0.0	0.4
	No	9.48	4.1	3.9	2.0	0.0	0.6	0.2	0.0	2.8
(10)	Yes	1.0	0.8	0.8	0.8	0.0	0.2	0.0	0.0	0.0
	No	2.0	0.0	0.0	0.0	0.0	0.5	0.0	0.0	0.5
(13)	Yes	35.2	9.7	8.6	2.2	0.7	0.7	0.0	0.0	1.6
	No	22.1	7.9	7.8	2.9	0.0	1.3	0.0	0.0	1.6
(17A)	Yes	9.5	5.8	2.5	2.5	0.0	0.5	0.0	0.0	1.8
	No	8.1	9.6	4.1	1.6	0.1	0.5	0.3	0.0	1.0
(17B)	Yes	6.2	6.8	3.8	3.2	0.0	0.7	0.0	0.0	0.0
	No	6.9	7.0	4.5	1.5	0.2	0.9	0.1	0.0	0.3

Table 5. Comparison of mean metrics of sets of classes

DP	#Class	ILOC	IC	CC	Methods			Attributes		
					public	prot.	priv.	public	prot.	priv.
Yes	112	**16.7**	**7.8**	**7.0**	2.5	0.44	0.6	0.3	0.06	1.7
No	721	**18.7**	**9.1**	7.1	2.8	0.30	1.2	0.7	0.03	1.6
No-all	2280	17.6	7.3	**5.4**	2.9	0.29	0.7	0.5	0.04	2.1

5.4   Discussion of Results of Metrics

Comparison of DP classes with remaining classes within the same projects (Table 4) does not indicate on unambiguous impact of DP on classes covered by them.

If we compare all DP classes (the first row in Table 5), with no DP classes of all considered programs (the last row), it could be observed that the cyclomatic complexity (CC) is higher in the first case. This could imply higher complexity of DP classes. However, comparison of classes within DP projects does not confirm the fact, as CC metrics are almost the same. We can only learn that in general projects with DP were more complex than projects without DP. Furthermore, in some projects, CC is higher for DP classes while in others vice versa (Table 4). This fact could be correlated with the number and types of DP applied in the programs. Though, this would have required more DP occurrences than those encountering in the programs of concern.

A situation is different when we compare the mean numbers of instructions (lLOC) and of interface complexity (IC). In these cases, DP and no DP classes within the DP projects have higher discrepancy than DP classes compared to all no DP classes (Table 5). The DP classes have less instructions (2 on average) and lower interface complexity (about 1.3) than the remaining classes. However, those discrepancies are not high, and furthermore, if we look at separate programs (Table 4), we can find different outcomes both for higher and lower such values.

Private and protected attributes constitute 85.6% of all attributes in classes with DP, while 70.3% in the remaining classes of the same set of projects. These results show that DP classes are more encapsulated than the remaining classes within the projects, which is consistent with the paradigm of object-oriented programming.

In general, usage of DP is associated with implementation a slightly more complex structures, in which a comparable amount of instructions is used preserving a similar interface complexity and encapsulation. This founding of DP class complexity has agreed with those of Java projects [13] although assessed with different metrics.

Maintainability Index, which takes 100 as maximum, is considerably high for all DP projects (Table 3). The average MI value of all DP projects equals to 87 and is higher than of projects without DP (80), testifying better ability of maintenance.

5.5   Threats to Validity

An external threat to validity is lack of guarantee that the results of the programs could be generalize. To alleviate this problem, we chose 20 programs from real software, written by different authors and from different domains. However, experiments on bigger sets could be valuable. A construct validity has been associated with measurements of software metrics and detection of DP. The metrics have been measured by commonly used tools; many of metrics have similar outcomes from different tools, hence their values could have been verified. Design patterns have been recognized according to the structural rules implemented in the DP detector [10]. The study was mainly bound by the set of 9 DP supported by the tool. Within the set of programs, no patterns that were detected as DP were classified by mistake. Analysis of three programs was confirmed by their authors in terms of all DP occurrences and DP types.

6   Conclusions

Development of the C# Analyzer tool, especially its DP detection module, has allowed us to conduct experiments on DP utilization in C# programs. As expected, usage of DP seems to have a positive impact on the code quality, although it could also have been influenced by other factors, including developer skills. According to project-level analysis, DP increase maintainability of software. Class-level analysis of programs with DP approved high encapsulation of DP code. However, it should be noted, that the detailed results might vary among different projects.

In general, differences between code with and without DP were more visible at the project level than at the class level. One of possible reasons could be the fact that the projects were developed by various programmers. Those developers who knew DP and applied them might have been able to implement high quality code that would be easy to comprehend and maintain.

References

1. Gamma, E., Helm, R., Johnson, R., Vlissides, J.: Design Patterns: Elements of Reusable Object-Oriented Software. Addison-Wesley, Boston (1995)
2. Ali, M., Elish, M.O.: A comparative literature survey of design patterns impact on software quality. In: Proceedings of International Conference on Information Science and Applications (ICISA), pp. 1–7 (2013). https://doi.org/10.1109/ICISA.2013.6579460
3. Ampatzoglou, A., Charalampidou, S., Stamelos, I.: Research state of the art on GOF design patterns: a mapping study. J. Syst. Softw. **86**(7), 1945–1964 (2013). https://doi.org/10.1016/j.jss.2013.03.063
4. Khomh, F., Gueheneuc, Y.-G.: Do design patterns impact software quality positively? In: 12th European Conference on Software Maintenance and Reengineering, pp. 274–278. IEEE Computer Society (2008). https://doi.org/10.1109/CSMR.2008.4493325
5. Mayvan, B.B., Rasoolzadegan, A., Yazdi, Z.G.: The state of the art on design patterns: a systematic mapping of the literature. J. Syst. Softw. **125**, 93–118 (2017). https://doi.org/10.1016/j.jss.2016.11.030
6. Metsker, S.J.: Design Patterns in C#. Addison-Wesley, Boston (2004)
7. Sarcar, V.: Design Patterns in C#. Apress, Berkeley (2018). https://doi.org/10.1007/978-1-4842-3640-6
8. Gatrell, M., Counsell, S., Hall, T.: Design patterns and change proneness: a replication using proprietary C# software. In: 16th Working Conference on Reverse Engineering (WCRE), pp. 160–164. IEEE Computer Society (2009). https://doi.org/10.1109/WCRE.2009.31
9. Gatrell, M., Counsell, S.: Faults and their relationship to implemented patterns coupling and cohesion in commercial C# software. Int. J. Inf. Syst. Model. Des. **3**(2), 69–88 (2012). https://doi.org/10.4018/jismd.2012040103
10. Derezińska, A., Byczkowski, M: Enhancements of detecting gang-of-four design patterns in C# programs. In: Borzemski, L., et al. (ed.) ISAT2018, AISC 852, pp. 277–286. Springer, Cham (2019). https://doi.org/10.1007/978-3-319-99981-4_26
11. Kan, S.H.: Metrics and Models in Software Quality Engineering, 2nd edn. Addison-Wesley Professional, Boston (2002)
12. Hussain, S., Keung, J., Khan, A.A., Bennin, K.E.: Correlation between the frequent use of Gang-of-Four design patterns and structural complexity. In: 24th Asia-Pacific Software Engineering Conference, pp. 189–198 (2017). https://doi.org/10.1109/APSEC.2017.25
13. Gravino, C., Risi, M.: How the use of design patterns affects the quality of software systems: a preliminary investigation. In: 43rd Euromicro Conference on Software Engineering and Advanced Applications (SEAA), pp. 274–277. IEEE Computer Society (2017). https://doi.org/10.1109/SEAA.2017.32
14. Ampatzoglou, A., Chatzigeorgiou, A., Charalampidou, S., Averiou, P.: The effect of GoF design patterns on stability: a case study. IEEE Trans. Softw. Eng. **41**(8), 781–802 (2015). https://doi.org/10.1109/TSE.2015.2414917
15. Mohammed, M., Elish, M., Qusef, A.: Empirical insight into the context of design patterns: modularity analysis. In: 7th International Conference on Computer Science and Information Technology (CSIT), pp. 1–6. IEEE (2016). https://doi.org/10.1109/CSIT.2016.7549474

16. Hsueh, N.-L., Chu, P.-H., Chu, W.: A quantitative approach for evaluating the quality of design patterns. J. Syst. Softw. **81**(8), 1430–1439 (2008). https://doi.org/10.1016/j.jss.2007.11.724
17. Derezińska, A.: Metrics in software development and evolution with design patterns. In: Silhavy, R. (ed.) CSOC2018, AISC, vol. 763, pp. 356–366. Springer, Cham (2019). https://doi.org/10.1007/978-3-319-91186-1_37
18. VS Docs: Code metrics values. https://docs.microsoft.com/en-us/visualstudio/code-quality/code-metrics-values?view=vs-2017. Accessed 14 Jan 2019
19. SourceMonitor. http://www.campwoodsw.com/sourcemonitor.html. Accessed 14 Jan 2019
20. Resource Standard Metrics (RSM). http://www.msquaredtechnologies.com/base/index.html. Accessed 14 Jan 2019
21. LocMetrics. http://www.locmetrics.com. Accessed 14 Jan 2019
22. Maintainability Index Range and Meaning. https://blogs.msdn.microsoft.com/codeanalysis/2007/11/20/maintainability-index-range-and-meaning/. Accessed 06 Mar 2019
23. Etl-pentaho-data-integration. https://www.etltool.com/etl-vendors/pentaho-data-integration-kettle/. Accessed 14 Jan 2019
24. Nagy, A., Kovari, B.: Programming language neutral design pattern detection. In: Proceedings of 16th IEEE International Symposium on Computational Intelligence and Informatics (CINTI), pp. 215–219, November (2015). https://doi.org/10.1109/CINTI.2015.7382925

Scheduling Tasks in a System with a Higher Level of Dependability

Dariusz Dorota$^{(\boxtimes)}$

Cracow University of Technology, 31-155 Cracow, Poland
ddorota@pk.edu.pl

Abstract. This paper presents a new method of scheduling which uses a modification of Muntz-Coffman algorithm. This novel method takes into account three-processors tasks. Acyclic directed graph is used as a specification of system, which involves one- and two-processors tasks, and attribute of divisibility/indivisibility of tasks. The graph constituting the input data for the created system is generated based on the TGFF algorithm. Scheduling of these tasks is prepared on NoC architecture which consists of three processors. In this paper, algorithm of scheduling tasks using new approach to prioritize and prepare ranking of tasks on chosen architecture, is presented.

Keywords: Multiprocessor tasks · Scheduling task · Divisibility of the task

1 Introduction

Modern computer development and the ever-growing trend of using mobile devices in the field of IoT and IoV enforces constant development of technology and equipment to meet the growing requirements [1, 12]. Such development will first of all enforce the efficiency of such systems to increase, i.e. the time of execution and reliability [17, 20], which is indispensably connected with the necessity to develop new and effective methods of scheduling tasks, especially multiprocessor tasks. These types of tasks are aimed at introducing and ensuring greater reliability of the system being created as well as complex processes in general. As a higher level of reliability it is possible to present mutual testing of result tasks on individual processors as presented in chapter 3. Another aspect motivating the use of multiprocessor tasks is to use them in the process of controlling and of walking in humanoid robots [2]. While performing each step of such a robot, it is necessary to simultaneously calculate many coordinates at the same time ensuring balance (stabilization) and alternate movements of individual limbs, so as to ensure that the distance is traveled with the set speed. One of the basic assumptions of the article is to propose the use of three-processor tasks in order to create a target system with an increased level of reliability. According to the current state of the present work, three-processor tasks were used for the first time in order to ensure the reliability of the system. The article proposes creating a system with a higher level of reliability using multiprocessor tasks. A new approach is to use these types of tasks and to propose an effective scheduling algorithm for the specified system specification.

© Springer Nature Switzerland AG 2020
W. Zamojski et al. (Eds.): DepCoS-RELCOMEX 2019, AISC 987, pp. 143–153, 2020.
https://doi.org/10.1007/978-3-030-19501-4_14

2   Problem of Tasks Scheduling

This chapter will present the scheduling problems and the most important concepts regarding their solutions.

2.1   Tasks Scheduling General Issues

To define the scheduling problem tasks and resources should be denominated. The set P was determined as executive elements, where $P \in (P_1, P_2, \ldots, P_n)$. The set T was determined as tasks, where $T \in (T_1, T_2, \ldots, T_n)$. Typically, the problem of scheduling is to assign tasks to resources taking into account defined constraints.

The multiplicity of scheduling problems and application areas forced them to be systematized. The most common classification is Graham's notation. In three-field notation, parameters α, β, γ can be distinguished [17, 20].

Considering the three-field notation, you can specify the type of problem (the so-called machine park) as the alpha parameter, specifying α_1 as:

– G - general problem
– O - open shop problem
– J - job shop problem
– F - flow shop problem
– FW - permutated flow-shop.

Then, the α_2 field, which is optional, determines the number of production machines. α_3 and α_4 field describe type and number of transport machines. Beta field means technological requirements in problem of scheduling tasks (tasks characteristic). When field β is empty it means there are no restrictions. Last parameter gamma determines goal function shape.

Assumption:

It is assumed that there is a function $f(x)$, which is associated with each task k, which depends on the moment of its ending C_k, where $k \in T$.

There are two types of optimization criteria:

1. Minimum criteria

$$f_{max} = \max_{k \in T} f_k(C_k) \tag{1}$$

2. Additive criteria

$$\sum f_k = \sum_{k \in T} f_k(C_k) \tag{2}$$

Multiprocessor systems allow increasing the speed of embedded systems, while applying other constraints such as the demand for system power or used area [14, 16, 19, 21]. Over the years, during the development of computer science and all kinds of computer systems, the scheduling methods have evolved and are constantly evolving in order to perform actions on the delivered equipment in the most effective way possible. Among the algorithms of scheduling, deterministic and stochastic algorithms can be distinguished [17, 20]. Both in the first and in the second case, it is possible to distinguish the scheduling on the single machine and on the parallel machines [17, 20].

Each scheduling problem can be described as a pair (X, F), where $X \subseteq X^0$ is set of acceptable solutions, while F: X → R is an accepted function of the goal (optimization criterion). The task scheduling problem boils down to determining a globally optimal solution $x^* \in X$ such that $F(x^*) = \min F(x)$, where x is the solution to the problem.

Regardless of what the system's input specification is, each of the tasks and the target system must have specific task parameters and resources to be used for scheduling.

Therefore, you can specify the characteristics of system parameters in which tasks must be ranked:

- availability (temporary availability intervals)
- quantity
- expense
- acceptable unit/resource load.

Due to the functions of the machine it is divided into:

- parallel (universal) - performing the same functions
- dedicated (specialized) - differing in functions fulfilled.

Typically, the scheduling process considers minimizing many parameters at the same time, which is called multi-criterial optimization. In this work, priority is first of all one parameter, time of system performance, which is one of the most important from of the perspective of the implementation of the entire system. A single task was defined as T_i. The task set can be defined as T = {T_1, T_2, ..., T_N}. A single processor is defined as P_j, while a set of processors is defined as P = {P_1, P_2, ..., P_M}. Embedded processors may use universal processors or hardware modules that can generally be described as a computational element (PE).

In the proposed method two basic types of PE can be distinguished:

- Universal programmable processor (PP),
- Specialized hardware modules (HC).

2.2   Tasks Representation in Embedded Systems

In scheduling methods we may talk about dependent and independent tasks. If we consider independent tasks then scheduler can map those independently, means without keeping any order. Opposite are tasks which we can schedule as dependent tasks, which are executed in order according to precedence.

This work presents an algorithm for dependent tasks. There are several ways to represent the second type of tasks, including the task graph, SystemC, etc. [11, 13]. In this work, an acyclic task graph was used for the system specification. The input tasks are specified by the graphs proposed by the author based on TGFF graphs [8]. TGFF provides a standard method for generating random allocation and scheduling problem. Randoming graphs which are generated based on this method have an additional task multiprocessor sign. The number of multiprocessor tasks and the multiprocessor designation is generated randomly for the entire graph. In this work designation for multiprocessor tasks are proposed as follows:

T_i^1 – signification for 1-processor tasks
T_i^2 – signification for 2-processor tasks
T_i^3 – signification for 3-processor tasks.

Acyclic directed graph G = {V, E} can be used to describe the proposed embedded system. Each node in the graph represents a task, while the edge describes the relationship between related tasks. Each of the edges in the graph is marked by a label d, where each of its indices defines the tasks that connect. The graphs of the tasks used for the system specification are presented below on Figs. 1, and 2.

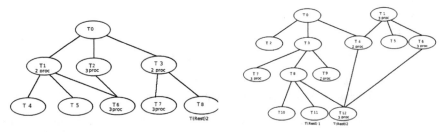

Fig. 1. Exemplary graph 1

Fig. 2. Exemplary graph 2

3   Three Processor Tasks

Currently presented tasks in the scheduling process are performed on a single machine and on parallel machines. A novelty in the approach presented in the article is the use of three-processor tasks, i.e. the need to perform a specific task/tasks on three processors at the same time.

The system's reliability is the motivation to introduce and use such tasks. Such reliability can be achieved by using redundancy for critical tasks and thus testing the correctness of the implementation of the selected task on the second processor. In the proposed approach, only selected tasks are three-processor tasks in order not to significantly increase the cost of the entire system. Reliability plays a particularly important role in systems requiring a high degree of operational correctness, especially in real-time systems, used in industries, where both temporal determinism and correct operation of the whole system play a significant role.

3.1   Motivation of Using Three-Processor Tasks

The introduction of mutual testing of three-processor tasks is one example of motivation [9]. This approach allows you to create a target system with a higher degree of reliability. The figure below (Fig. 3) presents the concept of mutual testing of tasks.

The T_6 task is presented here, which requires to run work simultaneously on 3 processors (P_1, P_2, P_3). It can be assumed that in the case under consideration, the T6 task consists of three identical tasks $T_6 = \{T_{6A}, T_{6B}, T_{6C}\}$ run at the same time, where T_{6B} and T_{6C} tasks run on P_2 processors and P_3 after completing tasks check each other results, while the T_{6A} task is an arbitrator and if the answers of these tasks are different from each other, they are tested in pairs with the T_{6A} task and then the correct answer is returned.

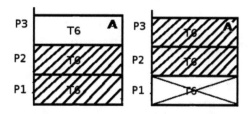

Fig. 3. Example of mutual testing of tasks

3.2   Representation of Tasks in Graph

As already mentioned in the work, the task graph has been adopted as input data for the system under consideration [9, 10]. The only change in relation to the one-processor tasks is to add annotations in the task graph indicating that the task must be completed as a two-processor or three-processor task. This was accomplished by adding the description "2 Proc" or "3-Proc" in the representation of a given task, which means the necessity of its simultaneous implementation on three machines. Assuming that T_i is a task with a number "i" in a task graph, then the task T_i^1, T_i^2, T_i^3, means that the task has a number and is one-, two-, and three-processor. Using the graphic symbol, such a task is presented as follows (Fig. 4):

Fig. 4. Graphical representation of three-, two- and one-processor task

3.3   Scheduling Process Using Three-Processor Tasks

The process of scheduling two-processor tasks follows the principles introduced in the Muntz-Coffman algorithm with the changes presented in this article. Task priorities are determined on the basis of the adopted algorithm with the reservation that one-, two- and three-processor tasks have the appropriate multiplier. The use of this approach allows for the promotion of three-processor tasks from other tasks, of course, the order conditions and time constraints must be met here. As an example, you can specify task prioritization according to the modified Munz-Coffman algorithm:

Table 1. Exemplary prioritization for tasks shown in Fig. 1.

Number	Time	Level of task/time to ending							
T0	10	130/50	0	0	0	0	0	0	0
T1	5	70/30	70/30	70/30	0	0	0	0	0
T2	20	120/40	120/40	0	0	0	0	0	0
T3	15	45/20	45/20	45/20	45/20	45/20	0	0	0
T4	15	15/15	15/15	15/15	15/15	15/15	0	0	0
T5	10	10/10	10/10	10/10	10/10	10/10	10/10	10/10	0
T6	20	60/20	60/20	60/20	60/20	0	0	0	0
T7	5	15/5	15/5	15/5	15/5	15/5	15/5	0	0
T8	10	10/10	10/10	10/10	10/10	10/10	10/10	10/10	0

4   Proposed Algorithm

In this chapter a Munt-Coffman algorithm will be discussed and proposed modification of this algorithm to resolve multiprocessors tasks scheduling.

4.1   Original Muntz-Coffman Algorithm

The Muntz-Coffman algorithm was proposed in [15] as the optimal scheduling of tasks on two processors. This algorithm is the result of the previous research of the authors. It assumes independent tasks in which the scheduling takes place depending on the priorities set. This approach allows for any scheduling of tasks so as not to exceed the predetermined time for the entire system. The issue of divisibility of tasks dealt with in earlier author's works is extremely interesting. Dividend in scheduling tasks for two processors was presented in [3]. The algorithm proposed by Muntz-Coffman is the optimal algorithm for scheduling tasks divided on two machines [15]. It is an extension of previous works by its authors. The proposed algorithm performs task scheduling according to the following procedure:

1. Setting levels of tasks not performed
2. Assignment of tasks with the highest levels to processors, assuming that if the number of tasks is greater than the processors - assign each task B = m/a of computing power, where is the number of tasks with the highest level. Performing task simulation until the task with the lowest level would be per formed earlier than the task with a higher level. Then go back to step 1. Repeat the procedure until all tasks have been completed
3. Apply the Mc Noughton algorithm to obtain optimal ranking.

4.2   Modification of Muntz-Coffman Algorithm

According to previous studies presented by the author, regarding the attribute of divisibility, also in this case, the use of the divisibility attribute resulted in beneficial effects. According to the current state of knowledge, apart from [4, 5], the Muntz-Coffman algorithm and its modifications have not been considered for the task scheduling process, especially for three-processor tasks.

In the proposed algorithm, tasks are specified in the form of a task graph, where 1-, 2- and 3- processors tasks are generated. The number of individual types of tasks is also randomly generated. Tasks are generated as described in Chapter 2. The following is a description of the algorithm used for scheduling tasks including three-processor tasks.

Based on the effects of the work presented above, as well as the author's previous experience with the issue of scheduling, three-processor tasks with the attribute of divisibility were proposed here [9, 10]. In the algorithm presented in the work, scheduling tasks at the beginning are determined paths in the graph using the A* algorithm. The algorithm proposed in [9] for scheduling two-processor tasks was used in the work. To streamline and speed up the operation of the algorithm, omitting the so-called multiprocessor ratio checking. As a result of the conducted research, it was proposed to use the coefficient for both single and multiprocessor tasks. This factor is equal to the number of processors necessary to execute the task. The modification was also proposed in terms of the use of multiprocessor tasks. This algorithm allows you to create a system with higher-level of reliability. The concept of a multiprocessor task comes from tests of scheduling in multiprocessor computer systems, testing of VLSI or other devices [22]. Testing processors by each other requires task action on at least two processors simultaneously [23]. Dependability for the entire system was calculated using the formula:

$$D_x = \frac{\sum_{i=x}^{n}(p_i)}{\sum_{i=x}^{n}(t_i)},$$

Where D_i denote level of dependability. According to the proposed formula, reliability levels were calculated for each of the examples for which the scheduling was performed. A summary of the reliability for individual systems is shown in Table 2.

The procedure of the modified Muntz-Coffman algorithm for three-processor tasks is given below:

1. Load the system specification in the form of a task graph
2. Set levels for all non-completed tasks:
    2.1 The task level px is the sum of the task execution times ti on the longest path for the selected task $P_x = A_x = max\sum_{i=x}^{n}(t_i)$, where $A_x = max(A_{i=1}^{K})$, is the time of the longest path for task x, for the initial task x is selected $max(A_{j=1}, \ldots, A_k)$
    2.2 If the selected T_i task at time t_i is a i-processor task:
        2.2.1 $p_x = t_x * i$
3. Select the task with the highest level
    3.1 If there are two (or more) tasks at the same level:
        3.1.1 The priority is a dual-processor task, (the exception are dependencies which condition the execution of subsequent tasks, i.e. enabling the implementation of successors, e.g. the need to perform a task/tasks not to exceed the time limits T (Rest) X, where x is the number of the next time limit)
    3.2 if there are two (or more) two-processor tasks at the same level, the first task should be higher in the hierarchy (or with the lower number of the task)
    3.3 Delete the selected task from the set (P_a, P_b, ..., P_z),

4. Simulate the task execution on the selected processor/processors per unit of time
5. If after the simulation the processor is available in the selected time unit:
    5.1 Select the next task (go to step 1)
6. If there are still unassigned tasks in the graph, go back to step 1
7. Exit the algorithm.

5   Result and Analysis

This chapter presents the results obtained during experiments that reach task scheduling with the attention of three-processor tasks. Experiments were carried out on a selected number of graphs for systems which included one-, two- and three processor tasks in three-processor and four-processor architecture. Exemplary ordering of tasks in the target system is shown in Fig. 5. The target system is implemented on a predefined multiprocessor architecture based on the NoC network. The target architecture is generated in the first step of creating the system, assuming that along with the specification given in the form of a graph of tasks, the number of computational elements on which the system is to be implemented is determined. The Table 1 presents the calculated priorities of tasks in individual steps that directly affect the way of ordering tasks in the target system, and thus have an impact on the duration of the entire system. The experiments were carried out on graphs generated on the basis of TGFF with the number of tasks from 10 to 50. For each number of tasks in the graph, several experiments were carried out to authenticate the scheduling results for graphs of different structure and identical number of tasks. The list of scheduling results has been placed in Table 2.

Fig. 5. Scheduling task in 3-processor architecture from specification in Fig. 1

The conducted research provides promising results for the application of three-processor tasks and optimal scheduling of tasks in the system, so that resources allocated to the implemented system will be used to the greatest possible extent. For the sake of simplicity, inter processor transmissions between tasks were omitted. Subsequent research will also focus on taking into account the transmission, as they can also affect the execution time of the entire system. The proposed algorithm allows obtaining satisfactory results for both three-processor and four-processor architectures.

Table 2. Exemplary task scheduling times for selected 3 and 4 processor architectures

Graph number	Count of tasks	One-Proc tasks	Two-Proc tasks	Three-Proc tasks	Count of Proc.	Time of system executions	Level of dependability
G1	9	4	2	3	3	85	2,29
G2	12	5	4	4	3	170	2,32
G3	14	5	5	5	3	185	2,35
G4	18	5	7	7	3	245	2,40
G5	21	6	6	10	3	305	2,66
G6	25	8	7	6	3	335	2,61
G7	31	11	8	13	3	440	2,68
G8	35	10	8	18	3	640	2,70
G9	42	11	10	22	3	780	2,60
G10	45	12	10	24	3	825	2,62
G11	50	12	14	25	3	925	2,70

6 Summary

The presented work will consider the problem of scheduling tasks in multiprocessor embedded systems based on the NoC network. The system specification is presented using the task graph, generated by author based on TGFF. Like the previous ones, this author's work also considers the attribute of divisibility of tasks. The scheduling process presented in the article is a modification of the approach used in the Muntz-Coffman algorithm. A novelty in relation to scheduling works is the consideration of three-processor tasks here. This is to ensure the reliability of the system. The obtained results of the experiments confirm that the introduction of three-processor tasks only affects the execution time of the entire system. The proposed coefficient for calculating the reliability of the system depending on the multiprocessor coefficient allows the targeted establishment of a system with a higher level of reliability. Because the level of credibility of the system depends here strictly on the multiprocessing of the task, using the three-processor tasks, the system is more reliable than the system using only two-processor tasks. What's more, the reliability of the system also closely depends on the percentage share of three-processor tasks. However, in areas that require reliability and correct operation of the system, the benefits are disproportionate to the longer time of the task. Certainly, more research will be carried out in this interesting direction. Future work will concern the introduction of n-processor tasks as well as consideration of inter processor transmissions in the NoC network for which the ordering is performed.

References

1. Ang, L.M., Seng, K.P., Ijemaru, G.K., Zungeru, A.M.: Deployment of IoV for smart cities: applications, architecture, and challenges (2019)
2. Bi, S., Zhuang, Z., Xia, T., Mo, H., Min, H., Luo, R.: Multi-objective optimization for a humanoid robot walking on slopes. In: 2011 International Conference on 2011 International Conference on Machine Learning and Cybernetics Machine Learning and Cybernetics (ICMLC), vol. 3, pp. 1261–1267, July 2011
3. Błażewicz, J., Cellary, W., Słowiński, R., Węglarz, J.: Badania operacyjne dla informatyków. WNT, Warszawa (1983)
4. Błażewicz, J., Drabowski, M., Węglarz, J.: Scheduling multiprocessor tasks to minimize schedule length. IEEE Trans. Comput. **5**, 389–393 (1986)
5. Błażewicz, J., Drozdowski, M., Guinand, F., Trystam, D.: Scheduling a divisible task in a two-dimensional toroidal mesch. Discret. Appl. Math. **94**, 35–50 (1999)
6. Błażewicz, J., Drozdowski, M., Węglarz, J.: Szeregowanie zadań wieloprocesorowych w systemie jenorodnych duoprocesorów. Zeszyty Naukowe Politechniki Śląskiej (1988)
7. Błądek, I., Drozdowski, M., Fuinand, F., Schepler, X.: On contiguous and non-contiguous parallel task scheduling (2015)
8. Dick, R.P., Rhodes, D.L., Wolf, W.: TGFF: task graphs for free. In: Proceedings of the 6th International Workshop on Hardware/Software Codesign (CODES/CASHE 1998). IEEE Computer Society, Washington, DC, USA, pp. 97–101 (1998)
9. Dorota, D.: Dual-processor tasks scheduling using modified Muntz-Coffman algorithm. In: Dependability and Complex Systems DepCoS-RELCOMEX, Brunów, Poland (2018)
10. Dorota, D.: Scheduling tasks in embedded systems based on NoC architecture using simulated annealing. In: Dependability and Complex Systems DepCoS-RELCOMEX, Brunów, Poland (2017)
11. Eles, P., Peng, Z., Kuchcinski, K., Doboli, A.: System level hardware/software partitioning based on simulated annealing and tabu search. Des. Autom. Embed. Syst. **2**(1), 5–32 (1997)
12. Gubbi, J., Buyya, R., Marusic, S., Palaniswami, M.: Internet of Things (IoT): a vision, architectural elements, and future directions. Future Gener. Comput. Syst. **29**(7), 1645–1660 (2013)
13. Khan, G.N., Iniewski, K. (eds.): Embedded and Networking Systems: Design, Software, and Implementation. CRC Press, Boca Raton (2013)
14. Kopetz, H.: Real-Time Systems: Design Principles for Distributed Embedded Applications. Springer, Heidelberg (2011)
15. Muntz, R.R., Coffman, E.G.: Optimal preemptive on two-processor systems. Trans. Comput. (1969)
16. Ost, L., Mandelli, M., Almeida, G.M., Moller, L., Indrusiak, L.S., Sassatelli, G., Moraes, F.: Power-aware dynamic mapping heuristics for NoC-based MPSoCs using a unified model-based approach. ACM Trans. Embed. Comput. Syst. (TECS) **12**(3), 75 (2013)
17. Pinedo, M.L.: Scheduling Theory, Algorithms and Systems. Springer, Heidelberg (2008)
18. Popieralski, W.: Algorytmy stadne w optymalizacji problem przepływowego szeregowania zadań, Ph.D. thesis (2013)
19. Rajesh, K.G.: Co-synthesis of hardware and software for digital embedded systems, Ph.D. thesis, 10 December 1993
20. Smutnicki, C.: Algorytmy szeregowania zadań. Oficyna Wydawnicza Politechniki Wrocławskiej, Wrocław (2012)
21. Tynski, A.: Zagadnienie szeregowania zadań z uwzględnieniem transportu. Modele, własności i algortmy, Ph.D. thesis, Wrocław (2008)

22. Tarun, K., Aryabartta, S., Manojit, G., Sharma, R.: Scheduling chained multiprocessor tasks onto large multiprocessor system (2017)
23. Drozdowski, D.: Selected problems of scheduling tasks in multiprocessors computer systems, Poznań (1997)

Comparison of Parallel and Non-parallel Approaches in Algorithms for CAD of Complex Systems with Higher Degree of Dependability

Mieczyslaw Drabowski[(✉)]

Department of Theoretical Electrical Engineering and Computing Science,
Cracow University of Technology, Warszawska 24, 31-155 Cracow, Poland
drabowski@pk.edu.pl

Abstract. The paper presents a comparison of results obtained from calculations of algorithms performing the following activities parallel: task scheduling, partition of resources, and allocation of tasks and resources, with results obtained from other - known from world literature - algorithms in which partition of resources, task scheduling and allocation of tasks and resources are not run in parallel. This parallelism is described in detail in [1] and is carried out in calculations in nested cycles of programs. Computational processes in the external cycle evolve in the space of solutions to the best (for example the most efficient or the most dependable) sets of resources, and in the internal cycle they evolve to e.g. of the shortest task schedules and best matching of tasks selected resources, whose collections can still be modified. This parallelism gives better - closer to optimal - results of calculations concerning the synthesis of complex systems.

Keywords: Parallel · Synthesis · Model · Approach · Algorithm · Digraph · Comparison

1 Introduction

The paper [2] presents a deterministic model of scheduling tasks and partitioning resources, which was applied in high-level synthesis of complex systems with a higher degree of dependability. This paper emphasizes the parallel of approaches in two areas this model: parallelism of the design of the resource structure (hardware) and design of the operational structure (software) and parallel in operation of multiprocessor tasks that form the basis for increasing credibility through self-testing and self-diagnosis, which are keys for building of fault tolerant systems, based of redundancy – Fig. 1.

The paper [1], in turn, presents the first algorithms based on a hybrid metaheuristic approach (genetic and simulated annealing) realizing these parallel ties and applicable in the CAD of such dependability and fault tolerant systems – Fig. 2.

© Springer Nature Switzerland AG 2020
W. Zamojski et al. (Eds.): DepCoS-RELCOMEX 2019, AISC 987, pp. 154–165, 2020.
https://doi.org/10.1007/978-3-030-19501-4_15

This article attempts to compare parallel approaches, whose first results are presented in the cited articles, with a classical (not parallel) approach presented in selected articles, available in the literature, other authors. In our computing tasks (implementations execute) used during the tests are generated as digraphs and they are received as:

- Graphs STG [3] http://www.kasahara.elec.waseda.ac.jp/schedule/
- Graphs TGFF [4] http://ziyang.eecs.umich.edu/ ~ dickrp/tgff/

These generators have been worked out in order to standardize random tests for research into common task scheduling and allocating resources, especially for system synthesis and for such implementations which need pseudo-random generating acyclic directed graphs.

In generators, the sort of graph, number of source and sink nodes, the length of maximal track, the node and edge weight, degree of graph, probability of predecessors and successors' number etc. should be determined. For example, time of tasks might be generated as follows: the average time value for the task (e.g. equal 5 units) and time of tasks determined by uniform distribution or regular distribution with a fixed standard deviation.

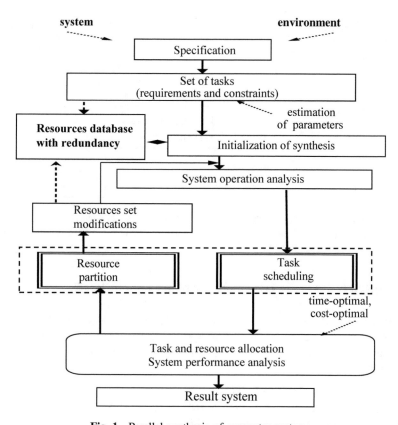

Fig. 1. Parallel synthesis of computer systems

For the tests, maximum number of tasks have been determined (e.g. equal 200 and independent tasks or equal 150 dependent tasks), about the sufficient number for the presentations of all algorithm features, and their comparisons as well (also to other algorithms) and is also the right number of operations for realistic system synthesis; obviously, system functions must be in its specification are given on the suitable level of granulation.

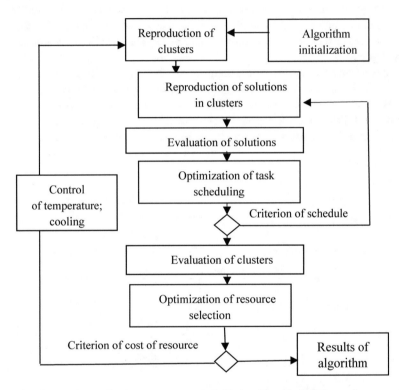

Fig. 2. Parallel synthesis of computer system – hybrid algorithms included genetic approach and simulated annealing, clusters included resources, solutions included schedules and evaluations were control with temperature and cooling.

2   Comparisons with Other Algorithms of Synthesis

The described algorithms were tested and compared to others, which known from literature of scope algorithms of system synthesis. The comparisons applied to the following approximate non-parallel, representative for system synthesis problems algorithms:

- Eles-Peng-Kuchcinski-Doboli Algorithm [5],
- Bianco-Auguin-Pegatoquet Algorithm [6],
- Dave-Lakshminarayana-Jha Algorithm [7],

- Yen-Wolf Algorithm [8],
- Eles-Peng-Kuchcinski-Doboli-Pop Algorithm [9], and
- Dick-Jha Algorithm [10],
- Prakash-Parker Algorithm [11].

We are going to present comparisons of coherent algorithms' solutions with these algorithms.

2.1   Comparisons with STG and TGFF Digraphs

Eles-Peng-Kuchcinski-Doboli Algorithm (EPKD Algorithm):
It is based on Tabu search and owns three ways of generating solution in each iteration. Firstly, neighboring solutions of smaller cost are considered – neighboring solutions are those which are created as a result of moving a task from one resource to the other. Secondly, solutions of cost modified by punishment function which prefers allocating tasks of the longest non-allocation are considered. Thirdly, solutions which are closest to the starting one and are not on the tabu list are generated. Iteration finishes when in the latest runs a global minimum cost has not been obtained. In short-term memory latest allocations are saved - tabu list, whereas in long-term memory statistics which determine the frequency of task occurrence in task allocations.
Idea of the algorithm:

1. *Construct initial configuration x_p*
2. ***for** each solution x' $\in N(x_p)$ **do***
   Compute change of cost function deltaC = C(x') - C(x_p);
3. **3.1. *for** each deltaC , 0, in increasing order of deltaC **do***
   *if not tabu(x') or tabu_aspirate(x') **then***
   $x_p=x';$
   goto 4
   3.2. *for each solution x' $\in N(x_p)$do Compute deltaC' = deltaC*
   + penalty(x')
   3.3. *for each deltaC' in increasing order of deltaC' **do***
   *if not tabu(x') **then***
   $x_p=x';$
   goto 4
   3.4. Generate x_p by performing the east tabu move
4. *4.1. **if** iterations since previous best solution < Nr **then goto**2*
   *4.2. **if** restarts **then***
   Generate initial configuration x_p considering frequencies
   ***goto**2*
5. ***return** solution corresponding to the minimum cost function*

Bianco-Auguin-Pegatoque Algorithm (BAP Algorithm):
This algorithm estimates the best and the worst case. In each step it calculates so called discrimination factor γ, which considers the increase in system structure cost.

Idea of the algorithm:

> **while** *all the tasks are not scheduled* **do**
>   *selected data ready tasks T_i;*
>   **for** *each implementation model R_j of T_i* **do**
>   *compute $\gamma(R_j, T_i)$;*
>   *place R_j in the corresponding list;*
> **endfor;**
>   **case of**
> $L^1_{reuse} \neq \varnothing$ *select $R_j \in L^1_{reuse}$ / $l_{max}(R_j, T_i)$ minimum;*
>   **break;**
> $L^2_{reuse} \neq \varnothing$ *select $R_j \in L^2_{reuse}$ / $\gamma(R_j, T_i)$ maximum;* **break;**
> $L_{new} \neq \varnothing$ *select $R_j \in L_{new}$ / area(R_j, T_i)minimum;* **break;**
> $L_{others} \neq \varnothing$ *select $R_j \in L_{others}$ / **if** $T_{new} + T_{reuse} < l$ **then** $\gamma($*
>   *$R_j, T_i)$ maximum **else** area(R_j, T_i)minimum* **endif;**
>   **break;**
>   *otherwise: error ("the partitioning is not feasible");*
>   **break;**
>   **end case;**
>   *schedule as soon as T_j on R_i according to available time slots of R_j;*
> **end while;**

Discrimination factor $\gamma\left(R_j, T_i\right) = 1 - \frac{e}{E} * \frac{pl}{PL}$
where:

e – This value represents the difference between the execution time of T_i by R_j and the lowest execution time of τ I by components of the library,
E – Maximum freedom of schedule. This value gives the maximum available time frame to schedule T_i (and successors of T_i) on R_j. Components with the largest values of E are those that impose the lowest constraints on the partitioning.
pl – Time of performing tasks which are located on the critical track containing these tasks.
PL – Time of performing tasks which are located on the critical track in the whole task graph

Dave-Lakshminarayana-Jha Algorithm (DLJ Algorithm):
This algorithm creates system structure by iterations through resource allocation from the pool to subsequent grouped tasks into clusters and minimization of system cost and power consumption. For each task, algorithm estimates the worst and the best case of allocation for the rest of tasks. If for the worst case no constraints are broken, it is taken; if for the best case not all the constraints are fulfilled, this solution is eliminated. Otherwise, algorithm chooses the solution of the smaller cost.

Idea of an algorithm:

assign randomly tasks from graphs to clusters;
__for__ each cluster __do__
   repeat
        take the next cluster allocation;
        schedule tasks from the clusters;
        calculate cost and power;
    until *all the constraints are fulfilled or all the allocations are considered;*
  __if__ not all the constraints are fulfilled, then eliminate the solution.

Yen-Wolf Algorithm (YW Algorithm):

In this algorithm speed and cost are optimized. The cost is constituted by a sum of processors.

Algorithm realizes 4 steps:

1. *Starting solution,*
2. *Elimination of the least loaded processors,*
3. *Balance of loading,*
4. *Ending.*

As a starting solution a structure where each task is performed by a different processor (of the highest proportion of the performed speed to the cost of the processor) is generated.

Next, processors of the lowest degree of use are eliminated by iteration and then the processors' loading is equalized in order to increase the speed of the system. Finally, algorithm tries to replace each of the selected processors with a cheaper one if such a change does not break constraints.

Eles-Peng-Kuchcinski-Doboli-Pop Algorithm (EPKDP Algorithm):

It is based on Simulated Annealing, in which new solutions are generated randomly and worse solutions are selected with diminishing proportion in subsequent algorithm runs. The parameters of the algorithm are: starting temperature TP, time of temperature duration TK, cooling coefficient a, stopping criterion.

Idea of the algorithm:

1.   *Construct initial configuration x_p*
2.   *Initial temperature $T = TP$*
3.   *3.1. __for i = 1 to TK do__*
      Generate randomly a neighboring solution $x' \in N(x_p)$
      Compute change of cost function $deltaC = C(x') - C(x_p)$
            __if__ $deltaC < 0$ __then__ $x_p = x'$
            __else__
                Generate $q = random(0,1)$
             __if__ $q < e^{-deltaC/T}$ __then__ $x_p = x$
    3.2. Set new temperature $T = a*T$
4.   *__if__ stopping criterion not met __then goto 3__*
5.   *__return__ solution corresponding to the minimum cost function*

Tests for graphs of files STG and TGFF which compare algorithms EPKD, BAP, DLJ, YW, EPKDP and coherent algorithms' of par-synthesis performance have been conducted. The subsequent columns present the name of the graph, number of tasks, time constraint and suitably for particular algorithms; cost, speed and power consumption of the obtained structures and computation time of algorithms.

In Table 1 input data are presented, in Tables 1a, 1b, 1c, 1d and 1e results of computation for suitable non-parallel algorithms, whereas in tables Table 2 a for suitable parallel algorithm for graphs STG. For graphs STG parallel algorithm definitely outperform, especially in achieving structure speed and scheduling; results better by around 15% than algorithm DLJ, 20% in proportion to algorithm BAP, 25% better than EPKD, 3% than EPKDP, and 15% than the results of YW. Cost is also better (from 5 to 15%), and power consumption is better (from 5 to 10%). We must indicate that algorithm DLJ is dedicated to minimization of power consumption. Parallel algorithm achieve their results at greater (20–70%) computation time.

Table 1. Input data with digraphs of type STG

Digraph	Number of tasks	Time constraints
rand0005	50	15
rand0020	50	15
rand0006	100	20
rand0166	166	25

Table 1a. Results actions of DLJ algorithm

DLJ algorithm

Cost	Power consumption	Time	Computation time
13,146	78,326	10,22	0.5
15,113	83,342	10,97	0.6
23,031	119,023	15,91	1,2
28,137	145,168	18,77	2.5

Table 1b. Results actions of BAP algorithm

BAP algorithm

Cost	Power consumption	Time	Computation time
13,812	82,236	10,78	0.6
16,003	89,175	11,63	0.8
23,951	126,164	16,71	1,7
29,825	152,468	19,72	3.2

Table 1c. Results actions of EPKD algorithm

EPKD algorithm

Cost	Power consumption	Time	Computation time
14,422	78,326	11,31	0.6
16,152	83,342	12,02	0.8
25,393	119,023	17,89	1,9
30,950	145,168	20,06	3.4

Table 1d. Results actions of EPKDP algorithm

EPKDP algorithm

Cost	Power consumption	Time	Computation time
12,094	83,026	9,91	0.8
14,657	88,175	10,61	0.9
21,188	125,688	15,27	1,9
26,449	153,878	18,01	4.3

Table 1e. Results actions of YW algorithm

YW algorithm

Cost	Power consumption	Time	Computation time
13,81	85,631	10,52	0.6
16,95	91,571	11,12	0.8
24,18	130,95	16,98	1,4
29,16	159,55	18,51	3.6

The solutions acquired by the parallel algorithms:

Table 2. Results actions of parallel genetic algorithm

Parallel genetic algorithm			
Cost	Power consumption	Time	Computation time
12,482	75,633	8,71	0.6
13,993	84,228	8,68	0.7
22,012	121,129	13,52	2,1
26,505	142,886	16,12	4.2

Table 3 presents the input data, in Tables 3a, 3b, 3c, 3d and 3e the results of computation experiments are given for suitable non-coherent algorithms, whereas in Table 4 of parallel algorithm for graphs TGFF.

Table 3. Input data with digraphs type TGFF

Digraph	Number of tasks	Time constraints
TG3	50	25
TG4	70	25
TG5	90	50
TG6	110	75

Table 3a. Results actions of DLJ algorithm

DLJ algorithm			
Cost	Power consumption	Time	Computation time
21,118	91,166	19,1	1,5
35,313	123,211	21,9	1,8
59,198	169,128	45,9	3,2
98,211	295,168	69,7	5,5

Table 3b. Results actions of BP algorithm

BP Algorithm			
Cost	Power consumption	Time	Computation time
21,751	95,125	19,9	1,6
36,714	129,514	23,1	2.1
60,855	173,885	47,9	3,4
101,255	299,312	72,49	5,9

Table 3c. Results actions of EPKD algorithm

EPKD algorithm			
Cost	Power consumption	Time	Computation time
20,421	93,621	18,24	1,7
32,921	126,123	20,96	2,2
57,023	171,219	43,83	3,5
95,264	296,227	66,91	5,8

Table 3d. Results actions of EPKDP algorithm

EPKDP algorithm			
Cost	Power consumption	Time	Computation time
20,062	94,812	17,7	1,7
33,441	129,001	20,4	1,9
56,297	176,062	41,8	3,6
93,301	306,918	64,7	6,3

Table 3e. Results actions of YW Algorithm

YW algorithm			
Cost	Power consumption	Time	Computation time
19,538	89,342	18,1	1,4
32,876	120,746	20,8	1,6
55,054	165,728	43,3	3,0
91,336	289,264	66,2	5,2

The solutions acquired by the parallel algorithm:

Table 4. Results actions of parallel genetic algorithm

Parallel genetic algorithm			
Cost	Power consumption	Time	Computation time
18,118	88,43	16,6	1,8
31,129	120,07	18,9	2,1
52,998	165,282	39,5	3,8
88,386	286,331	59,6	6,8

For graphs TGFF best results are achieved by coherent genetic algorithm: up to 15% better for time criterion and 10% for cost optimization, ant's algorithm better by around 12% for time and cost, whereas for criterion of power consumption around 12% - better than genetic, whose results are better only by around 3%. Parallel algorithm achieves around 9% better results than non-coherent algorithms for time and 7% for cost. Generally, one can see the superiority of parallel algorithm whose optimization criterion can be choose and this algorithm realize multi-criteria optimization, too.

2.2   Comparison with Hou-Wolf and Prakash-Parker Graphs

The following tests apply to the comparison between coherent algorithms to Dick-Jha and Prakash-Parker algorithms, when the input data are Hou-Wolf graphs [12] and Prakash-Parker graphs [11].

Dick-Jha Algorithm (DJ Algorithm)
Algorithm is based on a genetic attitude. Solution is represented by chromosomes:

1. *of universal processors' allocation,*
2. *of dedicated processors' allocation,*
3. *of communication channels' allocation,*
4. *of task allocation to resources.*

Solutions of the same allocation are grouped into clusters. Crossbreeding of chromosomes' allocations applies to solutions of different clusters. List algorithm of task scheduling is used. Cost and power consumption are estimated and solution ranking is created.

Prakash-Parker Algorithm (PP Algorithm)
For computation of task scheduling and allocation to processors, linear programming has been used. The set of equations describes total correctness of system performance (constraints) and allocation of processors to task performing. Algorithm is restrained to the small number of tasks and processors.

The example of Hou-Wolf graphs – Table 5

Parallel algorithms have found better solutions – as far as time is considered i.e. system speed – than algorithms DLJ and DJ (Tables 6a and 6b) and parallel algorithm (Table 6c) for Hou-Wolf graphs, though at greater computation time. Not in any of the algorithms structures have been changed and optimization applied only to their speed.

Table 5. Examples digraphs of Hou-Wolf

Digraph	Number of tasks
Hou1,2/u	20
Hou3,4/u	20
Hou1,2/c	8
Hou3,4/c	6

Table 6a. Results actions of DLJ algorithm

DLJ algorithm		
Cost	Time	Computation time
N.A.	N.A.	N.A.
N.A.	N.A.	N.A.
170	5,1	N.A.
N.A.	N.A.	N.A.

Table 6b. Results actions of DJ algorithm

DJ algorithm		
Cost	Time	Computation time
170	5,7	2,8
170	8,0	1,6
170	5,1	0,7
170	2,2	0,6

Table 6c. Results actions of parallel genetic algorithm

Parallel genetic algorithm		
Cost	Time	Computation time
170	5,5	5,0
170	7.2	3,1
170	4,9	1,9
170	2,0	1,5

Example of Prakash-Parker Graphs – Table 7

All the algorithms have achieved optimum costs, algorithm PP calculates (non-polynomial) an optimum for the criterion of cost optimization – Table 8a, whereas algorithms DJ – Table 8b – and parallel – Table 8c – for the same cost achieve better results for time optimization. Parallel algorithms achieve better results than algorithm DJ, though at greater computation time – Table 8c.

Table 7. Examples digraphs of Prakash - Parker

Digraph	Number of tasks
Prakash-Parker 1/1	4
Prakash-Parker 1/7	4
Prakash-Parker 2/8	9
Prakash-Parker 2/15	9

Table 8a. Results actions of PP algorithm

PP algorithm		
Cost	Time	Computation time
7	28	N.A.
5	37	N.A.
7	4,5	N.A.
5	N.A.	N.A.

Table 8b. Results actions of DJ algorithm

DJ algorithm		
Cost	Time	Computation time
7	3,3	0,2
5	2,1	0,1
7	2,1	0,2
5	2,3	0,1

Table 8c. Results actions of parallel genetic algorithm

Parallel genetic algorithm		
Cost	Time	Computation time
7	2,9	0,5
5	1,7	0,4
7	1,2	0,6
5	2,0	0,5

3 Conclusions

Parallel algorithms achieved better results of synthesis than the results of known algorithms of non-parallel synthesis.

For graphs STG and TGFF the results are better by a few percent and even up to 20% for selected algorithms, those for whose procedures of codes are given. For others and representative algorithms – exact, though ineffective (PP) and heuristic (DJ) the results of parallel algorithms' computations are also better.

Test results given in literature, especially for Prakash-Parker graphs have a considerably small number of tasks, because an algorithm of linear programming calculates, though optimally (for cost minimization), but generally ineffectively (exponentially). Unfortunately, the authors do not give computation time for this algorithm. We must indicate that for the given data, heuristic algorithm DJ, and parallel algorithms have found – at the same cost value – better results as a consequence of advanced procedures of task scheduling which play a key role in these algorithms and not only improve allocations of tasks to resources but also and more importantly of time realization of all system operations.

Data for synthesis problems are specified in the form of randomized digraphs, in which each edge and each node has assigned vector with of execution time, cost of execution and power consumption in relative units. Similarly, resources assigned to tasks have characteristics given in relative units. These comparisons are based on available databases that objective data input for calculations.

These results were achieved as a result the activities of the synthesis itself are performed in parallel; these activities are scheduling, partition and allocations.

References

1. Drabowski, M.: Boltzmann tournaments in evolutionary algorithm for CAD of complex systems with higher degree of dependability. In: Zamojski, W., Mazurkiewicz, J., Sugier, J., Walkowiak, T., Kacprzyk, J. (eds.) Tenth International Conference on Dependability and Complex Systems DepCos-RELCOMEX. AISVC, vol. 365, pp. 141–152. Springer (2015)
2. Drabowski, M., Wantuch, E.: Deterministic schedule of task in multiprocessor computer systems with higher degree of dependability. In: Zamojski, W., Mazurkiewicz, J., Sugier, J., Walkowiak, T., Kacprzyk, J. (eds.) Proceedings of the Ninth International Conference on Dependability and Complex Systems DepCos-RELCOMEX. AISC, vol. 286, pp. 165–175. Springer (2014)
3. http://www.kasahara.elec.waseda.ac.jp/schedule/
4. http://ziyang.eecs.umich.edu/∼dickrp/tgff/
5. Eles, P., Peng, Z., Kuchcinski, K., Doboli, A.: System level Hardware/Software partitioning based on simulated annealing and tabu search. J. Des. Automat. Embed. Syst. **2**, 5–32 (1997)
6. Bianco, L., Augin, M., Pegatoquet, A.: A path analysis based partitioning for time constrained embedded systems. In: Proceedings of the International Workshop on Hardware/Software Codesign, pp. 85–89 (1998)
7. Dave, B.P., Lakshminarayana, G., Jha, N.K.: COSYN: Hardware-Software co-synthesis of embedded systems. In: Proceedings of the Design Automation Conference, pp. 703–708 (1997)
8. Yen, T.Y., Wolf, W.H.: Communication synthesis for distributed embedded systems. In: Proceedings of the International Conference on CAD, pp. 288–294 (1995)
9. Eles, P., Kuchcinski, K., Peng, Z., Doboli, A., Pop, P.: Scheduling of conditional process graphs for the synthesis of embedded systems. In: Proceedings of the Design Automation and Test in Europe Conference, pp. 132–138 (1998)
10. Dick, R.P., Jha, N.K.: MOGAC: a multiobjective genetic algorithm for hardware-software synthesis of hierarchical heterogeneous distributed embedded systems. IEEE Trans. Comput. Aided Des. Integr. Circuits Syst. **17**(10), 920–935 (1998)
11. Rhode, D.L. Wolf, W.: Co-synthesis of heterogeneous multiprocessor systems using arbitrated communication. In: Proceedings of the International Conference on Computer Aided Design, pp. 339–342 (1999)
12. Hsiung, P.A., Lin, C.Y.: Synthesis of real-time embedded software with local and global deadline. In: Proceedings of the CODES + ISSS, pp. 114–119 (2003)

Examples of Applications of CAD Methods in the Design of Fault Tolerant Systems

Mieczyslaw Drabowski[✉]

Department of Theoretical Electrical Engineering and Computing Science,
Cracow University of Technology, Warszawska 24, 31-155 Cracow, Poland
drabowski@pk.edu.pl

Abstract. In CAD procedures for designing complex systems, in particular with increased dependability and fault tolerant, optimized algorithms for identifying and allocation resources and scheduling of tasks should be implemented. These problems are computational NP-complete (exact: their decision versions are NP-complete [1]), so CAD procedures usually calculate suboptimal but polynomial solutions. Such procedures are most often based on metaheuristic approaches, e.g. hybrid ones: genetic algorithms and simulated annealing. This paper presents an example of the application of such a solution: the procedure receives requirements and constraints on the proposed system - at the input - and calculates (on the output of procedure) - in accordance with the adopted criteria of system operation - selected types and numbers of resources and schedules of tasks, e.g. in the form of Gantt charts.

Keywords: Scheduling · Partition · Allocation · Dependable · Fault-tolerant

1 Introduction

In the paper [2], a system model was presented based on a modified model of a computer system for task scheduling. The model additionally includes the terms, in addition to the best task planning, also the best resource configurations taking into account, along with the time optimization criteria, also the cost and energy consumption criteria. In this model there, so-called multiprocessor tasks [3, 4] as tasks to test the system. These tasks enable checking the correct operation of the equipment and software, as well as fault analysis and fault diagnosis. Multiprocessor tasks together with transmission devices are a kind of "nerve" of the system responsible for detecting and immunity faults.

Based on this model, methods have been developed for computer-aided design of such complex computer systems (i.e. containing sets of processors and additional resources that process a large number of usable programs). The aim of these methods is in turn to find the optimal system implementation solution, in accordance with the requirements and limitations imposed by a specific system specification. Typically, the following optimality criteria are taken into account: speed of operation, implementation cost, power consumption, the degree of serviceability and the degree of dependability of the system and its ability to repair or fault tolerant.

© Springer Nature Switzerland AG 2020
W. Zamojski et al. (Eds.): DepCoS-RELCOMEX 2019, AISC 987, pp. 166–176, 2020.
https://doi.org/10.1007/978-3-030-19501-4_16

Identification and allocation of resources taking into account various techniques and implementation technologies are the basic issue of automatic system design. This partitioning is significant because each complex system must be implemented as a result of a hardware implementation for certain tasks and software implementation for others. In addition, scheduling problems are one of the most important problems occurring in the design of operational procedures responsible for controlling the allocation of tasks and resources in complex systems.

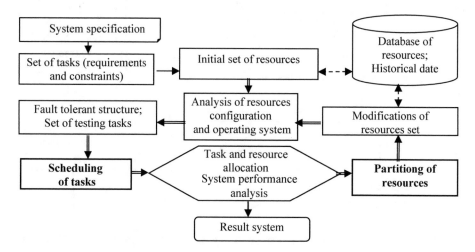

Fig. 1. The process of parallel synthesis of dependable computer system

The new model and new methods of building software and hardware elements are developed jointly connected with each other, which ultimately reduces costs and increases the speed of operation, unlike before. This model and methods for synthesis of systems with a high degree of dependability are presented in [2]. The process of their synthesis is show on Fig. 1. In this paper we will present examples illustrating the course of designing complex systems using methods of automatic support of synthesis.

2   Examples of Synthesis

We will present practical examples of the use of parallel approach in the design of computer systems. The method used in this parallel synthesis is a hybrid algorithm combining the genetic approach with Boltzmann tournaments in simulated annealing [5]. The scheme of this procedure is shown in Fig. 2. The procedure based on genetic algorithms that optimize the identification and selection of resources through the evolution of cluster populations and optimize the task schedule and their allocation to resources through the evolution of the population of solutions in clusters. All operations of the genetic algorithm are dependent on temperature and cooling parameters, which allows avoiding local extremes.

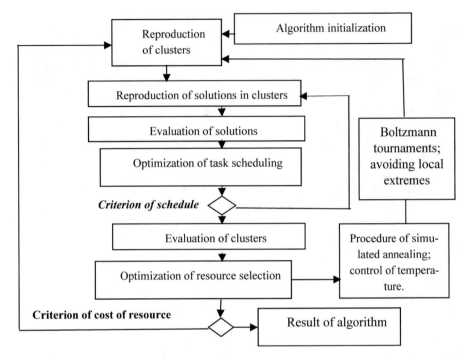

Fig. 2. Scheme of actions of the procedure of genetic with simulated annealing for parallel synthesis

2.1   Specification of System (Input for Procedure)

We assume, that algorithm of coherent synthesis has to design computer system, which functions and behavior described on this digraph of tasks – Fig. 3.

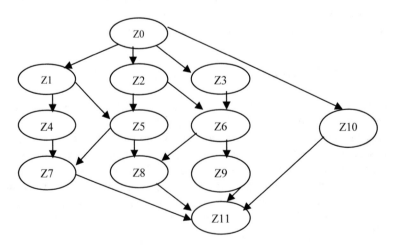

Fig. 3. Digraph of tasks

There are pools (container) of available resources – Table 1.

Table 1. Available resources

ID	Type	Cost	Cost of memory	Power consumption
P1	General	1	0.15	0.001
P2	General	1.5	0.2	0.002
A1	Dedicated	0.7	0	0.001
A2	Dedicated	0.9	0	0.002

Other requirements and constraints of the system are as following – Tables 2 and 3.

Table 2. Times of task processing if 0 - processor is not adapted to realization of such tasks

Tasks	Processors			
	P1	P2	A1	A2
Z0	2	1	3	2,5
Z1	4	2	1,5	0
Z2	2	1	3,5	3
Z3	4	2	0	0
Z4	5	2,5	0	0
Z5	6	3	2	1
Z6	3	1,5	0	0
Z7	2	1	0	0
Z8	1	0,5	0	0
Z9	4	2	0	1
Z10	2	1	0,5	0,4
Z11	2	1	0,5	0,4

Table 3. Power consumption of task processing if 0 - processor is not adapted to realization of such tasks

Tasks	Processors			
	P1	P2	A1	A2
Z0	1	1,5	1,5	1,5
Z1	2	3	1	0
Z2	1	1,5	2	2
Z3	2	3	0	0
Z4	2,5	3,5	0	0
Z5	3	4,5	2	1,5
Z6	1,5	3	0	0
Z7	1	2	0	0
Z8	0,5	1	0	0
Z9	2	3	0	1,5
Z10	1	2	0,7	0,6
Z11	1	2	0,7	0,6

Processor P2 is quicker than processor P1 but processor P2 takes more power and is more expensive than P1. Dedicated processor A1 can only realize tasks Z0, Z1, Z2, Z5, Z10 and Z11 but it is suitable for tasks: Z1, Z5, Z10 and Z11. Tasks Z0 and Z2 can be executed on this processor, but the time of their realization are longer than on universal processors. This processor is cheaper than dedicated processor A2, but its power consumption is greater. Dedicated processor A2 is suitable for execution of tasks Z5, Z9, Z10 and Z11. However, it is less suitable for tasks Z0 and Z1 than universal processors, but more suitable than dedicated processor A1. The cost of processor A2 is greater than of processor A1.

2.2   Results of Optimization (Output of Procedure)

To simplify the tests, we assume that four is the maximum number of processors in the resultant system structure.

Cost Minimization Not Including Memory Cost
For obvious reasons, in this case we receive system structure which consists of a single and cheapest universal processor. Dedicated processors, which are cheaper, do not have such possibilities of task execution in the system (Fig. 4 and Table 4).

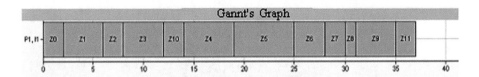

Fig. 4. Cost minimization not including cost of memory

Table 4. Cost minimization not including operating memory

Cost	Time	Power consumption
1	37	18,5

Cost Minimization Not Including the Cost of Memory
In this case we will receive system structure which consists of one universal processor (cheaper) and a single dedicated processor (cheaper). In this case it is better to apply two processors. The cost of realization of all tasks on the cheapest universal processor is 2, 8 (the cost of processor + 12 tasks × 0, 15 cost of memory unit) - Fig. 5 and Table 5.

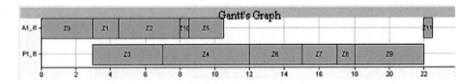

Fig. 5. Cost minimization including the cost of memory

Table 5. Cost minimization including the cost of operating memory

Cost	Time	Power consumption
2,6	23,5	17,4

Processing Time Minimization

In case of time minimization, system structure is generated, which consists of two faster processors A1, A2 and two dedicated processors A1, A2. Resultant scheduling was allocated by algorithm of synthesis to allow processors A1 and A2 to complete tasks to which they are dedicated (Fig. 6 and Table 6).

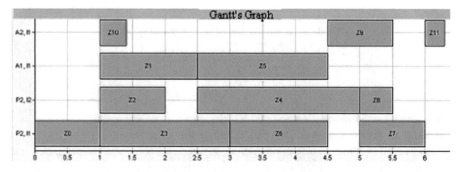

Fig. 6. Processing time minimization

Table 6. Minimization of processing time

Cost	Time	Power consumption
6	6,3	21,2

Minimization of Power Consumption

In the process of minimizing power consumption, the system which consists of two cheaper universal processors was received, power consumption of which is the lowest for most tasks. Task Z1 was executed on processor A1, as this processor was dedicated for effectively executed task. Task Z10, Z5, Z9 and Z11 are executed on dedicated processor A2, where power consumption is the lowest - Fig. 7 and Table 7.

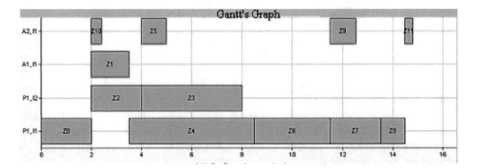

Fig. 7. Minimization of power consumption

Table 7. Minimization of power consumption

Cost	Time	Power consumption
4,65	14,8	14,7

Time and Cost Minimization for the Established Cost Constraint

The value of cost criteria was established during the tests as 4.

As a result of multi-criteria optimization, we receive a set of optimal solutions. The first and the second solution's costs are above minimum and the time larger than minimum time. However, there is a compromise between the two contradictory requirements. The first solution is cheaper than the other by *0,4*, but the execution time is longer by *2,7*. Power consumption is slightly lower than for more expensive architecture – Figs. 8 and 9, Tables 8 and 9.

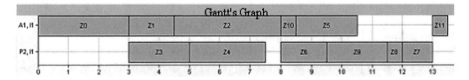

Fig. 8. Minimization of time and cost – cheaper system

Table 8. Minimization of time and cost – cheaper system

Cost	Time	Power consumption
3,4	13,5	23.4

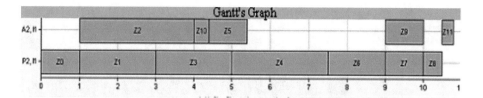

Fig. 9. Minimization of time and cost – quicker system

Table 9. Minimization of time and cost – quicker system

Cost	Time	Power consumption
3,8	10,8	23,2

The third case shows how multi-criteria optimization is capable of searching for the solution of the lowest cost, but apart from it can optimize the rest of criteria.

The third structure's cost is the same as the cheapest found during cost minimizing, but at the same time the length of allocation is smaller – Fig. 10 and Table 10.

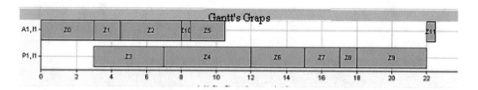

Fig. 10. Minimization of cost and time – cheapest system

Table 10. Minimization of cost and time – cheapest system; improvement of time

Cost	Time	Power consumption
2,6	22,5	17,4

Cost and Power Consumption Minimization for Established Time Constraints
The value of time criteria was established as 20.

The assumption of a resultant system was to limit the time; the system had to finish completing all the tasks within 20 time units. According to such assumption the rest of costs were optimized: time and power consumption. Two optimal, in sense of Pareto, structures were received. The first system completes its performance after 19, 8 time units and takes relatively more power than the other; however, the cost of the first system is smaller. The only difference is execution of task Z0. In the first system this task is completed by a dedicated processor which, in that case, takes more power, but does not need additional operating memory. Due to it the cost of the system decreases and power consumption rises. Completing this task by universal processor P1 causes taking less power, but needs additional operating memory. Executing time of all the tasks is shorter in the second system but both systems meet time restrictions – Figs. 11 and 12, Tables 11 and 12.

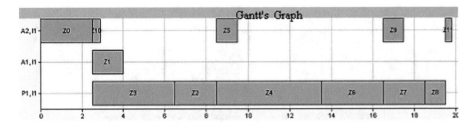

Fig. 11. Minimization of cost and power consumption – maximum time equal 20 – cheaper system

Table 11. Minimization of cost and power consumption – maximum time equal 20 – cheaper system

Cost	Time	Power consumption
3,5	19,8	15,2

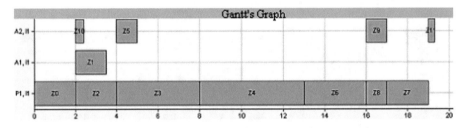

Fig. 12. Minimization of cost and power consumption – maximum time equal 20 – more expensive system

Table 12. Minimization of cost and power consumption – maximum time equal 20 – more expensive system

Cost	Time	Power consumption
3.65	19.3	14.7

Multi-criteria Optimization – Simultaneous Cost, Time and Power Consumption Minimization for the Established Time Limit

Maximum time value was established as 15.

As a result of three criteria optimization, a set of optimal solutions has been received (in sense of Pareto) which in a different way describes a compromise between contradictory criteria. Two structures have been taken to analyze whose time (14, 5) of task completing is close to the earlier established value of time criteria (15). The first consists only of two processors, due to which the cost is small. However, task completing on processor P2 causes significant power consumption.

Dedicated processor performs the tasks, execution of which causes smaller power consumption. Task Z2 is an exception. However, execution of this task on processor P2 could exceed time restrictions – Fig. 13 and Table 13.

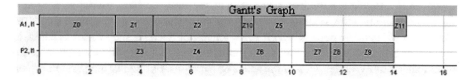

Fig. 13. Cost, time and power consumption minimization – cheaper system

Table 13. Cost, time and power consumption minimization – cheaper system

Cost	Time	Power consumption
3,4	14,5	23,4

The second structure consists of 4 processors. Such a structure is more expensive, but ensures meeting time restrictions of completing all the tasks and significantly decreases power consumption. – Fig. 14 and Table 14.

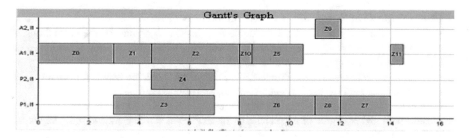

Fig. 14. Cost, time and power consumption minimization – cheaper system – system with optimized power consumption

Table 14. Cost, time and power consumption minimization

Cost	Time	Power consumption
4,9	14,5	17,9

3 Conclusions

The above presented examples show solutions of optimization for synthesis of computer systems. As a result of algorithms type CAD the designer receives a set of optimal or suboptimal solutions and results for multi-criteria optimization in sense of

Pareto. The designer has to decide which resource fulfills the requirements of the solution. Depending on the system requirements, it is possible to rely on one of the obtained results.

To learn the specification of the solution space for the given problem in-stance, it is important to provide a sufficiently long list of the best received solutions. It is also important to define slow cooling of the algorithm: parameters "temperature step" and "cooling coefficient" – depending on the numbers of tasks in the system. Thanks to this, is prevent the population from too big convergence [6]. Algorithm searches for the lower temperature a bigger area in the space of solutions. For higher temperatures there is a wider expansion of solutions and during cooling i.e. for lower temperatures, intensive exploration of solutions. The use of such a method of automatic synthesis of complex systems allows for optimization of solutions and ensure rapid prototyping of such systems.

References

1. Garey, M., Johnson, D.: Computers and Intractability: A Guide to the Theory of NP-Completeness. Freeman, San Francisco (1979)
2. Drabowski, M., Wantuch, E.: Deterministic schedule of task in multiprocessor computer systems with higher degree of dependability. In: Zamojski, W., Mazurkiewicz, J., Sugier, J., Walkowiak, T., Kacprzyk, J. (eds.) Advances in Intelligent Systems and Computing, Proceedings of the Ninth International Conference on Dependability and Complex Systems DepCos-RELCOMEX, vol. 286, pp. 165–175. Springer (2014)
3. Dick, R.P., Jha N.K.: MOGAC: a multiobjective genetic algorithm for the cosynthesis of hardware-software embedded systems. In: Proceedings of the International Conference on Computer Aided Design, pp. 522–529 (1997)
4. Błażewicz, J., Drabowski, M., Węglarz, J.: Scheduling multiprocessor tasks to minimize schedule length, IEEE Trans. Comput. **C-35**(5), 389–393 (1986)
5. Drabowski, M.: Boltzmann tournaments in evolutionary algorithm for cad of complex systems with higher degree of dependability. In: Zamojski, W., Mazurkiewicz, J., Sugier, J., Walkowiak, T., Kacprzyk, J. (eds.) Advances in Intelligent Systems and Computing, The Tenth International Conference on Dependability and Complex Systems DepCos-RELCOMEX, vol. 365, pp. 141–152. Springer (2015)
6. Elburi, A., Azizi, N., Zolfaghri, S.: A comparative study of a new heuristic based on adaptive memory programming and simulated annealing: the case of job shop scheduling. Eur. J. Oper. Res. **177**, 1894–1910 (2007)

The Concept of the ALMM Solver Knowledge Base Retrieval Using Protégé Environment

Ewa Dudek-Dyduch[1] ⓘ, Zbigniew Gomolka[2]([✉]) ⓘ,
Boguslaw Twarog[2] ⓘ, and Ewa Zeslawska[3] ⓘ

[1] Department of Biomedical Engineering and Automation,
AGH University of Science and Technology Cracow, Krakow, Poland
edd@agh.edu.pl
[2] Faculty of Mathematics and Natural Sciences, University of Rzeszow,
ul. Pigonia 1, 35-959 Rzeszow, Poland
{zgomolka,btwarog}@ur.edu.pl
[3] Faculty of Applied Informatics, Department of Applied Information,
University of Information Technology and Management in Rzeszow,
ul. Sucharskiego 2, 35-225 Rzeszow, Poland
ezeslawska@wsiz.rzeszow.pl

Abstract. The paper presents the concept of using a Protégé environment to extract knowledge about the properties and features of discrete problems described by ALMM technology. The concept of an Intelligent ALMM system was proposed firstly by Dudek-Dyduch E. The main purpose of the ALMM system is not only to solve problems with discreet optimization, but also to help the user in choosing the right method and algorithm for solving them. The authors, based on earlier work connected with component structure based solvers using ALMM technology, proposed the exploitation of description of the properties of individual problems in order to create the corresponding knowledge base. This paper presents an example of a knowledge base structure for the problem of scheduling tasks on a single machine that can be used to solve various optimization tasks. Using the OWL language convention, the architecture of classes and related properties of solved problems: axioms and assertions were proposed. An on-line knowledge base representing specific classes of problems together with its own query language can be a scenario for pre-modeling the architecture of the final application. For the ontological architecture, the examples of knowledge extraction scenarios to solve the optimization problem of the tasks scheduling on a single machine are presented too.

Keywords: ALMM technology · Knowledge base · Ontology approach · Protégé

1 Introduction

The Algebraic Logical Meta-Model (ALMM) of multi-stage decision-making processes provides a uniform template (pattern) for creating the formal notations of all constraints and information on deterministic problems that can be presented as a sequence of decisions. A suitable modeling method is mentioned in the bibliography

© Springer Nature Switzerland AG 2020
W. Zamojski et al. (Eds.): DepCoS-RELCOMEX 2019, AISC 987, pp. 177–185, 2020.
https://doi.org/10.1007/978-3-030-19501-4_17

for several issues of discrete optimization, production processes, logistic processes, project management, and more generally in planning and other areas [5–11, 18, 19]. Based on the meta-model, it is possible to create algebraic logic models (AL models) of individual problems of searching for an acceptable or optimal solution. Two basic classes of multistage decision processes can be distinguished: ordinary (Common Multi-stage Decision Process - CMDP) and dynamic (Dynamic Multistage Decision Process - MDDP). In MDDP a generalized state is defined as: $s = (x, t)$, in which x denotes the proper state, while t describes the moment of time.

The algebraic logical meta-model of a dynamic, multi-stage decision process is then defined as the discrete process P which is uniquely described by the six tuple:

$$P = (U, S, s_0, f, S_N, S_G) \tag{1}$$

In the process P the transition function is defined: $f : U \times S \to S$, the set U is called a set of control decisions or control signals, set $S = X \times T$ is called a set of generalized states and X is a set of proper states. The set $T \subset \mathbb{R}^+$ is referred to as a subset of non-negative real numbers representing the relevant time moments. The sets U, X, T are non-empty, and the transition function f is defined on the basis of the relationship defining the next proper state $f_x : U \times X \times T \to X$ and the relation defines the next moment in time $f_t : U \times X \times T \to T$, when the condition $[\Delta t = f_t(u, x, t) - t] > 0$ is fulfilled. The generalized initial state is denoted as: $s_0 = (x_0, t_0)$ and $s_0 \in S$ respectively. S_N is a set of inadmissible states ($S_N \subset S$) whereas a non-empty set of goal states is defined to as S_G ($S_G \subset S$). S_G describes a set of states in which the process should be found as a result of proper controls. In ALMM technology, states and decisions do not have to be defined by means of variables that take values in numerical spaces. Values of variables can be names of elements of certain sets, subsets, or other objects. As a consequence, the models can be defined with the help of both algebraic relations and logical formulas - hence the name of the Algebraic-Logic-Meta-Model. A problem where in admissible decision sequence can be sought is determined by defining a dataset as well as six basic components of the P process. An optimization problem Π is defined by a (P, Q) pair where Q denotes the criterion regulation. For specific data (provided explicitly) we receive a problem instance (P, Q) with P being a so-called individual process and Q denoting the criterion determined for a given instance. The dynamics of the individual process P is represented by its trajectory set. Thus we look for a decision sequence determining an admissible or optimal trajectory.

In the earlier works of Dudek-Dyduch [6–8] a system concept was presented, based on ALMM technology, which aims to optimize discrete problems. This system has been called the Intelligent ALMM System. It is distinguished from other solvers by the fact that the Intelligent ALMM System has a knowledge base that is used for:

- **storage** of problem models and their components represented in the ALMM technology;
- support for creating (**creation**) models of new problems through the use of stored components in the database;

- storing the **specification** of discrete optimization methods and algorithms, presented in the ALMM technology;
- storing **definitions** of general properties of discrete optimization problems and information about properties that are met by the individual problems.

Figure 1 shows a system structure that consists of three main parts: the Knowledge Base, which stores Knowledge on Problems and Knowledge on Solving Methods and Algorithms, the Intelligent User Interface that communicates with the user, and Solving Module which is responsible for calculations. A detailed description of the system is described in the works [6].

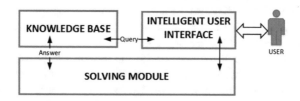

Fig. 1. Structure of intelligent ALMM system

2 Modelling an Ontology with Protégé Environment

Ontology is a formal, unambiguous description of the area of interest, which allows for the definition of the appropriate logical assertions about the types of a given area based on the definition of classes, properties and individual variables. Ontology in combination with Instances represents the Knowledge Base (KB). The starting point in designing ontologies is a deep, complete and appropriate analysis of a given problem. The knowledge about the problem can be represented/formalized using ontology. The main features of using ontologies are modeling, interoperability and data export. In practice, there are ontologies of a different degree of formalization - from predefined vocabulary to knowledge models based on descriptive logic (Description Logic - DL) on which the OWL language (Web Ontology Language) is based. OWL is the language of knowledge representation allowing for the expression, exchange and processing of knowledge about a given problem area. Ontologies defined with the use of OWL language creates classes, properties, individuals and data values. In order to distinguish properly object classes in the ontological architecture from individuals and express different relationships between them, the standard relationships were used:

- isA - denotes the relationship between the individual and his class;
- hasA - is a relation part-whole;
- aKo (a kind of) - is a relation between a subclass and a superclass, often also written as: SubClass or SS (SubSet).

Semantic webs are an ontological method of knowledge representation assuming that human memory best describes an associative model which is build of a set of objects connected with each other by various relations. It can be a graphically represented with a certain type of logic, where relations between objects can be presented in the form of a graph in which objects are nodes, and relations are branches. The advantage of semantic webs is their flexibility and the lack of restrictions on the number of nodes and arcs, as well as the possibility of treating information contained in the semantic network as a set (conjunction) of logical formulas directly expressed by the relationship between objects. In the designed ontology, all relations/properties (and their corresponding formulas) are binary (two-argument) relations. They express the relationship between facts, between facts and concepts, and between concepts. They can have the character of taxonomic dependencies, creating a hierarchy (network) of concepts and/or facts, as well as cause-and-effect relationships. By exploiting data in the Knowledge Base, described using a presumed language, we can obtain new information using the so-called the inference mechanism which, according to logical rules (facts, rules), proves a new hypothesis that is the target for the reasoning engine. The new scenario generated by the query with the use of DL Query technique will generate the chain of components necessary to run the solver with the appropriate parameters solving the selected problem. Currently, there are several tools and environments available that enable the modeling of ontological knowledge bases that use: OWL (Web Ontology Language), RDF (Resource Description Framework), RDF Schema (RDFS), XML and SPARQL (Protocol And RDF Query Language). Several tools for modeling, editing, visualization and validation of the structure and semantics can be distinguished: Protégé, OWL-S, SMORE, SNOBASE, OntoEdit, OntoStudio, SOE4W, SWOOP, ALTOVA SematicWorks, Top Braid, IBM IODT, COMA++ [1–4, 12–17].

The ALMM technology considered in earlier works used the mechanism of a file structure to construct the component of ALMM System structure. Such an approach enforces rigid programming rules for individual ALMM System modules, which are undoubtedly a crucial limitation of such of approach. In the simulation part of this work, the Protégé environment was adopted because it allows for:

- defining classes by means of properties (object properties), i.e. relations to other classes, by specifying the necessary conditions (SubClassOf, primitive classes) or necessary and sufficient conditions (EquivalentTo, defined classes);
- hierarchy of classes and properties including inheritance;
- application of functional, transitive, feedback and symmetrical features;
- using expressions of descriptive logic (DL - Description Logic) as synonyms of anonymous classes;
- advanced effects of reasoner's actions in the process of detecting inconsistencies of ontologies and detection of classified (implied) relationships between classes;
- ability to implement(Open World Assumptions/Reasoning) with closure axioms.

Figure 2 presents an ontological relationship example of classes defined by the system designer, within the definition of axioms and assertions and a set of logical relations between classes (A_1, B_1) and (A_2, B_2) generated respectively by a reasoner.

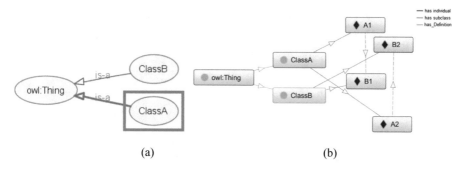

(a)   (b)

Fig. 2. The exemplary ontology structure and its dependencies

3 An Ontology Approach for ALMM Technology

In [7, 11] a software presentation of AL Models and Component-based structure of library have been described, in which the classes were proposed for the following separate categories: utility classes, atomic classes, elementary classes, composite classes and design classes. The analysis of problem models belonging to a certain area, i.e. tasks scheduling, graph problems and others, shows that the definitions of some model elements are identical. As a consequence, the same attributes and methods can be distinguished in classes modeling different problems. This fact implies the idea of creating a utility class from the appropriate atomic classes. The use and constructing method of the proposed utility classes can be presented on the example of the following scheduling problems addressed to one machine: $\prod_1 = 1||C_{max}$, $\prod_2 = 1|C_j \leq d_j|C_{max}$, $\prod_3 = 1|prec|C_{max}$, $\prod_4 = 1|r_j|C_{max}$, $\prod_5 = 1|C_{max} \leq \tilde{C}|C_{max}$, where C_j - denotes the completion time of $j - th$ job, C_{max} - total time of completing the whole task, $\tilde{C}$ - the maximum total time of the task completion, C_j - the sum of machine job end times, d_j and r_j - describes restrictions on working dates and time of release respectively, $prec$ - describes the existence of previous restrictions in the set of tasks. AL models of the above problems are characterized by identical definitions of state and decision structures and a definition of the initial state of the process. Definitions of a set of forbidden states, a set of final states, a transition function and a set of possible controls in a given state for a problem, allow us to assume them as default definitions. Some of the above definitions can be used without change for the four other problems. Regarding problems being analyzed, five specific definitions related to the constraints of the considered problem (in the beta notation) and three common definitions should be created. Each element definition of the problem is answered by a single atomic class implementing one or more methods. A total of twelve elementary classes will be created, which describes three definitions common to all problems, four default ones and five specific for individual problems. By building a representative ontology for a system using ALMM technology, we gain not only a deep insight into the basics of the modeled domain, but we can also determine the semantic space emerging from it (Table 1).

Table 1. Ontological model of SMSP problem described in the ALMM technology

Class	Problems	SubClass	SubClass	Individuals	Ontology annotation
Problem Π	$\Pi1, \Pi2,$ $\Pi3, \Pi5$	Process P	f	f_1	Transition_Function
	$\Pi4$			f_2	
	$\Pi1, \Pi2,$ $\Pi3, \Pi4,$ $\Pi5$		U	U	Decision
	$\Pi1, \Pi2,$ $\Pi4, \Pi5$		$U_p(s)$	$U_{p1}(s)$	UPossible_Decision
	$\Pi3$			$U_{p2}(s)$	
	$\Pi1, \Pi2,$ $\Pi3, \Pi4,$ $\Pi5$		S	S	State
	$\Pi1, \Pi2,$ $\Pi3, \Pi4$		S_G	S_{G1}	Goal_State
	$\Pi5$			S_{G2}	
	$\Pi1, \Pi3,$ $\Pi4$		S_N	S_{N1}	Non_Admissible_State
	$\Pi2$			S_{N2}	
	$\Pi3$			S_{N3}	
	$\Pi1, \Pi2,$ $\Pi3, \Pi4,$ $\Pi5$		s_0	s_{01}	Initial_State
	$\Pi1, \Pi2,$ $\Pi3, \Pi4,$ $\Pi5$	Criterion Q	Q	Q_1	Criterion_Q

In particular, we define how a given concept can be combined with each other and what types of statements may appear in a given model regarding particular elements. The generated result code uses the implemented semantic network structure which corresponds to the metamodel of the ALMM solver ontology (see Fig. 3).

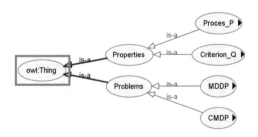

Fig. 3. Semantic space of the main concepts of the ALMM ontology

Changes in the ontology which are automatically propagated on its implementation, usually require smaller changes the procedures creating the model of meaning, and any possible discrepancies between them and the ontology can be detected at the stage of compiling the system. After analyzing the whole system, on the basis of the model of its created meaning, a class diagram is generated as well as a diagram showing the relations of properties corresponding to the structure of reality and the occurring processes in it Fig. 4.

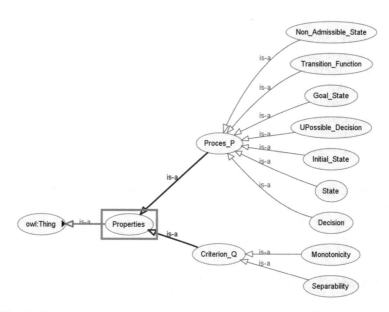

Fig. 4. The actual ontological structure and relations of basic metamodel processes

Querying ontologies using descriptive logic expressions (DL Query tab) allows for specifying the constraints by means of conditions limiting classes in the Description panel. At Protégé, they are printed in the so-called Manchester notation (also referred to as Manchester syntax). The Manchester OWL syntax is a simplified notation of constructors for descriptive logic used in the Protégè editor in order to specify complex concepts and roles (Listing 1 and Listing 2).

Listing 1. Manchester Syntax for individual Transition Function

```
Individual: F01
    Types:
        <#Transition_Function >
    Facts:
     has_Part  p:MS_1,
     p:_Machine_State   " "
    DifferentFrom:
        F02
```

Listing 2. Manchester Syntax for individual CMDP class

```
Individual: p:MS_1
    Annotations:
        rdfs:comment "Job-scheduling problems with a sin-
gle machine #1"@en
    Types:
        p:MS
    Facts:
     has_Definition  F01,
     has_Definition  S01,
     has_Definition  SG01,
     has_Definition  SN01,
     has_Definition  So01,
     has_Definition  U01,
     has_Definition  UP01,
```

The specific forms of responses generated on the basis of the descriptive logic will be the scenarios, which are necessary to implement the final layout of the assembled software components for the model that solves complex problems of discrete manufacturing processes.

4   Conclusion

The work presents the method of representation of an intelligent information system using an ontology and metamodeling technique. A dedicated ontology metamodel has been defined, which stands the grammar description for the Intelligent ALMM System. This approach gives a lot of freedom to model various semantic phenomena, while maintaining the accuracy of the specifications and transparency of the adopted solutions. The presented ontology reflects the semantic representation of the ALMM solver which allows for the solution of complex discrete issues regarding the scheduling of tasks on a single machine. In addition, the proposed solution provides an advanced knowledge base for generating various scenarios of the component implementation system of a dedicated application for the problems of management the production processes. The possibility of using properly selected queries in the DL-type descriptive logic allows for generate an input source for a parser that is an interface between the ontological knowledge base and the ALMM software representation architecture. The proposed method of knowledge extraction from the ontology of ALMM system can effectively assist the user in choosing methods for solving various optimization problems. As a continuation of the researches presented in this paper, it should be considered to include in the proposed ontology problems of tasks scheduling for parallel machines, with particular emphasis on the optimization criteria and dynamically changing resources of the system.

References

1. Bergman, M.K.: A Knowledge Representation Practionary, Guidelines Based on Charles Sanders Peirce. Springer, Cham (2018)
2. Bialas, A.: Common Criteria IT Security Evaluation Methodology – An Ontological Approach, Advances in Intelligent Systems and Computing, vol. 761. Springer, Cham (2018)
3. Bożejko, W., Gnatowski, A., Niżyński, T., Affenzeller, M., Beham, A.: Local Optima Networks in Solving Algorithm Selection Problem for TSP, Advances in Intelligent Systems and Computing, vol. 761. Springer, Cham (2019)
4. Ciskowski, P., Drzewiński, G., Bazan, M., Janiczek, T.: Estimation of Travel Time in the City Using Neural Networks Trained with Simulated Urban Traffic Data, Advances in Intelligent Systems and Computing, vol. 761. Springer, Cham (2019)
5. Dudek-Dyduch, E., Gomolka, Z., Twarog, B., Zeslawska, E.: Intelligent ALMM System - implementation assumptions for its Knowledge Base. ITM Web Conf. **21**, 00002 (2018)
6. Dudek-Dyduch, E.: Intelligent ALMM system for discrete optimization problems – the idea of knowledge base application. In: ISAT, vol. 657 (2017)
7. Dudek-Dyduch, E., Korzonek, S.: ALMM Solver for combinatorial and discrete optimization problems Idea of Problem Model Library. In: ACIIDS Part I. LNAI, vol. 9621 (2016)
8. Dudek-Dyduch, E.: Modeling manufacturing processes with disturbances a new method based on algebraic-logical meta-model. In: ICAISC, Part II. LNCS, vol. 9120 (2015)
9. Dudek-Dyduch, E.: Algebraic logical meta-model of decision processes new metaheuristics. In: ICAISC, Part 1. LNCS, vol. 9119, pp. 541–554 (2015)
10. Dudek-Dyduch, E.: Modeling manufacturing processes with disturbances – two-stage AL model transformation method. In: MMAR, pp. 782–787 (2015)
11. Korzonek, S., Dudek-Dyduch, E.: Component library of problem models for ALMM solver. J. Inf. Telecommun. **1**, 224–240 (2017)
12. Munir, K., Sheraz Anjum, M.: The use of ontologies for effective knowledge modelling and information retrieval. Appl. Comput. Inf. **14**(2), 116–126 (2018)
13. Protégé. https://protege.stanford.edu/. Accessed 30 Jan 2019
14. Drałus, G., Mazur, D., Gołębiowski, M., Gołębiowski, L.: One day-ahead forecasting at different time periods of energy production in photovoltaic systems using neural networks. In: 2018 International Symposium on Electrical Machines (SME), SEM0053, Andrychów, Poland, 10–13 June 2018, pp. 1–5. IEEE Xplore, 23 August 2018. https://doi.org/10.1109/ISEM.2018.8443044
15. Khattak, A.M., Latif, K., Lee, S.: Change management in evolving web ontologies. Knowl.-Based Syst. **37**, 1–18 (2013). https://doi.org/10.1016/j.knosys.2012.05.005. ISSN 0950-7051
16. Strzałka, D., Dymora, P., Mazurek, M.: Modified stretched exponential model of computer system resources management limitations - The case of cache memory. Phys. A **491**(1), 490–497 (2018). https://doi.org/10.1016/j.physa.2017.09.012
17. Drałus, G., Mazur, D.: Cascade complex systems - Global modeling using neural networks. In: 13th Selected Issues of Electrical Engineering and Electronics (WZEE), pp. 1–8 (2016). https://doi.org/10.1109/wzee.2016.7800198, http://ieeexplore.ieee.org/document/7800198/ 2016-12-29
18. Walkowiak, T., Mazurkiewicz, J.: Soft computing approach to discrete transport system management. In: Lecture Notes in Computer Science. Lecture Notes in Artificial Intelligence, vol. 6114, pp. 675–682 (2010). https://doi.org/10.1007/978-3-642-13232-2_83
19. Walkowiak, T., Mazurkiewicz, J.: Analysis of critical situations in discrete transport systems. In: Proceedings of DepCoS - RELCOMEX 2009, Brunów, Poland, 30 June–02 July 2009, pp. 364–371. IEEE (2009). https://doi.org/10.1109/depcos-relcomex.2009.39

Graph-Based Vehicle Traffic Modelling for More Efficient Road Lighting

Sebastian Ernst, Konrad Komnata$^{(\boxtimes)}$, Marek Łabuz, and Kamila Środa

Department of Applied Computer Science, AGH University of Science
and Technology, Al.Mickiewicza 30, 30-059 Kraków, Poland
{ernst,kkomnata,mlabuz,ksroda}@agh.edu.pl

Abstract. Road traffic is one of the primary characteristics of modern
cities. It affects the travel time, which is used by navigation and route
planning systems. However, traffic flow can also be modelled with regard
to the number of vehicles. This approach can be applied e.g. to dynamic
adjustment of street lighting intensity, provided the data is available and
comes from a reliable source. This paper proposes a new model – the
Traffic Flow Graph – which can be used to represent measurable flows
of vehicles. It can be used to verify the reliability of sensor data and to
broaden area of dynamic street lighting, to streets without precise traffic
detectors. The obtained values allowed the application of dynamic street
lighting control, which resulted in 13.8% of energy savings in the road
under consideration.

Keywords: Traffic flow · Graph · Road traffic · Smart city

1 Introduction and Motivation

Road traffic is one of the most significant issues in modern cities, in two aspects.
On one hand – obviously – traffic congestions hamper travel, which affects com-
muters, companies and tourists. On the other hand, knowledge of traffic characteristics can – and should – be part of the decisions made by the city authorities
and technical departments.

One real-life example, which is not so obvious, is related to road and street
lighting. Energy needed for lighting constitutes almost a fifth of the power
consumed by the urban infrastructure, and therefore is a significant contributor to the city's regular expenses [7]. This issue can be optimised using many
approaches [9], but one of them – dynamic control of luminaire intensity using
sensor data – is not yet widely exploited. Research has shown that introduction
of such systems can reduce energy consumption up to 34% [10].

However, as traffic intensity is one of the most important factors used by such
systems, it is crucial to provide accurate and dependable data. The traffic sensor
infrastructure in cities is often sparse, which means that only a small portion of
streets can be equipped with dynamic lighting control. For instance, due to this

© Springer Nature Switzerland AG 2020
W. Zamojski et al. (Eds.): DepCoS-RELCOMEX 2019, AISC 987, pp. 186–194, 2020.
https://doi.org/10.1007/978-3-030-19501-4_18

problem, in a pilot project in a large city in Poland, carried out on an area of two districts and including almost 4,000 fixtures, dynamic control was possible in only 15% of the streets.

One might argue, of course, that traffic data is widely used in online maps and navigation devices. However, there are two issues. First, while the data is usually accurate and very complete, the sources for the fused data is usually not disclosed – and one may not wish to use it to control the city infrastructure. The data must be reliable and dependable, which means it should originate from official sensors, and also provide means for *validation* of their accuracy. Second, the data is profiled for route planning – and therefore focuses on the *travel time*, not on the *number of vehicles*, which is the basis for lighting class determination according to standards [5].

Of course, the scope of application of traffic intensity data is not limited to street lighting. Knowledge of traffic data can be used to optimise the traffic itself, by adjusting traffic signal cycles and making strategic decisions. Moreover, it can be used to identify commuter flows and optimise the public transport system, which – in turn – may lead to improved living comfort and reduced carbon dioxide emissions.

In this paper, we propose a graph formalism used to model traffic flows on urban roads. Its goal is to provide formal means of identifying the possible manoeuvres of vehicles and binding them to real-life sensor data. It constitutes the basis for estimation of traffic intensity on roads not covered by sensors, as well as for the aforementioned accuracy validation.

The practicality of the proposed model is proved using the example of real-life junction in one of major Polish cities, where the graph is used to identify potential anomalies in traffic intensity data recorded over a period of 12 months. Simulation of real street lighting dynamic control has been also performed, to prove energy saving potential.

The paper is organised as follows. Section 2 presents other approaches to modelling and estimation of road traffic. Section 3 provides a description of the proposed model, followed by an example of its practical application in Sect. 5.

2   State of the Art

Traffic flow measurements often use historical data as the source for volume prediction. Such data is usually insufficient and requires additional effort to make it complete enough for prediction algorithms. Therefore, creation of an exact model of traffic flow is essential.

Literature contains a variety of approaches on how to model traffic flow. One of the attempts is described in [1]. Authors use an origin-destination matrix to express the number of vehicles passing between zones. Link-to-link dividing ratios are calculated using traffic flow data completion methodology upon initially estimated and historical data from San Francisco, CA, USA.

On the other hand, many researchers take advantage of neural networks. As studies [3,6,8] show, artificial intelligence-based approach yields good results when it comes to traffic prediction. Due to the nature of the problem and the usually large volume of time series traffic data, the usage of neural networks seems to be a reasonable choice.

In [2], the traffic flow model is built focusing on the geometry of intersection and positioning of places that are equipped with sensor devices. On the contrary, in our paper we are trying to create a formalised traffic flow graph. Additionally, in [2] the main effort is put upon the discovery of unknown traffic volumes with the use of a neural network.

Authors in [3] take into consideration incident and atypical conditions apart from normal, non-incident conditions due to the fact that traffic flow forecasting is most needed in case of an accident. This is a different approach than dynamic street lighting, which focuses on typical traffic flow levels. However, the study shows that the variety of use cases for traffic flow modeling is very extensive.

In complex street lighting systems, it is crucial to have properly defined formalisms and methods to manage them. Dynamic street lighting control is becoming an important aspect of street lighting systems, which needs proper methodology [11]. Research has proven that introducing dynamic lighting control can reduce energy consumption up to 34% [10]. Dynamic control is possible on streets equipped with traffic intensity measurements. Other research shows, that picking proper time window of measurements can improve energy saving even by 20% [12].

3   Solution Proposal

3.1   General Description

As mentioned in Sect. 2, there are many models of traffic flow modelling. So why create a new model? For dynamic control of road lighting we need to strongly rely on real traffic intensity to make lighting safe for all traffic participants. While other traffic flow models just describe the travel time and delays, our model can be used for calculation of the number of vehicle, based on traffic detector data. In addition, choosing a graph as the formal model makes it easy to integrate with existing dynamic lighting decision systems, such as those using the Control Availability Graph (CAG) [13].

Each node in the Traffic Flow Graph represents a place with deterministic, measurable traffic flow, usually one road lane. Edges represent possible traffic flows between nodes. Node labels identify streets, e.g. nodes $a_1...a_3$ belong to one street. Edge labels are numeric values showing possible division of traffic originating from a given node.

3.2   Traffic Flow Graph Formal Model

Definition 1. *Traffic Flow Graph is a directed graph described by a graph grammar:*

$$\Psi = (V, E, \Sigma, \Upsilon, \delta, \lambda, I, \Theta) \tag{1}$$

where:

- *V is a set of nodes, distinguished by indexing function I,*
- $E \subset V \times \Upsilon \times V$ *is a set of edges,*
- *Σ is a set of node labels,*
- *Υ is a set of edge labels,*
- *$\delta : V \longrightarrow \Sigma$ is the node labeling function,*
- *$\lambda : E \longrightarrow \Upsilon$ is the edge labeling function,*
- *I is a function from V to δ indexing nodes,*
- *Θ is an ordering relation in the set of edge labels,*
- *$e = (a,\ b)$, where $e \subset E$ and $a, b \in V$, understood as a directed edge from a to b.*

and following condition is fulfilled: $V, E \neq \emptyset$.
Nodes represent places with deterministic, measurable traffic flow, edges repre-sent possible traffic flow between nodes. Node labels identify precisely part of street. Node attributes are holding current traffic intensity values. Edge labels carry information about traffic flow division with exact values of that division. This definition is enhanced general graph definition, allowing synchronization with other graph structures.

4   Practical Example

We have chosen one intersection located on one of main streets in a major Polish city as an example. This intersection is part of one of the city's ring roads, so it is significant for cross-city traffic. A diagram showing possible traffic flow is presented in Fig. 1.

4.1   Graph Structure

The graph presented in Fig. 1 shows possible traffic flows in the examined inter-section. As mentioned in Sect. 3.2, each node represents a lane, where edges are possible traffic flows between them. The graph clearly shows the possible traffic flows. The Traffic Flow Graph can support precise calculation of weights repre-senting the division of flows between nodes. Equations describing possible flows and their usage are presented in the following sections.

4.2   Traffic Flow Equations

Traffic flow in the considered intersection can be described by the following equations:

$$e_1 = a_1 \cdot w_1 + f_1$$
$$b_1 = a_1 \cdot (1 - w_1)$$
$$b_2 = a_2 + e_3 \cdot w_4 + f_2 \cdot w_2$$

$$b_3 = a_3 \cdot w_3 + e_3 \cdot (1 - w_4) + f_2 \cdot (1 - w_2)$$
$$d_1 = c_2 + f_3$$
$$d_2 = c_3 + f_4$$
$$d_3 = c_4$$
$$f_1 = c_1 \cdot w_5$$
$$f_2 = c_1 \cdot (1 - w_5)$$
$$f_3 = e_2 \cdot w_6 + a_3 \cdot (1 - w_3) \cdot w_7$$
$$f_4 = e_2 \cdot (1 - w_6) + a_3 \cdot (1 - w_3) \cdot (1 - w_7)$$
$$c_1 = f_1 + f_2$$

Where $a_i, b_i, c_i, d_i, e_i, f_i$ are values of traffic at certain points and w_i are weight values of how traffic is divided.

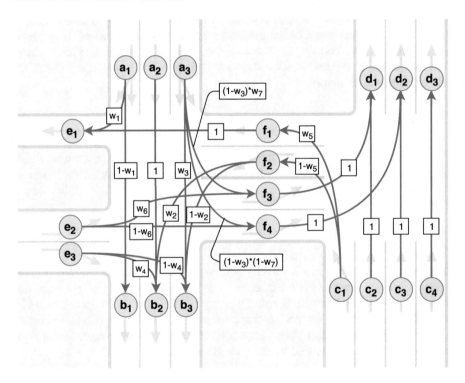

Fig. 1. Intersection model with Traffic Flow Graph elements

5   Application Examples

5.1   Validation Using Historical Data

One simple application of the proposed model is verification of the sensor accuracy using historical data from traffic detectors located on individual lanes. For every tested time period, we have taken traffic values and verified the relation between incoming and outgoing traffic. Since traffic light cycles can affect the results, the validation has been performed using several different time window sizes. Results have been presented in Sect. 6.1.

5.2   Dynamic Control Simulation

Another example is related to estimation of traffic intensity on streets with no sensors. For that purpose, a quasi-blind test has been developed.

We have assumed that the street with lanes marked as e_i in Fig. 1 (further referred to as road e) does not have any traffic sensors. In that street, there are 46 lamps, located on both of its sides.

They are equipped with LED luminaires with following parameters:

- power – 99 W
- luminous flux – 13 721.80 lm
- symmetry – none

Each of them has nominal power equal to 99 W, so every hour, when all of lamps are on without any dimming changes, 4,554 kWh are used. The goal is to enable the use of dynamic lighting control, which is based on real traffic intensity. Knowing the intensity, we are able to change the dimming of lamps, so they use less energy where lighting is still compliant with standards and safe for traffic participants. In this experiment we will perform simulation of traffic intensity prediction and compare it to real-life data. With that calculated traffic intensity, simulation of dynamic lighting control will be performed to assess energy savings.

The main Lighting Class (ME3c) for this road has been chosen in compliance with CEN/TR 13201-1:2004 requirements [4]. This norm allows to lower the lighting class on the road when traffic intensity drops. Photometric calculations have been performed for the main class as well as for lower classes. For every lighting class we have calculated the dimming values so that all photometric requirements are fulfilled with minimum energy consumed by the lamp.

For this scenario, the following performance values have been calculated: the accuracy of prediction compared to real-life data (Sect. 6.2), the accuracy of lighting class selection according to the standards (Sect. 6.3) and the energy savings resulting from the performed calculations (Sect. 6.4).

6   Experiments and Results

6.1   Sensor Reliability Validation

To have enough data and to make model reliable, we have taken traffic measurements from the whole year (06.2015–06.2016). For that period, taking half-hour time periods, the mean error is 15,76%, which shows that model is valuable. For shorter time periods (i.e. 5 min) error rate is much higher.

6.2   Traffic Intensity Calculation from Traffic Flow Graph

Having calculated the weight values from the Traffic Flow Graph equations (described in Sect. 4.2), we are able to calculate traffic values on road e, assuming that there are no traffic detectors. Equations to calculate traffic intensity on road e are as follows:

$$e_1 = a_1 \cdot w_1 + c_1 \cdot w_5$$
$$e_2 = (d_1 - c_2 - a_3 \cdot (1 - w_3) \cdot w_7)/w_6$$
$$e_3 = (b_2 - a_2 - c_1 \cdot (1 - w_5) \cdot w_2)/w_4$$

For each timestamp, we have calculated the expected values of e_i and compared them to real, measured values. The mean error is equal to 14.14%.

6.3   Calculation of Lighting Class Configuration

With calculated predicted traffic on road e, we were able to determine, which lighting class should be applied for each half-hour window. For real traffic, we were able to choose the lighting class with 100% confidence. Then, we calculated the lighting class based on predicted traffic intensity. Precision of such prediction of lighting class is 93% which is a very good result. The accuracy of prediction was obtained by comparing lighting classes determined by an expert with lighting classes which were calculated using approximated traffic flow data from presented model.

6.4   Energy Savings Gained by Our Solutions

Having predicted the lighting class configuration for each time window, we were able to calculate how much energy can be saved by implementing our solution on the street. Without any dynamic control, during the entire year, street lighting on street e uses 16,697 MWh. Introducing dynamic control using real-life data saves energy usage by 15,1%. Our calculated values gave possibility to gain energy savings of 13,8%, which is only 1,3% less than using real-life data. The energy usage without and with control is presented in Fig. 2.

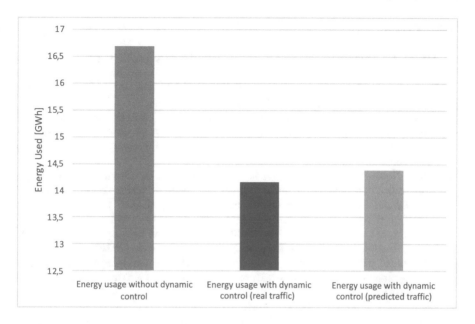

Fig. 2. Energy usage.

7   Conclusions

The paper introduces a formal, graph-based model for vehicle flows – defined as the number of vehicles passing in a given time – in urban road systems. Its goal is to provide a clear representation of points where traffic intensity can be measured and means to define the dependencies between them in the form of equations. Its usefulness manifests itself especially in applications based on the utilisation of the design road capacity (as opposed to those based on travel time).

One of such applications is the area of dynamic road lighting control. The proposed model allows for easy representation of the existing sensor infrastructure and for calculation of probable (or almost certain) traffic flow values to broaden area to which dynamic street lighting control can be applied, which leads to higher energy savings. Verification of the model in a real-life scenario has proven energy savings by 13,8%.

An important issue in this context is the *scalability* of the solution. The presented examples are limited to a small part of the city road network for clarity of presentation. However, in practical situations, the modelled area will often be an entire city or one of its districts. Using a graph formalism (see Sect. 3.2) allows us to develop the model and the transformations on a conceptual level, which does not determine the implementation. As the model itself is theoretically capable of handling an arbitrarily complex road structure, its implementation may utilise horizontally-scalable, distributed systems, such as graph databases.

References

1. Abadi, A., Rajabioun, T., Ioannou, P.A.: Traffic flow prediction for road transportation networks with limited traffic data. IEEE Trans. Intell. Transp. Syst., 1–10 (2014). https://doi.org/10.1109/TITS.2014.2337238, http://ieeexplore.ieee.org/lpdocs/epic03/wrapper.htm?arnumber=6878453

2. Bielecka, M., Bielecki, A., Ernst, S., Wojnicki, I.: Prediction of traffic intensity for dynamic street lighting. In: 2017 Federated Conference on Computer Science and Information Systems (FedCSIS), pp. 1149–1155 (2017). https://doi.org/10.15439/2017F389

3. Castro-Neto, M., Jeong, Y.S., Jeong, M.K., Han, L.D.: Online-SVR for short-term traffic flow prediction under typical and atypical traffic conditions. Expert Syst. Appl. **36**(3), 6164–6173 (2009). https://doi.org/10.1016/j.eswa.2008.07.069. https://linkinghub.elsevier.com/retrieve/pii/S0957417408004740

4. CEN: CEN/TR 13201-1:2004, Road lighting. Selection of lighting classes. Technical report, European Committee for Standardization, Brussels (2004)

5. CEN: CEN/TR 13201-1:2014, Road lighting – Part 1: Guidelines on selection of lighting classes. Technical report, European Committee for Standardization, Brussels (2014)

6. Chan, K.Y., Dillon, T.S., Singh, J., Chang, E.: Neural-network-based models for short-term traffic flow forecasting using a hybrid exponential smoothing and Levenberg-Marquardt algorithm. IEEE Trans. Intell. Transp. Syst. **13**(2), 644–654 (2012). https://doi.org/10.1109/TITS.2011.2174051. http://ieeexplore.ieee.org/document/6088012/

7. European Commission: Green Paper: Lighting the Future. Accelerating the deployment of innovative lighting technologies (2011)

8. Hu, W., Liu, Y., Li, L., Xin, S.: The short-term traffic flow prediction based on neural network. In: 2010 2nd International Conference on Future Computer and Communication, vol. 1, pp. V1-293–V1-296 (2010). https://doi.org/10.1109/ICFCC.2010.5497785

9. Sędziwy, A.: A new approach to street lighting design. LEUKOS **12**(3), 151–162 (2016). https://doi.org/10.1080/15502724.2015.1080122

10. Wojnicki, I., Ernst, S., Kotulski, L.: Economic impact of intelligent dynamic control in urban outdoor lighting. Energies **9**(5), 314 (2016). https://doi.org/10.3390/en9050314

11. Wojnicki, I., Ernst, S., Kotulski, L., Sedziwy, A.: Advanced street lighting control. Expert Syst. Appl. **41**(4), 999–1005 (2014). https://doi.org/10.1016/j.eswa.2013.07.044

12. Wojnicki, I., Kotulski, L.: Empirical study of how traffic intensity detector parameters influence dynamic street lighting energy consumption: a case study in Krakow, Poland. Sustainability **10**(4), 1221 (2018). https://doi.org/10.3390/su10041221

13. Wojnicki, I., Kotulski, L.: Improving control efficiency of dynamic street lighting by utilizing the dual graph grammar concept. Energies **11**(2), 402 (2018). https://doi.org/10.3390/en11020402

A Fuzzy Approach for Evaluation of Reconfiguration Actions After Unwanted Events in the Railway System

Johannes Friedrich[1] and Franciszek J. Restel[2(✉)] (iD)

[1] Technische Universität Berlin, Salzufer 17-19, 10587 Berlin, Germany
JFriedrich@railways.tu-berlin.de
[2] Faculty of Mechanical Engineering, Wroclaw University of Science
and Technology, 27 Wybrzeze Wyspianskiego Str., Wroclaw, Poland
franciszek.restel@pwr.edu.pl

Abstract. Due to infrastructure constraints in railway transportation system, each disruption can have a very important influence on the system operation. Sometimes it is possible to affect the disrupted traffic by dispatching actions. Traffic reconfiguration may lead to a quick system recovery, but on the other hand it may lead also to cascade failures and safety infringement. The examination of a recovery strategy is hard to perform due to parameters that should be taken into account. The structure of timetable is a key factor, that has effect on the transportation system robustness. A large number of possible cases of train scheduling in the timetable makes it currently impossible to deal with this factor. Therefore, the goal of the paper is to show a method which supports the dispatching decision making process. The paper starts with a literature review, that includes the decision making process and the system recovery. Afterwards, important indicators of correct train traffic were identified, such as: number of delayed trains, average train delay, average stop loses and the probability of further disruptions. Due to the challenge to combine the indicators, the fuzzy logic was chosen to build an evaluation method. The membership functions and the decision rules were established together with experts from the railway industry. A case study was performed to show the usability of the method. Finally, a discussion about the further research about the analyzed topic was carried out.

Keywords: Railway · Evaluation method · Reconfiguration

1 Introduction

As a fundamental dilemma in the restriction of railway infrastructure arises the question of possible reconfiguration of disrupted trains. It is to be separated according to the type of unwanted event: is the capacity of the track limited (speed reduction, single-track sections, etc.), or is a track not usable at all. If at least a partial unavailability of the infrastructure occur, than some trains should be theoretically redirected. There are various consequences of a rescheduling - apart from self-effects such as delays due to a

© Springer Nature Switzerland AG 2020
W. Zamojski et al. (Eds.): DepCoS-RELCOMEX 2019, AISC 987, pp. 195–204, 2020.
https://doi.org/10.1007/978-3-030-19501-4_19

longer journey time, there are also external effects on other trains and the use of other infrastructures.

In principle, estimating the resulting effects before the start of a reconfiguration requires scheduling and validation. However, there is no time for this at short-term events, so a manageable alternative needs to be developed. Especially for systems with regular train traffic such as rapid transit systems exist for different incidents graduated incident concepts to possess a tangible disposition concept with suspected or proven feasibility and be able to respond as quickly as possible. For systems with a variety of train implementation and the integration of different train types, especially mixed passenger and freight traffic, other solutions must be found.

Basically, the problem of finding alternatives is based on the fact that timetable studies are carried out for planned timetables. Robustness is either neglected by adding flat-rate margins or by carrying out timetable studies which usually validate results using simulation (cf. Guideline 405 of DB Netz AG).

The relevant methodologies for assessing timetables relate to the degree of infrastructure capacity utilization and the additional delay after an unwanted event. Only selected characteristics and scenarios are taken into account. With the simulation of a large number of cycles it is attempted to generate quasi stationary result and therefore to get valid results. Nevertheless, this procedure can only map a part of the resulting operating situations. However, for any spontaneous reaction such a procedure is neither expedient (it would suffice a single simulation run as soon as the concrete delay data are known) nor resource-efficient (to employ a team of investigators of a railway operational investigation for each incident seems oversized).

In reference to the above, the purpose of the work arises, to elaborate a method for evaluation of reconfiguration scenarios after unwanted events. The method should give results in a short time period to be able to be used as support for the dispatching decision making process.

2   Literature Review

The occurrence of undesirable events in the railway network is linked to the specific subsystems, which at the same time constitute the sources of failures. Mainly three failure sources can be identified: train, infrastructure and environment (external influences). Human failures, technical and organizational failures as well as random events form the set of unwanted events [13, 15]. The failure intensity changes depending on the characteristics of the train, the infrastructure and the environment. Furthermore, the consequences also change. Operational failure consequences can be considered as damage to the system in terms of reliability [3, 14], in example as temporary failures. Since the railway system is repairable, it can be determined that the damage period is finite. In this sense, a longer passenger changeover time can be considered as a system failure as well as a train disaster - with the difference that the consequences of the first event, take five minutes, the second turn a few days.

The combination of system reliability and recovery of traffic is also reflected in the literature [5]. Part of the traffic disruption is compensated by time reserves in the timetable [17, 24]. However, a timetable cannot fully compensate any upcoming delays

[21]. Otherwise this would lead to a massive increase in time reserves, thus a significant reduction of network capacity and significant increase of travel times. Therefore, in addition to the timetable robustness, the resilience of the system must also be considered. This property describes how fast the normal operation of the technical system can be restored [7].

Railway operators use a well-defined set of dispositive methods in case of delayed trains or broken vehicles. As the sheer number of disposition possibilities in each case is way too big to evaluate even parts of the solution set entirely, fixed-rule setups help to find a reasonable solution with a decent outcome without triggering the complexity problem. As for suburban railways and metros, public transit providers tend to develop fixed solutions based on the frequency and scheduled timetable of the running lines for predefined events [4, 10, 20, 22].

Depending on the type of the event the remaining traffic is to be rerouted, held back or the service has to be dismissed partially. Event classes can therefore be used, differing by the impact the events cause, to prepare the most efficient disposition solution. Another advantage of using fixed-rule setups is the ease of the information flow in case of an incident. As all involved teams such as information teams for platforms, electronic timetable adjustments etc. can refer to the fixed-rule setup they know the content and correct amount of information to distribute [12]. Also, personnel and vehicle scheduling can be prepared for these incidents and therefore easily adapt the new conditions [22].

With the EU directive 91/440/EG and its successor, the EU directive 2012/34/EU, railway infrastructure managers are forced to grant non-discriminatory network access to railway operators. In case of major service disruptions, a fixed-rule setup can help to prove that this requirement is met. In cases of mixed traffic routes that hold traffic of different operators and different train types (e.g., long distance trains and suburban railways), fixed-rule solutions can mostly be used as a draft and need to be adjusted manually by controllers [4, 20].

In many cases, best-practice solutions are being developed by controllers as time goes by and similar events happen to occur more often. Controllers know pretty well, how a solution set will work out or not ("historical knowledge"), but especially in open setups of mixed-traffic lines their power is limited as there are far too much possible solutions to be evaluated in the limited time [22, 23, 25, 26].

Robust timetables [6, 11, 17] and dispatching actions after failure occurrence [1, 16], build the core of a good working railway system.

The definition of normal operation is not simple. Basing on the main functional objective of the railway system, normal operation can be considered as the correct realization (in time and space) of the timetable. This means that theoretically all organizational assumptions (e.g., vehicle and personnel planning) can be discarded to achieve the desired stops in the planned time and to allow all connections. The effective restoration of the planned rail operation is largely linked to the correct assessment of the traffic situation and the chosen evaluation criteria [9].

3 Evaluation Qualities

Before dispatching reconfiguration, the evaluation criteria have to be identified. The fundamental assumption was made that a reconfigured train is such a train whose implemented schedule is different than the theoretical one. In other words, a train rescheduled by the dispatcher will be called as reconfigured as well as a delayed train without rescheduling actions.

Due to the operation process, the basic quality is punctuality or delays. It is hard to define borders of cumulated delays, which can be named as good or bad. The problem stems from changing traffic volume. In other words, sixty minutes cumulated delay of ten trains seems to be acceptable, while sixty minutes delay of one train is not acceptable. Therefore the average delay will be used as first quality. The average value is calculated only for the delayed trains, thus not taking into account punctual trains. For the average delay, according to the literature review shown in [19], the delay level which can be named as punctual varies from 2.5 min up to 5 min. A delay from 5 to 15 min can be sometimes accepted while delays higher than 20 min are not acceptable.

The second quality for reconfiguration solution evaluation is the number of missed stops. Such a value will also depend on the traffic volume, therefore the average number of missed stops will be taken into account. The average will be calculated as before only in relation to the reconfigured trains. The most difficult situation in terms of that quality can take place for the regional railway system. It is there possible that a rerouting action can lead to missing of a half of all stops on the course. It means that, a scenario where more than six stops are missed can be seen as bad.

The third quality is the number of reconfigured trains. A good situation is such a one in which the unwanted event changes the schedule of at most two trains. Conditionally can be accepted a number of reconfigured trains not higher than six, while a higher number indicates a worse scenario.

Finally the fourth quality. It describes how probable is a further delay increasing of the reconfigured trains due to their own unwanted events that may occur in the new configuration. Stochastic functions can be used in railway operations to predict the show-up probability of a train for a given moment. To determine operational issues of different trains trying to occupy the same track node the interference probability has to be determined.

The punctuality of trains and the train meeting probability must be set in relation for any specific traffic situation. Thus, not only the punctuality of trains for specific time corridors is needed, but also the probability distribution of the arrivals or departures for any combination of train and infrastructure in time.

The cumulated waiting time of subordinate trains can be used to distinct and assess the different disposition alternatives for a given situation. This waiting time approach is just an approximation of the best solution; as larger-scale operational impacts cannot be directly determined, the solution is not definite. For given few-train setups this method may be used to imply the best solution for the current situation and needs to be evaluated under the condition of resulting larger-scale impacts (e.g., marshalling yard track occupancies that result from the current operation).

For rerouted trains, different reroute alternatives can be evaluated and their resulting system waiting time can be used as benchmark to rate the different solutions. In this case, the functions need to be split into two (or more) different functions (one per route) with links between the functions so that their usage is limited to the time frames where the best result is met. The further delay probability greater than 0.5 can be defined as bad, while from zero to 0.5 as intermediate and conditionally good. A more detailed description of the probability calculation can be found in [2, 8].

4 The Fuzzy Approach

The proposed decision making process bases on solution evaluation in terms of four criteria described in the previous chapter. Each of them is hard to define in terms of reached value as good or not good. Therefore, the combined evaluation of a scenario in terms of all them is hard to perform. A possible way to evaluate such a multi criteria decision making process, is the fuzzy logic approach. For the built fuzzy model, the following input variables were assumed:

X_{RT} – number of reorganized trains,
X_{AD} – average trains delay with reference to the theoretical schedule,
X_{LS} – number of lost stops,
X_{FD} – probability of further delays.

Evaluation of the dispatching scenario Y_{DS} is the linguistic output variable of the model. The general outlook of the model is shown in Fig. 1.

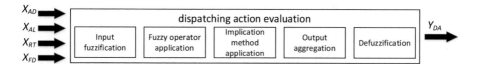

Fig. 1. Concept of the dispatching scenario evaluation model.

A Mamdani fuzzy inference system was built according to basics of the approach [18, 27]. For each input variable and the output variable were assumed membership functions, which describe the linguistic variables: good, intermediate, bad. The function shapes are shown in Fig. 2. Membership functions were established in cooperation with experts from the railway industry and basing on the literature review [19].

The output variable evaluation of dispatching actions was settled by a range from zero to five. Five is the best grade, zero is the worst one. For the output variable four membership functions were established. Two fuzzy sets describe the linguistic evaluation intermediate (P-poor, G-good).

The fuzzy operator is used to define implementing methods of logical operations. The AND and OR operators are used. For the action A AND B the function min (A, B) is used, to find the lowest value of membership function A and B. For the OR operator the maximum membership function value is estimated max (A, B).

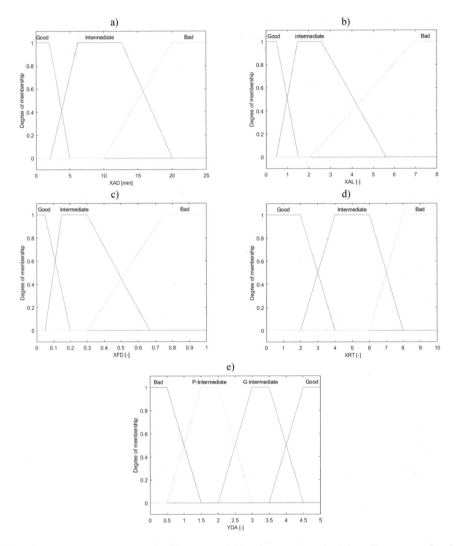

Fig. 2. Membership functions for the input variables: (a) average train delay, (b) average missed stops, (c) further delay of reconfigured trains, (d) number of reconfigured trains, and the output variable (e) dispatching action evaluation.

The implication method application is implemented in two steps. The first one is the rule elaboration. Each rule can be weighted, depending on its importance. In the presented model each rule has the same weight. A set of thirty seven rules was established, that is shown in Table 1.

Basing on the elaborated rules with respect to the input membership functions, the output membership function value is calculated respectively for all rules $\mu_{out}^{R1}(z), \mu_{out}^{R2}(z), \ldots, \mu_{out}^{R37}(z)$. For the implication step, the minimum function was used.

Table 1. Rule set for the elaborated evaluation method; notation of the linguistic variables for the input variables: 1 – "Good", 2 – "Intermediate", 3 – "Bad", 0 – none; notation of the output linguistic variables: 1 – "Good", 2 – "Good-Intermediate", 3 – "Poor-intermediate", 4 – "Bad"

Rule	Operator	X_{AD}	X_{AL}	X_{FD}	X_{RT}	Y_{DA}	Rule	Operator	X_{AD}	X_{AL}	X_{FD}	X_{RT}	Y_{DA}
R1	AND	1	1	1	1	1	R20	AND	2	1	1	1	2
R2	AND	1	1	1	2	2	R21	AND	2	1	1	2	3
R3	AND	1	1	1	3	3	R22	AND	2	1	1	3	4
R4	AND	1	1	2	1	1	R23	AND	2	1	2	1	2
R5	AND	1	1	2	2	2	R24	AND	2	1	2	2	3
R6	AND	1	1	2	3	3	R25	AND	2	1	2	3	4
R7	AND	1	1	3	1	2	R26	AND	2	1	3	1	3
R8	AND	1	1	3	2	3	R27	AND	2	1	3	2	3
R9	AND	1	1	3	3	4	R28	AND	2	1	3	3	4
R10	AND	1	2	1	1	1	R29	AND	2	2	1	1	2
R11	AND	1	2	1	2	2	R30	AND	2	2	1	2	3
R12	AND	1	2	1	3	3	R31	AND	2	2	1	3	4
R13	AND	1	2	2	1	3	R32	AND	2	2	2	1	2
R14	AND	1	2	2	2	3	R33	AND	2	2	2	2	3
R15	AND	1	2	2	3	4	R34	AND	2	2	2	3	4
R16	AND	1	2	3	1	2	R35	AND	2	2	3	1	3
R17	AND	1	2	3	2	3	R36	AND	2	2	3	2	4
R18	AND	1	2	3	3	4	R37	AND	2	2	3	3	4
R19	OR	3	3	0	0	4							

The maximum function was used during the output aggregation for the algebraic connection of all output membership functions.

$$\mu_{out}(z) = max\{\mu_{out}^{R1}(z), \mu_{out}^{R2}(z), \ldots, \mu_{out}^{R8}(z)\} \tag{1}$$

Finally according to the most common used technique, the defuzzification process basses on the centroid estimation of the output membership function.

$$z^* = \frac{\int \mu_{out}(z) \cdot z dz}{\int \mu_{out}(z) dz} \tag{2}$$

The described dispatching action evaluation model was implemented in the MATLAB tool.

5   Case Study

Figure 3 shows an exemplary operating situation. The stations A and B are connected to each other by a double-track main line (via station U) and a branch line (via Z). The branch line has a lower maximum speed and a lower traffic volume. The regional train

RB 101 is subject to disruption on the track and blocks the track. This disturbance has no direct influence on the opposite traffic, but depends on the duration of the following trains on the same track (here the EC 103). The red colour marks delayed trains, and the green one marks rescheduled trains.

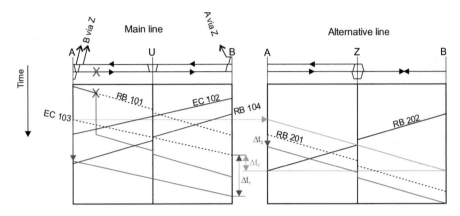

Fig. 3. Traffic situation after an unwanted event with marked scenarios: "no rerouting" and "rerouting to the alternative line".

From the dispatching point of view, there are three possible scenarios: no external intervention, EC 103 rerouting to the track next to the blocked one, EC 103 rerouting to the alternative line. Each one was analysed to calculate the input variable values. The results are shown in Table 2.

Table 2. Scenario evaluation according to the input variable values.

Scenario	X_{AD}	X_{AL}	X_{RT}	X_{FD}	Y_{DA}
No rerouting	29.75	0	2	0.508	0.53
Rerouting to the track next to	14.60	0	3	0.639	1.62
Rerouting to the track next to	9.75	0	4	0.735	1.75
Rerouting to the alternative line	7.60	0.33	3	0.597	2.19

The first solution, EC 103 waits until RB 101 starts the journey again. In this case, EC 103 would be delayed by Δt_1. The next solution, EC 103 can be routed to the track next to the basic one. As consequence EC 102 and/or RB 104 would be delayed.

Finally the built model evaluated the four analyzed solutions. As it was expected, the worst solution is the first one without dispatching actions. Despite the missed station in the fourth solution, it is the best noted due to the lowest average delay, an intermediate probability of further delays and the second lowest number of reconfigured trains.

6   Conclusions

The paper shows a multi-criteria evaluation method for dispatching actions. Four important criteria are used in the presented decision making support method. The delays and missed represent the most important transportation system qualities from the customer's point of view. Both qualities are represented in the model as average values. It allows to compensate the scale effect for systems with different load.

The number of reconfigured trains is an indicator of possible cascade consequences which may occur a few moments after the decision making. Finally the further delay probability allows to analyse how possible are new delays due to a new compilation of rolling stock and infrastructure.

The fuzzy logic model seems to be the best solution for the combination of such different qualities. The presented method allows to analyse effectively possible reconfiguration solutions and the gathered results are promising. Therefore the implementation of the method in the real system will be the further work. Despite that, a next research field in the topic is a fast solution for calculating the input variables in case of complicated timetable structures.

References

1. Acuna-Agost, R., et al.: SAPI: statistical analysis of propagation of incidents. A new approach for rescheduling trains after disruptions. Eur. J. Oper. Res. **215**, 227–243 (2011)
2. Beck, C., Briggs, K.: Modelling train delays with q-exponential functions (2008)
3. Bertsche, B., Lechner, G.: Zuverlässigkeit im Fahrzeug- und Maschinenbau. Springer (2004)
4. Böhme, A., Chu, F., Wolters, A.: Störfallprogramme betrieblich umsetzen. Deine Bahn June 2013, S. 20–25 (2013)
5. Chen, H.-K.: New models for measuring the reliability performance of train service. In: Safety and Reliability. Swets & Zeitlinger, Lisse (2003)
6. Dicembre, A., Ricci, S.: Railway traffic on high density urban corridors: capacity, signalling and timetable. J. Rail Transp. Plan. Manag. **1**, 59–68 (2011)
7. Enjalbert, S., Vanderhaegen, F., Pichon, M., Ouedraogo, K.A., Millot, P.: Assessment of transportation system resilience. In: Human Modelling in Assisted Transportation. Springer (2011)
8. Friedrich, J., Restel, F.J., Wolniewicz, Ł.: Railway operation schedule evaluation with respect to the system robustness. In: Zamojski, W., Mazurkiewicz, J., Sugier, J., Walkowiak, T., Kacprzyk, J. (eds.) Contemporary Complex Systems and Their Dependability. DepCoS-RELCOMEX 2018. Advances in Intelligent Systems and Computing, vol. 761. Springer, Cham (2019)
9. Guze, S.: Business availability indicators of critical infrastructures related to the climate-weather change. In: Zamojski, W., Mazurkiewicz, J., Sugier, J., Walkowiak, T., Kacprzyk, J. (eds.) Contemporary Complex Systems and Their Dependability. Advances in Intelligent Systems and Computing. Springer, vol. 761 (2018). https://doi.org/10.1007/978-3-319-91446-6_24
10. Guze, S.: An application of the selected graph theory domination concepts to transportation networks modelling. Sci. J. Marit. Univ. Szczecin **52**(124), 97–102 (2017)

11. Hansen, I., Pachl, J.: Railway timetable and traffic: analysis-modelling - simulation. Eurailpress, Zagreb Kliewer. A note on the online nature of the railway delay management problem. Networks **57**, 403–412 (2011)

12. Jodejko-Pietruczuk, A, Werbińska-Wojciechowska, S.: Block inspection policy for non-series technical objects. In: Safety, Reliability and Risk Analysis: Beyond the Horizon, Proceedings of 22nd Annual Conference on European Safety and Reliability (ESREL) 2013, Amsterdam 2014, pp. 889–898 (2013)

13. Kierzkowski, A., Kisiel, T.: Simulation model of security control system functioning: a case study of the Wroclaw Airport terminal. J. Air Transp. Manag. (2016). http://dx.doi.org/10.1016/j.jairtraman.2016.09.008

14. Kierzkowski, A.: Method for management of an airport security control system. In: Proceedings of the Institution of Civil Engineers - Transport (2016). http://dx.doi.org/10.1680/jtran.16.00036

15. Kisiel, T., Zak, L., Valis, D: Application of regression function - two areas for technical system operation assessment. In: CLC 2013: Carpathian Logistics Congress - Congress Proceedings, pp. 500–505 (2014). WOS: 000363813400076

16. Kroon, L., Huisman, D.: Algorithmic support for railway disruption management. In: Transitions Towards Sustainable Mobility Part 3. Springer (2011)

17. Liebchen, C., et al.: Computing delay resistant railway timetables. Comput. Oper. Res. **37**, 857–868 (2010)

18. Mamdani, E.H., Assilian, S.: An experiment in linguistic synthesis with a fuzzy logic controller. Int. J. Man Mach. Stud. **7**(1), 1–13 (1975)

19. Restel, F.J.: Reliability and safety models of transportation systems - a literature review. In: Probabilistic Safety Assessment and Management, PSAM 12, s. 1–12 (2014)

20. Schranil, S., Weidmann, U.: Betrieblicher Umgang mit Störereignissen in der Bahnproduktion. Eisenbahningenieur, July 2012, S. 44–48 (2012)

21. Schwanhäußer, W.: Die Bemessung der Pufferzeiten im Fahrplangefüge der Eisenbahn. Schriftenreihe des verkehrswissenschaftlichen Instituts der RWTH Aachen, Heft 20 (1974)

22. Straube, S.: Neue dispositive Konzepte und deren Auswirkungen bei der Ringbahn der S-Bahn Berlin. Technische Universität Berlin (2015)

23. Tubis, A., Werbińska-Wojciechowska, S.: Operational risk assessment in road passenger transport companies performing at Polish market. In: European Safety and Reliability, ESREL 2017, 18–22 June 2017, Portoroz, Slovenia (2017)

24. UIC Code 451-1. In den Fahrplänen vorzusehende Fahrzeitzuschläge, um die pünktliche Betriebsabwicklung zu gewährleisten – Fahrzeitzuschläge

25. Walkowiak, T., Mazurkiewicz, J.: Soft computing approach to discrete transport system management. LNCS. LNAI, vol. 6114, pp. 675–682 (2010). https://doi.org/10.1007/978-3-642-13232-2_83

26. Walkowiak, T., Mazurkiewicz, J.: Analysis of critical situations in discrete transport systems. In: Proceedings of DepCoS - RELCOMEX 2009, Brunów, Poland, 30 June–02 July 2009, pp. 364–371. IEEE (2009). https://doi.org/10.1109/depcos-relcomex.2009.39

27. Zadeh, L.A.: Outline of a new approach to the analysis of complex systems and decision processes. IEEE Trans. Syst. Man Cybern. **3**(1), 28–44 (1973)

On Some Computational and Security Aspects
of the Blom Scheme

Alexander Frolov$^{(\boxtimes)}$

National Research University Moscow Power Engineering Institute,
Krasnoka-zarmennaya, 14, Moscow 111250, Russian Federation
abfrolov@mail.ru

Abstract. We study some computational and security aspects of the (k, m)-Blom scheme. This key pre-distribution scheme is defined by a symmetric polynomial $f(x_1, x_2, \ldots, x_k)$ of degree m in each variable over a field F_p and a set of user identifiers $r_i, r_i \in F_p, i = 1, \ldots, n$. The pre-shared keys that are distributed to users are the coefficients of the polynomials $f(r_i, x_2, \ldots, x_k)$ with distinct users identifiers r_i, $i = 1, \ldots, n$. Implementing Stinson's proof of unconditional security of $(2, m)$-Blom scheme we get the proof of unconditional security of (k, m)-Blom scheme. We establish that for disclosing of the (k, m)-Blom scheme we do not need to use all coefficients of $m + 1$ compromised polynomials. We show that for disclosure, it is enough to use just $\binom{m+k}{m}$ certain coefficients - exactly as many coefficients as there are present in a non-redundant representation of the original polynomial. Additionally, we estimate the number of multiplications in the field F_p sufficient to compute a pre-shared key and the shared key $f(r_i, r_2, \ldots, r_k)$ for communication within a privileged group of k users and to find the initial polynomial as well.

Keywords: Symmetric polynomial over finite field ·
Non-redundant representation · Blom key pre-distribution scheme ·
Disclosing condition · Key calculation complexity

1 Introduction

Blom scheme was the first proposed key pre-distributing scheme for computer networks [1, 2] and it is successfully implemented in contemporary systems especially in sensor wireless computer networks [3–5]. It is defined by a symmetric polynomial $f(x_1, x_2, \ldots, x_k)$ of degree m in each of k variables over a field F_p. The coefficients of the polynomial are kept secret. Distinct n elements $r_1, \ldots, r_n$ of the field F_p are defined as user identifiers. These elements are public knowledge and protected against modification. Each user i knows the secret coefficients of polynomial $f(r_i, x_2, \ldots, x_k), i = 1, \ldots, n$, as a personal pre-shared key. Any k privileged users can compute the shared key $K = f(r_1, r_2, \ldots, r_k)$ to communicate with each other. At the same time, any other m forbidden users, by combining their pre-shared keys $f(r_i', x_2, \ldots, x_k), i = 1, \ldots, m$, cannot obtain

© Springer Nature Switzerland AG 2020
W. Zamojski et al. (Eds.): DepCoS-RELCOMEX 2019, AISC 987, pp. 205–214, 2020.
https://doi.org/10.1007/978-3-030-19501-4_20

any information about the key K. Indeed, one can show, that for each possible key K^* for communication between k privileged users it is possible to define implicitly symmetric polynomial $f^*(x_1, x_2, \ldots, x_k)$, such that $f^*(r_1, r_2, \ldots, r_k) = K^*$ whereas $f^*(r_i', x_2, \ldots, x_k) = f(r_i', x_2, \ldots, x_k), i = 1, \ldots, m$. This polynomial $f^*(x_1, x_2, \ldots, x_k)$ can be defined as follows [6]:

$$
f^*(x_1, x_2, \ldots, x_k) = f(x_1, x_2, \ldots, x_k)
$$
$$
+ (K^* - K) \prod_{1 \leq i \leq m} \frac{(x_1 - r_i')(x_2 - r_i')\ldots(x_k - r_i')}{(r_1 - r_i')(r_2 - r_i')\ldots(r_k - r_i')}
$$

This indicates the unconditional (independent of computing resources) security of the (k, m)-Blom scheme with respect to the compromise of polynomials of any m participants.

But any $m + 1$ users, by combining their pre-shared keys $f(r_i, x_2, \ldots, x_k), i = 1, \ldots, m + 1$ and applying the Lagrange interpolation formula for a polynomial in many variables can calculate the coefficients of the original polynomial $f(x_1, x_2, \ldots, x_k)$ [6]:

$$
f(x_1, x_2, \ldots, x_k) = \sum_{j=1}^{m+1} f(r_j, x_2, \ldots, x_k) \prod_{\substack{i=1 \\ i \neq j}}^{m+1} \frac{x_1 - r_{i+1}}{r_{j+1} - r_{i+1}}.
$$

In this chapter, we use a non-redundant representation of symmetric polynomials $f(x_1, x_2, \ldots, x_k)$, as linear combination of $\binom{m+k}{m}$ homogeneous polynomials containing all monomials of a given type.

We show that to disclose the (k, m)-Blom scheme, it is sufficient to use the $\binom{m+k}{m}$ certain coefficients from the coefficients of the indicated $m + 1$ polynomials. We define also the number of multiplications in the field F_p to be performed to compute pre-shared key $f(r_i, x_2, \ldots, x_k)$ and the shared key $f(r_i, r_2, \ldots, r_k)$ for communication within a privileged group of k users.

The rest of this chapter is organized as follows: In the second section, we consider non-redundant representations of symmetric polynomials $f(x_1, x_2, \ldots, x_k)$ and $f(r_i, x_2, \ldots, x_k)$, $i = 1, \ldots, m + 1$ allowing to reveal the relations between the coefficients of these polynomials. These relations expressed in the form of systems of linear algebraic equations over the field F_p. The third section is devoted to the description and complexity estimation of the (k, m)-Blom scheme disclosure algorithm based on the systems of equations of the previous section and to the complexity estimation for the calculation of both the pre-shared key and the shared key. In the conclusion remarks, we give a summary describing the proposed method for the fast disclosure of a compromised (k, m)-Blom scheme.

2 Polynomials Used in the (k, m)-Blom Scheme

Monomials with the same sets of powers of variables will be called *similar monomials*. It is clear that the coefficients for similar monomials of a symmetric polynomial $f(x_1, x_2, \ldots, x_k)$ of degree m in each of k variables over a field F_p are the same. Let us designate them as $a_{i_1, i_2, \ldots, i_k}$ where the set of indices corresponds to the set of degrees of monomial variables in the non-decreasing order. Such sets of indexes will be considered as *types of monomials* and interpreted as decimal equivalents of numbers in a numeration system with base $m + 1$. We denote homogeneous polynomials containing all monomials of type $(i_1, i_2, \ldots, i_k)$ of a symmetric polynomial $f(x_1, x_2, \ldots, x_k)$ by $p_{i_1, i_2, \ldots, i_k}(x_1, x_2, \ldots, x_k)$. Now this symmetric polynomial can be represented as a linear combination over the field F_p of such homogeneous polynomials, listed within it in order of increasing the decimal equivalents of their indices:

$$f(x_1, x_2, \ldots, x_k) = \sum_{\substack{(0,0,\ldots,0) \\ i_1 \le i_2 \le \ldots \le i_k}}^{(m,m,\ldots,m)} a_{i_1,i_2\ldots i_k} p_{i_1,i_2\ldots i_k}(x_1, x_2, \ldots, x_k). \qquad (1)$$

Such a representation of a symmetric polynomial is non-redundant.

Consider the set of indices of the elements of this linear combination enumerated in mentioned order:

$$A(k, m) = ((0, \ldots, 0), \ldots, (i_1, i_2, \ldots, i_k), \ldots, (m, \ldots, m)).$$

It is easy to check that the length of this set is $\binom{m+k}{m}$. This set $A(k, m)$ is partitioned into $\binom{m+k-1}{m}$ subsets $A_{i_1, i_2, \ldots, i_{k-1}}(k, m)$ indices with the same first $k - 1$ values. Denoted by $S(k, m)$ will be the set of lengths $S_{i_1, i_2, \ldots, i_{k-1}}(k, m)$ of these subsets. It is easy to see that the sum of elements of the set $S(k, m)$ is equal to $\binom{m+k}{m}$.

Example 1. Some concrete numbers and sets are the following ($k = 3$, $m = 2$):

$$\binom{2+3}{2} = 10; \quad \binom{2+3-1}{2} = 6;$$

$A(3,2) = ((0,0,0),\ (0,0,1),\ (0,0,2),\ (0,1,1),\ (0,1,2),\ (0,2,2),\ (1,1,1),\ (1,1,2),\ (1,2,2),$ $(2,2,2));$

$A_{00}(3,2) = ((0,0,0),\ (0,0,1),\ (0,0,2)),\ A_{01}(3,2) = ((0,1,1),\ (0,1,2)),\ A_{02}(3,2) = ((0,2,2)),$ $A_{11}(3,2) = ((1,1,1),\ (1,1,2)),\ A_{12}(3,2) = ((1,2,2)),\ A_{22}(3,2) = ((2,2,2)).$ $S(3,2) = (S_{00}(3,2),\ S_{01}(3,2),\ S_{02}(3,2),\ S_{11}(3,2),\ S_{12}(3,2),\ S_{22}(3,2)) = (3,2,1,2,1,1).$

Next, consider the polynomial

$$f(r, x_2, \ldots, x_k) = \sum_{\substack{(0,0,\ldots,0) \\ i_1 \le i_2 \le \ldots \le i_k}}^{(m,m,\ldots,m)} a_{i_1,i_2\ldots,i_k} P_{i_1,i_2\ldots,i_k}(r, x_2, \ldots, x_k), r \in F_p.$$

It is also symmetric and can be represented as

$$f(r, x_2, \ldots, x_k) = \sum_{\substack{(0,\ldots,0) \\ i_2 \le \ldots \le i_k}}^{(m,\ldots,m)} c_{i_2,\ldots,i_k} P_{i_2\ldots,i_k}(x_2, \ldots, x_k).$$

Equating the coefficients of polynomials we get

$$c_{i_2,\ldots,i_k} P_{i_2\ldots,i_k}(x_2, \ldots, x_k) = \sum_{i_1=0}^{m} a_{i_1,i_2\ldots,i_k} P_{i_1,i_2\ldots,i_k}(r, x_2, \ldots, x_k)$$

$$= \sum_{i_1=0}^{m} a_{i_1,i_2\ldots,i_k} r^{i_1} P_{i_2\ldots,i_k}(x_2, \ldots, x_k).$$

So we obtain the relation between the coefficients of two symmetric polynomials

$$c_{i_2,\ldots,i_k} = \sum_{i_1=0}^{m} a_{i_1,i_2\ldots,i_k} r^{i_1}. \tag{2}$$

$$c_{i_2,\ldots,i_k;j} = \begin{cases} a_{j,i_2,\ldots,i_k}, & \textit{if } j \le i_2, \\ a_{i_2,\ldots,i_t,j,i_{t+1},\ldots,i_k}, & \textit{if } i_t \le j \le i_{t+1} \end{cases}. \tag{3}$$

For given different $m + 1$ elements $r_1, \ldots, r_{m+1}$ of the field F_p and $\binom{m+k-1}{m}$ corresponding coefficients $c^{(i)}_{i_2,\ldots,i_k}$ of polynomials $f(r_i, x_2, \ldots, x_k)$, we obtain $\binom{m+k-1}{m}$ systems of linear algebraic equations (SLAE) for unknown coefficients $c_{i_2\ldots,i_k;j}$ of the original polynomial $f(x_1, x_2, \ldots, x_k)$

$$c^{(i)}_{i_2,\ldots,i_k} = \sum_{j=0}^{m} c_{i_2,\ldots,i_k;j} r_i^j, i = 1, \ldots, m+1. \tag{4}$$

Such SLAE's are of rank $m + 1$, since they are filled with Vandermonde matrices.

3 Blom's Scheme Fast Disclosure Algorithm

Let's consider and solve SLAE (4) in an ascending order of numerical equivalents of index sets $(i_2, \ldots, i_k)$. If we solve them independently of one another, then some of the coefficients of the original polynomial will be calculated many times and it is possible to use some already calculated values to simplify the next SLAE.

The solution of the first SLAE

$$c_{0,\ldots,0}^{(i)} = \sum_{j=0}^{m} c_{0,\ldots,0;j} r_i^j, \, i = 1, \ldots, m+1 \tag{5}$$

of rank $m + 1$ contains the values $\dot{c}_{0,\ldots,0;j} = \dot{a}_{0,\ldots,0,j}, j = 0, \ldots, m$ of $S_{0,\ldots,0}(k, m) = m + 1$ initially unknown coefficients $c_{0,\ldots,0;j} = a_{0,\ldots,0,j}, j = 0, \ldots, m$ with indices from the subset $A_{0,\ldots,0}(k, m)$. One of them $\dot{c}_{0,\ldots,1;0} = \dot{a}_{0,0,\ldots,0}$ replaces the corresponding variable $c_{0,\ldots,1;0}$ of the second SLAE

$$c_{0,\ldots,1}^{(i)} = \sum_{j=0}^{m} c_{0,\ldots,1;j} r_i^j, \, i = 1, \ldots, m+1,$$

that can be simplified decreasing the rank and the number of variables and equations:

$$c_{0,\ldots,1}^{(i)} - \dot{c}_{0,\ldots,1;0} r_i^0 = \sum_{j=1}^{m} c_{0,\ldots,1;j} r_i^j, \, i = 2, \ldots, m+1.$$

Denoted by $C(k, m)$ is the sequence of the index sets $(i_2, \ldots, i_k)$ listed in order of increasing their decimal equivalents.

For simplicity, the elements $c_{i_2 \ldots i_k}$ and $c_{i_2 \ldots i_k;j}$ in (4) will be denoted as c_t and $c_{t;j}$, where $t = 0, 1, \ldots, \binom{m+k-1}{m} - 1$ is the ordinal number of the set $(i_2, \ldots, i_k)$ in the sequence $C(k,m)$. After solving the t-th SLAE, the values of the coefficients with indices from the subsets $A_0(k, m), \ldots A_{t-1}(k, m)$ will be known and the unknown of the $t + 1$-st SLAE will correspond to $S_t(k, m)$ indices from the subset $A_t(k, m)$. Hence, $t + 1$-st SLAE can be simplified up to rank $S_t(k, m)$:

$$c_t^{(i)} - \sum_{j=0}^{m-S_t(k,m)} c_{t;j} r_i^j = \sum_{m-S_t(k,m)+1}^{m} c c_{t;j} r_i^j, \tag{6}$$

$$i = m - S_t(k, m) + 1, \ldots, m+1.$$

Thus, the algorithm for restoring the secret coefficients $a_{i_1,i_2...,i_k}$ of the symmetric polynomial (1) according to the list L containing lists $c^{(i)}$ of $\binom{m+k-1}{m}$ coefficients $c_{i_2...,i_k}^{(i)}$, $i = 0, \ldots, m$, of m + 1 compromised polynomials is the following:

1. Define the sets of indices $A(k, m)$, $C(k, m)$, subsets $A_{t-1}(k, m)$, $t = 0, \ldots,$ $\binom{m+k-1}{m} - 1$, and the set $S(k, m)$ of SLAE ranks.

2. Construct and solve SLAE (5), obtaining the values of secret coefficients with indices from a subset $A_0(k, m)$.

3. For t in range $(1, \binom{m+k-1}{m} - 1)$: using already obtained secret coefficients with indices from the subsets $A_0(k, m), \ldots, A_{t-1}(k, m)$ construct and solve SLAE (6), obtaining values of secret coefficients with indices from a subset $A_t(k, m)$.

It follows that to disclose initial polynomial it is sufficient to use $\binom{m+k}{m}$ equations that correspond to certain coefficients of compromised m + 1 polynomials. To disclose initial polynomial, it is necessary to perform at most $(m^2 - 1) + \sum_{t=0}^{|S(k,n)|} (M(s_t) - (s_0 - s_t)s_t)$ multiplications in the field F_p, where $s_t = S_t(k, m)$, $M(s_t)$ is the number of multiplications for solution of SLAE of rank s_t.

Example. Let $k = 3$, $m = 2$. Let three compromised polynomials be given by list

$$L = [c^{(1)}, c^{(2)}, c^{(3)}]$$
$$= [[c_{00}^{(1)}, c_{01}^{(1)}, c_{02}^{(1)}, c_{11}^{(1)}, c_{02}^{(1)}, c_{22}^{(1)}], [c_{00}^{(2)}, c_{01}^{(2)}, c_{02}^{(2)}, c_{11}^{(2)}, c_{02}^{(2)}, c_{22}^{(2)}], [c_{00}^{(3)}, c_{01}^{(3)}, c_{02}^{(3)}, c_{11}^{(3)}, c_{02}^{(3)}, c_{22}^{(3)}]],$$

and corresponding identifiers of participants be given by list $r = [r_1, r_2, r_3]$.

1. $A = A(k,m) = [[0, 0, 0], [0, 0, 1], [0, 0, 2], [0, 1, 1], [0, 1, 2], [0, 2, 2], [1, 1, 1], [1, 1, 2], [1, 2, 2], [2, 2, 2]]$.
   $C = A(k-1, m) = [[0, 0], [0, 1], [0, 2], [1, 1], [1, 2], [2, 2]]$; $S = S(k,m) = [3, 2, 1, 2, 1, 1]$.

2. The first SLAE is the following:

$$\begin{pmatrix} c_{00}^{(1)} = c_{00;0}r_1^0 + c_{00;1}r_1^1 + c_{00;2}r_1^2 \\ c_{00}^{(2)} = c_{00;0}r_2^0 + c_{00;1}r_2^1 + c_{00;2}r_2^2 \\ c_{00}^{(3)} = c_{00;0}r_3^0 + c_{00;1}r_3^1 + c_{00;2}r_3^2 \end{pmatrix}.$$

Its solution consists of the set of values $(\dot{c}_{00;0}, \dot{c}_{00;1}, \dot{c}_{00;2}) = (\dot{a}_{000}, \dot{a}_{001}, \dot{a}_{002})$ of polynomial (1) coefficients with indices from the set $A_{00}(3,2)$.

3. (a) The second SLAE is the following:

$$\left(\begin{matrix} c_{01}^{(2)} = c_{01;0}r_2^0 + c_{01;1}r_2^1 + c_{01;2}r_2^2 \\ c_{01}^{(3)} = c_{01;0}r_3^0 + c_{01;1}r_3^1 + c_{01;2}r_3^2 \end{matrix} \right)$$

Replacing designations according to (3) and using the result of the previous step we get

$$\left(\begin{matrix} c_{01}^{(2)} - \dot{a}_{001}r_2^0 = a_{011}r_2^1 + a_{012}r_2^2 \\ c_{01}^{(3)} - \dot{a}_{001}r_3^0 = a_{011}r_3^1 + a_{012}r_3^2 \end{matrix} \right)$$

The solution of this SLAE contains the values $\dot{a}_{011}, \dot{a}_{012}$ of the coefficients of the original polynomial with indices from the set $A_{01}(3,2)$.

(b) The third SLAE consists of one equation:

$$\left(c_{02}^{(3)} = c_{02;0}r_3^0 + c_{02;1}r_3^1 + c_{02;2}r_3^2 \right)$$

which is converted to an equation

$$\left(c_{02}^{(3)} - \dot{a}_{002}r_3^0 - \dot{a}_{012}r_3^1 = a_{022}r_3^2 \right)$$

and gives the value $\dot{a}_{022}$ of the coefficient of the original polynomial with the index from $A_{02}(3,2)$.

(c) The fourth SLAE consists of $S_{11}(3,2) = 2$ equations:

$$\left(\begin{matrix} c_{11}^{(2)} = c_{11;0}r_2^0 + c_{11;1}r_2^1 + c_{11;2}r_2^2 \\ c_{11}^{(3)} = c_{11;0}r_3^0 + c_{11;1}r_3^1 + c_{11;2}r_3^2 \end{matrix} \right).$$

Using the already obtained values it can be converted to a SLAE

$$\left(\begin{matrix} c_{11}^{(2)} - \dot{a}_{011}r_2^0 + a_{111}r_2^1 + a_{112}r_2^2 \\ c_{11}^{(3)} - \dot{a}_{011}r_3^0 + a_{111}r_3^1 + a_{112}r_3^2 \end{matrix} \right)$$

and gives the values $\dot{a}_{111}, \dot{a}_{112}$ of the coefficient of the original polynomial with the index from $A_{11}(3,2)$.

(d) The fifth SLAE contains $S_{12}(3,2) = 1$ equation:

$$\left(c_{12}^{(3)} = c_{12;0}r_3^0 + c_{12;1}r_3^1 + c_{12;2}r_3^2 \right)$$

which is converted to an equation

$$\left(c_{12}^{(3)} - \dot{a}_{012}r_3^0 - \dot{a}_{112}r_3^1 = a_{122}r_3^2 \right)$$

and gives the value $\dot{a}_{122}$ of the coefficient of the original polynomial with the index from $A_{12}(3,2)$.

(*e*) The sixth SLAE contains $S_{22}(3,2) = 1$ equation:

$$\left(c_{22}^{(3)} = c_{22;0} r_3^0 + c_{22;1} r_3^1 + c_{22;2} r_3^2 \right)$$

which is converted to an equation

$$\left(c_{22}^{(3)} - \dot{a}_{022} r_3^0 - \dot{a}_{122} r_3^1 = a_{222} r_3^2 \right)$$

and gives the value $\dot{a}_{222}$ of the coefficient of the original polynomial with the index from $A_{22}(3,2)$.

Thus, all coefficients of the original polynomial are restored, while $\binom{2+3)}{2} = 10$ coefficients $c_{00}^{(1)}, c_{00}^{(2)}, c_{00}^{(3)}; c_{01}^{(2)}, c_{01}^{(3)}; c_{02}^{(3)}; c_{11}^{(2)}, c_{11}^{(3)}; c_{12}^{(3)}; c_{22}^{(3)}$ of the compromised polynomials were used.

For numerical illustration, consider this example for a field of small characteristic $p = 127$. Let the original polynomial be given by a list of coefficients

$$[\dot{a}_{000}, \dot{a}_{001}, \dot{a}_{002}, \dot{a}_{011}, \dot{a}_{012}, \dot{a}_{022}, \dot{a}_{111}, \dot{a}_{112}, \dot{a}_{122}, \dot{a}_{222}]$$
$$= [104, 69, 69, 99, 100, 92, 114, 123, 80, 92]$$

and identifiers $r_1 = 1, r_2 = 2, r_2 = 3$ of $m + 1 = 3$ compromised computer network participants are given. Using the pre-calculated degrees

$$\left(r_1^0, r_1^1, r_1^2 \right) = (1, 1, 1), \left(r_2^0, r_2^1, r_2^2 \right) = (1, 2, 4), \left(r_3^0, r_3^1, r_3^2 \right) = (1, 3, 9)$$

of the latter one can calculate the coefficients of the compromised polynomials:

$$L = [\mathbf{c}^{(1)}, c^{(2)}, c^{(3)}]$$
$$= [[115, 14, 7, 82, 49, 10], [10, 32, 2, 57, 31, 112], [43, 123, 54, 24, 46, 17]].$$

The first SLAE looks like this:

$$\begin{pmatrix} 115 = a_{000} \cdot 1 + a_{001} \cdot 1 + a_{002} \cdot 1 \\ 10 = a_{000} \cdot 1 + a_{001} \cdot 2 + a_{002} \cdot 4 \\ 43 = a_{000} \cdot 1 + a_{001} \cdot 3 + a_{002} \cdot 9 \end{pmatrix}.$$

Its solution gives three initial coefficients $\dot{a}_{000} = 104, \dot{a}_{001} = 69, \dot{a}_{002} = 69$.

The solutions of the remaining five SLAEs complement the list of the desired coefficients of the original polynomial.

Concluding this section, let's note that a non-redundant representation of pre-shared key $f(r_i, x_2, \ldots, x_k)$ takes $\binom{m+k-1}{m}$ field elements and computing of this key requires at most $m + m \binom{m+k-1}{m} - 1$ multiplications in the field F_p, whereas computing of shared key requires at most $\sum_{i=2}^{k}(m + m\binom{m+k-i}{m} - 1)$ multiplications in this field.

It was emphasized in the first section, the unconditional (independent of computational resources) security of (k, m)-Blom scheme with respect to the compromise of polynomials of any m participants.

So, in sections one and three the following theorem has been proven:

Theorem. *The (k, m)-Blom scheme is unconditionally secure regarding the compromise of the pre-shared keys of any m users, and for its disclosure it is sufficient to use* $\binom{m+k}{m}$ *certain coefficients from the coefficients of the pre-shared keys of $m + 1$ participants of the alienated group, for example, coefficients with indices from the list A (k,m). Each participant must receive and store a pre-shared key of* $\binom{m+k-1}{m}$ *field elements and perform* $\sum_{i=2}^{k}(m + m\binom{m+k-i}{m} - 1)$ *multiplication operations to calculate a shared key.*

Conclusion remarks

In this chapter, we describe a non-redundant representation of symmetric polynomials of degree m in each of k variable over prime field. We established that for disclosing of (k, m)-Blom scheme it is enough to use just $\binom{m+k}{m}$ coefficients of $m + 1$ compromised polynomials- exactly as many coefficients as there are present in a non-redundant representation of the original polynomial. These coefficients are used to build a sequence of $\binom{m+k-1}{m}$ SLAE of ranks at most $m + 1$ whose sequential solutions are all coefficients of the original polynomial. The union of these SLAE's is a SLAE of rank $\binom{m+k}{m}$. This suggests the following fast method of disclosing a compromised Blom scheme. Given sets of coefficients $c_{i_2,\dots,i_k}^{(i)}$, $i = 0,\dots, m$ of $m + 1$ compromised pre-shared keys, it is necessary to select from each of them the coefficients corresponding to the indices $(i_2, \dots, i_k; i)$ from the corresponding sets $A_{i_2,\dots,i_k}(k, m)$ of the set $A(k, m)$. For the obtained groups of coefficients, one should compile and consistently solve the SLAEs following the algorithm proposed in the third chapter. Additionally, we determined the number of field elements for non-redundant representation of a pre-shared key $f(r_i, x_2, \dots, x_k)$ and estimated the numbers of multiplications in the field F_p sufficient to compute this key and the shared key and to find initial secret polynomial as well.

Acknowledgement. This research has been supported financially by Russian Foundation of Basic Research, project 19-01-00294a. The complexity estimates discussed in this chapter were experimentally confirmed by Natalia Davydova.

References

1. Blom, R.: Nonpublic key distribution. Advances in Cryptology. In: Proceedings of EUROCRYPT'82, pp. 231–236. Plenum. New York (1983)
2. Stinson, D.R.: Cryptography. Theory and Practice, 3rd edn. CRC Press, Boca Raton (2006)
3. Akhbarifar, S., Rahmani, A.: A survey on key pre-distribution schemes for security in wireless sensor networks. Int. J. Comput. Netw. Commun. Secur. 2(12), 423–442 (2014)
4. Li, X., Li, D., Wan, J., Vasilakos, A.V., Lai, C.-F., Wang, S.: A review of industrial wireless networks in the context of Industry 4.0. Wireless Netw. **23**, 23–41 (2017). https://doi.org/10.1007/s11276-015-1133-7
5. Di-Pietro, R., Guarino, S., Verde, N.V., Domingo-Ferrer, J.: Security in wireless ad-hoc networks: a survey. Comput. Commun. **51**, 1–20 (2014)
6. Stinson, D.: Overview of Attack Models and Adversarial Goals for SKDS and KAS. CS 758. http://cgi.di.uoa.gr/~halatsis/Crypto/Bibliografia/Crypto_Lectures/Stinson_lectures/lec09.pdf

Registration and Analysis of a Pilot's Attention Using a Mobile Eyetracking System

Zbigniew Gomolka[1]([⊠]) [iD], Boguslaw Twarog[1] [iD],
Ewa Zeslawska[2] [iD], and Damian Kordos[3]

[1] Faculty of Mathematics and Natural Sciences, University of Rzeszow,
ul. Pigonia 1, 35-959 Rzeszow, Poland
{zgomolka,btwarog}@ur.edu.pl
[2] Faculty of Applied Informatics, Department of Applied Information,
University of Information Technology and Management in Rzeszow,
ul. Sucharskiego 2, 35-225 Rzeszow, Poland
ezeslawska@wsiz.rzeszow.pl
[3] Rzeszow University of Technology,
al. Powstańców Warszawy 8, 35-959 Rzeszów, Poland
d_kordos@prz.edu.pl

Abstract. This paper presents the next stage of research into the registration and analysis of a pilot's attention during the take-off and landing procedures. This is a continuation of the research, which the authors conducted using the static SensoMotoric Instruments (SMI) Red500 eyetracker on a group of pilots applying for a pilot's license. Because static eyetracking introduces a series of restrictions on the properties of the observed scene, a system using Tobii Glasses Pro mobile eye tracker was proposed. Using the mobile system and a training simulator, a measurement stand was prepared for the flight of the aircraft configured as follows: Seneca II airplane with a three-point hidden chassis, location: EPRZ airport - Rzeszów Jasionka. An aviation task was prepared, which was carried out by participants in the experiment. A Tobii project was developed to define several Area of Interests (AOI) for key instruments in the cockpit of the used aircraft. It was observed that for the implementation of an exemplary task, there is no possibility of smoothly modifying coordinates of areas of interest of AOI in subsequent frames of the recorded video stream. Authors designed in the Matlab environment, the Smart Trainer application for a smooth analysis of attention using the characteristic points tracking mechanism. A fuzzy AOI contour is proposed using a modified form of the Butterworth 2D filter for individual instruments, which allows more effective registration of fixations. During the measurement it is possible to observe fixation histograms for a defined set of instruments.

Keywords: Eye tracking · Pilot attention · Tobii Glasses Pro

1 Introduction

Eyetracking research is an area of the modern computing that deals with eye movement measurements in response to visual, auditory or cognitive stimuli. This technique of the registration and analysis of people's attention is used in various areas, i.e. education,

© Springer Nature Switzerland AG 2020
W. Zamojski et al. (Eds.): DepCoS-RELCOMEX 2019, AISC 987, pp. 215–224, 2020.
https://doi.org/10.1007/978-3-030-19501-4_21

sports, entertainment marketing, ergonomics. The most important parameters describing the process of seeing are fixations, saccades, chasing for traffic and the path of sight. Based on the analysis of the emerging trajectory of attitudes, it is possible to draw conclusions regarding the process of perceiving key areas of the analyzed scenes. Modern eyetracking methods are associated with electrooculography, the coil method, the system of light and dark pupils [3, 4, 6–9, 11–14]. The Tobii Glasses Pro mobile eye tracker used in the experimental part of the work uses a method based on a system of dark pupils. The system of dark pupils allows eye tracking through the use of infrared light in the eyes and a high-sensitivity recording camera, which results in very accurate measurements. The research results presented in this paper are a continuation of a research project concerning the pilot's visual activity during key flight and takeoff procedures using a flight simulator. The analysis of obtained results of registered eye attachments of the pilot using the stationary eye tracer SMI RED 500 has been described in [5, 10].

2   Research Methods

In order to carry out research the assumptions defining the problem and information acquisition were introduced in order to facilitate the observations. In the preliminary experimental studies, the following assumptions were made (see Fig. 1): the test stand consists of a mobile eye tracker and instruments for controlling the airplane. The tests were conducted among two groups of people with different aviation experience: IFR PILOT (people who have passed only instrument training) and VFR PILOT with basic training for visibility flights. The flight task was performed on the Seneca II aircraft, and the piloting was only able to use the throttle, control column and control pedals. Analysis of the observed flight attempts was carried out using Tobii Pro Studio and dedicated application to analyze the distribution of saccades and fixations over time in the Matlab environment [1, 2, 15–18].

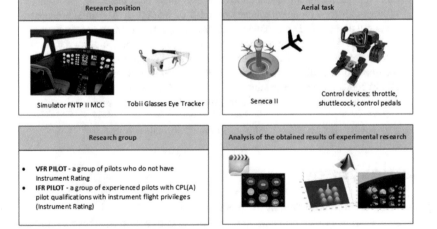

Fig. 1. The assumed research methodology

In the simulation tests, the FNTP II MCC flight simulator was used, which was configured as a twin-engine airplane with a dropped chassis. A mobile eye tracker was used to monitor the scene, allowing for a non-invasive, convenient and quick observations of pilot attention trajectory. Additionally, in order to facilitate the analysis, in the strategic area, due to the aviation task performed, four red markers mounted in the background of the observation area - were introduced. The measuring stand together with the markers are shown in Fig. 2(a) and (b) shows the analyzed on-board instruments used during the flight task.

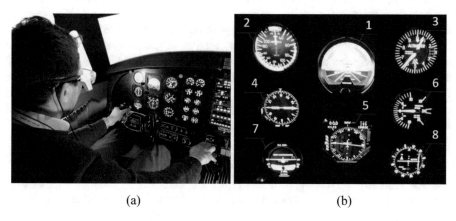

(a)   (b)

Fig. 2. Measuring stand (a) and a view of the cockpit instruments used during the tests (b)

The group of surveyed pilots participating in the experiment had one task to accomplish, which involved making a precise instrumental approach according to the Instrument Landing System (ILS) guidelines, with horizontal and vertical guidance. The approach was performed for visibility along the RVR 550 m runway, with clouds base at the height of 200 ft AGL (Above Ground Level). The flight took place under windless conditions, on a configured Seneca II airplane with a three-point hidden chassis. The flights were made at the EPRZ airport - Rzeszów Jasionka, on the 27th strip with a landing length of 3192 m and a width of 45 m. The elevation of the airport is 693 ft and the magnetic course of the belt was 265°. The pilots started the flight at a distance of 7 nm from the aircraft runway start point and at a barometric altitude of 3000 ft AMSL (Above Medium Sea Level). The task was accomplished when the decision height of DH (Decision Height) equal to 200 ft was reached [1, 2]. The pilots were provided with the indications of the following instruments during the air flight (see Fig. 2(b)): Attitude indicator (1), Speed indicator (2), Altimeter (3), RMI indicator (4), VOR/ILS indicator (5), Vertical speed indicator (6), Turn coordinator (7), CDI indicator (8).

3   Results

Tobii Pro Studio's dedicated environment has the Area Of Interest (AOI) option that allows you to create areas to analyze the number of fixations that appear. The creation and modelling of the moving AOI's for the analyzed video stream allowed to identify a problem that made it impossible to continue the research in its basic form using the Tobii environment. For the initial manually created AOI, the program, even despite the lack of changes in the location of the mobile eyetracker in relation to the subject, made it impossible to track the areas of interest for the interior of the defined instruments. With small changes in the position of the head, the AOIs created from the first frame of the film were not properly aligned with the position of the instruments, as a result of the imprecise AOI position with regard to the motion of the observed instruments, the results with errors were obtained. Therefore, one of the possible solutions was a manual analysis of the film frame with the modification of AOI. Such a solution proved to be extremely time-consuming and inefficient, and the analysis of fixation for eight defined simultaneously areas became impossible. Therefore, a separate analysis was carried out for each of the studied areas, and the results obtained are summarized in Table 1.

Table 1. The obtained number of fixations on the AOI obtained from Tobii Pro Studio

Attitude indicator	Speed indicator	Altimeter	RMI indicator	VOR/ILS indicator	Vertical speed indicator	Turn coordinator	CDI indicator
3293	370	130	8	1437	8	1	1

Due to the inefficient analysis of results in the dedicated Tobii Pro Studio environment, the authors designed the Smart Trainer application in the Matlab environment (see Fig. 3), which allows fixations in the individual areas of pilot's attention to be counted during the performance of a flight task. After loading the frame from the saved video, the user could define markers that enable tracing the coordinates of the simulator cockpit. The application generates characteristic points for markers and starts tracking the attributes for the video stream, which is displayed along with the information about the attention and AOI. Depending on the choice of the tracking method (sharp contour method and the method using the modified Butterworth filter shape), it is possible to generate a histogram with the amount of fixation occurrences in the set AOI areas. In case of using a smoothed mask option, the program also generates 3D mask model and the appropriate cross section of the particular mask.

In order to compare the effectiveness of the designed application, the authors analyzed the same obtained research material with the analogically defined AIO set in the Tobii Pro Studio program. The results obtained are shown in Fig. 4.

The designed application has been used to analyze the received video sequences for individuals in the IFR PILOT group and VFR PILOT respectively. The flight was completed by all persons tested, with varying execution time. As a result of the analysis, the number of fixations was obtained depending on the defined AOI (Table 2).

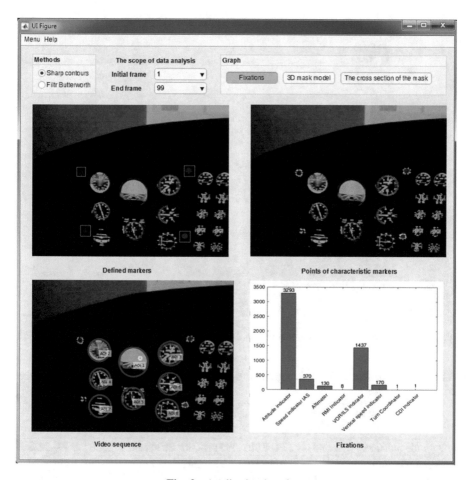

Fig. 3. Application interface

Taking into account different duration times in the recorded materials for the individual experiments and the number of fixations received for particular instruments, an analysis was made of the average percentage share of fixation at the given AOI. The obtained results are summarized in Table 3.

Based on the results obtained, it appears that the most important instruments in the VFR PILOT group were: artificial horizon, VOR/ILS indicator, speedometer, variometer and altimeter. However, in the IFR PILOT Group was an artificial horizon, VOR/ILS indicator, variometer and a high-altimeter. During the analysis of the results, the authors noticed that a significant part of the fixations was located outside the defined areas of AOI (see Fig. 5(a)). Therefore, the obtained results may be subject to errors related to the non-fulfillment of the condition of the fixation being found in the defined area of the defined AOI (see Fig. 5b).

The emergence of such situations is related to a person's movements. The occurring situations and assumptions adopted during the calibration of the eyetracker allowed for

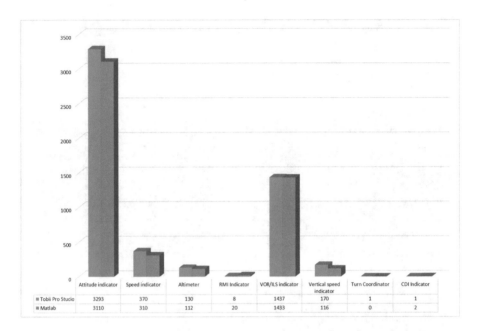

Fig. 4. The number of fixations obtained in Tobii Pro Studio and *Smart Trainer* application

Table 2. The number of fixations registered using the Smart Trainer program, detailing their areas of occurrence

Cockpit instruments	VFR PILOT					IFR PILOT				
	P01	P02	P03	P04	P05	P06	P07	P08	P09	P10
Attitude indicator	4374	532	3110	1751	3030	2230	2489	2250	2370	3333
Speed indicator	198	0	310	22	341	137	180	46	53	24
Altimeter	180	8	112	104	3	159	135	149	26	109
RMI indicator	0	45	20	0	3	6	0	1	0	0
VOR/ILS indicator	1450	962	1433	320	1023	2282	3038	2754	1616	2326
Vertical speed indicator	96	79	116	12	94	423	530	447	29	315
Turn coordinator	0	0	0	0	28	15	0	18	0	0
CDI indicator	1	18	2	0	0	0	0	0	0	0

interpretation of the instruments indications, even in the time when there is a lack of exact appearance of the pilot attention in a given area of AOI. An additional coordinate axis was introduced bringing the layout of the image function to the 3D system. In order to analyze this occurrence of attention, the authors used a modified form of the Butterworth filter:

$$Mask(x, y) = \frac{1.0}{1.0 + c\left[\left((x - x_0)^2 - (y - y_0)^2 / r_0\right)\right]^n} \qquad (1)$$

Table 3. Average percentage of time for the individual instruments

Cockpit instruments	Value (%)	
	VFR PILOT	IFR PILOT
Attitude indicator	61.80	46.75
Speed indicator	3.55	1.59
Altimeter	2.06	2.03
RMI indicator	0.64	0.03
VOR/ILS indicator	29.34	43.46
Vertical speed indicator	2.25	6.03
Turn coordinator	0.12	0.12
CDI indicator	0.23	0.00

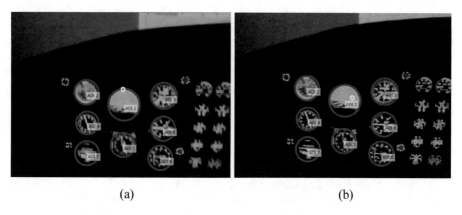

(a)   (b)

Fig. 5. Fixations that do not meet the condition defined (a) and fulfill the condition (b)

where: x_0, y_0 – mask center coordinates, r_0 – instrument mask radius, n – assumed degree of the modified Butterworth filter, $c = 0.414$ – filter constant, for which the profile of the instrument mask achieves the so-called half point. By adopting the above model of constructing individual areas of interest, it was found that their fuzzy contours can enter into mutual coincidence. If a large scale of contour dilution is used, particular areas of interest can simultaneously introduce parallel registration of the attentions. An example of the appearance of such an event is shown in Fig. 6.

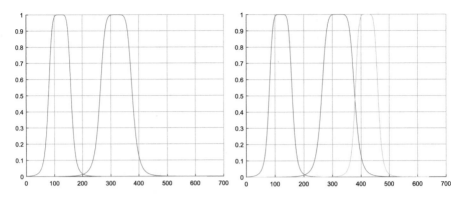

Fig. 6. Possible forms of coincidence of individual AOI areas

Using the above methods of pilot attention analysis, an average percentage share of fixations was obtained for particular pilot groups (see Table 4). Based on the results, it was found that for the task performed, the use of fuzzy gradations of attention local-ization causes the acquisition of more fixations located within the AOI, which results in more accurate results of the analysis of the attributes.

Table 4. Percentage share of fixation for the research methods used

Methods	Fixation	Value (%)	
		VFR PILOT	IFR PILOT
Sharp contours	Inside of the AOI	54.25	77.73
	Outside of the AOI	45.75	22.27
Fuzzy contours	Inside of the AOI	82.18	91.90
	Outside of the AOI	18.82	8.10

4   Conclusion

The experiments presented in this work constitute the preliminary stage of research that will be conducted for pilot students applying for the acquisition of a line pilot license. Based on the experiments carried out in previous studies, the limitations of static eyetracking have been demonstrated. As part of the pre-performed tasks, it was shown that the high complexity of the standard Tobii Pro project makes it impossible to analyze long video sequences, making it practically impossible to use this tool to analyze the attention of real aerial tasks under real conditions. Using the Tobii Glasses Pro mobile system, the authors proposed an application designed to dynamically determine the so-called markers that can be tracked by the special module in order to solve the scenery perspective and find out the coordinates of the instruments being observed during the flight. As part of preliminary measurements, it was shown that there is a group of instruments that play a key role in the course of the aviation task being performed. It has been shown also that a group of experienced pilots acquire better information about the indications of these instruments, which have significant implications for flight safety. A comparison of the histograms of observing the pilot's attention with the use of Tobii Pro and Smart Trainer was made. It was shown that the designed application allows for effective counting of the observer's attention trajectory while not limiting the complexity of the AOI of individual instruments for the observed cockpit. In the presented form, the designed tool will be used to measure the validity and further analysis of the selection of the shape and size of the different AOIs. The designed system will be used as a tool supporting pilots during the implementation of aviation training.

Acknowledgment. The research and experiments were conducted in the Laboratory of Cognitive Science Research, Computer Graphics and Digital Image Processing Laboratory and Real Time Diagnostic Systems Laboratory at the Centre for Innovation and Transfer of Natural Sciences and Engineering Knowledge, University of Rzeszow as a result of EU project "Academic Centre of Innovation and Technical-Natural Knowledge Transfer" based on "Regional Operational Program for Subcarpathian Voivodship for years 2007-2013" Project No. UDA-RPPK.01.03.00-18-001/10-00.

References

1. Aeronautical Information Regulation And Control: AIRAC 1607-EPRZ Airport (2016)
2. Cessna Aircraft Company, Pilot's Operating Handbook for Cessna 172 N Skyhawk, Kansas (1978)
3. Duchowski, A.T.: Eye Tracking Methodology: Theory and Practice. Springer, London (2007)
4. Ellis, K.K.E.: Eye tracking metrics for workload estimation in flight deck operations. MS (Master of Science) thesis, University of Iowa (2009)
5. Gomolka, Z., Twarog, B., Zeslawska, E.: Cognitive investigation on pilot attention during take-offs and landings using flight simulator. In: ICAISC 2017, vol. 10246 (2017)
6. Glaholt, M.G.: Eye tracking in the cockpit: a review of the relationships between eye movements and the aviators cognitive state. Defence Research and Development Canada (2014)
7. Przybyło, J., Kańtoch, E., Augustyniak, P.: Eyetracking-based assessment of affect-related decay of human performance in visual tasks. Future Gener. Comput. Syst. **92**, 504–515 (2019). ISSN 0167-739X
8. Mokatren, M., Kuflik, T., Shimshoni, I.: Exploring the potential of a mobile eye tracker as an intuitive indoor pointing device: a case study in cultural heritage. Future Gener. Comput. Syst. **81**, 528–541 (2018). ISSN 0167-739X
9. Orlosky, J., Toyama, T., Sonntag, D., Kiyokawa, K.: Using eye-gaze and visualization to augment memory: a framework for improving context recognition and recall. In: Streitz, N., Markopoulos, P. (eds.) Proceedings of the 16th International Conference on Human-Computer Interaction. LNCS, vol. 8530, pp. 282–291. Springer (2014)
10. Pawlak, A., Gomółka, P., Kordos, D., Gomółka, Z.: Badanie mózgu pilota podczas lotów na symulatorze. Mechanika, 89 (2/17) (2017). https://doi.org/10.7862/rm.2017.18
11. DeFanti, T.A., Dawe, G., Sandin, D.J., Schulze, J.P., Otto, P., Girado, J., Kuester, F., Smarr, L., Rao, R.: The StarCAVE, a third-generation CAVE and virtual reality OptIPortal. Future Gener. Comput. Syst. **25**(2), 169–178 (2009)

12. Seppelt, B.D., Lee, J.D.: Keeping the driver in the loop: dynamic feedback to support appropriate use of imperfect vehicle control automation. Int. J. Hum.-Comput. Stud. **125**, 66–80 (2019)
13. Causse, M., Lancelot, F., Maillant, J., Behrend, J., Cousy, M., Schneider, N.: Encoding decisions and expertise in the operator's eyes: using eye-tracking as input for system adaptation. Int. J. Hum.-Comput. Stud. **125**, 55–65 (2019)
14. Thomas, P.R.: Performance, characteristics, and error rates of cursor control devices for aircraft cockpit interaction. Int. J. Hum.-Comput. Stud. **109**, 41–53 (2018)
15. Bouzekri, E., Canny, A., Fayollas, C., Martinie, C., Palanque, P., Barboni, E., Deleris, Y., Gris, C.: Engineering issues related to the development of a recommender system in a critical context: application to interactive cockpits. Int. J. Hum.-Comput. Stud. **121**, 122–141 (2019)
16. Gomolka, Z., Twarog, B., Zeslawska, E., Gomolka, P.: Cognitive perception of scenes of complex in the process of research eye-tracking studies. Elektronika **10**, 5–9 (2018)
17. Gomolka, Z., Twarog, B., Krutys, P.: Prediction mechanism for the face tracking algorithm. Przeglad Elektrotechniczny **89**(7), 202–205 (2013)
18. Krutys, P., Gomolka, Z., Twarog, B., Zeslawska, E.: Synchronization of the vector state estimation methods with unmeasurable coordinates for intelligent water quality monitoring systems in the river. J. Hydrol. (2019) https://doi.org/10.1016/j.jhydrol.2019.02.038. ISSN 0022-1694

Minimization Problem Subject to Constraint of Availability in Semi-Markov Reliability Models

Franciszek Grabski[✉]

Chair of Mathematics and Physics, Polish Naval Academy,
Śmidowicza 69, 81-127 Gdynia, Poland
F.Grabski@amw.gdynia.pl

Abstract. Semi-Markov decision processes theory delivers methods which allow to control an operation processes of the systems. Semi-Markov decision processes theory was developed by Jewell [10], Howard [7–9], Main and Osaki [13], Gercbakh [4]. Those processes are also discussed in [5] and [6]. We investigate the infinite duration SM decision processes. The cost minimization problem subject to an availability constraint for the infinite duration Semi-Markov reliability model is discussed in the paper. The problem is transformed on some linear programing minimization problem.

Keywords: Semi-Markov decision processes · Minimization · Linear programing

1 Introduction

Semi-Markov decision processes theory delivers methods which give the opportunity to control an operation processes of the systems. We investigate the infinite duration SM decision processes. Semi-Markov decision processes theory was developed by Jewell [10], Howard [7–9], Main and Osaki [13], Gercbakh [4]. Those processes are also discussed in [5] and [6]. The cost minimization problem subject to an availability constraint for Semi-Markov model of the operation in reliability aspect is discussed in the paper. It should be mention that Boussemart and Limnios [2] presented Markov decision processes with a constraint on the average asymptotic failure rate. In this paper the problem is transformed on some minimization problem of linear programing.

2 Semi-Markov Decision Processes

Semi-Markov decision processes theory delivers methods which give the opportunity to control an operation processes of the systems. In such kind of problems we choose the process that brings the smallest loss among some alternatives available in the operation. We investigate the infinite duration Semi-Markov decision processes in the paper.

W. Zamojski et al. (Eds.): DepCoS-RELCOMEX 2019, AISC 987, pp. 225–234, 2020.
https://doi.org/10.1007/978-3-030-19501-4_22

The cost minimization problem under availability limitation for the infinite duration Semi-Markov reliability model is considered in the paper. The problem is transformed on some linear programing problem.

A number $u_i^{(k)} = m_i^{(k)} r_i^{(k)}$ is an expected value of the cost that is generated by the process in the state i at one interval of its realization for the decision $k \in D_i$. We assume that considered semi-Markov decision process with a finite state space $S = \{1, ..., N\}$ satisfy assumption of the limiting theorem.

The criterion function

$$C(\delta) = \frac{\sum_{j \in S} \pi_j(\delta) u_j^{(k)}}{\sum_{j \in S} \pi_j(\delta) m_j^{(k)}} = \frac{\sum_{j \in S} \pi_j(\delta) m_j^{(k)} r_j^{(k)}}{\sum_{j \in S} \pi_j(\delta) m_j^{(k)}} \tag{1}$$

means an expected value of the cost that is generated by the process in the state i at one interval of its realization for the decision $k \in D_i$ as a result of a long operating system. The numbers $\pi_j(\delta)$, $j \in S$ represent the stationary distribution of the embedded Markov chain of the semi-Markov process defined by the kernel $Q^{(\delta)}(t) = \left[Q_{ij}^{(k)}(t) : t \geq 0, \right.$ $i, j \in S, k \in D_i]$. A number $m_i^{(k)}$ denotes mean of waiting time in state i for decision k. The function

$$K(\delta) = \frac{\sum_{j \in S_+} \pi_j(\delta) m_j^{(k)}}{\sum_{j \in S} \pi_j(\delta) m_j^{(k)}} \tag{2}$$

denotes the limiting availability parameter.

Our aim is finding strategy δ that manimises loss $L(\delta)$ subject to availability constraint $K(\delta) > \alpha$ for given $0 < \alpha \leq 1$.

3   Linear Programing Method

Mine and Osaki [13] presented linear programing method for solving the problem of optimization without additional constraints. The problem of optimization with a system availability constraint is investigate in this paper.

Stationary probabilities $\pi_j(\delta)$, $j \in S$ for every decision $k \in D_i$ satisfy the following linear system of equations

$$\sum_{i \in S} \pi_i(\delta) p_{ij}^{(k)} = \pi_j(\delta), \ j \in S, \ \sum_{i \in S} \pi_i(\delta) = 1, \ \pi_j(\delta) > 0, \ j \in S \tag{3}$$

where

$$p_{ij}^{(k)} = \lim_{t \to \infty} Q_{ij}^{(k)}(t), \quad i, j \in S, \tag{4}$$

Let $a_j^{(k)}$ be a probability that in the state $j \in S$ has been taken decision $k \in D_j$. It is obvious that

$$\sum_{k \in D_j} a_j^{(k)} = 1, \quad 0 \le a_j^{(k)} \le 1, \, j \in S. \tag{5}$$

The criterion function (1) and constrain can be written as

$$C(\delta) = \frac{\sum_{j \in S} \sum_{k \in D_j} a_j^{(k)} \pi_j(\delta) m_j^{(k)} r_j^{(k)}}{\sum_{j \in S} \sum_{k \in D_j} a_j^{(k)} \pi_j(\delta) m_j^{(k)}}, \tag{6}$$

$$K(\delta) = \frac{\sum_{j \in S_+} \sum_{k \in D_j} a_j^{(k)} \pi_j(\delta) m_j^{(k)}}{\sum_{j \in S} \sum_{k \in D_j} a_j^{(k)} \pi_j(\delta) m_j^{(k)}} > \alpha. \tag{7}$$

The Eqs. (3) are equivalent to the system of equations

$$\sum_{i \in S} \sum_{k \in D_i} a_i^{(k)} \pi_i(\delta) p_{ij}^{(k)} = \pi_j(\delta), \quad \sum_{i \in S} \pi_i(\delta) = 1, \pi_j(\delta) \ge 0, \quad j \in S. \tag{8}$$

Substituting

$$a_i^{(k)} \pi_i(\delta) = x_j^{(k)} \ge 0, \quad j \in S, k \in D_j \tag{9}$$

and taking into account that

$$\sum_{k \in D_i} a_i^{(k)} \pi_i(\delta) = \pi_i(\delta) = \sum_{k \in D_i} x_j^{(k)} \ge 0, \quad j \in S, \, k \in D_j, \tag{10}$$

from (5), (6) and (7) we get a following optimization problem

$$\min \left[\frac{\sum_{i \in S} \sum_{k \in D_i} m_i^{(k)} r_i^{(k)} x_i^{(k)}}{\sum_{i \in S} \sum_{k \in D_i} m_i^{(k)} x_i^{(k)}} \right] \tag{11}$$

subject to constraint

$$\frac{\sum_{j \in S_+} \sum_{k \in D_j} m_j^{(k)} x_j^{(k)}}{\sum_{j \in S} \sum_{k \in D_j} m_j^{(k)} x_j^{(k)}} > \alpha, \tag{12}$$

From (8), (9) and (10) we get

$$\sum_{k \in D_j} x_j^{(k)} - \sum_{i \in S} \sum_{k \in D_i} p_{ij}^{(k)} x_i^{(k)} = 0, \quad j \in S \tag{13}$$

$$\sum_{j \in S} \sum_{k \in D_j} x_j^{(k)} = 1, \tag{14}$$

$$x_j^{(k)} \ge 0, \, j \in S, \, k \in D_j \tag{15}$$

Notice that $\sum_{i \in S} \sum_{k \in D_i} m_i^{(k)} x_i^{(k)} > 0$. We introduce new variables

$$y_j^{(k)} = \frac{x_j^{(k)}}{\sum_{i \in S} \sum_{k \in D_i} m_i^{(k)} x_i^{(k)}} \tag{16}$$

$$y = \frac{1}{\sum_{i \in S} \sum_{k \in D_i} m_i^{(k)} x_i^{(k)}} \tag{17}$$

Now the rules (10)–(14) take the form

$$\min \left[\sum_{i \in S} \sum_{k \in D_i} m_i^{(k)} r_i^{(k)} y_i^{(k)} \right], \tag{18}$$

under constraints

$$\sum_{j \in S_+} \sum_{k \in D_j} m_j^{(k)} y_j^{(k)} \geq \alpha, \tag{19}$$

$$\sum_{k \in D_j} y_j^{(k)} - \sum_{i \in S} \sum_{k \in D_i} p_{ij}^{(k)} y_i^{(k)} = 0, \quad j \in S, \tag{20}$$

$$\sum_{j \in S} \sum_{k \in D_j} m_j^{(k)} y_j^{(k)} = 1, \tag{21}$$

$$y_j^{(k)} \geq 0, \quad j \in S, k \in D_j. \tag{22}$$

$$\sum_{j \in S} \sum_{k \in D_j} y_j^{(k)} = y, \tag{23}$$

From theorem 5.5 [13] it follows, that for every $j \in S$ exists exactly one $k \in D_j$ such that $y_j^{(k)} > 0$. Finally we obtain linear programing problem defined by (18)–(22). A constrain (19) is equivalent to inequality

$$\sum_{j \in S_-} \sum_{k \in D_j} m_j^{(k)} y_j^{(k)} \leq 1 - \alpha, \tag{24}$$

where $S_- = S - S_+$ denotes subset of "down" states.
From (15) and (16) we have

$$x_j^{(k)} = y y_j^{(k)} \quad \text{for all } j \in S \text{ and } k \in D_j. \tag{25}$$

The optimal stationary strategy consists of decisions determined by probabilities

$$a_j^{(k)} = \frac{x_j^{(k)}}{\sum_{k \in D_j} x_j^{(k)}} = \frac{y y_j^{(k)}}{\sum_{k \in D_j} y y_j^{(k)}} = \frac{y_j^{(k)}}{\sum_{k \in D_j} y_j^{(k)}}, \tag{26}$$

independent of y.

From theorem 5.5 [13] it follows, that for the optimization problem without subject to constraint of availability (24), for each state $j \in S$ exists exactly one $k \in D_j$ such that $y_j^{(k)} > 0$.

4 Minimization Problem in Semi-Markov Model of the Operation

4.1 Description and Assumptions

The object (device) works by performing the two types of tasks 1 and 2. Duration of task r is a nonnegative random variable $\xi_r, r = 1, 2$ governed by a CDF $F_{\xi_r}(x)$, $x \geq 0$, $r = 1, 2$. Working object may be damaged. Time to failure of the object executing a task r is a nonnegative random variable $\zeta_r, r = 1, 2$ with probability density function $f_{\zeta_r}(x)$, $x \geq 0$, $r = 1, 2$. A repair time of the object performing task r is a nonnegative random variable $\eta_r, r = 1, 2$ governed by probability density function $f_{\eta_r}(x)$, $x \geq 0$, $r = 1, 2$. Each repair is a renewal of the object. After each operation or maintenance the object waits for the next task like new. After completing the task, the renovation is also carried out in a negligible short time. A duration of inspection after task r is a nonnegative random variable γ_r, having PDF $f_{\gamma_r}(x), x \geq 0, r = 1, 2$. After the inspection is completed, the object starts execution of task 1 with the probability p_1 or task 2 with the probability $q_1 = 1 - p_1$. Furthermore we assume that all random variables and their copies are independent and they have the finite and positive second moments.

4.2 Model Construction

We start the construction of the model with the determination of the operation process states.

  1 – control of the object technical condition and renewal after the task executing
  2 – object operation - performing of the task 1
  3 – object operation - performing of the task 2
  4 – repair after failure during executing of the task 1
  5 – repair after failure during executing of the task 2.

To construct a decision stochastic process we have to determine sets of decision (alternatives) for every states.

D_1:   1 – normal inspection after task 1 performing
      2 – expencive inspection after task 1 performing
      3 – normal inspection after task 2 performing
      4 – expencive inspection after task 2 performing
D_2:   1 – normal profit per unit of time for the task 1 executing,
      2 – higher profit for the task 1 executing,
D_3:   1 – normal profit per unit of time for the task 2 executing,
      2 – higher profit for the task 2 executing
D_4:   1 – normal repair after failure during the task 1 executing
      2 – expencive repair after failure during the task 1 executing

D_5:   1 – normal repair after failure during the task 2 executing
        2 – expencive repair after failure during the task 2 executing

Possible state changes of the process appears as Fig. 1.

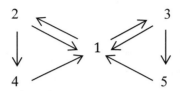

Fig. 1. Possible state changes of the process.

A model of an object operation is a decision semi-Markov process with a state space $S = \{1,2,3,4,5\}$, sets of actions (decisions) D_1, D_2, D_3, D_4, D_5 and the family of kernels

$$
Q^{(\sigma)}(t) = \begin{bmatrix}
0 & Q_{12}^{(k)}(t) & Q_{13}^{(k)}(t) & 0 & 0 \\
Q_{21}^{(k)}(t) & 0 & 0 & Q_{24}^{(k)}(t) & 0 \\
Q_{31}^{(k)}(t) & 0 & 0 & 0 & Q_{35}^{(k)}(t) \\
Q_{41}^{(k)}(t) & 0 & 0 & 0 & 0 \\
Q_{51}^{(k)}(t) & 0 & 0 & 0 & 0
\end{bmatrix}.
$$

The model is constructed if all kernel elements are determined. According to assumptions we calculate elements of the matrix $Q^{(\delta)}(t)$.
Finally we obtain:

$$
Q_{12}^{(k)}(t) = p_1 \int_0^t f_{\gamma_1}(x)dx, \ Q_{13}^{(k)}(t) = q_1 \int_0^t f_{\gamma_1}(x)dx
$$

$$
Q_{21}^{(k)}(t) = \int_0^t \left[1 - F_{\zeta_1^{(k)}}(x)\right]dF_{\xi_1^{(k)}}(x), \ Q_{24}^{(k)}(t) = \int_0^t \left[1 - F_{\xi_1^{(k)}}(x)\right]dF_{\zeta_1^{(k)}}(x)
$$

$$
Q_{31}^{(k)}(t) = \int_0^t \left[1 - F_{\zeta_2^{(k)}}(x)\right]dF_{\xi_2^{(k)}}(x), \ Q_{35}^{(k)}(t) = \int_0^t \left[1 - F_{\xi_2^{(k)}}(x)\right]dF_{\zeta_2^{(k)}}(x)
$$

$$
Q_{41}^{(k)}(t) = F_{\eta_1^{(k)}}(t), Q_{51}^{(k)}(t) = F_{\eta_2^{(k)}}(t).
$$

The transition probability matrix of the embedded Markov chain $\{X(\tau_n) : n \in \mathbb{N}_0\}$ we obtain using well known equality

$$
p_{ij}^{(k)} = \lim_{t \to \infty} Q_{ij}^{(k)}(t) :
$$

$$P^{(\delta)} = \begin{bmatrix} 0 & p_{12}^{(k)} & p_{13}^{(k)} & 0 & 0 \\ p_{21}^{(k)} & 0 & 0 & p_{24}^{(k)} & 0 \\ p_{31}^{(k)} & 0 & 0 & 0 & p_{35}^{(k)} \\ 1 & 0 & 0 & 0 & 0 \\ 1 & 0 & 0 & 0 & 0 \end{bmatrix},$$

$$p_{21}^{(k)} = \int_0^\infty \left[1 - F_{\zeta_1^{(k)}}(x)\right] dF_{\zeta_1^{(k)}}(x), \quad p_{24}^{(k)} = \int_0^\infty \left[1 - F_{\zeta_1^{(k)}}(x)\right] dF_{\zeta_1^{(k)}}(x)$$

$$p_{31}^{(k)} = \int_0^\infty \left[1 - F_{\zeta_2^{(k)}}(x)\right] dF_{\zeta_2^{(k)}}(x), \quad p_{35}^{(k)} = \int_0^\infty \left[1 - F_{\zeta_2^{(k)}}(x)\right] dF_{\zeta_2^{(k)}}(x).$$

4.3   Numerical Example

In our problem we have following decision variables

$$y_1^{(1)}, y_1^{(2)}, y_1^{(3)}, y_1^{(4)}, y_2^{(1)}, y_2^{(2)}, y_3^{(1)}, y_3^{(2)}, y_4^{(1)}, y_4^{(2)}, y_5^{(1)}, y_5^{(2)}$$

Expected durations of one step of operation in the state i for the decision $k \in D_i$ are

$$m_1^{(1)}, m_1^{(2)}, m_1^{(3)}, m_1^{(4)}, m_4^{(2)}, m_3^{(1)}, m_3^{(2)}, m_4^{(1)}, m_4^{(2)}, m_5^{(1)}, m_5^{(2)}.$$

Expected costs of one step of operation in the state i for the decision $k \in D_i$ are

$$u_1^{(1)}, u_1^{(2)}, u_1^{(3)}, u_1^{(4)}, u_2^{(1)}, u_2^{(2)}, u_3^{(1)}, u_3^{(2)}, u_4^{(1)}, u_4^{(2)}, u_5^{(1)}, u_5^{(2)}$$

The criterion function is

$$\begin{aligned} C = & u_1^{(1)}y_1^{(1)} + u_1^{(2)}y_1^{(2)} + u_1^{(3)}y_1^{(3)} + u_1^{(4)}y_1^{(4)} + u_2^{(1)}y_2^{(1)} + u_2^{(2)}y_2^{(2)} \\ & + u_3^{(1)}y_3^{(1)} + u_3^{(2)}y_3^{(2)} + u_4^{(1)}y_4^{(1)} + u_4^{(2)}y_4^{(2)} + u_5^{(1)}y_5^{(1)} + u_5^{(2)}y_5^{(2)} \end{aligned}$$

The constraints take the form (Table 1)

$$u_4^{(1)}y_4^{(1)} + u_4^{(2)}y_4^{(2)} + u_5^{(1)}y_5^{(1)} + u_5^{(2)}y_5^{(2)} \le 1 - \alpha$$

$$y_1^{(1)} + y_1^{(2)} + y_1^{(3)} + y_1^{(4)} - \left(p_{21}^{(1)}y_2^{(1)} + p_{21}^{(2)}y_2^{(2)} + p_{31}^{(1)}y_3^{(1)} + p_{31}^{(2)}y_3^{(2)} + y_4^{(1)} + y_4^{(2)} + y_5^{(1)} + y_5^{(1)}\right) = 0,$$

$$y_2^{(1)} + y_2^{(2)} - \left(p_{12}^{(1)}y_1^{(1)} + p_{12}^{(2)}y_1^{(2)} + p_{12}^{(3)}y_1^{(3)} + p_{12}^{(4)}y_1^{(4)}\right) = 0,$$

$$y_3^{(1)} + y_3^{(2)} - \left(p_{13}^{(1)}y_1^{(1)} + p_{13}^{(2)}y_1^{(2)} + p_{13}^{(3)}y_1^{(3)} + p_{13}^{(4)}y_1^{(4)}\right) = 0,$$

$$y_4^{(1)} + y_4^{(2)} - (p_{24}^{(1)}y_2^{(1)} + p_{24}^{(2)}y_2^{(2)}) = 0,$$

$$y_5^{(1)} + y_5^{(2)} - (p_{35}^{(1)}y_3^{(1)} + p_{35}^{(2)}y_3^{(2)}) = 0,$$

$$m_1^{(1)}y_1^{(1)} + m_1^{(2)}y_1^{(2)} + m_1^{(3)}y_1^{(3)} + m_1^{(4)}y_1^{(4)} + m_2^{(1)}y_2^{(1)} + m_2^{(2)}y_2^{(2)} + m_3^{(1)}y_3^{(1)} + m_3^{(2)}y_3^{(2)} + m_4^{(1)}y_4^{(1)}$$

$$+ m_4^{(2)}y_4^{(2)}m_5^{(1)}y_5^{(1)} + m_5^{(2)}y_5^{(2)} = 1,$$

$$y_i^{(k)} \geq 0, \ i = 1,2,\ldots,5, \quad k = 1,2,3,4.$$

We assume

$$\alpha = \alpha = \mathbf{0.941}.$$

Using MATHEMATICA computer system we obtain solution of the problem

Table 1. The transition probabilities and cost parameters of the SM decision process

State i	Decision k	$p_{i1}^{(k)}$	$p_{i2}^{(k)}$	$p_{i3}^{(k)}$	$p_{i4}^{(k)}$	$p_{i5}^{(k)}$	$m_i^{(k)}$	$r_i^{(k)}$	$u_i^{(k)}$
1	1	0	0.76	0.24	0	0	2.5	120	300
	2	0	0.35	0.65	0	0	3	150	450
	3	0	0.80	0.20	0	0	3	130	390
	4	0	0.28	0.72	0	0	3.5	150	525
2	1	0.99	0	0	0.01	0	10.2	−250	−2550
	2	0.98	0	0	0.02	0	9.4	−380	−3572
3	1	0.98	0	0	0	0.02	9.2	−340	−3128
	2	0.96	0	0	0	0.04	8.8	−460	−4048
4	1	1	0	0	0	0	72	120	8640
	2	1	0	0	0	0	66	135	8910
5	1	1	0		0	0	80	122	9760
	2	1	0		0	0	70	140	9800

$$y_1^{(1)} = 0.008125, \quad y_1^{(2)} = 0,$$
$$y_1^{(3)} = 0.064596, \quad y_1^{(4)} = 0,$$
$$y_2^{(1)} = 0.057852, \quad y_2^{(2)} = 0,$$
$$y_3^{(1)} = 0.014869, \quad y_3^{(2)} = 0,$$
$$y_4^{(1)} = 0, \quad\quad\quad\ y_4^{(2)} = 0.0005785,$$
$$y_5^{(1)} = 0, \quad\quad\quad\ y_5^{(2)} = 0.000297,$$

From (26) we obtain probabilities

$$a_1^{(1)} = \mathbf{0.000908}, \quad a_1^{(2)} = 0$$
$$a_1^{(3)} = \mathbf{0.999092}, \quad a_1^{(4)} = 0$$

$$a_2^{(1)} = \mathbf{1}, \quad a_2^{(2)} = 0, \quad a_3^{(1)} = \mathbf{1}, \quad a_3^{(2)} = 0,$$
$$a_4^{(1)} = 0, \quad a_4^{(2)} = \mathbf{1}, \quad a_5^{(1)} = 0, \quad a_5^{(2)} = \mathbf{1},$$

The vector of optimal action in each step is

$$(\gamma, 1, 1, 2, 2),$$

where

$$\gamma = \begin{cases} 1 & \text{with prob.} \quad 0.000908, \\ 3 & \text{with prob.} \quad 0.999092 \end{cases}$$

Practically optimal decision vector has the form

$$(3, 1, 1, 2, 2).$$

It should be added that for the availability parameter $\alpha \geq \mathbf{0.942}$ **no solution** subject to above constraint.

The minimum expected cost of one step of the operation in this case is

$$\begin{aligned} C &= 300 * 0.008125 + 390 * 0.000908 - 2550 * 0.057852 - 3128 \\ &\quad * 0.014869 + 8910 * 0.0005785 + 9800 * 0.000297 \cong -\mathbf{183.278}. \end{aligned}$$

This means that **the object operation brings profit.**

5   Conclusion

Semi-Markov decision processes theory provides the possibility to formulate and solve the optimization problems that can be modelled by SM processes. In such kind of problems we choose the process that brings the lowest cost or the largest profit among some decisions available for the operation. If the semi-Markov process describing the evolution of the real system in a long time satisfies the assumptions of the limit theorem, we can use the results of the infinite duration SM decision processes theory. The presented algorithm that allows to find the best strategy is equivalent to the special problem of linear programing.

From theorem 5.5 [13] for problem without additional constraint it follows, that for every $j \in S$ exists **exactly one** $k \in D_j$ such that $y_j^{(k)} > 0$. For the cost minimization problem subject to constraint of availability this theorem **is not true**. The optimal stationary strategy can consist of the vectors containing mixed decisions. This fact extends the previously known results.

References

1. Barlow, R.E., Proschan, F.: Mathematical Theory of Reliability. Wiley, New York (1965)
2. Boussemart, M., Limnios, N.: Markov decision processes with a constraint on the average asymptotic failure rate. Commun. Stat. Theory Methods **33**(7), 1689–1714 (2004)
3. Feller, W.: An Introduction to Probability Theory and Its Applications, vol. 2. Wiley, New York (1966)
4. Gertsbakh, I.B.: Models of Preventive Service. Sovetskoe radio, Moscow (1969). (in Russian)
5. Grabski, F.: Decision problem for infinite duration Semi-Markov process. J. Pol. Saf. Reliab. Assoc. Summer Saf. Reliab. Semin. **5**(1–2) (2014)
6. Grabski, F.: Semi-Markov Processes: Applications in Systems Reliability and Maintenance. Elsevier, Amsterdam (2014). (in preparation)
7. Howard, R.A.: Dynamic Programing and Markov Processes. MIT Press, Cambridge (1960)
8. Howard, R.A.: Research of semi-Markovian decision structures. J. Oper. Res. Soc. Jpn. **6**, 163–199 (1964)
9. Howard, R.A.: Dynamic probabilistic system. In: Semi-Markov and Decision Processes, vol II. Wiley, New York (1971)
10. Jewell, W.S.: Markov-Renew. Program. Oper. Res. **11**, 938–971 (1963)
11. Korolyuk, V.S., Turbin, A.F.: Semi-Markov Processes and Their Applications. Naukova Dumka, Kiev (1976). (in Russian)
12. Limnios, N., Oprisan, G.: Semi-Markov Processes and Reliability. Birkhauser, Boston (2001)
13. Mine, H., Osaki, S.: Markovian Decision Processes. AEPCI, New York (1970)
14. Silvestrov, D.C.: Semi-Markov Processes with a Discrete State Space. Sovetskoe Radio, Moscow (1980). (in Russian)

Method for Railway Timetable Evaluation in Terms of Random Infrastructure Load

Szymon Haładyn, Franciszek J. Restel$^{(\boxtimes)}$ (ID),
and Łukasz Wolniewicz (ID)

Faculty of Mechanical Engineering, Wroclaw University of Science
and Technology, 27 Wybrzeze Wyspianskiego Str., Wroclaw, Poland
222908@student.pwr.edu.pl, {franciszek.restel,
lukasz.wolniewicz}@pwr.edu.pl

Abstract. The traffic schedule is planned to keep distances between trains which allow a conflict-free train operation. For this reason the load of the power supply system is distributed in time. In consequence, the power limit is not exceeded. After unwanted event occurring, the traffic structure may change in such a way that an accumulation of section load can occur. As a result, the power limit can be exceeded and the electricity system may be damaged. Therefore, the main goal of the paper is to elaborate a method for railway timetable evaluation due to random power infrastructure load, caused by undesirable events. To reach the goal, a literature review was performed in terms of timetable evaluation, buffer times, undesirable events, system resilience and robustness. A method based on transportation mean movement theory with probability issues was proposed. The method was verified on a timetable for a selected railway line in Poland. Tractive calculations in terms of the electric power consumption were carried out. Using the method, critical situations were identified and changes were proposed. The paper ends with a discussion of the method and further research.

Keywords: Railway infrastructure · Timetable · Undesirable events

1 Introduction

Railway lines with a high capacity usage are often equipped with electricity supply systems, which transfer the energy to electric vehicles through the contact line. The maximum load of these devices is limited. Exceeding of the limit can lead to temporary unavailability, long lasting failures or accelerated degradation.

The train timetable is planned to keep distances between trains which allow a conflict-free train operation. For this reason the load of the power supply system is distributed in time. In consequence, the power limit is theoretically not exceeded. After unwanted event occurring, the traffic structure may change in such a way that an accumulation of section load can occur. As a result, the power limit can be exceeded and the named consequences may occur. Starting from this, the main goal of the paper is to elaborate a method for railway timetable evaluation due to random power infrastructure load, caused by undesirable events. The proposed bases on motion theory and a probabilistic approach which will allow to find the probability of electric device overload.

© Springer Nature Switzerland AG 2020
W. Zamojski et al. (Eds.): DepCoS-RELCOMEX 2019, AISC 987, pp. 235–244, 2020.
https://doi.org/10.1007/978-3-030-19501-4_23

2   Literature Review

According to [4, 5, 23], a flexible timetable structure allows to absorb train delays, which is a key issue in the timetable evaluation process. In the papers [3, 4, 9, 19] buffer times and rearrangement of train sequence are used in a flexible timetabling. The articles [3, 8, 10, 13, 16] include problems of long-term strategic decision-making process in passenger railway and detailed planning of operations. Operational research methods have an increasing role in the planning process. Recently, more attention has been paid to the resilience of planning issues. In the context of the railway system resilience, it is often defined as the ability to continue operations at a certain level in the event of disturbances such as delays or failures. This has led to the need to pay attention to ways of ensuring continuity of operation after disturbance occurrence. In the approach of integrated routing and passenger exchange at the station, the basic timetable measure [3] is defined as WTTE (Weighted travel time extension). *WTTE* is the ratio of realised passenger travel time and nominal travel time.

Another approach described in [11] includes the evaluation of buffer times in the timetable. The time buffer included in the operation schedule is disadvantageous for passengers. However it does increase the system robustness to operational disturbances. Therefore, a *WAD* (Weighted Average Distance) coefficient has been introduced. It is the weighted average distance of buffer times added to the travel time. This measure is intended to indicate the extent to which the buffer affects the train access to the final station and departure from the first station. If there are $N + 1$ trips on a line, from trip $t = 0$ to trip $t = N$, and each trip has supplement time (s_t). In this case, a *WAD* value of 0 would mean that all buffer is assigned to trip t = 0 and a WAD value of 1 would mean that all buffer is allocated to trip t = N, and *WAD* of 0.5 would mean that the buffer is equal for both, the first and second half of the trip. In other cases, the authors of the article [11] use the revised *WAD* index to calculate the relative distance to the execution margin from the beginning of the line to take the allocation. Dividing the line into N sections and letting s_t denote the amount of margin related to section t.

Studies in the article [18] show the robustness evaluation at a critical point. This article presents the application of the optimization method: ex-post evaluation by means of a microscopic simulation. The aim is to understand the increase of the Robustness at critical point (*RCP*) to a certain level within the schedule for the pairs of trains analysed. A case study is presented in which the initial schedule and schedule with increased RCP are reviewed. Articles [17, 20–22] present risk evaluation approaches. They can be used for timetable critical points.

The article [1] presents the results of experimental studies, in which the benchmark of critical points was made. The conclusion was, that attention should be paid to timetable points where trains enter the railway line or overtake. From the robustness point of view, the time difference between critical point is important. It is also proposed to delay trains that are already delayed by extending station stops in order to allow the remaining trains to run on time. The measures of resilience of train schedules can be divided into two groups: ex-ante measures related to timetable characteristics and ex-post measures which are based on rail line operational conditions (see also [6, 7]).

During the literature research it was found, that the randomness of load distribution is not analysed for the timetable designing process.

3   Deterministic Energy Load

Sometimes propagation of delays in rail system cannot be reduced due to limitation of maximum value of amperage charged from contact line. That kind of situation may occur especially in systems working with 3000 V DC voltage (applied in Poland) where rail carriers use vehicles with engines assuring large accelerations. Facility of reducing delays in that systems have to be checked by execution of tractive calculations reliant on the equation of motion solution [12, 14, 15, 24]:

$$F = m \cdot k \cdot \frac{d^2x}{dt^2} \tag{1}$$

where: F – resultant force, m – mass; k – coefficient of swirling mass.

$$F(v, S) = Z(v) - W(v) - I(S) - \text{for accelerating} \tag{2}$$

$$F(v, S) = B(v) + W(v) + I(S) - \text{for braking} \tag{3}$$

$$F(v, S) = W(v) + I(S) - \text{for ride with constant velocity} \tag{4}$$

Where: Z – tractive effort, B – braking force, W – movement resistance force and I – resistance forces dependent from tenor of railway line.

Equation of motion (1) can be noted as:

$$S = m \cdot k \int_{v_1}^{v_2} \frac{v dv}{F} \tag{5}$$

$$t = m \cdot k \int_{v_1}^{v_2} \frac{dv}{F} \tag{6}$$

We also know that:

$$P = F \cdot \frac{dx}{dt} \tag{7}$$

And:

$$P = U \cdot I \tag{8}$$

After transformation of Eq. (8) and subrogation of Eqs. (1) and (7), the formula for train run amperage can be obtained:

$$I = \frac{m \cdot k \cdot \frac{d^2x}{dt^2} \cdot \frac{dx}{dt}}{U} \, [A] \qquad (9)$$

The railway line is divided into sections of power units. Each section has one-side powering by electricity substation. In case of powering by 3000 V DC voltage, one section involves up to 30 km of the railroad [2].

Whereas it is necessary to take into account voltage drops on overhead line. For DC voltage, the voltage drop is directly proportional to the distance between supplying point and train position on the railroad (Fig. 1).

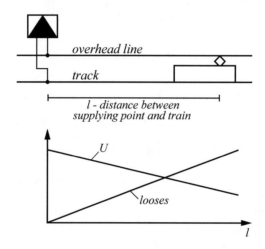

Fig. 1. Voltage changing depending on distance between supplying point and train.

In order to preserve equal power of train for dropping values of voltage, load should raise in accordance with Eq. (7). It also must be taken into account in the formula (9).

4   Description of a System in the Aspect of Infrastructure Load

For a selected railway line in Poland, basing on timetable were identified the most heavily loaded places. The route has 15 stations or stops. The timetable for a working day consists of 119 trains (including 22 freight trains). The shape of the selected railway line in terms of power sections is featured in Table 1.

At the same time in one section may be located several trains, which may accelerate, move with constant velocity, decelerate or stay in a station. Due to the shape of typical curve of tractive effort, different stages of movement determine sundry amperage charged from the contact line. In general, accelerating train requires the most amperage. In first stage of acceleration the tractive effort is constant, lower than

Table 1. Composition of power sections for the selected railway line

Section	Start point [km]	End point [km]	Stations placed in section
A	0.00	8.01	A, B
B	8.01	30.83	C, D, E, F
C	30.83	49.87	G, H
D	49.87	71.43	I, J, K, L
E	71.43	82.14	M, N, O

resultant from Eq. (7). Amperage increases up to passing the area of constant power hyperbola. At this stage amperage is the biggest and accomplishes maximal values (Fig. 2). When train moves with constant velocity, value of resultant force is equal to sum of movement resistance force and resistance forces dependent from gradient of railway line and it is smaller than during acceleration. Train uses least energy when it is not in movement or when breaks are activated.

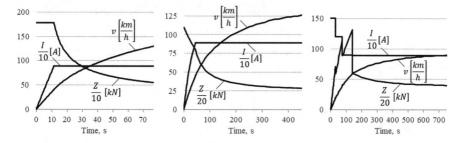

Fig. 2. Tractive effort, amperage charged from contact line and velocity of: short-distance passenger train, long-distance passenger train and freight train depending of time (for $U = 3000$ V DC and $I(S) = 0$)

Entire demand for amperage at section varies in time and it is sum of several uptakes:

$$I(t) = \sum_{i=1}^{n} I_i(t)[A] \tag{10}$$

Estimating of amperage at section requires to count wastage and other consumption [15], so value from Eq. (10) ought to be escalate by additional indicators:

$$I_{ent} = I \cdot \frac{1}{\eta_1} \cdot \frac{1}{\eta_2} \cdot \frac{1}{\eta_3} \cdot \frac{1}{\eta_4} \cdot \frac{1}{\eta_5}[A] \tag{11}$$

Where: η_1 – efficiency counted other energy consumption: to power apparatus located in locomotive or cars (0.96), η_2 – efficiency counting train's interior heating (0.93), η_3 – efficiency of contact line (0.91), η_4 – efficiency counted aberrant movement work (0.98), η_5 – efficiency of electricity substation (0.94).

Analyses of traffic volume based on timetables in each power section gives information about the allocation of increased energy request in time and space.

A probability analysis of each moment in time, for all train combinations will be time-consuming and for some situations np-hard. Therefore, traffic situations were found, when the load is highly increased. The traffic situations were modelled using the MATLAB Simulink package. The results are detailed in Table 2.

Table 2. Biggest strain of rail power grid in sections

Number of trains on section	Section	Number of traffic situation (Time of occurrence)
5		1 (13:55–13:56), 2 (16:51), 3 (17:04–17.06)
4	B	4 (9:54–10:01), 5 (13:57–14:01), 6 (15:05–15:06) 7 (16:52–17:03), 8 (17:07–17:12), 9 (18:54 – 18:59)
	C	10 (5:40–5:43), 11 (12:43), 12 (13:39–13:43), 13 (17:43)
	D	14 (6:10), 15 (15:37–15:40), 16 (17:58–17:59)

The identified traffic situations were also presented in Fig. 3. The red line shows the load limit (in Poland 4 kA [2]).

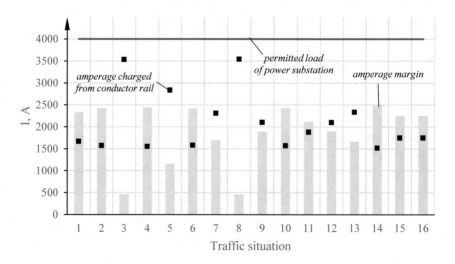

Fig. 3. Amperage charged from contact line for the identified traffic situations

The identification of the traffic situation simplifies the further probabilistic analysis to a finite set of possible scenarios.

5   Random Infrastructure Loads and Timetable Evaluation

In fact each traffic situation can escalate if one of two events will take place. The first, one or more earlier trains will be delayed and their acceleration will overlap with the given traffic situation. The second, the whole traffic situation will move for later, and due to that it will overlap with following trains. For that reason, delays of trains have been analysed for four months.

The timetable includes freight, regional and long-distance passenger trains. Thus, this three types were analysed in terms of punctuality. After the punctuality analysis were fitted probability density functions for each train type. The goodness of fit was verified by the Chi-squared test at the significance level 0.05.

It was found, the probability for passenger trains can be described by the two-parameter exponential distribution. The second parameter moves the function to the left, beside zero. Therefore the probability of punctual arrivals is taken into account by calculation of the integral for zero. For freight trains was the normal distribution fitted. The integral of the function from minus infinity to zero represents the probability of punctual train arrivals. The probability density functions are shown in Table 3, where were also placed the maximum amperages and their duration during acceleration.

Table 3. Train type characteristics

Train type	Probability density function for deviation from scheduled time	Max. amperage [A]	Load duration [s]
Passenger – long distance	$f(x) = 0.11 \cdot \exp(-0.11 \cdot (x + 12))$	890	418
Passenger – short distance	$f(x) = 0.19 \cdot \exp(-0.19 \cdot (x + 9))$	890	130
Freight train	$f(x) = \frac{1}{24 \cdot \sqrt{2 \cdot \pi}} \cdot \exp\left(\frac{-(x + 9)^2}{1152}\right)$	1050	363

Combining the knowledge from Tables 2 and 3 with the timetable, the overload probability for each traffic situation can be estimated. It has to be analysed how many more trains should overlap one section to cause an overload. For the situations three and eight only one more train will be enough to induce the overload. For the remaining situations two or more than two extra trains. Having the information about delay scenarios which lead to an overload for a given traffic situation, the Boolean algebra can be used to estimate the probability that one of the scenarios will occur. The results for the analysed case are shown in Fig. 4.

The highest probabilities, for the third, fifth and eight traffic situation, result from a close allocation of freight trains in relation to passenger train accumulations in the timetable. For the situations three and eight there is also more than one long-distance passenger train in the accumulation. Knowing the traffic situation overload probability, the section overload probability can be calculated from the formula:

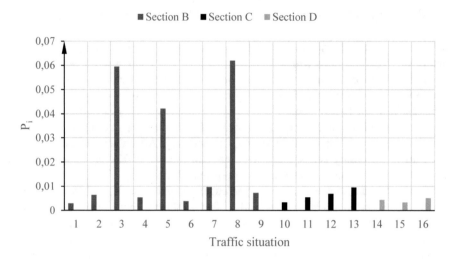

Fig. 4. Overload probability P_i for the identified traffic situations

$$P_{over} = 1 - \prod_{i=1}^{n} (1 - P_i) \tag{12}$$

Where: P_{over} – overload probability for the analysed timetable, P_i – overload probability for i-th traffic situation, n – umber of traffic situations.

The overload probability for the analysed sections is shown in Fig. 5. According to the traffic situation probabilities, the highest overload probability is related to section B. Due to the performed analysis, weak points in the timetable were found. For the section B the main reason of a high overload probability are three traffic situations where the passenger train accumulation is combined with freight trains.

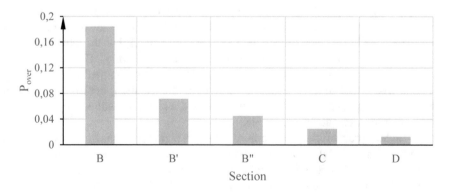

Fig. 5. Overload probability P_{over} for the given sections

Therefore, freight trains in this weak points were rescheduled to other free time windows ± one hour. That intervention decreased strongly the overload probability, what is shown as section B' in Fig. 5.

After that, also the long-distance passenger trains were replaced by 10–26 min. In result, the overload probability was once again decreased, what is shown as section B" in Fig. 5.

6 Conclusions

Trains with large level of current consumption (e.g. freight trains or trains during increasing velocity) may have an impact on exceedance permissible values so it is necessary to predict buffer times which are safeguards against occurrence of collapses, especially in the case of railway 3000 V DC power systems.

After literature review it was found, that there is lack of methods for timetable evaluation taking into account the load of energy supply systems. It was also found that the issue of robustness evaluation in terms of the energy equipment can be solved by simulation, which is time-consuming.

The presented method combines the well-known motion theory with the probabilistic approach in terms of punctuality. As a result, the method allows to evaluate timetables in terms of possible overload situations.

The results are promising, but follow-up research is necessary. The main reason for that are the time-consuming calculations of the amperage on sections. Also finding of the set of trains which may cause the escalation of a traffic situation has not been automated yet. Thus, for further studies it is planned to create an automated tool for the shown method. Moreover, further research on optimization of timetables has to be connected to the results gathered from the method.

References

1. Andersson, E., Peterson, A., Krasemann, T.J.: Quantifying railway timetable robustness in critical points. J. Rail Transp. Plan. Manag. **3**, 95–110 (2013). https://doi.org/10.1016/j.jrtpm.2013.12.002
2. Attachment to Resolution of 16 July 2018 No. 566/2018, PKP Polish Railway Lines
3. Burggraeve, S., Bull, S.H., Vansteenwegen, P., Lusby, R.M.: Integrating robust timetabling in line plan optimization for railway systems. Transp. Res. Part C **77**, 134–160 (2017). https://doi.org/10.1016/j.trc.2017.01.015
4. Burggraeve, S., Vansteenwegen, P.: Robust routing and timetabling in complex railway stations. Transp. Res. Part B **101**, 228–244 (2017). https://doi.org/10.1016/j.trb.2017.04.007
5. Chao, L., Jinjin, T., Leishan, Z., Yixiang, Y., Zhitong, H.: Improving recovery to optimality robustness through efficiency-balanced design of timetable structure. Transp. Res. Part C **85**, 184–210 (2017). https://doi.org/10.1016/j.trc.2017.09.015
6. Guze, S.: Business availability indicators of critical infrastructures related to the climate-weather change. In: Zamojski, W., Mazurkiewicz, J., Sugier, J., Walkowiak, T., Kacprzyk, J. (eds.) Contemporary Complex Systems and Their Dependability. Advances in Intelligent Systems and Computing, vol. 761. Springer (2018). https://doi.org/10.1007/978-3-319-91446-6_24

7. Guze, S.: An application of the selected graph theory domination concepts to transportation networks modelling. Sci. J. Marit. Univ. Szczecin **52**(124), 97–102 (2017)
8. Ghaemia, N., Zilko, A., Yan, F., Cats, O., Kurowicka, D., Goverde, R.: Impact of railway disruption predictions and rescheduling on passenger delays. J. Rail Transp. Plan. Manag. **8**, 103–122 (2018). https://doi.org/10.1016/j.jrtpm.2018.02.002
9. Kierzkowski, A., Kisiel, T.: A model of check-in system management to reduce the security checkpoint variability. Simul. Model. Pract. Theory **74**, 80–98 (2017). https://doi.org/10.1016/j.simpat.2017.03.002
10. Kierzkowski, A., Kisiel, T.: Simulation model of security control system functioning: a case study of the Wroclaw Airport terminal. J. Air Transp. Manag. Part B **64**, 173–185 (2017). https://doi.org/10.1016/j.jairtraman.2016.09.008
11. Kroon, L.G., Dekker, R., Vromans, M.J.C.M.: Cyclic railway timetabling: a stochastic optimization approach. In: Algorithmic Methods for Railway Optimization, vol. 4359, pp. 41–66. Springer, Berlin (2007). https://doi.org/10.1007/978-3-540-74247-0_2
12. Kwaśnikowski, J.: Elementy teorii ruchu i racjonalizacja prowadzenia pociągu. Wydawnictwo Naukowe Instytutu Technologii Eksploatacji, Radom (2013)
13. Lusby, R., Larsen, J., Bull, S.: A survey on robustness in railway planning. Eur. J. Oper. Res. **266**, 1–15 (2017). https://doi.org/10.1016/j.ejor.2017.07.044
14. Madej, J.: Teoria ruchu pojazdów szynowych. Oficyna Wydawnicza Politechniki Warszawskiej, Warszawa (2012)
15. Podoski, J., Kacprzak, J., Mysłek, J.: Zasady Trakcji Elektrycznej. Wydawnictwa Komunikacji i Łączności, Warszawa (1980)
16. Salido, M.A., Barber, F., Ingolotti, L.: Robustness in railway transportation scheduling. In: 7th World Congress on Intelligent Control and Automation, Chongqing, China, pp. 2880–2885 (2008). https://doi.org/10.1109/wcica.2008.4594481
17. Skupień, E., Tubis, A.: The use of linguistic variables and the FMEA analysis in risk assessment in inland navigation. Int. J. Mar. Navig. Saf. Sea Transp. **12**, 143–148 (2018). https://doi.org/10.12716/1001.12.01.16
18. Solinen, E., Nicholson, G., Peterson, A.: A microscopic evaluation of railway timetable robustness and critical points. J. Rail Transp. Plan. Manag. **7**, 207–223 (2017). https://doi.org/10.1016/j.jrtpm.2017.08.005
19. Takeuchi, Y., Tomii, N.: Robustness indices for train rescheduling. In: Proceedings of the 1st International Seminar on Railway Operations Modelling and Analysis, Netherlands (2005)
20. Tubis, A., Gruszczyk, A.: Measurement of punctuality of services at a public transport company. In: Carpathian Logistics Congress, Jesenik, Czech Republic, pp. 512–517 (2015)
21. Tubis, A.: Route risk assessment for road transport companies. In: Contemporary Complex Systems and their Dependability, vol. 761, pp. 492–503. Springer, Cham (2019). https://doi.org/10.1007/978-3-319-91446-6_46
22. Walkowiak, T., Mazurkiewicz, J.: Analysis of critical situations in discrete transport systems. In: Proceedings of DepCoS - RELCOMEX 2009, Brunów, Poland, pp. 364–371 (2009), https://doi.org/10.1109/depcos-relcomex.2009.39
23. Walkowiak, T., Mazurkiewicz, J.: Soft computing approach to discrete transport system management. LNCS. LNAI, vol. 6114, pp. 675–682 (2010). https://doi.org/10.1007/978-3-642-13232-2_83
24. Wyrzykowski, W.: Ruch kolejowy. Wydawnictwa Komunikacji, Warszawa (1954)

Representing Process Characteristics to Increase Confidence in Assurance Case Arguments

Aleksander Jarzębowicz$^{(\boxtimes)}$ ⓘ and Szymon Markiewicz

Department of Software Engineering, Faculty of Electronics,
Telecommunications and Informatics, Gdańsk University of Technology,
Narutowicza 11/12, 80-233 Gdańsk, Poland
olek@eti.pg.edu.pl

Abstract. An assurance case is a structured, evidence-based argument demonstrating that a safety or other quality objective of a high integrity system is assured. Assurance cases are required or recommended in many industry domains as a means to convince the regulatory bodies to allow commissioning of such system. To be convincing, an argument should address all potential doubts and thus cover numerous additional issues, including the processes that led to development of the considered system. It is however not obvious, which elements of processes (and which characteristics of them) should be documented and how to include them in the argument without making it too large and complex. In this paper we provide description structures for essential process elements. The structures were developed on the basis of literature search and reviews of publicly available assurance cases. We also show how to include such information within the overall assurance case in a way that reduces the complexity and allows to distinguish process-related elements from the primary argument.

Keywords: Assurance case · Safety case · Confidence argument · Defeater

1 Introduction

Assurance cases are developed for systems considered as safety-critical or expected to demonstrate other high integrity attribute (e.g. security, reliability). An assurance case is defined as "A reasoned and compelling argument, supported by a body of evidence, that a system (…) will operate as intended for a defined application in a defined environment" [1].

Assurance case is a structured, tree-like argument which starts with high-level claims about considered system's attribute(s) like safety or security. Then a supporting argument for each such claim is provided. The supporting argument will include more detailed, lower-level claims, which in turn need to be supported. This process continues iteratively, until claims need no further decomposition, but can be addressed by providing facts and evidence demonstrating their validity. A simplified example depicting the frequently used schema of arguing safety by addressing particular hazards

© Springer Nature Switzerland AG 2020
W. Zamojski et al. (Eds.): DepCoS-RELCOMEX 2019, AISC 987, pp. 245–255, 2020.
https://doi.org/10.1007/978-3-030-19501-4_24

(situations that can potentially lead to an accident) is shown in Fig. 1. The example does not conform to any particular assurance case notation, to avoid the need to introduce it here.

Assurance cases have been adopted in several industry domains, which require development of an assurance case before the system can be commissioned and used (e.g. railway [2], flight control [3], automotive [4]). Also, a cross-domain ISO/IEC standard dedicated to assurance cases was published [5]. Moreover, for several other standards or guidelines, which do not explicitly encourage nor mention using an assurance case, it was shown that development of such argument can help to demonstrate conformance to standard's requirements – examples include ISO 61508 [6, 7], ISO 15408 [8] or safety and quality management of healthcare services [9, 10].

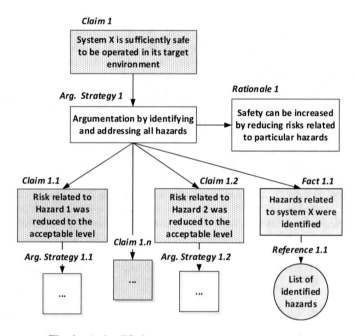

Fig. 1. A simplified assurance case argument example.

Despite the presence of several graphical notations dedicated to assurance case representation, all of them are based on the underlying argument model by Toulmin [11]. This model defines how to express so-called defeasible reasoning, which, in contrary to deductive reasoning (known from e.g. formal logic), cannot be proved with absolute certainty. Defeasible arguments are probably most common in real-world discussions (including e.g. law or politics). Assurance cases are defeasible arguments as well, perhaps with some exceptions when formal proofs are used to e.g. demonstrate consistency between referenced models. The confidence in a defeasible argument can be decreased by pointing out its weaknesses like questionable reasoning or insuffi-ciently reliable evidence. Such weaknesses are represented in the original Toulmin's model as "rebuttals" [11], also alternative terms "deficit" [12] and "defeater" [13] are used in the literature dedicated to assurance cases. In this paper we will henceforth use the term "defeater".

In our example from Fig. 1, one could easily identify defeaters, both related to the reasoning (Argumentation Strategy 1 + Rationale 1) and the evidence (Fact 1.1 + Reference 1.1) e.g.:

- Reasoning – Is handling each hazard separately sufficient? What about the possibility of two or more hazards occurring simultaneously?
- Evidence – What confidence can we have that the list of identified hazards is complete? What hazard identification methods were used and why should they be considered adequate? Were the people responsible for hazard identification task sufficiently qualified and experienced?

Both kinds of defeaters are important, but in this paper we will focus on the second kind (i.e. defeaters related to evidence) only.

To address potential defeaters, additional elements are added to the assurance case. In our example, attempts to strengthen the argument with respect to the evidence could be made by providing claims (together with their supporting arguments) and facts documenting the process characteristics of hazard identification, including its participants, activities, methods and tools used. Another reason for inclusion of such elements in an assurance case is the fact that some standards and regulations demand process-related evidence concerning activities, artefacts, roles (part 1, p. 35 of [5], p. 20 of [3]).

Including such additional process-related elements in the assurance case has its consequences. The positive result is that it would (hopefully) lead to the increase of argument's confidence and cover requirements of some standards. There are however negative consequences as well: the assurance case becomes larger, more complex and harder to understand, as some parts argue that the main objective like system safety is achieved (assurance argument), while others focus on dealing with defeaters and increasing confidence (confidence arguments) [12]. Similarly, the set of evidence grows and includes two kinds of evidence items: those referred to by the main assurance argument (direct evidence) and those used to show that some other evidence is reliable (backing evidence) [3].

Despite such drawbacks, it is still necessary to include confidence arguments and process-based backing evidence in order to make assurance case a "compelling argument", as its definition states. It however raises the following questions:

1. Which process elements should be considered as evidence items and what characteristics should be specified for each of them?
2. How to structure the assurance case to include confidence arguments and process-based evidence but to distinguish them from the main assurance argument?

In this paper we attempt to provide answers to these questions. Hence, its main contributions are:

1. A tabular description structure specifying essential process elements and their characteristics, that are expected to be required to include in an assurance case. It can be used as a checklist when developing an assurance case and documenting evidence. It is not the first such proposal published, but it is more comprehensive, as it is based on the previous proposals as well as on assurance case reviews.

2. A distinction between generic and context-specific attributes of process elements. The former are closely associated with the evidence item, while the latter should be provided within a context of a particular confidence argument.
3. A way of representing (1) and (2) in the assurance case argument structure.

The remainder of the paper is structured as follows. In Sect. 2 we outline the related work. Section 3 presents our proposal of process elements to be used in assurance cases. Section 4 explains how to represent process-based evidence in confidence arguments. Finally, the paper is concluded in Sect. 5.

2   Related Work

The main areas of related work include: (1) representing defeaters and confidence arguments in assurance cases; and (2) defining process elements to be included in assurance case argument.

2.1   Defeaters and Confidence Arguments

As mentioned, defeaters (under a different name) were already included in the original Toulmin's model [11]. The idea and categorization of defeaters was further researched both for general arguments [14, 15] and specifically for assurance case arguments [16, 17], also a concept of additional confidence argument was introduced to cope with them.

Hawkins et al. [12] developed a proposal of dividing an assurance case into two separated but interrelated parts: safety argument (which can be generalized into assurance argument) and confidence argument. The resulting confidence argument is a large structure gathering arguments addressing various unrelated defeaters. Goodenough et al. [13] introduced so called confidence maps, which extends the known notation with additional diagram to represent defeaters and confidence arguments addressing them. Ayoub et al. [18] describe the approach which includes identifying potential defeaters and developing a separate ("contrapositive") confidence argument. Jarzębowicz and Wardziński [19] proposed that a given step of the main assurance argument should be associated with a corresponding "local" confidence argument. Such confidence argument can be attached to the rationale/justification element that explains the overall validity of the reasoning used in such step.

We consider our work as the further step in the research on representing defeaters and confidence arguments, based on two observations: (1) confidence arguments should be compact and focused on addressing particular defeaters, instead of being enumeration of all good practices used; (2) the existing, Toulmin-based notations are sufficient to represent confidence aspects without the need to introduce additional elements or diagrams.

2.2   Process Elements in Assurance Cases

Graydon et al. [20] analyzed a software assurance argument and pointed out that software dependability can be compromised by several process-based causes including fallible humans, immature tools and unsuitable techniques. Ayoub et al. [18] proposed a list of process elements used in assurance cases and a pattern of a confidence argument addressing such elements. An alternative list of process-based elements and their characteristics (contributing to trustworthiness and appropriateness) as well as associated confidence argument pattern was given by Nair et al. [21]. Hawkins et al. [22] went one step further, by using process models (documented in software tools) to automate instantiation of confidence argument patterns.

The lists of process elements and their characteristics given by particular sources differ. As several recent papers express the need for better representations and ways to include confidence aspects of process-based elements [23, 24] and for evidence [25], we believe that it is useful to aggregate the existing body knowledge and to create a comprehensive list on such basis.

3   Process Elements and Their Characteristics

Our aim was to identify process elements and their characteristics that are relevant to increase confidence in an assurance case argument. We intended to do it in a thorough manner, so we used diversified sources and divided this task into several steps:

1. Studying related papers that included lists or models covering process-related defeaters and process elements which should be addressed in confidence arguments. We used Common Characteristics Map from [18], the list of factors influencing assessor confidence from [21] and defeater checklist from [19].
2. Reviewing publicly available assurance cases and related documents from various domains: medical devices [26, 27], airspace control [28] and railway [29] to identify processes and resources used by them and referred to in assurance arguments.
3. Analyzing the experience-based knowledge available in GSN standard [1] about assurance case reviews and problems commonly found during them.
4. Combining the inputs from steps 1–3 into a consistent list, removing potential duplicates, unifying the language and designing the structure to represent the results.

Our proposal is based on the observation, that the key, central entity is the artefact used as an evidence to support assurance case claims. Other process elements (like activities, people, tools or input resources) appear in the confidence argument only as a means to increase the confidence in that evidence item. The artefacts include e.g.:

- Reports of hazard and risk analyses;
- Requirements including safety/security requirements;
- Technical documentation describing system's architecture, implementation, components used, interlocks applied etc.;
- Test specifications and reports;
- Formal proofs of correctness;

- Field reports;
- Procedures addressing operation and maintenance.

Not all such artefacts have to be authored by the organization responsible for the development of the high-integrity system under consideration and its assurance case, as for example, when a third party solution is used, a relevant artefact describing this solution could be included as evidence. We however exclude from our area of interest items like: regulations, norms and standards, guidelines used in a particular industry domain etc. Such artefacts are referenced by assurance cases, but as a rationale for argument decomposition, not as evidence. Besides, they are considered trustworthy – assurance case developer is not expected to identify and address weaknesses of an international safety standard. Of course, one can challenge the decision of using a given standard or guideline as the basis for the development of an assurance case or its part, but it would be a defeater against reasoning, not against evidence.

Our research study described above resulted in the identification of the process elements listed below. The references given for each element indicate the sources that contributed to developing its description structure.

1. Artefact – the basic information (necessary e.g. for the auditors) that allows to identify the artefact and its version, summarize its contents and retrieve its source [27–29].
2. Author – the characteristics of people responsible for the development of the artefact, their background and involvement [1, 21].
3. Process – description of the process, method or technique used as part of artefact development [18, 21].
4. Tool – the characteristics of the tool used in the process of artefact development e.g. compiler, proof checker, automated testing tool [21, 27, 28].
5. Contents – considerations of the quality and adequacy of artefact's contents, as well as the role it should play in the assurance case argument [1, 21].
6. Language – the description of the language used in the artefact, including conventions, abbreviations and symbols. It is important to prevent communication errors when the artefact is referenced from the argument [27–29].
7. Associations – the explicit list of associated artefacts together with considerations related to their quality. Such characteristics allow to eliminate defeaters implying that an artefact is based on some other unreliable artefacts or that some hidden interdependencies exist [18, 27–29].
8. Reviews – details about conducted reviews of the artefact, including incorporation of reviewers' remarks [18, 27–29].

Each of the listed process elements is described through a number of its characteristics. We make a distinction whether a characteristic is generic and associated with the element regardless of context or context-specific i.e. important in the context of arguing against a particular defeater. For example, a name and contact information of artefact's author are certainly generic and should always be provided together with the artefact. On the other hand, description of author's experience and technical competence can form a long resume and only a part of such information would matter in the context of a given confidence argument. Our example from Fig. 1. would likely require confidence argument that each member of the team responsible for hazard identification task is sufficiently qualified and experienced. If so, it would be important to document each person's experience in hazard identification, his/her courses, trainings, certificates etc., but the information that someone has a substantial job experience in software testing and is a certified Scrum Master would not be relevant here. Thus, in accordance to our postulate that confidence arguments should be compact and focused on addressing particular defeaters, we insist on providing only such process-based evidence that is relevant for a given context.

The example description structure including characteristics for "Author" element is shown in Table 1. The meaning of the "G/CS" column is whether a given characteristic is generic or context-specific. Due to space limitations we are not able to present any more description structures here, however a full report is available online[1].

Table 1.   Description structure for "Author" element.

Characteristic	Description	G/CS
2.1. Personnel	The personal data of the person participating in the development of an artefact. In case an artefact is a product of teamwork, all team members should be listed (and characterized in the following rows). Contact info like phone number and e-mail should also be provided	G
2.2. Domain knowledge and experience	The confidence that the author has a knowledge about the industrial domain and experience in working on similar projects. Job history and projects conducted are the most important information to be included here. In addition: reports, scientific papers or presentations given can be mentioned	CS
2.3. Technical competence	The confidence about technical skills and competencies, based on education history (diplomas), certificates obtained, courses finished etc.	CS
2.4. Independence	The confidence that the work conducted by the author on the artefact is free from unwanted dependencies e.g. the same person develops and tests a software module or a subordinate verifies the document created by his/her superior. A description explaining the separation of task assignment and lack of conflict of interests	CS
2.5. Team organization	In case of teamwork, team structure and responsibilities should be described. Also the leader who is responsible for the artefact should be explicitly determined	G

---

[1] https://drive.google.com/open?id=1uo-PlUhPLJ2KIY1ck9kBkBfmifCl8c67.

4   Representing Process-Based Evidence in Confidence Arguments

In our earlier work [19], we proposed that confidence arguments should be strongly context-related and associated with a particular step of the main assurance argument. In order to avoid introducing additional concepts and notational elements, we proposed a solution where an element which explains the reasoning used in a given step of argumentation (called rationale or justification in existing notations) is the "root" of a confidence argument regarding this step. It makes sense, as such rationale should convince a reader that, given the premises (lower-level claims, facts etc.), the conclusion (higher-level claim) is valid – therefore it should also address any defeaters targeting evidence, reasoning or any other part of this argumentation step. Moreover, it is consistent with the original Toulmin's model.

In this paper we build on the mentioned previous work and extend it with the approach of handling process-based evidence. Based on the available sources and on reviews of available assurance cases, we compiled a list of essential process elements and we defined a set of characteristics for each element. We also introduced the distinction between generic and context-specific characteristics. We suggest that such distinction should be reflected in the way how evidence elements are defined and referenced from the assurance case argument.

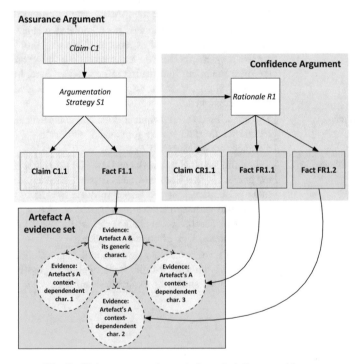

Fig. 2. Using process element characteristics as evidence.

First, let us consider the generic characteristics of process elements associated with a given artefact used as evidence. Such characteristics mostly provide the basic information about the artefact (following the earlier example: the list of identified hazards has its identifier, document version, identification of its authors, name and version of the tool it was stored in etc.), which is important whenever an artefact is referred to, regardless of the context. Such information should therefore be always stored together with the artefact itself. The artefact, together with its generic characteristics, will be referenced as direct evidence for the primary assurance argument.

Now, let us consider the context-dependent characteristics. One could store them together with the artefact, but it would result in a large portion of data that would never be used as a whole, instead only particular selected information would be needed. For example, when dealing with defeaters regarding "list of identified hazards", we would need evidence confirming team leader's experience in hazard analysis, proficiency in using FMEA method etc., while other qualifications he/she may possess (e.g. being an experienced system architect or a certified tester) would not be relevant here. On the other hand, evidence confirming such qualifications may be crucial when system architecture or test reports are referenced by some other parts of the assurance case. The evidence documenting context-dependent characteristics should be considered a backing evidence and is to be referenced from confidence arguments. It should be kept in the form of several evidence items, separately from the artefact, but linked to it.

The proposed solution is presented in Fig. 2. Please note that evidence management is not treated here as integral part of assurance case. Evidence set concerning Artefact A includes the artefact itself together with its generic characteristics. The other, context-dependent characteristics are kept as separate evidence items within that set. Interrelationships between the artefact and related evidence items are maintained. The artefact is referenced by a fact from the primary assurance argument. The use of this artefact is addressed in the associated confidence argument – if defeaters concerning the trustworthiness of the artefact are identified, the confidence argument references appropriate evidence items describing relevant context-dependent characteristics (e.g. author's competence, integrity of tools used).

5   Conclusions

In this paper we discussed the existing approaches to represent confidence arguments and to model process elements that are used as part of assurance case argumentation. By analyzing existing published proposals and by reviewing assurance cases and related reports, we identified process elements together with their characteristics that we consider as important information for assurance case development and review. We distinguished generic and context-specific characteristics and recommended that the former should be used in the primary assurance argument and stored together with the evidence artefact, while the latter is intended for confidence arguments and should be kept in the form of several, fine-grained evidence items. Finally, we demonstrated how an assurance case using both kinds of evidence can be structured.

We are aware that the approach using explicit confidence arguments is a demanding task, but on the other hand, process-based evidence and other evidence e.g. explaining the applied reasoning are still included in assurance cases, while, according to the literature mentioned in Sect. 1, assurance and confidence arguments should not be mixed as it results in a number of drawbacks.

Our proposed description structures include a significant number of characteristics that would have to be documented during the project of high-integrity system development, which means additional effort of project participants. Our proposal is however of similar scale as the other proposals we described as related work in Sect. 2. Also, considering what is at stake – auditors refusing to accept the system or worse: the system endangering people's health and lives – it does not seem a steep price.

We conducted a preliminary validation case study, in which an existing assurance case argument was refactored. The refactoring included extracting the process-related parts of the argument to create confidence arguments and applying the approach to manage and reference evidence items described in Sect. 4. The case study confirmed that the proposed approach can be used, however we are aware that more sound validation should be provided in future.

A promising direction of further research is to use process elements' description structures to define confidence patterns (schemes of building confidence arguments for frequently encountered defeaters). As we are currently implementing tool support for automated patterns instantiation, it may result in a valuable assistance to assurance case developers.

References

1. The Assurance Case Working Group (ACWG): Goal Structuring Notation community standard version 2 (2018)
2. CENELEC: EN 50126. Railway Applications: The Specification and Demonstration of Reliability, Availability, Maintainability and Safety (RAMS) (1999)
3. Eurocontrol: Safety Case Development Manual (2006)
4. International Organization for Standardization (ISO): ISO/DIS 26262: Road Vehicles - Functional Safety (2011)
5. International Organization for Standardization (ISO): 15026-2:2011: Systems and Software Engineering – Systems and Software Assurance – Part 2: Assurance Case (2011)
6. Stensrud, E., Skramstad, T., Li, J., Xie, J.: Towards goal-based software safety certification based on prescriptive standards. In: First International Workshop on Software Certification (WoSoCER), pp. 13–18 (2011)
7. Sklyar, V., Kharchenko, V.: Assurance case driven design based on the harmonized framework of safety and security requirements. In: 13th International Conference on ICT in Education, Research and Industrial Applications (ICTERI 2017), pp. 670–685 (2017)
8. Yamamoto, S., Kaneko, T., Tanaka, H.: A proposal on security case based on common criteria. In: Information and Communication Technology-EurAsia Conference. LNCS, vol. 7804, pp. 331–336. Springer (2013)
9. Sujan, M., Spurgeon, P., Cooke, M., Weale, A., Debenham, P., Cross, S.: The development of safety cases for healthcare services: practical experiences, opportunities and challenges. Reliab. Eng. Syst. Saf. **140**, 200–207 (2015)
10. Górski, J., Jarzębowicz, A., Miler, J.: Validation of services supporting healthcare standards conformance. Metrol. Measur. Syst. **19**(2), 269–284 (2012)

11. Toulmin, S.: The Uses of Argument. Cambridge University Press, Cambridge (2003). Updated Edition
12. Hawkins, R., Kelly, T., Knight, J., Graydon, P.: A new approach to creating clear safety arguments. In: Advances in Systems Safety, pp. 3–23. Springer (2011)
13. Goodenough, J., Weinstock, C., Klein, A.: Eliminative induction: a basis for arguing system confidence. In: Proceedings of the 2013 International Conference on Software Engineering, pp. 1161–1164. IEEE Press (2013)
14. Pollock, J.L.: Defeasible reasoning. Cogn. Sci. **11**, 481–518 (1987)
15. Verheij, B.: Evaluating arguments based on Toulmin's scheme. Argumentation **19**(3), 347–371 (2005)
16. Kelly, T.: Reviewing assurance arguments - a step-by-step approach. In: Proceedings of Workshop on Assurance Cases for Security, Dependable Systems and Networks Conference (2007)
17. Grigorova, S., Maibaum, T.: Argument evaluation in the context of assurance case confidence modeling. In: Proceedings of 25th IEEE International Symposium on Software Reliability Engineering Workshops, Naples, Italy, pp. 485–490. IEEE Computer Society (2014)
18. Ayoub, A., Kim, B., Lee, I., Sokolsky, O.: A systematic approach to justifying sufficient confidence in software safety arguments. In: Proceedings of 31st International Conference on Computer Safety, Reliability and Security. LNCS, vol. 7612, pp. 305–316 (2012)
19. Jarzębowicz, A., Wardziński, A.: Integrating confidence and assurance arguments. In: 10th IET System Safety and Cyber Security Conference (2015)
20. Graydon, P., Knight, J., Yin, X.: Practical limits on software dependability: a case study. In: International Conference on Reliable Software Technologies, pp. 83–96 (2010)
21. Nair, S., Walkinshaw, N., Kelly, T., de la Vara, J.L.: An evidential reasoning approach for assessing confidence in safety evidence. In: 26th IEEE International Symposium on Software Reliability Engineering (ISSRE), pp. 541–552. IEEE (2015)
22. Hawkins, R., Richardson, T., Kelly, T.: Using process models in system assurance. In: Skavhaug, A., Guiochet, J., Bitsch, F. (eds.) Computer Safety, Reliability, and Security. LNCS, vol. 9922, pp. 27–38. Springer, Heidelberg (2016)
23. Asplund, F., Törngren, M., Hawkins, R., McDermid, J.A.: The need for a confidence view of CPS support environments. In: 16th International Symposium on High Assurance Systems Engineering (HASE), pp. 273–274 (2015)
24. Retouniotis, A., Papadopoulos, Y., Sorokos, I., Parker, D., Matragkas, N., Sharvia, S.: Model-connected safety cases. In: International Symposium on Model-Based Safety and Assessment. LNCS, vol. 10437, pp. 50–63. Springer, Cham (2017)
25. Holloway, C.M., Graydon, P.: Evidence under a magnifying glass: thoughts on safety argument epistemology. In: Proceedings of 10th IET System Safety and Cyber Security Conference, Bristol, UK (2015)
26. Larson, B.R., Hatcliff, J., Chalin, P.: Open source patient-controlled analgesic pump requirements documentation. In: 5th International Workshop on Software Engineering in Health Care (SEHC), pp. 28–34 (2013)
27. Larson, B.R.: Open PCA Pump Assurance Case, SAnToS Research Group, Kansas State University (2014). http://openpcapump.santoslab.org/
28. Civil Aviation Office: Poland & Civil Aviation Administration, Lithuania: Baltic FAB Safety Case (2012). https://www.pansa.pl/aap/Safety_Case_2.0.pdf
29. London Underground Limited: London Underground Safety Certificate and Safety Authorisation, ver. 5.1 (2017). http://content.tfl.gov.uk/london-underground-safety-certificate-and-safety-authorisation.pdf

Dependability of Multichannel Communication System with Maintenance Operations for Air Traffic Management

Igor Kabashkin[✉]

Transport and Telecommunication Institute,
Lomonosova 1, Riga 1019, Latvia
kiv@tsi.lv

Abstract. Air Traffic Management (ATM) systems are one of the key element of critical for flight safety infrastructure. Among ATM systems communications play a special role as a vital part in information exchange between air traffic controllers and pilots, especially in hazardous situations. In the modern ATM systems each controller has independent direct communication channel with separate radio frequency. For reliability increasing of such ATM communication channels the redundancy with common set of standby radio stations is used. Traditionally, maintenance is additional tool to increase the efficiency of radio stations operation. This paper examines the effectiveness of the ATM communication channels with the simultaneous using both of the above methods—redundancy and periodic maintenance of radio stations. Mathematical model of the channel reliability for this case is developed.

Keywords: Reliability · Redundancy · Air Traffic Management · Communication network · Maintenance

1 Introduction

Air Traffic Management (ATM) systems are one of the key element of critical for flight safety infrastructure. Among ATM systems communications play a special role as a vital part in information exchange between air traffic controllers and pilots, especially in hazardous situations. Breakdown in effective human communication is one of the main factors in the majority of aviation accidents [1].

The modern ATM system provides independent communication with the aircrafts for all controllers and has for each of them independent direct communication channel (CC) with separate radio frequencies f_i, $i = \overline{1, m}$, where m is number of CC. Technical support of controller-pilot communication carried out by means of radio stations (RS). Interoperability of RS and controllers in ATM communication network is provided by voice communications system (VCS) which is a state-of-art solution for air traffic control communication. The high reliability of ATM communication channels provided by use of reliable equipment with different redundancy strategies.

Efficiency of redundancy in the systems with identical elements can be increased by use of the common group of standby elements.

© Springer Nature Switzerland AG 2020
W. Zamojski et al. (Eds.): DepCoS-RELCOMEX 2019, AISC 987, pp. 256–263, 2020.
https://doi.org/10.1007/978-3-030-19501-4_25

In this paper repairable redundant communication network of ATM communication system is studied. There are $N = m + n$ radio stations in communication system, m of which are the main radio stations (MRS). The system has set of n redundant radio station (RRS) as universal identical standby pool (Fig. 2). All reconfiguration of reserve architecture and switching radio frequencies are carried out by VCS.

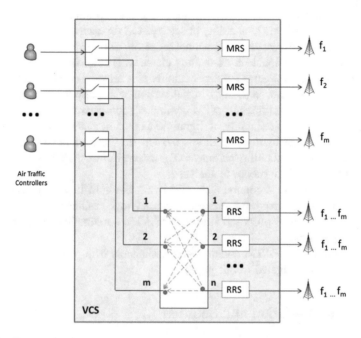

Fig. 1. Communication network of ATM system with common set of standby elements

Traditionally, maintenance is additional tool to increase the efficiency of radio stations operation. This paper examines the effectiveness of the ATM communication channels with the simultaneous using both of the above methods—redundancy and periodic maintenance of radio stations.

The content of this paper is organized as follows. In Sect. 2 some important works in the area of reliability in communication systems with redundancy are reviewed. In Sect. 3 the main definitions and assumptions are described and the model of reliability for study system is proposed. In Sect. 4 the conclusions are presented.

2 Related Works

One of the approaches for better efficiency of redundancy in the architecture with identical elements is k-out-of-n model. The k-out-of-n model is used to indicate an n-component system that functionality is good if and only if at least k of the n components are not failed. This system is called a k-out-of-n:G system. The papers [2–4] study the *k-out-of-n:G* system for different case studies. The analysis of the *k-out-of-n:G* system with non-exponential distributions of components lifetime are evaluated in paper [5].

In paper [6] an *k-out-of-n:F* architecture of *n*-component system that fails if and only if at least k of the n components fail The *k-out-of-n* model is a popular architecture in fault-tolerant systems. It is used as redundancy in different telecommunication systems [7, 8].

In ATM communication system it is important to know not the only dependability of the network at whole but availability of selected CC for each controller which carries out the communication session. The channel reliability of the architecture with identical elements and with additional set of standby elements is studied in [9].

The paper [10] investigates the dependability of a selected communication channel with common set of reserve stations in the system with periodical sessions of communications for real conditions of ATM communication network.

In the paper [11] the reliability of digital ATM communication system on the base of embedded cloud technology is discussed. On the base of developed model the boundary value of dependability parameters for automatics of VCS based on embedded cloud for different failure modes is analyzed.

In paper [12] for ATC communication network the model with dynamic structure reconfiguration is proposed for increasing of network resilience, and comparative analysis of reliability for proposed strategies of ATM communication network reconfiguration is performed.

The reliability of the ATM communication channels with periodic maintenance of radio stations is investigated in this paper.

3   Model Formulation and Solution

The following symbols have been used to develop equations for the models:

λ - Failure Rate for MRS and RRS
μ - Repair Rate for MRS and RRS
m - Number of communication channels and number of MRS
n - Number of RRS in common set of redundant radio stations
A - Channel Availability
U - Channel Unavailability, $U = 1 - A$
λ_M - Intensity of maintenance
μ_M - Intensity of the performance of maintenance work
A_0 - Required Availability of the CC
l - Number of repair bodies
$\gamma = \lambda/\mu$
$\omega = \gamma/l$

The repairable redundant communication network of ATM system includes $N = m + n$ radio stations, m of which are MRS and n radio stations are used as a common set of redundant elements (Fig. 1). All reconfiguration of the architecture and frequency setting change of reserve radio stations are carried out by VCS. Time of all failures and repairs of radio stations has exponential distributions [7].

The behavior of the examined system with $1 \leq l \leq n$ repair bodies and active backup mode of redundant elements is described by the Markov Chain state transition diagram (Fig. 2), where: H_i – state with i failed RS, but in the selected channel there is a workable RS; H_{il} – state with $i + 1$ failed RS, in the selected channel there is no a workable RS; H_M – state without failed RS and with one RS at the maintenance.

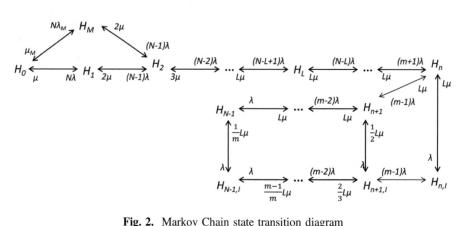

Fig. 2. Markov Chain state transition diagram

On the base of this diagram the system of Chapman–Kolmogorov's equations can be writing in accordance with the general rules [13].

We can determine the values of the stationary probabilities h_i of states H_i of the considered system using the computational method proposed in [14]. The channel availability will be defined by equation

$$A = 1 - U = 1 - \sum_{i=n}^{N-1} h_{i,l} \tag{1}$$

To do this, we carry out the decomposition of the transition graph of the original system

$$H = H'UH'' = \{H_0, H_M, H_1, H_2\}U\{H_{ij}, H_{j,l} : i = \overline{2, N-1}; j = \overline{n, N-1}\}$$

into two independent subsystems R and S.

Graphs of these subsystems form a sets of states

$$R = \{R_0, R_M, R_1, R_2\};$$

$$S = \{S_{ij}, S_{jl} : i = \overline{2, N-1}; j = \overline{n, N-1}\}$$

which have the same form and are characterized by the same parameters as the corresponding part of the original graph

$$R \equiv H'; S \equiv H''$$

Graph of subsystem R is shown at the Fig. 3; graph for subsystem S is shown at the Fig. 4.

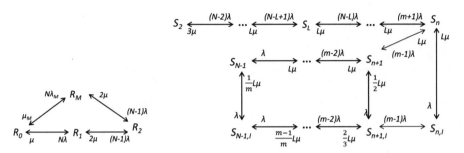

Fig. 3. Markov Chain state transition diagram for subsystem R

Fig. 4. Markov Chain state transition diagram for subsystem S

In this case, according to the method proposed in [14], we can obtain

$$h_i/h_2 = r_i/r_2 : \forall i \tag{2}$$

$$h_{\alpha\beta}/h_2 = s_{\alpha\beta}/s_2 : \forall \alpha, \beta \tag{3}$$

$$h_2 = r_2 s_2/(r_2 + s_2 - r_2 s_2) \tag{4}$$

where $h_i, h_{\alpha\beta}, r_i, s_{\alpha\beta}$ - stationary probabilities for the system in the states $H_i, H_{\alpha\beta}, R_i, S_{\alpha\beta}$ of related subsystems.

On the base of the state transition diagram for a Markov's process shown on Fig. 3, we can write the system of Chapman-Kolmogorov's equations according to the general rule for a stationary process [13]:

$$s_0'(t) = -N(\lambda + \lambda_M)s_0(t) + \mu s_1(t) + \mu_M s_M(t)$$

$$s_1' = N\lambda s_0(t) - [(N-1)\lambda + \mu]s_1(t) + 2\mu s_2(t)$$

$$s_2'(t) = (N-1)\lambda s_1(t) - 4\mu s_2(t) + (N-1)\lambda s_M(t)$$

$$s_M'(t) = N\lambda_M s_0(t) - [(N-1)\lambda + \mu_M]s_M(t) + 2\mu s_2(t)$$

Solving this equation, we can determine the stationary probability s_2, which for highly reliable systems ($\lambda \ll \mu$) will be described by the expression

$$s_2 = \frac{N(N-1)(\lambda + \lambda_M)\gamma}{a[(N-1)\gamma\mu + 2N\lambda_M + 2\mu_M]} \tag{5}$$

where $a = \begin{cases} 1, l = 1, \\ 2, 1 < l < n, \end{cases}$ $\gamma = \lambda/\mu$

For the subsystem R the formula for stationary probability r_2 can be borrowed from [11] and in case of the system with ideal switch VCS will be described by the expression

$$r_2^{-1} = 1 + \frac{2}{N(N-1)} \sum_{i=3}^{l} \binom{i}{N} \gamma^{i-2}$$

$$+ \frac{2(N-2)! l^{l-2} \omega^{N-2}}{l!} \left\{ \sum_{i=m}^{N-l-1} \frac{1}{i! \omega^i} + \frac{1}{m} \left[\frac{1}{(m-1)! \omega^{m-1}} + \sum_{i=0}^{m-2} \frac{(m-i))}{i! \omega^i} \right] \right\} \tag{6}$$

The stationary probability h_2 of the original system (Fig. 2) can be obtained by substituting expressions (5) and (6) into (4).

Taking into account the formulas (2) and (3), the expression for the availability of selected communication channel (1) takes the form

$$A = h_2 \left[\frac{1 - s_2}{s_2} + r_2^{-1} \sum_{i=2}^{N-2} r_i \right] \tag{7}$$

where in accordance with [11]

$$\sum_{i=0}^{N-1} r_i = r_2 \left\{ 1 + \frac{2}{N(N-1)} \sum_{i=3}^{l} \binom{i}{N} \gamma^{i-2} \right.$$

$$+ 2(N-2)! l^{l-2}/l! \left[\sum_{i=l+1}^{n} \frac{\omega^{i-2}}{(N-i)!} + \sum_{i=n+1}^{N-1} \frac{\omega^{i-2}}{(N-i-1)! m} \right] \right\}$$

Thus, expressions (4)–(6) when substituting them into (7) completely determine the value of the channel availability of the original system (Fig. 2).

Numerical Example
The effect of radio station downtime during maintenance can be evaluated by using the channel availability degradation factor

$$V = \frac{1 - A}{1 - A_M}$$

where A – channel availability in the model excluding downtime of RS during maintenance [11], A_M – channel availability in the model taking into the account downtime of RS during maintenance.

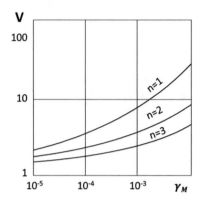

Fig. 5. The reliability degradation factor

At the Fig. 5 the availability degradation factor V shown as function of maintenance parameter $\gamma_M = \lambda_M/\mu_M$ for number of communication channels $m = 10$ with different number n of radio stations in common redundant set of RS with reliability parameter $\gamma = 10^{-4}$.

Analysis of Fig. 5 shows that for communication network with small value of parameter γ_M that is typical for condition-based maintenance, the influence of RS maintenance downtime is weak on channel availability. For large value of the parameter γ_M that is typical for periodical schedule-based maintenance, the influence of RS maintenance downtime results in significant deterioration of channel availability.

4 Conclusions

Mathematical model for reliability of the ATM communication channel in the real conditions of radio stations operation is developed.

Expression for channel availability of the ATM communication network with common set of redundant radio stations and periodic maintenance of radio stations to increase efficiency of the communication network is developed.

It is shown, that for communication network with highly reliable radio stations and condition-based maintenance, the influence of RS maintenance downtime is weak on channel availability and may not be taken into account. For periodical schedule-based maintenance, the influence of RS maintenance downtime results in significant deterioration of channel availability. For such case the model for calculation of communication channel availability is developed.

Acknowledgment. This work was supported by Latvian state research project "The Next Generation of Information and Communication Technologies (Next IT)".

References

1. Wiegmann, D., Shappell, S.: Human error perspectives in aviation. Int. J. Aviat. Psychol. **11** (4), 341–357 (2001)
2. Barlow, R., Heidtmann, K.: On the reliability computation of a k-out-of-n system. Microelectron. Reliab. **33**(2), 267–269 (1993)
3. Misra, K.: Handbook of Performability Engineering, 1315 p. Springer (2008)
4. McGrady, P.: The availability of a k-out-of-n:G network. IEEE Trans. Reliab. **R-34**(5), 451–452 (1985)
5. Liu, H.: Reliability of a load-sharing k-out-of-n:G system: non-iid components with arbitrary distributions. IEEE Trans. Reliab. **47**(3), 279–284 (1998)
6. Rushdi, A.: A switching-algebraic analysis of consecutive-k-out-of-n:F systems. Microelectron. Reliab. **27**(1), 171–174 (1987)
7. Ayers, M.: Telecommunications System Reliability Engineering, Theory, and Practice, 256 p. Wiley-IEEE Press, Hoboken (2012)
8. Chatwattanasiri, N., Coit, D., Wattanapongsakorn, N., Sooktip, T.: Dynamic k-out-of-n system reliability for redundant local area networks. In: Proceedings of the 9th International Conference on Electrical Engineering/Electronics, Computer, Telecommunications and Information Technology (ECTI-CON), 16–18 May, pp. 1–4, 16-18 (2012)
9. Kozlov, B., Ushakov, I.: Reliability Handbook (International series in decision processes), 416 p. Holt, Rinehart & Winston of Canada Ltd., New York (1970)
10. Kabashkin, I.: Effectiveness of redundancy in communication network of air traffic management system. In: Proceedings of the Eleventh International Conference on Dependability and Complex Systems DepCoS-RELCOMEX. 27 June–1 July 2016, Brunów, Poland. In the book "Dependability Engineering and Complex Systems". AISC, vol. 470, pp. 257–265. Springer, Switzerland (2016)
11. Kabashkin, I.: Analysing of the voice communication channels for ground segment of air traffic management system based on embedded cloud technology. In: Information and Software Technologies. CCIS, vol. 639, pp. 639–649. Springer International Publishing, Switzerland (2016)
12. Kabashkin, I.: Dynamic reconfiguration of architecture in the communication network of air traffic management system. In: Proceedings of the 17th IEEE International Conference on Computer and Information Technology (IEEE CIT-2017), Helsinki, Finland, 21–23 August, 2017. CPS, pp. 345–350. IEEE Computer Society, IEEE Xplore (2017)
13. Rubino, G., Sericola, B.: Markov Chains and Dependability Theory, 284 p. Cambridge University Press, Cambridge (2014)
14. Kabashkin, I.: Computational method for reliability analysis of complex Systems based on the one class Markov models. Int. J. Appl. Math. Inform. **10**, 98–100 (2016)

Modelling and Safety Assessment of Programmable Platform Based Information and Control Systems Considering Hidden Physical and Design Faults

Vyacheslav Kharchenko[1,2(✉)] ⓘ, Yuriy Ponochovnyi[3] ⓘ,
Anton Andrashov[2] ⓘ, Eugene Brezhniev[1,2] ⓘ,
and Eugene Bulba[1,2] ⓘ

[1] National Aerospace University KhAI, Kharkiv, Ukraine
V.Kharchenko@csn.khai.edu, e.brezhnev@csis.org.ua,
evhenb@gmail.com
[2] Research and Production Company Radiy, Kropyvnytskyi, Ukraine
a.andrashov@radiy.com
[3] Poltava State Agrarian Academy, Poltava, Ukraine
yuriy.ponch@gmail.com

Abstract. The information and control system (I&CS) of Nuclear Power Plant (NPP) is considered as a set of three independent hardware channels including on-line testing system. NPP I&C system's design on programmable platforms is rigidly tied to the V-model of the life cycle. Functional safety and availability during its life cycle are assessed using Markov models. Markov models are used to assess availability function and proof test period. The basic single-fragment model MICS01 contains an absorbing state in case of hidden faults and allows to evaluate risks of "hidden" unavailability. The MICS02 model simulates "migration" of states with undetected failures into states with detected faults. The results of Markov modeling (models MICS01 and MICS02) are compared to evaluate proof test period taking into account requirements for SIL3 level and limiting values of hidden fault probabilities.

Keywords: Functional safety modeling · Information and control system · Availability functions · Undetected failure

1 Introduction

Nuclear Power Plant (NPP) information and control systems (I&CSs) such as reactor trip systems are safety critical systems. Very stringent requirements have been developed to such systems characteristics and life cycle processes. During the development cycle, it is possible to change the architecture of the NPP I&CS project, correct the parameters of its elements, program code and other changes. This affects the final characteristics of the system (quality, reliability, availability, safety, etc.).

The software architecture is modified on functions changing or on the detected software faults elimination. This affects code characteristics and failures and recovery

© Springer Nature Switzerland AG 2020
W. Zamojski et al. (Eds.): DepCoS-RELCOMEX 2019, AISC 987, pp. 264–273, 2020.
https://doi.org/10.1007/978-3-030-19501-4_26

flow parameters. In papers [1–3], the mathematical apparatus of Markov and semi-Markov processes was used to study systems with variable indicators. In [4], a systematic approach to the construction of multi-fragment models was considered, which allows to structure complex systems models development. NPP I&CS design on programmable platforms such as RadICS [5] is rigidly tied to the V-model of the life cycle. During the development, verification and validation phases of the project, the results are recorded at intermediate stages. During the execution of the stage, it is possible to change both the system architecture and its parameters. Therefore, the primary developed safety assessment models need to be adjusted for changes and linked to the stages of the V model.

I&CS design of one of the NPP emergency protection system subsystems, previously considered in [5, 6], is investigated in this paper. Block diagram of the I&CS is shown in Fig. 1. It's one-version I&CS which implements logic of voting "2 out of 3". It includes three independent channels basing on programmable (FPGA) modules of the platform, in particular digital and analog input and output modules, logic module. Each channel is checked by the on-line testing systems for the dangerous failures presence. The hardware channels use the similar software (VHDL). If VHDL design faults occur it is equivalent to a common cause failure (CCF). To minimize CCF risk diversity approach (hardware-software components and process version redundancy) is applied [7, 8].

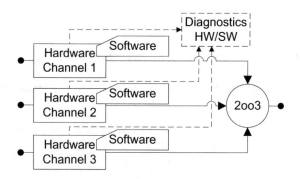

Fig. 1. I&CS reliability structural diagram [2].

Control system is characterized by the parameter of on-line test diagnostics coverage (DC). In the designed system, there are distinctions between the diagnostics of the hardware and software with the parameters DC_{HW} and DC_{SW}. Control is carried out continuously and the detected failures are eliminated immediately after detection.

The class of I&CS on programmable platforms RadICS [5, 8], which are used in NPP reactors emergency protection systems, is studied in this paper. Emergency protection systems belong to the group of complex objects, which are provided with regular proof tests. These tests are performed according to the regulations and should be treated as a separate process.

Multiphase models are used to simulate such tests as separate processes with a periodic component according to IEC61508-6 [7]. They are represented by a set of labeled graphs, each of which models a separate phase - the time interval between proof tests.

Within one phase, the system is simulated by a Markov model with discrete states and continuous time. Unfortunately, the dimension of Markov models for one phase of designed I&CSs is tens to hundreds of states, which makes it impossible to examine the full multiphase model in the framework of this study.

For the NPP I&CS, a single-fragment or multi-fragment Markov model is placed on the time interval between proof tests. The assigned tasks are limited to the responsibility zone of one "phase" (the gap between the proof tests). In this case, from the I&CS model availability function resulting diagram, it is possible to determine the time interval at which the requirements for functional safety $A > A_{req}$ are accomplished. In approximation, it can be considered the interval between the proof tests.

Thus, the purpose of the paper is to develop and investigate models of such systems, taking into account two types of hidden failures for hardware (physical faults) and software (design faults). Considering these failures is very important to assess risk of such faults and to minimize risk of overstated availability and safety assessment. The paper is structured as followed. Next Sect. 2 describes two NPP I&CS functional safety Markov models, their assumptions, states and transitions between them (Subsects. 2.1 and 2.2). The results of the I&CS modeling and safety assessment are analyzed in Sect. 3. Section 4 concludes and describes future steps.

2   ICS Functional Safety Assessment Models Development

2.1   I&CS Functional Safety Assessment Model with Absorbing State (MICS01)

At the initial design stages, the V-models use the initial (basic) I&CS functional safety evaluating model. It is based on the results of the similar systems' analysis, existing functional safety standards and a minimal assumptions set. Model MICS01 (marked graph) is presented in Fig. 2. Developing of the Markov graph is based on the approach described in [7]. It contains one absorptive state with undetected dangerous failures of the hardware and software S17. In the classical model it is impossible to leave the undetected dangerous failure state without carrying out additional measures (proof tests). In spite of the fact that a single-fragment model is considered, its graph is divided into three parts. They simulate three state groups by the signs:

- design/software fault didn't occur,
- design/software fault occurred, but wasn't detected (hidden),
- design/software fault detected by the diagnostic system.

In the first approximation, after an obvious software fault occurrence the system is restored without its elimination (the software is restarted with a rate μ_s). After the latent software fault occurrence, the system continues functioning in the inoperable state.

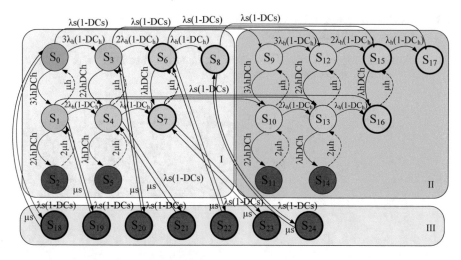

Fig. 2. Marked graph I&CS functional model with absorbing state

The first group of states (software is intact) contains:

(a) healthy states: S0, S1 and S3;
(b) inoperable states, they are detected by the majority authority: S2 and S5;
(c) states with undetected dangerous software failures, the majority body is unable to fend them off: S4, S6, S7, S8.

After dangerous hardware failure detection, the inactive channel is disabled. Then it is restored with the rate μ_h. This is modeled by the transitions S1 → S0, S2 → S1, S4 → S3, S5 → S4, S7 → S6.

The second group of states (software with hidden failure). The system continues to function, but all S9 ... S17 states are inoperative. After detecting a dangerous hardware failure, the inactive channel is turned off and restored with the rate μ_h. This is modeled by the transitions S10 → S9, S11 → S10, S13 → S12, S14 → S13, S16 → S15.

The third group of states (obvious software failure detected, the system stops, the software failure is eliminated by restarting without the defect liquidation). The group contains inoperable states S18 ... S23. In them, the software is restored (restarted) with the rate μ_s. This is modeled by the transitions S18 → S0, S19 → S1, S20 → S3, S21 → S4, S22 → S6, S23 → S7, S24 → S8.

The availability function taking into account dangerous failures is defined as (1):

$$A(t) = P_0(t) + P_1(t) + P_3(t) \tag{1}$$

Baseline conditions: t = 0, $P_0(t) = 1$.

2.2   I&CS Functional Safety Assessment Model with the Migration of Undetected Failures (MICS02)

The basic single-fragment model MICS01 contains an absorbing state. It causes a decrease in system availability to zero.

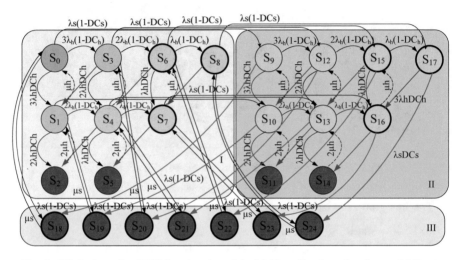

Fig. 3. Marked graph of ICS functional model with the migration of undetected failures

This behavior of the availability function is a typical for most Markov models and weakly correlates with real practical systems. Critical systems development engineers do not allow the possibility of such situations. A possible coordination option may be the approach from [6]. It uses the assumption that after the transition of the system to the absorbing state the conduction of a new failure, which will be revealed by the diagnostic system, is likely over time. After detection, a reaction to failure procedure will be initiated and missed hidden failures may be detected. In this paper, this situation is designated as "migration" of hidden failures into obvious ones.

On the MICS02 model graph (Fig. 3) the arrows-transitions number is increased. This graph is a one-fragment Markov model. Like the last graph (Fig. 2), it is conventionally divided into three groups of states. In the model with the migration of hidden failures, there are exits from the state of undetected dangerous failure without carrying out additional measures (proof tests). This model also considers the restoration of the system after the occurrence of an obvious software defect through the restarting of the software (with rate μ_s).

The number and nature of the MICS02 model graph states is identical to the previous model. In addition to the MICS01 model, transitions were considered that simulate the migration of hidden hardware failures: S3 → S1, S4 → S2, S6 → S4, S7 → S5, S8 → S7, S12 → S10, S13 → S11, S15 → S13, S16 → S14, S17 → S6; and transitions that simulate the migration of hidden software failures: S9 → S18, S10 → S19, S12 → S20, S13 → S21, S15 → S22, S16 → S23, S17 → S24.

3 Simulation and Comparative Analysis

The primary input parameters of Markov models were determined on the basis of certification data [8] for the previous I&CS versions samples. Their values are presented in Table 1.

Table 1. Values of simulation processing input parameters

#	Parameter	Base value
1	λ_{Dh}	46.04622e−6 (1/hour)
2	DCh	0.9989
3	$\mu_h = 1/MRTh$	1/8 = 0.125 (1/hour)
4	λ_{Ds}	6.27903e−6 (1/hour)
5	DCs	0.9902
6	$\mu_s = 1/MRTs$	10 (1/hour)
7	μ_{sr}	1/24 = 0.04167 (1/hour)
8	$\Delta\lambda_{Ds}$	1.5697575e−06 (1/hour)

According to the operational standards [11], I&CS on programmable platforms boundary exploitation term is 15 years, or 131490 h. But since this research involves the study of availability function behavior, the time interval in question will be extended by several orders of degree.

To build the matrix of Kolmogorov-Chapman system of differential equations (SDU) in Matlab, matrix A function was used [12]. The SDU solution is obtained using the ode15s function [13]. The simulation results are shown in Fig. 4. In the MICS01 model there is an absorbing state. Therefore, its availability gradually decreases to zero during $4 * 10^7$ h. In the MICS02 model, there is no absorbing state. It has a typical availability function behavior - approach to stationary value of 0.9901.

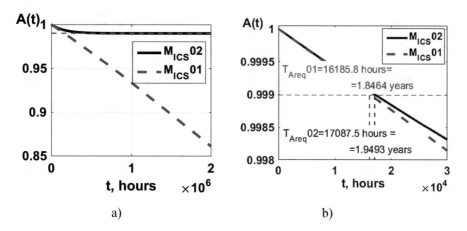

a)   b)

Fig. 4. Simulation results for availability function (a) and determining the interval T_{Areq} with calculation error $\xi = 1e-6$ (b) for models MICS01 and MICS02

The availability function level reduction below 0.999 occurs after 16186 h (1.85 years) in the MICS01 model and after 17087 h (1.95 years) for the MICS02. This does not satisfy the typical standards for industrial systems (3 years or 26298 h). Therefore, we conduct additional research. It is necessary to determine the input parameters values at which $T_{Areq} \geq 26298$ h.

The intervals of input parameters change are defined as follows:

- on the one hand, they are based on the possible options for the architecture [14],
- on the other hand, the DC change limit is extended to 1 for research purposes.

The variable parameters values series are presented in Table 2.

Table 2. Variable input parameters of the ICS model

#	Variable parameter	Designation	Values series
1	The rate of dangerous hardware failures	$\lambda_{D\,H}$	[0.05...5]e−5 (1/hour)
2	Diagnosing dangerous hardware failures control completeness	DC_H	[0..1]
3	The rate of dangerous software failures	$\lambda_{D\,S}$	[0.05...7]e−6 (1/hour)
4	Diagnosing dangerous software failures control completeness	DC_S	[0..1]

Cyclic scripts for Matlab were built to calculate the models. The results of the research are shown as graphical dependences in Figs. 5 and 6.

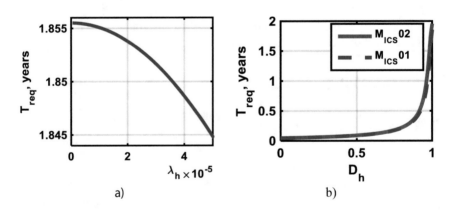

a)   b)

Fig. 5. T_{Areq} interval diagrams (a) from the parameter $\lambda_{D\,H}$ MICS01, (b) for different values of the input parameter DC_H

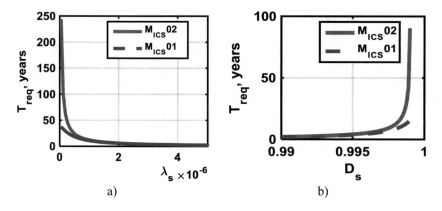

Fig. 6. T_{Areq} interval diagrams (a) for different values of the input parameter $\lambda_{D\ S}$, (b) for different values of the input parameter DC_S

The input parameter $\lambda_{D\ H}$ affects the behavior of the MICS01 model availability function. If it is reduced by an order of degree, the availability will be longer reduced to zero (the time of decrease increases from $4 * 10^7$ to $8 * 10^7$ h, that is, twice). However, this result does not matter for practical application. The ICS will already be decommissioned in such time interval. More important is the result presented in Fig. 5(a). It illustrates the insignificant influence of the parameter $\lambda_{D\ H}$ on the length of the interval $[0 \ldots T_{Areq}]$ (more precisely, on its upper limit T_{Areq}). After the simulation, it becomes obvious that a change of $\lambda_{D\ H}$ by two orders of degree does not allow to ensure the $T_{Areq} \geq 26298$ h condition.

The value of the input parameter DC_H almost does not affect the speed of the availability function transition to the stationary mode. On the other hand, if DC_H is changed (from 0 to 1), the Aconst MICS02 model will change (within $[0 \ldots 0.990199]$). The result, presented in Fig. 5(b), is also important for practice. After the simulation, it becomes obvious that increasing the DC_H to 1 does not allow to ensure the $T_{Areq} \geq 26298$ h condition (as in the MICS01 model).

A decrease by an order of degree in the rate of MICS01 model dangerous software failures $\lambda_{D\ S}$ does not affect the speed of reducing the availability level to zero. For practical application it is not essential. After all, for the time interval of $1.3 * 10^5$ h, the ICS will be decommissioned. The result presented in Fig. 6(a) is more interesting. The parameter $\lambda_{D\ S}$ significantly affects the upper limit of the interval $[0 \ldots T_{Areq}]$. When $\lambda_{D\ S} = 3.9 * 10^{-6}$ (1/hour), the condition $T_{Areq} \geq 26298$ h is satisfied.

With a decrease in the rate of dangerous software failures, $\lambda_{D\ S}$ A_{const} of the MICS02 model increases linearly, but not significantly (6 decimal places). The result of determining the T_{Areq} value is presented in Fig. 6(a) in comparison with the result of the model MICS01. If $\lambda_{D\ S} = 4.11 * 10^{-6}$ (1/hour), the condition $T_{Areq} \geq 26298$ h is provided.

By reducing the test coverage of dangerous software failures by an order of degree (from $DC_S = 0.999$ to $DC_S = 0.99$, etc.), the MICS01 model availability function goes to zero level several times faster (from $6 * 10^7$ to $3 * 10^7$ h). The result presented in

Fig. 6(b) is important for practice. Starting from $DC_S = 0.9939$, the condition $T_{Areq} \geq 26298$ h is satisfied.

In the MICS02 model, the input parameter DC_S linearly affects A_{const}. It is important to note that with $DC_S = 1 \rightarrow A_{const} = 0.999992$. The value for the SIL3 requirements ($A_{const} = 0.99901$) is achieved at $DC_S = 0.9991$. Theoretically, this allows us to talk about systems without proof test. But from a practical point of view, it is very difficult and expensive to achieve such a level of control completeness.

The result presented in Fig. 6(b) is important for practice: starting from the value $DC_S = 0.9934$, the condition $T_{Areq} \geq 26298$ h is satisfied.

4 Conclusion

The necessity of the development of adequate models, linked to the I&CS development life cycle V model stages and errors of embedded or on-line testing subsystems, is substantiated and confirmed basing on investigation of industrial safety critical system. Two models MICS01 and MICS02 of the NPP I&CS safety have been developed and analyzed taking into account two types of hidden failures of programmable platform components.

In the MICS01 model, the availability function is reduced to zero (due to the absorbing state in the digraph). For typical input parameters values (Table 1), the fulfillment of SIL3 requirements is guaranteed in the [0 … 1.86 years] interval. An increase in the interest interval $T_{proof\ test}$ up to 3 years is possible if the rate of dangerous software failures $\lambda_{D\ S} \leq 3.9 * 10^{-6}$ (1/hour), or if the control completeness of dangerous software failures $DC_S \geq 0.9939$.

In the MICS02 model, the availability function is reduced to the stationary A_{const} value. For typical input parameters values (Table 1), the fulfillment of SIL3 requirements is guaranteed in the [0 … 1.95 years] interval. An increase in the interest interval $T_{proof\ test}$ up to 3 years is possible if $\lambda_{D\ S} \leq 4.11 * 10^{-6}$ (1/hour), or if $DC_S \geq 0.9934$. Starting from $DC_S = 0.9991$, the fulfillment of SIL3 requirements is guaranteed without additional test checks.

The developed models and technique of considering hidden failures allows to minimize over estimate risk and adds metric-based technique of choice of tools and settings for tools to improve accuracy of safety assessment [17].

Further research should be directed to the adequate consideration of the parameters for software faults eliminating processes and reducing the failure rate, as well as the development of the automatic correction of the structure and parameters of the I&CS availability and functional safety models technology.

References

1. Ghosh, R., Longo, F., Frattini, F., Russo, S., Trivedi, K.: Scalable analytics for IaaS cloud availability. IEEE Trans. Cloud Comput. **2**, 57–70 (2014)
2. Trivedi, K., Kim, D., Roy, A., Medhi, D.: Dependability and security models. In: 7th International Workshop on Design of Reliable Communication Networks, pp. 11–20 (2009)

3. Kharchenko, V., Ponochovnyi, Y., Boyarchuk, A.: Availability assessment of information and control systems with online software update and verification. In: Ermolayev, V., Mayr, H., Nikitchenko, M., Spivakovsky, A., Zholtkevych, G. (eds.) Information and Communication Technologies in Education, Research, and Industrial Applications, ICTERI 2014. Communications in Computer and Information Science, vol. 469, pp. 300–324 (2014)
4. Kharchenko, V., Ponochovnyi, Y., Boyarchuk, A., Brezhnev, E., Andrashov, A.: Monte-Carlo simulation and availability assessment of the smart building automation systems considering component failures and attacks on vulnerabilities. In: Zamojski, W., Mazurkiewicz, J., Sugier, J., Walkowiak, T., Kacprzyk, J. (eds.) Contemporary Complex Systems and Their Dependability, DepCoS-RELCOMEX 2018. Advances in Intelligent Systems and Computing, vol. 761, pp. 270–280 (2018)
5. Bulba, Y., Ponochovny, Y., Sklyar, V., Ivasiuk, A.: Classification and research of the reactor protection instrumentation and control system functional safety Markov models in a normal operation mode. CEUR Workshop Proc. **1614**, 308–321 (2016)
6. Ponochovniy, Y., Bulba, E., Yanko, A., Hozbenko, E.: Influence of diagnostics errors on safety: Indicators and requirements. In: 2018 IEEE 9th International Conference on Dependable Systems, Services and Technologies (DESSERT), pp. 54–58 (2018)
7. IEC 61508-6:2010: Functional safety of electrical/electronic/programmable electronic safety related systems, Part 6: Guidelines on the application of IEC 61508-2,3 (2010)
8. D7.24-FSC(P3)-FMEDA-V6R0. Exida FMEDA Report of Project: Radiy FPGA-based Safety Controller (FSC) (2018)
9. Langeron, Y., Barros, A., Grall, A., Berenguer, C.: Combination of safety integrity levels (SILs): a study of IEC61508 merging rules. J. Loss Prev. Process Ind. **21**(4), 437–449 (2008)
10. The function for drawing graphs and digraphs using MATLAB. http://iglin.exponenta.ru/All/GrMatlab/grPlot.html. Accessed 24 Feb 2019
11. IEC 61513:2011, Nuclear power plants - Instrumentation and control important to safety - General requirements for systems (2011)
12. Kharchenko, V., Ponochovnyi, Y., Boyarchuk, A., Gorbenko, A.: Secure hybrid clouds: analysis of configurations energy efficiency. In: Zamojski, W., Mazurkiewicz, J., Sugier, J., Walkowiak, T., Kacprzyk, J. (eds.) Theory and Engineering of Complex Systems and Dependability, DepCoS-RELCOMEX 2015. Advances in Intelligent Systems and Computing, vol. 365, pp. 195–209 (2015)
13. Solve stiff differential equations and DAEs; variable order method - MATLAB ode15s. https://www.mathworks.com/help/matlab/ref/ode15s.htmll. Accessed 24 Feb 2019
14. Sklyar, V.V.: Elements of the information and control systems functional safety analysis methodology. Radioelectron. Comput. Syst. **6**(40), 75–79 (2009)
15. Kharchenko, V., Ponochovnyi, Y., Abdulmunem, A., Andrashov, A.: Availability models and maintenance strategies for smart building automation systems considering attacks on component vulnerabilities. In: Zamojski, W., Mazurkiewicz, J., Sugier, J., Walkowiak, T., Kacprzyk, J. (eds.) Advances in Dependability Engineering of Complex Systems, DepCoS-RELCOMEX 2017. Advances in Intelligent Systems and Computing, vol. 582, pp. 186–195 (2017)
16. Kharchenko, V., Ponochovnyi, Y., Boyarchuk, A., Brezhnev, E.: Resilience assurance for software-based space systems with online patching: two cases. In: Zamojski, W., Mazurkiewicz, J., Sugier, J., Walkowiak, T., Kacprzyk, J. (eds.) Dependability Engineering and Complex Systems, DepCoS-RELCOMEX 2016. Advances in Intelligent Systems and Computing, vol. 470, pp. 267–278 (2016)
17. Kharchenko, V., Butenko, V., Odarushchenko, O., Sklyar, V.: Multifragmentation markov modeling of a reactor trip system. ASME J. Nucl. Eng. Radiat. Sci. **1**(3), 031005-1–031005-10 (2015)

Dependability of Service of Substation Electrical Equipment: Estimation of the Technical Condition State with the Use of Software and Information Tools

Alexander Yu. Khrennikov[1], Nikolay M. Aleksandrov[2]([✉]),
and Pavel S. Radin[3]

[1] Scientific and Technical Center, Federal Grid Company of United Energy
System, Moscow, Russia
Hrennikov_AY@ntc-power.ru
[2] SPE "Dynamics", Cheboksary, Russia
nickdynamics@gmail.com
[3] Federal Grid Company of the Unified Energy System, Yakutsk, Russia

Abstract. This paper presents an experience in application of electrical equipment condition estimation. With the transition to the system of the electrical equipment repair to condition-based maintenance (c) system and the introduction of the system of planning the resources of enterprise in FGC UES Company planning measures for maintenance and repair of electrical equipment rose to qualitatively different level and, as consequence increased the requirements, presented to the estimation of equipment state.

Diagnostics specialists faced a complex of issues, without solution of which was impossible the realization of all effective tools of the productive system automatic control system (ASU) of maintenance and repair and final transition to CBM repair system.

The task lies in the fact that it is possible to decrease, or exclude faults with the use of information tools. By information tool in this work is implied software, located on a stationary PC or a laptop. As a whole by information tool it is possible to understand the specialized automated working place (further SAWP).

To make decisions such as to evaluating controlling influences can be only based on the estimation of technical condition state (TCS).

In the article the general relationship of errors is examined in those influencing the quality of the estimation of the technical condition state of electrical equipment and so the way of their decrease and of the eliminations of reasons, which cause errors. The use of special information tools (software) will make it possible to considerably decrease the subjective errors, which appear in the process TCS electrical equipment.

Making a decision about development and introduction in FGC Company of the tools of similar SAWP is extremely necessary, since TCS has a significant effect on the total effectiveness of company.

W. Zamojski et al. (Eds.): DepCoS-RELCOMEX 2019, AISC 987, pp. 274–283, 2020.
https://doi.org/10.1007/978-3-030-19501-4_27

Keywords: Maintenance · Repair · Condition-based maintenance ·
Error of measurement · The automated working place ·
The estimation of technical state · Information tools

1 Introduction

The electrical objects of Federal Grid Company of United Energy System (FGC Company) are located in 73 regions of the Russian Federation by the total area of more than 13,6 millions of km^2. Company exploits 125,3 thousand km of overhead transmission lines ensures the functioning of 856 electrical substations with the total installed transformer capacity of more than 322,6 thousand MVA and the voltage from 35 to 1150 kV [1].

With the transition to the system of the electrical equipment repairs to "condition-based maintenance" and the introduction of the system of planning the resources of enterprise FGC Company planning measures for maintenance and repair (M&R) of electrical equipment rose to qualitatively different level and, as consequence increased the requirements, presented to the estimation of equipment state. Before the specialists of divisions, who carry out diagnostics of electrical equipment, faced with the complex of questions, without solution of which was impossible the realization of all effective tools of the productive system of M&R and final passage to "condition-based maintenance" repair system [2].

The component part of IEPS AAN is the business-process "diagnostics of electrical equipment", which is achieved within the framework the activity of FGC Company and is one of the most important business processes. This occurs because the quality of the estimation of technical condition state (TCS) directly influences on the reliability of the work of electrical equipment, the effectiveness of the functioning of company and, as a result, to the investment FGC Company as discussed elsewhere [1–5].

In entire TCS allows to get information in objective and sufficient way about the state of the active company equipment and by other ways, except transition to TCS, this information cannot be obtained. To make decisions such as to evaluating controlling influences can be only based on the estimation of technical condition state (TCS) [4, 6–10].

2 Technical Condition State of the Transformer-Reactor Oil-Filled Equipment

The TCS and the technical diagnosis of electrical equipment can be divided into two basic segments: (a) the diagnosis of the basic oil-filled equipment of substations and (b) the diagnosis of the elements of air electrical lines. The classification of reasons and forms of damages and also the block diagram of the technical diagnosis of the transformer-reactor oil-filled equipment are presents in Fig. 1.

Designations in the diagram: DGA-the chromatographic analysis of the dissolved gases of transformer oil; PCA-the physical chemistry analysis of transformer oil, Ktr-transformation ratio, Pol-the estimation of the degree of polymerization of paper

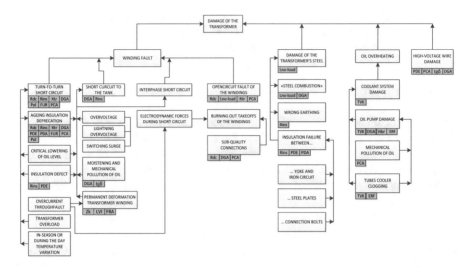

Fig. 1. The classification diagram of damage's reasons and forms. schematic of the technical diagnosis of the transformer-reactor oil-filled equipment

insulation, Rins-the measurement of the insulation resistance, RDC-the measurement of direct-current resistance, FUR-the determination of the content of furan derivatives, tgδ-the determination of the dielectric power factor, PDA-level measurement of partial discharges by acoustic method, PDE-level measurement of partial of discharges by electrical method, LVI-diagnosis by the method of low-voltage pulses (deformation of windings), FRA-the determination of the mechanical state (deformations of windings) of transformer (reactor) by frequency response analysis, Zk-the determination of short-circuit impedance, Lno-load-the determination of the no-load losses, TVK-thermal-vision control, Vibr-measurement of vibration characteristics, ERF-estimates of the rate of oil flow as discussed elsewhere [3–7, 11].

The quantity and the variety of the oil-filled equipment diagnosis methods several times exceed the amount of the methods of diagnosis of overhead transmission lines. This is caused historically due to paying more attention to the reliability of the oil-filled equipment; by the same reasons there are a large quantity of effective "on-line" monitoring systems of the transformer equipment, for example, bushings control systems R-1500, systems of the transformer oil monitoring Hydran, TRANSFIX, TIM, TDM.

The guarantee of failure-free operation of electrical stations and networks of Russia, under the conditions of aging of electrotechnical equipment and also the improvements of TCS system, optimization of expenditures for modernization and technical ree-quipping of the objects of power engineering requires both the creation of the new methods of diagnosis and improvement of the already used means of the inspection of the flaw detection of transformer and reactor electrical equipment (TREE) [3, 11]. Development and introduction of the new methods of diagnosis makes it possible to calculate the value of the worn resource taking into account of actual conditions and modes of operation TREE as discussed by Lech and Tyminski [11].

The application of contemporary methods of diagnostics is directed toward the timely detection of the appearing defects, the adoption of measures for their elimination, the detection of the most worn units TREE and the determination of the priority of the modernization of power facilities. The development of the methods of diagnosis, especially used during the TREE service, will make it possible to switch over from the system of regular overhauls to repair system due to the technical state (TS) [4, 11].

3 Metrological Estimation of Error in the Measurements with the Determination of the Technical State of the Electrical Equipment

The general percentage of errors influencing on the quality of the estimation of the technical condition state of electrical equipment is given in Fig. 2 [10, 12, 13].

The instrument error - is the error, which is determined by errors in the means of measurements used, and it is caused by the imperfection of operating principle, by an inaccuracy in the calibration of the scale, by the lack of clarity of instrument. At present in practice it is excluded by means of the application of reliable and sufficiently precise means of measurements.

The systematic (methodological) error - is the error, caused by the imperfection of method and also by the simplifications, assumed as the basis of procedure. At present it is considerably decreased by the way of refining theoretical knowledge of the assumed as basis procedures.

Subjective error - is the error, caused by the degree of attentiveness, concentration, preparedness and by other qualities of operator. In the recent decades it is not practically accepted effective measures for reduction in the data of errors; on the contrary, an increase in the influence of data of errors under the action of the following factors occurred:

- A notable increase in the quantity of equipment for that being undergoing TCS;
- A notable increase in the parameters of control TCS;
- An increase in the tempo of the fulfillment of works;
- An increase in information traffic of that processed by operator;
- An increase in the psychological load of operator as discussed elsewhere [10, 12, 13].

By subjective error in this paper is understood the influence of the errors, which appear, beginning from the stage, when the operator readouts the readings of instrument to the stage of the formation of conclusion about the state of object. The utilized methods and tools with working of the values, obtained with the direct measurements on the equipment, practically did not change (they were not improved) in the last 20 years [4, 8–10, 12, 13].

If we accept the percentage of errors with TCS at the beginning of the 90's for the starting point, then at present the percentage of errors with TCS electrical equipment will be analogous with the given percentage in Fig. 2 [4, 6, 7, 14, 15].

In the different subdivisions the Company the percentage of subjective errors to the instrument and by systematic are different, but general tendency corresponds to that given in Fig. 2. Being based on this it is possible to draw the conclusion that after removing the reasons for the appearance of subjective errors (further errors) it is possible to considerably increase quality TCS.

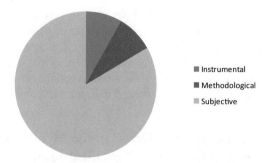

Fig. 2. The percentage of errors with TCS at present

Let us examine what stages TCS and what factors lead to the appearance of error (Fig. 3).

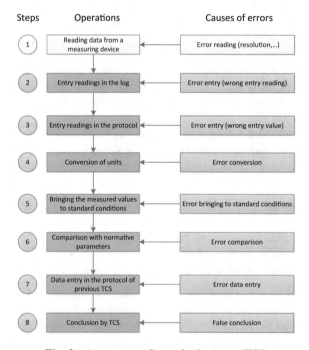

Fig. 3. Appearance of error in the stages TCS

Factors, indicated in Fig. 3 are specified on following basic reasons [9, 10, 12]:

– The repeated transfer of data of one data carrier in another, what leads to entering of incorrect values-stage 2–7;

- "Manual" calculation of values-stages 4, 5; so bringing a number of the parameters of those utilized with TCS to the standard conditions is conducted with the aid of the nomograms (visual method), which so entering additional error;
- The absence of the reliable system of the verification both of the entered, calculated, corrected values and conclusion TCS-stages 1–8.

To remove the reasons, which cause error in stages 2–8, possibly and this will be shown below. Concerning the first stage it is possible to say that with the standard measurements and the assumption about final attentiveness of operator the today this error is practically unavoidable (exception-control of counted data by the second operator) as discussed elsewhere [5–10, 12].

4   Results and Discussion

4.1   The Temperature Drift of the Measuring Channel Parameters

For example, the drift of temperatures of the bus arrangement of transformer, which has the defective contact connection, determined according to the results of thermal-vision inspection makes it possible to determine the temperature distribution along the busbar (Fig. 4).

Fig. 4. Defect of the contact connection of the disconnector of 330 kV substation at the south of Russia

Additive error-this is the error, which appears because of the summing up of numerical values and which does not depend on the value of the measured quantity, undertaken on the module (absolute).

Multiplicative error - this is the error, which is changed together with a change in the values of the value, which is undergone measurements.

It is possible to compensate multiplicative and additive error by the calibration of instrument. The multiplicative error includes an error in the input divider, an error in the complete scale in the analog-to-digital converter (ADC). The additive error includes the error, caused by electromotive force (EMF) of displacement and by input currents of operational amplifier (OA) of infrared device, the error in the zero drift of ADC, the error, caused by the instability of the feeding voltage.

By calibration is compensated the systematic value of additive and multiplicative error under normal conditions, the accuracy of the result of measurement will be determined by the temperature drift of the parameters, the determining metrological properties of measuring channel.

Let us determine a multiplicative error in the input divider from the temperature drift of resistors, with a change in the temperature for $\Delta\theta = 65\ °C$

$$\delta Uvh := \left| \frac{R1}{(R1 + R2) \cdot R2} \right| \cdot \Delta R2 + \left| \frac{-1}{R1 + R2} \right| \cdot \Delta R1 \qquad (1)$$

where: Uvh - input phase voltage,
R1, R2 - resistances of input divider,
$\Delta Ri = \Delta Ri/\Delta\theta \cdot \Delta\theta$ - the deviation of resistance from the nominal value with a change in the temperature on $\Delta\theta$.

$$\delta Uvh = 0{,}032\%$$

Additive error OA, determined by the drift of input currents and EMF of displacement from the temperature with a change in the temperature on $\Delta\theta = 65\ °C$

$$\Delta Uoutput2 = 173\ mcV$$

4.2   The Estimation of a Maximum Error in the Measuring Channel

Maximum additive error is equal to the sum of additive errors OA and error in the instability of the power source.

$\Delta add = 174$ mcV
Maximum multiplicative error is equal
$\Delta mul = 0.0323\%$

To considerably descend or more precisely to in practice exclude error is possible with the use of information tools. By information tool in this work is implied software, located on stationary computer or on laptop. As a whole by information tool it is possible to understand the specialized automated working place (further SAWP) (Fig. 5).

The principle of use SAWP is extremely simple and consist only in the initial introduction by the operator of data of those obtained in stage 1 (Fig. 3), further transport and data processing is for the most part achieved by program; the function of operator-expert, in responsibility of whom enters only the administration of SAWP (Fig. 5). Since the errors, which appear in stages 2–8, are caused only by the subjective factor (emotional and mental condition of operator), with the exception of the direct influence of the state of operator on the process of decision making are excluded data of error, that leads to a substantial increase in the quality TCS.

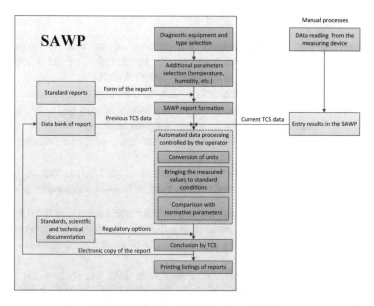

Fig. 5. Diagram of functioning SAWP

With the use SAWP with TCS on stationary computer are excluded errors in stages 3–8 (Fig. 3). The probability of the occurrence of error in stage 2 remains and is caused by the need for the transfer in SAWP of data of the measurements, fixed in the periodical on the spot of the installation of equipment. With the use of laptop the periodical (Fig. 3, stage 2) is excluded and data are introduced by operator directly in SAWP, that makes it possible to exclude the probability of the appearance of an error by the caused specific character of stage 2 (Fig. 3) [4–8, 11].

5 Conclusions

The TCS and the technical diagnosis of electrical equipment is divided into two basic segments: The diagnosis of the basic oil-filled equipment of substations and the diagnosis of the elements of air electrical lines.

To considerably descend or more precisely to in practice exclude error is possible with the use of information tools by the specialized automated working place (further SAWP).

The principle of use SAWP is extremely simple and consist only in the initial introduction by the operator of data of those obtained in stage 1, further transport and data processing is for the most part achieved by program; the function of operator-expert, in responsibility of whom enters only the administration of SAWP.

Basic condition of developed SAWP:

- Progressive ergonomic interface-for the purpose of solution of basic problem-reduction in the load on operator and as the consequence of the decrease of the probability of the occurrence of subjective errors with TCS;
- Simplicity of the utilized algorithms of TCS-in the stage of the formation of conclusion about equipment state according to the results of tests and measurements not rationally to begin to operate bulky prognostic algorithms;
- Modular structure SAWP-for increasing the flexibility of system during its further modernization;
- The distributed structure of SAWP-with the use of an WEB-interface, for guaranteeing the condition of the access into the system from any computer of Company.

As a whole it is possible to make the preliminary conclusion that the use of special information tools (software) will make it possible to considerably decrease the subjective errors, which appear in the process of TCS of electrical equipment.

Acknowledgment. Mr. Richard Malewski, Poland, Mr. John Lapworth, Great Britain, are greatly acknowledged for supporting this study. Cooperation of Universities and Innovation Development, Doctoral School project "Complex diagnostic modeling of technical parameters of power transformer-reactor electrical equipment condition" has made publishing of this article possible.

References

1. Fogelberg, T., Girgis, R.S.: ABB power transformers-a result of merging different technologies with prospects for significant future advancements. In: International Symposium "Electrotechnics-2010", Moscow, no. 1 (1994)
2. Khrennikov, A.Yu.: Power transformer's fault diagnostics at SAMARAENERGO Co, including FRA/LVI method. Reports from School of Math. and System Engineering, Vaxjo University, Sweden N43 (2000). ISSN: 1400-1942
3. Malewski, R., Khrennikov, A.Yu.: Monitoring of winding displacements in HV transformers in service, pp. 4–9. CIGRE Working Group 33.03, Italy (1995)
4. Lech, W., Tyminski, L.: Detecting transformer winding damage-the low voltage impulse method. Electr. Rev. **179**, 768–772 (1966)
5. Khrennikov, A.Yu.: Diagnostics of electrical equipment's faults, defects and weaknesses. In: Reports of Conference on Condition Monitoring and Diagnosis (CMD 2006), Korea (2006)
6. Khrennikov, A.Yu.: New "intellectual networks" (Smart Grid) for detecting electrical equipment faults, defects and weaknesses. Smart Grid Renew. Energy **3**(03), 159 (2012)
7. Khrennikov, A.Yu.: Smart Grid technologies for detecting electrical equipment faults, defects and weaknesses. In: Workshop on Mathematical Modelling of Wave Phenomena with Applications in the Power industry, Linnaeus University, Växjö (2013). http://lnu.se/mmwp
8. Khrennikov, A.Yu., Shakaryan, Y.G., Dementyev, Y.A.: Shortcurrent testing laboratories. Short-circuit performance of power transformers, transformer testing experience. Int. J. Autom. Control Eng. (IJACE). **2**(3), 120–127 (2013). http://www.seipub.org/ijace/AllIssues.aspx

9. Khrennikov, A.Yu., Mazhurin, R.V., Radin, P.S.: Infra-red and ultraviolet control, LVI-testing, partial discharges and another diagnostic methods for detection of electrical equipment's faults, defects. J. Multidiscip. Eng. Sci. Technol. (JMEST) 1(4) (2014). http://www.jmest.org/vol-1-issue-4-november-2014/. ISSN: 3159-0040

10. Khrennikov, A.Yu.: Diagnostics of Electrical Equipment Faults and Power Overhead Transmission Line Condition by Monitoring Systems (Smart Grid): Short-Circuit Testing of Power Transformers, 174 p. Nova publishers, New York (2016). https://www.novapublishers.com/catalog/product_info.php?products_id=56561

Reliability Modeling of Technical Objects in the Airport Security Checkpoint

Tomasz Kisiel$^{(\boxtimes)}$ ⓘ and Maria Pawlak

Faculty of Mechanical Engineering, Department of Maintenance and Operation
of Logistics, Transportation and Hydraulic Systems,
Wroclaw University of Science and Technology, Wroclaw, Poland
{tomasz.kisiel,maria.pawlak}@pwr.edu.pl

Abstract. The paper presents a simulation model that proposes to supplement
the approach currently used in structure planning of a security control system.
The current solutions presented in scientific works focus only on the system
structure without taking into account the disturbances caused by failures of
technical objects. By using the Monte-Carlo simulation, a method for estimating
the resilience of a security control system to technical facilities damage under
varying load conditions was proposed. The article presents various configurations of security control lanes and their reliability structures. Next, the simulation model was described and its advantages were demonstrated. The model was
validated for the selected structure of the security control system. The validation
was carried out for 5 parallel security control lanes, equipped with Explosive
Detection System EDS and Explosive Trace Detection ETD devices. The paper
analyzes the impact of ETD failures on the waiting time of passengers in the
system. Sensitivity analysis has been performed.

Keywords: Simulation model · Security control · Airport

1 Introduction

For several years, there has been a dynamic increase in air traffic. This fact is a
challenge in guaranteeing punctuality during air operations. The current infrastructure
of air transport is insufficient to handle air traffic planned for the coming years. The
existing disturbances are connected with the capacity of the airspace and airports.

Airport capacity may be defined as the maximum number of passengers or operations that an airport is able to handle in a given unit of time. However, it should be
mentioned that this parameter is only a mathematical formula, therefore the capacity
should be planned much higher so that the airport can handle the actual number of
passengers with any disturbances. This is due to the high seasonality of air traffic
throughout the year, as well as changes occurring within individual days and even
hours. Therefore, the key role of the proper functioning of the airport is played by
proper adjustment of the throughput of its individual elements according to the demand
for the air service (much less for a short period of time than a day, month or year).

Selected hubs and regional airports were analyzed. One hour was selected as the
time interval. The analysis shows that the capacity is utilized not uniformly. The

W. Zamojski et al. (Eds.): DepCoS-RELCOMEX 2019, AISC 987, pp. 284–292, 2020.
https://doi.org/10.1007/978-3-030-19501-4_28

coefficient of variation can be as much as 0.77 for hubs and 0.95 for regional airports. This means that there are both rush hours and time periods that do not fully utilize the system.

When the capacity is fully utilized during rush hours, any failure of the technical objects can cause very long delays, while in the remaining time interval the system is resilient because it will have a structural excess.

Taking into account the presented research problem, the aim of the work was formulated. This article presents a simulation model, which aims to take into account the failure of technical objects that may reduce the throughput of the security control system at the airport. Security control is considered because it is a bottleneck in the passenger service process.

The next parts of the article present in turn: literature review, theoretical aspects of reliability of security control lanes, simulation model and its validation. The simulation model is a continuation of the previous scientific work of the authors and will be partially referred to another source. In this work, additional functionality was introduced, which complements the previous work with interferences caused by damage to technical devices.

2 State of the Art

In the scientific literature there are lots of scientific articles that are devoted to the planning and management of the airport security checkpoint. It is not necessary to put many of them here and give examples because they have been reviewed in [1]. The important thing is that these works assume that the system is reliable. The purpose of this works is to adapt the structure of the system to the required capacity [2–4]. Another goal of this articles is the dynamic resource planning issues [5].

Looking more broadly at the research areas, issues related to the reliability of security controls may also be found. Here, this is understood in a different way. Reliability is focused on security-related aspects. The purpose of such work is to check if the system works properly and prohibited articles are eliminated from the system [6–8].

However, there are many works from other branches of transport systems [9–12] or industries [13–15] that indicate the importance of the reliability of the technical system. The robustness of the rail transport system to various types of failures is described in [16, 17]. An intermodal container terminal [18] was also considered whether road transport [19]. The applications are very different. The technical system may be in different states [20]. His availability is very often considered [21, 22]. In addition, there are issues related to risk in such cases [23]. But security control has not yet been considered in such aspects.

This article also takes into account the significance of the field of reliability and has extended the approach in modeling such a system.

3   Security Checkpoint Reliability

The process of security control at airports is carried out in accordance with the regulations [1]. These regulations allow for the possibility of implementing the process using various methods of detection of prohibited items. Passengers are subjected to security checks using the following methods [24]:

- manual control;
- metal detection gates (WTMD);
- dogs for detecting explosives;
- devices for detecting traces of explosives (ETD).
- devices for screening persons not using ionizing radiation;
- devices for detecting traces of explosives (ETD) in combination with a hand-held metal detector (HHMD).

Cabin baggage is subjected to a security check using the following methods [24]:

- manual control;
- X-ray devices;
- explosive detection systems (EDS);
- dogs for detecting explosives in combination with manual control,
- ETD devices.

The method of implementation and the use of particular detection methods are specified in Commission Decision C (2015) 8005. This article considers the reliability of the security control system whose lanes are equipped with EDS, WTMD and ETD devices. Passenger control uses the WTMD device in combination with manual control and ETD device. The cabin baggage check is carried out with the EDS device in conjunction with manual control. In the case of failure to selected technical objects, it is possible to use alternative methods, however this article focuses on taking into account the reliability of the basic security control system.

Two types of desks can be distinguished in the security control system. For a single-lane desk, technical devices are dedicated exclusively to the control of passengers moving through one control lane. Double-lanes desk is equipped with devices that can be part of the two control lanes. Typically, a WTMD device is a common part. However, sometimes an ETD device is also shared.

It was assumed that the system is in the up-state when at least one lane can perform its functions. The reliability structure of a single-lane security control desk is shown in Fig. 1. Failure of one of the technical devices causes the station to go into the down-state.

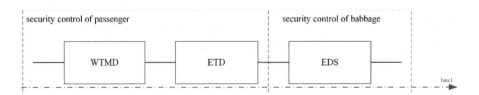

Fig. 1. The reliability structure of a single-lane security control desk

The double-lanes desk, in which the common element is a WTMD device (see Fig. 2), is in a down-state when the WTMD device, two ETD devices, two EDS devices or one ETD and EDS device are failed for two different control lanes. In the event of failure of one of the ETD or EDS devices, the double-lanes desk is in partially down-state, which limits the possibility of performing security control for only one of two lanes.

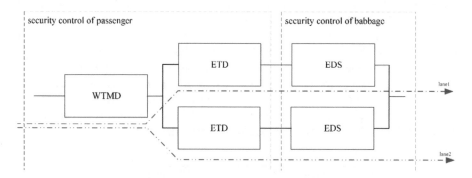

Fig. 2. The reliability structure of a double-lanes security control desk (with common WTMD)

The double-lanes desk, in which the common element are WTMD and ETD devices (see Fig. 3), is in a down-state when the WTMD, ETD or 2 EDS devices are failed. In the event of failure of one of the EDS devices, the double-lanes desk is in partially down-state, which limits the possibility of per-forming security control for only one of two lanes.

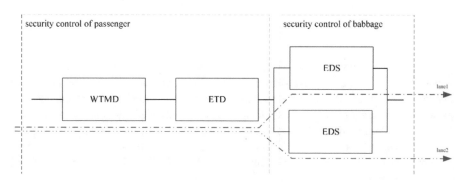

Fig. 3. The reliability structure of a double-lanes security control desk (with common WTMD & ETD)

A security control system consisting of several parallel desks will be in a down-state if all positions are failed. It is acceptable to implement the process in the condition of in partially down-state. Next chapter presents a proposal for modeling a security control system that allows estimating the impact of failures to technical devices on the basic functional indicators of the system.

4   Simulation Model of the Security Checkpoint

The developed model consists of two modules that depend on each other (Fig. 4). This is a continuation of the previous scientific work, in which Module I has already been described in detail in [25]. This module is responsible for the implementation of the process in accordance with the assumed work schedule. A simplified passenger service algorithm is presented in Fig. 5. The passenger reports to the queuing system in accordance with the probability density function of the notification time before the scheduled departure time $f(t_r)$. Then the passenger following the passengers in front of him, also expecting a vacant security control lane. The service time is carried out as a random variable described by the probability density distribution $f(t_s)$.

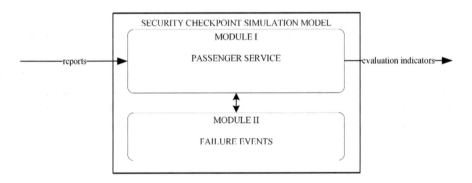

Fig. 4. Simulation model concept

In this work, Module II complements the presented approach in [25]. Module II is responsible for introducing disturbances in the system, which are caused by failures of technical objects. Module II is based on the discrete event method. The time between consecutive failures of a given technical object is expressed by a random variable, described by the function $f(i_{bf})$, expressing the distribution of the number of passenger services carried out between the failures. The time of objects recovery is expressed by a random variable described by the function $f(t_{tr})$ expressing the distribution of the recovery duration.

The Module II algorithm is implemented in parallel for each device from the moment of starting the simulation experiment. In the first step, it determines the moment of device failure i_{bf}. Then, after each passenger service, the value of i from i_u is compared. If there is compliance, the device goes to down-state and the selected lane is closed.

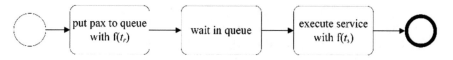

Fig. 5. Simplified passenger service algorithm (prepared in BPMN notation)

Then, in accordance with the variable t_{tr}, the recovery process is carried out, after which the algorithm is carried out again. A simplified algorithm scheme is shown in Fig. 6. The algorithm stops working only after the simulation experiment is finished.

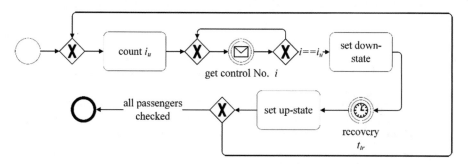

Fig. 6. Simplified technical objects failures algorithm (prepared in BPMN notation)

5 Simulation Model Validation

Model validation was performed for a 5 parallel single security control lanes. Input data in accordance with formulas (1–4) was assumed.

$$f(t_r) = \frac{\left((t_r - 229.80)/25553.35\right)^{7.89}}{25553.35 \cdot B(8.89, 53.24)\left(1 + (t_r - 229.80)/25553.35\right)^{62.13}} \, [s] \qquad (1)$$

$$f(t_s) = 0.03 \left(\frac{t_s}{22.81}\right)^{-0.17} \exp\left(-\left(\frac{t_s}{22.81}\right)^{0.83}\right) [s] \qquad (2)$$

$$f(i_{bf}) = 0.002 \cdot \exp(-0.002 \cdot i_{bf}) [pas.] \qquad (3)$$

$$f(t_{tr}) = \frac{\exp\left(-\frac{1}{2}\left(\frac{t_{tr} - 1800}{300}\right)^2\right)}{300\sqrt{2\pi}} [s] \qquad (4)$$

Formulas (1) and (2) were determined on the basis of preliminary research on the real system carried out at Wroclaw Airport. These functions were estimated with the EasyFit tool. Kolmogorov-Smirnov's compliance test was carried out at the materiality level $\alpha = 0.05$. Formulas (3) and (4) were assumed theoretically (based only on the knowledge of experts from Wroclaw airport) and represent the distribution of random variables of time between failures and repair time of ETD devices. It was assumed that the reliability of other devices is equal to 1.

The Monte-Carlo simulation was carried out for a given example of the flight and security lanes schedule using the FlexSim software. The average waiting time of passengers in the queue was determined. The sensitivity analysis of the system was performed depending on the parameter λ affecting the random variable i_{bf}.

For the changed parameter λ in the exponential distribution function (3) it can be seen that the system is robust (it has a sufficient structural excess) if the average time between failures of ETD devices is greater than 130 passengers checked $\lambda > 0.0077$ (see Fig. 7). If it turns out that λ is less than 130, it will be necessary to expand the system structure with additional ETD devices. These devices will be used as a structural stock.

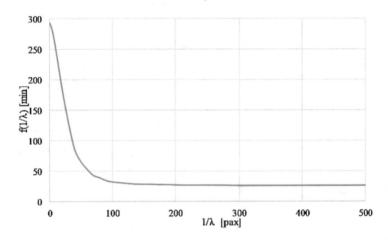

Fig. 7. Sensitivity analysis of waiting time in the queue from the ETD failures intensity

The conducted analysis can thus provide relevant information about the structure of the analyzed system for any stream of reports and at any given service schedule.

6 Summary

The article presents the concept of a new approach in modeling the airport security checkpoint. In contrast to the current approach, the proposed method predicts the possible effects caused by technical objects failures.

The presented concept can be used at the stage of system planning as well as during its operation. The process manager can adjust the flight schedule in such a way as to ensure an adequate structural excess of the technical system. This will make the system resilient to interference caused by technical objects failures.

The presented concept assumes the development of a discrete event simulation model and requires the introduction of input data representing random variables of passenger reports, service times as well as times between failures of technical objects and times of recovery them.

References

1. Wu, P., Mengersen, K.: A review of models and model usage scenarios for an airport complex system. Transp. Res. Part A **47**, 124–140 (2013)
2. Kierzkowski, A.: Method for management of an airport security control system. In: Proceedings of the Institution of Civil Engineers - Transport, vol. 170, no. 4, pp. 205–217 (2017). https://doi.org/10.1680/jtran.16.00036
3. Neufville, R.: Building the next generation of airports systems. Transp. Infrastruct. **38**(2), 41–46 (2008)
4. Wilson, D., Roe, E.K., So, S.A.: Security checkpoint optimizer (SCO): an application for simulating the operations of airport security checkpoints. In: Winter Simulation Conference, pp. 529–535 (2006)
5. Kierzkowski, A., Kisiel, T.: A model of check-in system management to reduce the security checkpoint variability. Simul. Modell. Pract. Theory **74**, 80–98 (2017). https://doi.org/10.1016/j.simpat.2017.03.002
6. Kierzkowski, A.: Model of reliability of security control operation at an airport. Tehnički vjesnik. **24**(2), 469–476 (2017). https://doi.org/10.17559/TV-20150812153802
7. Skorupski, J.: The simulation-fuzzy method of assessing the risk of air traffic accidents using the fuzzy risk matrix. Saf. Sci. **88**, 76–87 (2015). https://doi.org/10.1016/j.ssci.2016.04.025
8. Skorupski, J., Uchroński, P.: A fuzzy model for evaluating airport security screeners' work. J. Air Transp. Manage. **48**, 42–51 (2015). https://doi.org/10.1016/j.jairtraman.2015.06.011
9. Krzykowska, K., Siergiejczyk, M., Rosiński, A.: The safety level analysis of the SWIM system in air traffic management. TransNav **10**(1), 85–91 (2016). https://doi.org/10.12716/1001.10.01.09
10. Restel, F.J., Wolniewicz, Ł.: Tramway reliability and safety influencing factors. Procedia Eng. **187**, 477–482 (2017). https://doi.org/10.1016/j.proeng.2017.04.403
11. Walkowiak, T., Mazurkiewicz, J.: Soft computing approach to discrete transport system management. Lecture Notes in Computer Science. Lecture Notes in Artificial Intelligence, vol. 6114, pp. 675–682 (2010). https://doi.org/10.1007/978-3-642-13232-2_83
12. Walkowiak, T., Mazurkiewicz, J.: Analysis of critical situations in discrete transport systems. In: Proceedings of DepCoS - RELCOMEX 2009, Brunów, Poland, 30 June–02 July 2009, pp. 364–371. IEEE (2009). https://doi.org/10.1109/DepCos-RELCOMEX.2009.39
13. Vintr, Z., Valis, D.: Modeling and analysis of the reliability of systems with one-shot items. In: 2007 Annual Reliability and Maintainability Symposium, pp. 380–385 (2007). https://doi.org/10.1109/RAMS.2007.328106
14. Giel, R., Plewa, M.: Analysis of the impact of changes in the size of the waste stream on the process of manual sorting of waste. In: Świątek, J., Wilimowska, Z., Borzemski, L., Grzech, A. (eds.) Information Systems Architecture and Technology: Proceedings of 37th International Conference on Information Systems Architecture and Technology – ISAT 2016 – Part III. Advances in Intelligent Systems and Computing, vol. 523. Springer, Cham (2017)
15. Giel, R., Plewa, M., Młyńczak, M.: Analysis of picked up fraction changes on the process of manual waste sorting. Procedia Eng. **178**, 349–358 (2017). https://doi.org/10.1016/j.proeng.2017.01.063
16. Friedrich, J., Restel, F.J., Wolniewicz, Ł.: Railway operation schedule evaluation with respect to the system robustness. In: Contemporary Complex Systems and Their Dependability: Proceedings of the Thirteenth International Conference on Dependability and Complex Systems, DepCoS-RELCOMEX. Springer, vol. 761, pp. 195–208 (2019). https://doi.org/10.1007/978-3-319-91446-6_19

17. Restel, F.J.: Defining states in reliability and safety modelling. In: Zamojski, W., Mazurkiewicz, J., Sugier, J., Walkowiak, T., Kacprzyk, J. (eds.) Theory and Engineering of Complex Systems and Dependability. DepCoS-RELCOMEX 2015. Advances in Intelligent Systems and Computing, vol. 365. Springer, Cham (2015). https://doi.org/10.1007/978-3-319-19216-1_39

18. Zajac, M., Swieboda, J.: An unloading work model at an intermodal terminal. In: Zamojski, W., Mazurkiewicz, J., Sugier, J., Walkowiak, T., Kacprzyk, J. (eds.) Theory and Engineering of Complex Systems and Dependability. DepCoS-RELCOMEX 2015. Advances in Intelligent Systems and Computing, vol. 365. Springer, Cham (2015)

19. Tubis, A., Gruszczyk, A.: Measurement of punctuality of services at a public transport company. In: Carpathian Logistics Congress, CLC 2015, Jesenik, Czech Republic, 4th–6th November 2015, pp. 512–517. Tanger (2016)

20. Werbińska, S.: Interactions between logistic and operational system - an availability model. In: Risk, Reliability and Societal Safety, vol. 2, pp. 2045–2052. Taylor and Francis, Leiden (2007)

21. Jodejko-Pietruczuk, A., Werbińska-Wojciechowska, S.: Development and sensitivity analysis of a technical object inspection model based on the delay-time concept use. Maint. and Reliab. **19**(3), 403–412 (2017)

22. Tubis, A.: Route risk assessment for road transport companies. In: Zamojski, W., Mazurkiewicz, J., Sugier, J., Walkowiak, T., Kacprzyk, J. (eds.) Contemporary Complex Systems and Their Dependability. DepCoS-RELCOMEX 2018. Advances in Intelligent Systems and Computing, vol. 761. Springer, Cham (2019). https://doi.org/10.1007/978-3-319-91446-6_46

23. Restel, F.J., Zając, M.: Reliability model of the railway transportation system with respect to hazard state. In: International Conference on Industrial Engineering and Engineering Management (IEEM), pp. 1031–1036 (2015). https://doi.org/10.1109/ieem.2015.7385805

24. Commission Implementing Regulation (EU) 2015/1998 of 5 November 2015 laying down detailed measures for the implementation of the common basic standards on aviation security

25. Kierzkowski, A., Kisiel, T.: Simulation model of security control system functioning: a case study of the Wroclaw airport terminal. J. Air Transp. Manage. **64**, Part B, 173–185 (2017). https://doi.org/10.1016/j.jairtraman.2016.09.008

Multi-clustering Used as Neighbourhood Identification Strategy in Recommender Systems

Urszula Kużelewska[(⊠)]

Bialystok University of Technology, Wiejska 45A, 15-351 Bialystok, Poland
`u.kuzelewska@pb.edu.pl`

Abstract. This article describes clustering approach to neighbourhood calculation in collaborative filtering recommender systems. Precise identification of neighbours of an active object (a user to whom recommendations are generated) is very important due to its direct impact on quality of generated recommendation lists. Clustering techniques, although improving time effectiveness of recommender systems, can negatively affect quality (precision) of recommendations. In this article it is proposed a new algorithm based on multi-clustering, as well as author's description of this term. Despite of various definitions in papers, a common distinctive feature of multi-clustering is its multiple point of view of one dataset. Various views discover their different aspects, selecting the most appropriate data model to solution of a current problem. The article contains experiments confirming advantage of multi-clustering approach over the traditional method based on single-scheme clustering. The results include recommendation quality and time effectiveness comparison, as well.

Keywords: Recommender systems · Multi-clustering · Collaborative filtering

1 Introduction

Nowadays people frequently encounter a problem of information overload on the internet. Recommender Systems (RSs) are computer applications designed to help users in searching for items (music, books, news, goods, etc.) as well as delivering unexpected ones while just hunting the internet for something interesting. They base on the information about users or items, which are in user's search history, visited web sites including time spent on them, as well as item's description, similarity to other items, etc. and predict a level of interest of users on new, never seen, items [2,4,6]. On one hand, Recommender Systems are great decision making support in an excess of internet data, on the other hand, they maximize profit for e-commerce and information-based companies such as Google, Netflix, Twitter [11].

© Springer Nature Switzerland AG 2020
W. Zamojski et al. (Eds.): DepCoS-RELCOMEX 2019, AISC 987, pp. 293–302, 2020.
https://doi.org/10.1007/978-3-030-19501-4_29

Although the most common are hybrid, Recommender Systems can be classified into the following categories [2]: collaborative filtering (CF), content-based (CB) and knowledge-based systems (KB). The first methods search for similar users or items, and assume that users with corresponding interests prefer the same items. Collaborative filtering approach have been very successful due to its precise prediction ability. Content-based techniques look for similarity in items by analysis of their text data e.g. description. They form a term vectors model for documents and process them with respect to term frequency. Content-based algorithms do not require lengthy behaviour history from users, therefore they can cope with a new items problem. The last category is the knowledge-based approach. The algorithms take into account specific domain knowledge and its relation to users' needs and specifications [12].

The article is organised as follows: Sect. 2 presents selected collaborative filtering with focus on clustering based solutions in domain of Recommender Systems, including problems they solve and encounter. The following section, Sect. 3 describes the proposed multi-clustering algorithm on the background of alternative clustering techniques, whereas Sect. 4 contains results of performed experiments with the aim to compare multi-clustering to kNN as well as single-clustering approach. The last section concludes the paper.

2   Related Work

Collaborative filtering methods base on either user-based or item-based similarity to make recommendations. User-based models assume that similar users have the same taste, therefore they search for users who have similar ratings to active users. Item-based approach bases on the principle, that similar items are evaluated in the same way by the same user [1]. A method to evaluate a prediction rating $\hat{r}(x_i)$ is similar for both of the mentioned approaches, and (1) defines it for user-based model composed of k users:

$$\hat{r}(x_i) = \mu_{x_i} + \frac{\sum_{j=1, i \neq j}^{k} sim(x_i, x_j) \cdot (r(x_j) - \mu_{x_j})}{\sum_{j=1, i \neq j}^{k} |sim(x_i, x_j)|} \tag{1}$$

where μ_{x_i} is a mean rating (see (2)) for user x_i on a vector of ratings $V = \{v_1, \ldots, v_c\}$ and $sim(x_i, x_j)$ is a similarity value between users: x_i and x_j. As a similarity value it can be used one of several common measures, e.g based on Euclidean distance, cosine value, Pearson correlation, LogLikehood based, Tanimoto, adopted from mathematical applications [12].

$$\mu_{x_i} = \frac{\sum_{q \in V(x_i)} r(x_{iq})}{|V(x_i)|} \tag{2}$$

An example similarity formula based on Pearson correlation is as follows (3):

$$sim_P(x_i, x_j) = \frac{\sum_{k \in V_{ij}} (r(x_{ik}) - \mu_{x_i}) \cdot (r(x_{jk}) - \mu_{x_j})}{\sqrt{\sum_{k \in V_{ij}} (r(x_{ik}) - \mu_{x_i})^2} \cdot \sqrt{\sum_{k \in V_{ij}} (r(x_{jk}) - \mu_{x_j})^2}} \tag{3}$$

where $V_{ij} = V(x_i) \cap V(x_j)$ is a set of ratings present in both user's vectors: i and j.

The item-based approach usually generates more relevant recommendations due to the fact that it uses user's own ratings - there are identified similar items to a target item, and the user's own ratings on those items are used to extrapolate the ratings of the target. This approach is more resistant to changes of the ratings, as well, because usually a number of users is much larger than a number of items and new items are less frequently added to the dataset [1].

The main problem with collaborative filtering methods is their time complexity. Usually, a set of items and users is extremely big, therefore to make them in reasonable time it is appropriate to reduce a search space for candidate objects. The most commonly used method for this issue is k Nearest Neighbors (kNN) [2,15]. It calculates all user-user or item-item similarities and identifies the most k similar objects (users or items) to the active object as its neighbourhood. Then, prediction is performed only on objects from the neighbourhood reducing the time of calculations. The kNN algorithm is a reference method for determining neighbourhood of an active user for the collaborative filtering recommendation process [4]. Its advantages are: simplicity and reasonably accurate results; its disadvantages - low scalability and vulnerability to sparsity in data.

An efficient solution to this problem can be clustering algorithms, that identify clusters for further use as a pre-defined neighbourhood [14]. Recently, clustering algorithms have drawn much attention of researchers and there were proposed new algorithms, particularly developed for recommender systems application [8,16]. The efficiency of clustering techniques is related to fact, that a cluster is a neighbourhood that is shared by all the cluster members, in contrast to kNN approach determining neighbours for every object separately [1]. The disadvantage of this approach is usually loss of prediction accuracy. The level of reduction depends on clusters quality and granularity [14], however there are many cluster-based recommender system that are precise and scalable [7,13].

The explanation for decreasing recommendations accuracy is in the way how clustering algorithms work. A typical approach bases on a single partitioning scheme, which is generated once and then not updated significantly. There are two major problems related to quality of clustering. The first is the clustering results depend on the input algorithms parameter, and additionally, there is no reliable technique to evaluate clusters before on-line recommendations process. The second issue is imprecise neighbourhood modelling of data located on borders of clusters.

To improve quality of the neighbourhood modelling the author decided to use multiple clustering schemes, instead of a single partitioning, and to select the most suitable one to the particular data object. This way leads to elimination inconvenience of decreased recommendations quality keeping high time effectiveness.

The mentioned method has one drawback: it is resources-consuming solution. However, nowadays there are servers very rich in memory and CPU resources, therefore the drawback is not a great limitation.

3   Description of M-CCF Algorithm on the Background of Alternative Clustering

The role of a multi-clustering technique in recommendations generation process is to determine the most appropriate neighbourhood for an active user. It means that the algorithm selects the best cluster from a set of clusters prepared previously. The general algorithm $M - CCF$ (Multi-Clustering Collaborative Filtering) is presented in Algorithm 1. The input set contains data of n users, who rated a subset of items - $A = \{a_1, \ldots, a_k\}$. The set of possible ratings - V - contains values $v_1, \ldots, v_c$. The input data are clustered ncs times into nc clusters every time giving as a result a set of clustering schemes CS. Finally, the algorithm generates a list of recommendations R_{x_a} for the active user.

Algorithm 1. A general algorithm $M - CCF$ of a recommender system based on multi-clustering used in the experiments

---

Data:

- $U = (X, A, V)$ - matrix of clustered data, where $X = \{x_1, \ldots, x_n\}$ is a set of users, $A = \{a_1, \ldots, a_k\}$ is a set of items and $V = \{v_1, \ldots, v_c\}$ is a set of ratings values,
- $\delta : v \in V$ - a similarity function,
- $nc \in [2, n]$ - a number of clusters,
- $ncs \in [2, \infty]$ - a number of clustering schemes,
- $CS = \{CS_1, \ldots, CS_{ncs}\}$ - a set of clustering schemes,
- $CS_i = \{C_1, \ldots, CS_{nc}\}$ - a set of clusters for a particular clustering scheme,
- $CS_r = \{c_{r,1}, \ldots, c_{r,nc \cdot ncs}\}$ - the set of cluster centres,

Result:

- A_{Rx_a} - a list of recommended items for an active user x_a,

begin
  $\delta_1..\delta_{ncs} \longleftarrow$ calculateSimilarity(CS_r, CS_i, δ);
  $C_{best_{x_a}} \longleftarrow$ findTheBestCluster$(x_a, CS_r, \delta_1..\delta_{ncs \cdot ncs}, CS_r, CS_i)$;
  $R_{x_a} \longleftarrow$ recommend$(x_a, C_{best_{x_a}}, \delta_1..\delta_{nc \cdot ncs})$;

The set of groups is identified by clustering algorithm which is run several times with the same or different values of its input parameters. In the experiments described in this article, as a clustering method, $k - means$ was used. The set of clusters delivered for the collaborative filtering process was generated with the same parameter k (a number of clusters). This step, although time consuming, has a minor impact on overall system scalability, because it is performed rarely and in off-line mode.

After neighbourhood identification, the following step, appropriate recommendation generations, is executed. This process requires, despite of a great precision, high time effectiveness. Multi-clustering satisfies these two conditions, because it can select the most suitable neighbourhood area of an active user for candidates searching and the neighbourhood of all objects is already determined, as well.

One of the most important issues of this approach is to generate a wide set of input clusters that is not very numerous, however provides a high similarity for every user or item. The second matter concerns matching users with the best clusters as their neighbourhood. It can be obtained in the following ways. The first of them compares active user's ratings with the cluster centers' ratings and searches for the most similar one using a certain similarity measure. The other way, instead of cluster centers, can compare an active user with all cluster members and select the one with the highest overall similarity. The both solutions have their pros and cons, e.g. the first one will work well for clusters of spherical shapes, whereas the second one requires higher time consumption. In the experiments presented in this paper, the clusters for active users are selected basing on their similarity to centers of groups (see Algorithm 2).

Algorithm 2. Algorithm of cluster selection of $M - CCF$ recommender system used in the experiments

---

Data:

- $U = (X, A, V)$ - matrix of clustered data, where x_a is an active user, $A = \{a_1, \ldots, a_k\}$ is a set of items and $V = \{v_1, \ldots, v_c\}$ is a set of ratings values,
- $\delta : v \in V$ - a similarity function,
- $CS = \{CS_1, \ldots, CS_{ncs}\}$ - a set of clustering schemes,
- $CS_i = \{C_1, \ldots, CS_{nc}\}$ - a set of clusters for a particular clustering scheme,
- $CS_r = \{c_{r,1}, \ldots, c_{r,nc \cdot ncs}\}$ - the set of cluster centres,

Result:

- $C_{best_{x_a}}$ - the best cluster for an active user x_a,
- δ_{best} - a matrix of similarity within the best cluster

begin
    $\delta_1..\delta_{ncs \cdot ncs} \longleftarrow$ calculateSimilarity(x_a, CS_r, δ);
    $\delta_{best} \longleftarrow$ selectTheHighestSimilarity$(\delta_1..\delta_{ncs})$;
    $C_{best_{x_a}} \longleftarrow$ findTheBestCluster$(\delta_{best}, CS, CS_i)$;

Afterwards, a recommendations generation process works typically basing on a collaborative filtering approach, however the candidates are searched only within the selected cluster of neighbourhood.

The described algorithm is an author's technique, not found in other publications. The presented approach defines a multi-clustering process as generation a set of clustering results obtained from an arbitrary clustering algorithm with the same data on its input. An advantage of this approach is better quality of the neighbourhood modelling leading to high quality of predictions, keeping real time effectiveness provided by clustering methods.

There are some other methods, which can be generally called as alternative or multi-view clustering, that find partitioning schemes on different data (e.g. ratings and text description) combining results after all ([3,9]). The aim of a multi-view partitioning is to generate distinct aspects of the data and to search for the mutual link information among the various views, finally leading to the same cluster structure [5].

Examples of alternative clustering applications in recommender systems are the following. A method described in [10] combines both content-based and collaborative filtering approaches. The system uses multi-clustering, however it is interpreted as clustering of a single scheme on both techniques. It groups the ratings, to create an item group-rating matrix and a user group-rating matrix. As a clustering algorithm it uses $k - means$ combined with a fuzzy set theory to represent the level of membership of an object to the cluster. Then a final prediction rating matrix is calculated to represent the whole dataset. In the last step of pre-recommendation process $k - means$ is used again on the new rating matrix to find a group of similar users. The groups represent neighbourhood of users to limit a search space for a collaborative filtering method.

Another solution is presented in [17]. The authors observed, that users might have different interests over topics, thus might share similar preferences with different groups of users over different sets of items. The method $CCCF$ (Co-Clustering For Collaborative Filtering) first clusters users and items into several subgroups, where the each subgroup includes a set of like-minded users and a set of items in which these users share their interests. The groups are analysed by collaborative filtering methods and the result recommendations are aggregated over all the subgroups.

4   Experiments

This section contains results of experiments with the new multi-clustering recommender system $M - CCF$. There were examined precision of generated recommendation lists as well as time effectiveness and scalability. It was taken into consideration various similarity indices (LogLikehood - $LogLike$, cosine coefficient - $Cosine$, Pearson correlation - $Pearson$, Euclidean distance - $Euclid$, CityBlock metrics - $CityBl$ and Tanimoto - $Tanimoto$) during the recommendation process, as well as two types of distance measures: Euclidean and cosine-based, during the clustering phase. All results were compared with traditional collaborative filtering method based on single-scheme clustering $k - means$ for neighbourhood identification.

The clustering method, similarity and distance measures were taken from Apache Mahout environment [18]. The multi-clustering algorithm was implemented basing on classes, data models and structures derived from Apache Mahout, as well. The dataset examined in the experiments is a subset of benchmark LastFM data [19] containing 10 million ratings. The subset ($100\,kdata$) consisted of 100 000 entries - 2 032 users and 22 174 artists.

Quality of recommendations was calculated with $RMSE$ measure (a classical error measure - Root Mean Squared Error) in the following way. For every user from the input set their ratings were divided into the following parts: training (70%) and testing. The values from a testing part were removed and estimated by the recommender system. Difference between the original and calculated value is taken for evaluation. For a test set containing N ratings for evaluation, this measure is calculated using 4, where $r_{real}(x_i)$ is a real rating of user x_i for a particular item i and $r_{est}(x_i)$ is a rating estimated by a recommender system for this user. Although the lower value of $RMSE$ denotes a better prediction ability, there is no maximal value for this measure.

$$RMSE = \frac{\sum_{i=1}^{N} |r_{real}(x_i) - r_{est}(x_i)|}{N} \tag{4}$$

Time effectiveness (t_{av}) has been measured in the following way. A set of M (in the experiments M was equal 100) users was constructed by random users selection from a whole dataset. Then, for each of them it was generated a list of propositions consisted H (in the experiments H was equal 5) elements. The process was repeated K times (in the experiments K was equal 10) and final value is average time of recommendations generation per one user (see 5).

$$t_{av} = \frac{\sum_{i=1}^{K} \frac{\sum_{j=1}^{M} \sum_{k=1}^{H} t_{rec}(x_{jk})}{M}}{K} \tag{5}$$

The first experiment compared the previous results and performance of single-scheme clustering as a technique of modelling of neighbourhood. The clustering algorithm was $k - means$ and the distance measure - Euclidean metrics. Tables 1 and 2 contain results of $RMSE$ values and time (in s) of execution on clusters. The shortest values of time are in bold. The clustering schemes were generated for the following values of k parameter: 20, 50, 200 and 1000. It is evident, that precision of recommendations decreased - $RMSE$ changes from 0.62 to 0.68. This statistics are slightly connected with a number of clusters: for cases of 50 and 200 groups the error is the greatest.

Table 1. RMSE of item based collaborative filtering recommendations with neighbourhood determined by a single clustering $k - means$. The best values are in bold.

Number of clusters	Similarity measure					
	LogLike	Cosine	Pearson	Eucl	CityBl	Tanimoto
20	0.65	0.65	**0.64**	**0.64**	0.65	0.66
50	0.67	0.67	0.67	**0.66**	0.67	0.68
200	0.68	0.67	0.67	**0.66**	0.68	0.67
1000	0.66	0.64	0.65	**0.62**	0.65	0.66

In the following experiment, it was tested $M - CCF$ multi-clustering recommender system. The dataset as well as a distance measure were the same. It was examined an impact of a number of clusters, as well - the dataset was split separately 4 times - for 20, 50, 200 and 1000 clusters. By analogy with the previous experiments, different measures were used to determine a level of similarity among vectors of items in the appropriate recommendations generation process.

Table 2. Time [s] of item based collaborative filtering recommendations with neighbourhood determined by a single clustering $k - means$. The best values are in bold.

Number of clusters	Similarity measure					
	LogLike	Cosine	Pearson	Eucl	CityBl	Tanimoto
20	0.017	0.019	0.018	0.019	**0.015**	0.016
50	0.026	0.027	0.027	0.027	**0.024**	**0.024**
200	0.011	0.012	0.011	0.011	**0.010**	**0.010**
1000	**0.010**	0.020	0.020	0.020	**0.010**	**0.010**

For every case of a number of clusters, the $k - means$ algorithm was executed 3 times with the same value of k and all generated schemes were used in CF method for its input as it was described in Algorithm 1.

Tables 3 and 4 contain $M - CCF$ algorithm performance. The time of recommendations generation is greater even than in the experiments where neighbourhood was modelled by kNN method, however the value of $RMSE$ is tremendously lower. The only problematic case was Pearson similarity metrics, which often generated not measurable results.

This approach is also slightly a number of cluster dependent - the greater its value the worse precision. It means, that in case of greater number of clusters the optimal number of clustering schemes is greater, as well. The neighbourhood in this case has limited area and is very adjusted to the specific cluster center and the greater set of schemes offers higher probability that an active user is closer to any cluster's center.

Table 3. RMSE of item based collaborative filtering recommendations with neighbourhood determined by multi-clustering $M - CCF$ method. The best values are in bold.

Number of clusters	Similarity measure					
	LogLike	Cosine	Pearson	Eucl	CityBl	Tanimoto
20	0.15	0.15	-	**0.11**	0.15	0.15
50	0.15	0.15	-	**0.10**	0.16	0.16
200	0.20	0.22	-	**0.16**	0.22	0.21
1000	0.34	0.35	**0.00**	**0.31**	0.34	0.34

Table 4. Time [s] of item based collaborative filtering recommendations with neighbourhood determined by multi-clustering $M - CCF$ method. The best values are in bold.

Number of clusters	Similarity measure					
	LogLike	Cosine	Pearson	Eucl	CityBl	Tanimoto
20	**2.21**	2.65	-	2.62	2.91	2.30
50	**3.11**	3.41	-	3.21	3.26	3.14
200	13.50	15.70	-	14.26	16.28	**13.15**
1000	40.62	46.65	43.68	**39.95**	40.50	46.00

Taking into consideration all experiments presented in this article, it can be stated, that $M - CCF$ multi-clustering recommender system and the technique of dynamic selection the most suitable clusters offers very valuable results with respect to precision of recommendations. However, there is a disadvantage that needs improvement - time effectiveness.

5   Conclusions

This article presented a new approach to collaborative filtering recommender systems that focuses on a problem of an active user's neighbourhood identification. The algorithm $M - CCF$ is based on multi-clustering, that is it requires on its input to have a set of clustering schemes generated á priori on the same dataset regardless of the values of clustering algorithm input parameters and even regardless of the type of clustering method.

Application of the algorithms based on a single-scheme clustering is characterised by high time efficiency and scalability. However, this benefit usually involves decreased accuracy of prediction feature. It results from inaccurate modelling of object neighbourhood in case of data located on borders of clusters. The aim of $M - CCF$ algorithm was to model an active user's neighbourhood precisely through selection the most appropriate group from the set of clustering schemes.

Multi-clustering approach has proved its benefit as it eliminated inconvenience of decreased quality of predictions. It can be stated, that $M - CCF$ generates results, that are particularly valuable with respect to precision of recommendations. However, the step of the best cluster selection negatively impacts the time effectiveness and the further effort must be put to lead the method to reasonable time efficiency.

Acknowledgment. The present study was supported by a grant S/WI/1/2018 from Bialystok University of Technology and founded from the resources for research by Ministry of Science and Higher Education.

References

1. Aggrawal, C.C.: Recommender Systems: The Textbook. Springer, Switzerland (2016)
2. Tuzhilin, A., Adomavicius, G.: Toward the next generation of recommender systems: a survey of the state-of-the-art and possible extensions. IEEE Trans. Knowl. Data Eng. **17**, 734–749 (2005)
3. Bailey, J.: Alternative Clustering Analysis: A Review, Intelligent Decision Technologies: Data Clustering: Algorithms and Applications, pp. 533–548. Chapman and Hall/CRC, Boca Raton (2014)
4. Bobadilla, J., Ortega, F., Hernando, A., Gutiérrez, A.: Recommender systems survey. Knowl. Based Syst. **46**, 109–132 (2013)
5. Guang-Yu, Z., Chang-Dong, W., Dong, H., Wei-Shi, Z.: Multi-view collaborative locally adaptive clustering with Minkowski metric. Expert Syst. Appl. **86**, 307–320 (2017)
6. Jannach, D.: Recommender Systems: An Introduction. Cambridge University Press, New York (2010)
7. Kużelewska, U.: Clustering algorithms in hybrid recommender system on MovieLens data. Stud. Logic gramm. Rhetor. **37**, 125–139 (2014)
8. Logesh, R., Subramaniyaswamy, V., Vijayakumar, V., Gao, X.-Z., Indragandhi, V.: A hybrid quantum-induced swarm intelligence clustering for the urban trip recommendation in smart city. Future Gener. Comput. Syst. **83**, 653–672 (2017)
9. Mitra, S., Banka, H., Pedrycz, W.: Rough-fuzzy collaborative clustering. IEEE Trans. Syst. Man Cybern. Part B (Cybern.) **36**(4), 795–805 (2006)
10. Puntheeranurak, S., Tsuji, H.: A multi-clustering hybrid recommender system. In: Proceedings of the 7th IEEE International Conference on Computer and Information Technology, pp. 223–238 (2007)
11. Portugal, I., Alencar, P., Cowan, D.: The use of machine learning algorithms in recommender systems: a systematic review. Expert Syst. Appl. **97**, 205–227 (2017)
12. Ricci, F., Rokach, L., Shapira, B.: Recommender Systems: Introduction and Challenges Recommender Systems handbook, pp. 1–34. Springer, Boston (2015)
13. Rongfei, J.: A new clustering method for collaborative filtering. In: Proceedings of the International IEEE Conference on Networking and Information Technology, pp. 488–492 (2010)
14. Sarwar, B.: Recommender systems for large-scale e-commerce: scalable neighborhood formation using clustering. In: Proceedings of the 5th International Conference on Computer and Information Technology (2002)
15. Schafer, J.B., Frankowski, D., Herlocker, J., Sen, S.: Collaborative filtering recommender systems. In: Brusilovsky, P., Kobsa, A., Nejdl, W. (eds.) The Adaptive Web, pp. 291–324 (2007)
16. Sheugh, L., Alizadeh, S.: A novel 2D-Graph clustering method based on trust and similarity measures to enhance accuracy and coverage in recommender systems. Inf. Sci. **432**, 210–230 (2017)
17. Wu, Y., Liu, X., Xie, M., Ester, M., Yang, Q.: CCCF: improving collaborative filtering via scalable user-item co-clustering. In: Proceedings of the Ninth ACM International Conference on Web Search and Data Mining, pp. 73–82 (2016)
18. Apache Mahout. http://mahout.apache.org/. Accessed 14 Dec 2018
19. A Million Song Dataset. https://labrosa.ee.columbia.edu/millionsong/lastfm/. Accessed 02 Nov 2018

Assessment of the Potential of the Waterway in the City Using a Fuzzy Inference Model

Michał Lower[(✉)] and Anna Lower

Wroclaw University of Science and Technology, ul. Wyb. Wyspianskiego 27, 50-370 Wroclaw, Poland
{michal.lower,anna.lower}@pwr.edu.pl

Abstract. The paper involves the assessment of the potential of the waterway in the city as a complement to the city's communication infrastructure network. The evaluation is made using the fuzzy inference model built by the authors. In practice, it happens that strategic decisions are made incorrectly without a deeper analysis. Our model can be useful in making such strategic decisions. Research using the model can replace the work of a team of experts whose work is time-consuming and expensive. Therefore, such detailed analyzes are not always carried out in strategic decisions. The analysis using the model is easy to carry out, gives the effect in a short time at a low cost. Fuzzy logic is a good tool in such cases because the factors to be assessed are often ambiguous and fuzzy, difficult to be precisely determined but possible for the evaluation by an expert at the level of the linguistic variable values. The constructed model takes into account geographical, technical and urban criteria. The result of the inference determines the assessment of the river's potential in terms of use for tourist and business transport. The proposed fuzzy inference model has been validated on the example of the city of Wroclaw in Poland. The obtained results are convergent with the current decision of the city authorities made in 2018 regarding the possibility of using the structure of the water network in Wroclaw and with the results of previous analysis carried out and published by the authors.

Keywords: Fuzzy inference · Water transport

1 Introduction

Contemporary cities experience significant communication problems in the field of vehicular traffic. The slowdown of traffic, especially strong during peak hours, is acute for traffic participants, residents and pedestrians alike. The phenomenon of congestion is attempted to be minimized in various ways. One of the methods is introducing P&R systems, widely discussed in the literature, e.g. in publications [4,5] presenting the method of assessing the location of system facilities. In the case of waterside cities, the potential of the river as a waterway can be used.

© Springer Nature Switzerland AG 2020
W. Zamojski et al. (Eds.): DepCoS-RELCOMEX 2019, AISC 987, pp. 303–310, 2020.
https://doi.org/10.1007/978-3-030-19501-4_30

The launch of water transport can significantly reduce the congestion. Some researchers believe that it is possible to reduce traffic by as much as 10%. Authors of [9] also point out that the introduction of water transport can be a competitive method to reduce congestion and solve the transport problem in the city.

For cities with an extensive network of water structures, the alternative in the form of waterbus appears to be very attractive, but in reality it will not always bring real benefits. The functional assessment of the waterway potential is a complex process and requires the analysis of a team of experts. Such studies are time-consuming and expensive, so decisions made at the strategic level are often based on general premises and the enthusiasm of city authorities. An example is the planning process in Wroclaw. Due to the extensive infrastructure of water connections in the city, waterbus communication seemed a natural need. In planning documents, for many years there was a record of the introduction of a waterbus on the Odra river, even designating the location of stops. It was only in the current Study [1] of 2018 this notation has been changed and the recreational role of the river as the leading one in the city structure was indicated. It is also an important aspect of river functioning in the city, widely considered in literature. The authors of [2] note that tourism on rivers has increased in recent years. Research is being carried out on both the geographical conditions of Europe's rivers and the level of available infrastructure serving tourists on the waterway. Also in [8] the authors point out that equipping the river with tourist facilities and amenities may affect the attractiveness of the river. In the literature studies that describe functional elements of a waterbus based solution can be found. An important topic is the need to build a network structure connecting different modes of transport [6]. The role of waterbus makes the most sense in a comprehensive system. The aspect of network interconnection is also pointed out by the authors [7] who introduce a quantitative analysis of the criteria for assessing transport networks. Our proposal is a fuzzy inference model that will allow to comprehensively assess the potential of a river in a city to introduce water transport. The model was developed on the basis of expert knowledge and knowledge collected in literature studies. The proposed model will make it possible to quickly assess the potential of the waterway by people who are not highly qualified experts, without the significant costs of the work of team of experts.

2   Water Transport in the City

Water transport in the city has been divided into two functional categories:

- public transport for residents (business) (PuTR)
- tourist transport, tourist traffic around the city (TuTR)

The first category is associated with everyday travels of the city's residents and is characterized by a small annual variability. A characteristic reduction in traffic, associated with a favorable vacation period, occurs in the summer months, i.e. those that are most advantageous for water transport. In this category, the transport system should present features similar to land transport, in order to meet the expectations of travellers [9]. It should be integrated with land transport, providing a convenient transfer to other modes of transport. It should also offer the shortest possible journey time, which is not always possible due to the specificity of the waterway, where various obstacles may appear to slow down navigation.

The second category is an offer of even regular water transport lines, addressed to tourists or residents on occasional recreational trips. It is the possibility of moving between attractive tourist places accessible from the waterway. It often takes the form of cognitive cruises as a form of guided city sightseeing. Cities have historically often developed on the basis of the river, which nowadays results in the fact that many historic buildings can be found by the river or even with river frontage.

3   Fuzzy Inference Model of the Waterway Potential Assessment in the City

The fuzzy inference model of the waterway potential assessment has been shown in Fig. 1.

It can be observed that the presented model has 11 input linguistic variables and 2 output linguistic variables. 31 linguistic values were assigned to the input linguistic variables, the analysis of such number of variables within one inference model would be complex and difficult for the expert due to too many rules of inference, therefore it was decided to divide the problem into local inference models. Linguistic variables were grouped thematically.

The result of inference are two values separately for public transport for residents (Pu) and for tourists (Tu). The final results are determined by local inference models (PuTR and TuTR). Values of output linguistic variables have been determined as bad, unsufficient, mediocre, good, very good.

The next separation of local models of inference was made on the basis of the following features which were divided into three groups:

- geographical (GK) - resulting from the initial conditions, e.g. climate, (lay of the land - water current), watercourse size, length of the shipping season, water richness (its stability), the shape of the river,
- technical (TH) - the result of many years of human activity within the watercourse, the effect of the transformation process, e.g. weirs, locks, bridges, closed sections, water protection zones, etc.
- urban planning (UR) - a result of many years of human activity within the transformation of the city's spatial structure, e.g. land infrastructure (city - water contact, levees, riverside infrastructure, availability of river water from the wharfs), functional city structure (linking potential travel sources and destinations by water, links to land transport nodes).

In the context of the output linguistic variables, it was found that UR requires a separate analysis for PUTR and TuTR. Therefore, UR wa divided into two

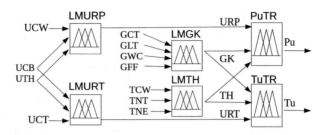

Fig. 1. The fuzzy inference model of the waterway potential assessment in the city.

separate local LMURP inference models (for PuTR) and LMURT (for TuTR). On the basis of the TH and UR features, local inference models LMTH and LMGK were distinguished. The above division allowed us to implement the entire inference model in six local Mamdani inference models. In all local inference models triangular membership functions and center of mass defuzzification methods were defined.

3.1   The First Local Inference Model (LMGK) of Input Parameters - Indicator of Geographical Conditions

According to the expert assessment and earlier studies [3], four parameters being linguistic variables in the model are specified.

The first parameter, defined as GCT, is the spatial structure of the river within the city. The more complex the structure, the greater the chance of linking important functions by waterway, and thus the greater sense of the existence of such communication.

The second factor that has been taken into account is the difference in ground levels throughout the whole area planned for navigation. As a consequence water accumulation is possible. In the model it has been defined as GLT.

Another factor affecting the possibility of using the waterway is the speed of the current, defined as GWC.

The last, fourth parameter resulting directly from the climatic features of the analyzed site is the length of the waterway shipping period. It has been defined as GFF parameter which determines the proportion between the closing time of a waterway or its fragment in relation to the length of the period in which the tasks of PuTR and TuTR can be carried out.

The following parameters have been adopted as linguistic variables of the first local inference model:

- GCT - network layout of waterways: linguistic variable values: very good, good, bad
- GLT - differences in levels of terrain: linguistic variable values: big, medium, small
- GWC - the speed of the current: linguistic variable values: big, small

- GFF - shipping period limited by climate: linguistic variable values: big, medium, small

The result of inference (GK) has six values of linguistic variables.

3.2   The Second Local Inference Model (LMTH) of Input Parameters - Indicator of Technical Criteria

This group includes all obstacles and delays in overcoming the waterway resulting from the development of the waterway structure in the city. In the technical criteria group, three parameters have been distinguished.

The first is continuity of the waterway, defined as TCW. This parameter is used to determine the length of collision-free sections, the degree of collision - whether weirs are accompanied by locks enabling the continuation of navigation or not, etc.

The second parameter represents the nuisance on the waterway which limits the possibility of shipping or slow down the navigation. It has been defined as TNT.

The last parameter is aesthetic nuisance, defined as TNE. It is not an impediment to mobility, but it is a group of features that can discourage users to some extent, such as noise, an unpleasant smell from a neighbouring sewage treatment plant, an ugly view, etc.

The following parameters have been adopted as linguistic variables of the second local inference model:

- TCW - continuity of the waterway: linguistic variable values: big, medium, small
- TNT - nuisance time on the waterway: linguistic variable values: big, medium, small
- TNE - Aesthetic nuisance: linguistic variable values: big, small

The result of inference (TH) has six values of linguistic variables.

3.3   The Last Two Local Inference Models (LMUR) of Input Parameters – Indicator of Urban Criteria

In this group of criteria, the functional and spatial structure of the city is important. The group has been divided into two functional categories, PuTR and TuTR. That is the reason of the division of LMUR into two separate models -for public transport (LMURP) and for tourists (LMURT).

The first category is water communication adapted to the needs of tourists (allowing to reach the main tourist attractions and business centers by water). The second one is adapted to the needs of residents - it connects large housing estates with the destinations of everyday travel - business centers, service centers, etc. In both categories, three input parameters have been distinguished, two of which are common for both groups.

The first common parameter is the layout of the river and its branches in connection with selected business centers, defined as UCB.

The second common parameter is linking the waterway with land-based communication nodes, defined as UTH.

The third parameter has been diversified for different user groups and their needs. In the case of tourist traffic services, it is the connection of a waterway with tourist attractions, the parameter defined as UCT.

For residents and their daily travel, a parameter has been introduced, defined as UCW, illustrating links between housing estates and the waterway.

The following parameters have been adopted as linguistic variables of the third local inference model:

- UCB - the shape and system of the river and its branches in connection with selected business centers: linguistic variable values: big, medium, small
- UTH - the shape and system of the river and its branches in connection with selected multi-modal transport hub: linguistic variable values: big, medium, small
- UCT - the shape and system of the river and its branches in connection with selected tourist attractions, tourist centers: linguistic variable values: big, medium, small
- UCW - direct contact of housing estates with river water: linguistic variable values: big, medium, small

The result of both inferences (URP and URT) has six values of linguistic variables.

4   Validation of the Method - The Inference Results

Validation of the method was done on the example of the city of Wroclaw. Parametric analysis of Wroclaw.

GCT - In Wroclaw the difference between the water level and the terrain level is balanced in many places. In the area of the city, the Odra river is divided into two main branches. Both waterways merge below the city center area. Additionally, they are connected by a side channel, which is continued by the City Canal, accessible only to small units. The GCT parameter gets the value 60.

GLT - On the entire length of the Odra river in Wroclaw, the difference in water levels is around 11 m. The GLT parameter gets the value 30.

GWC - The water current in some parts reaches the speed up to 3 km/h. This means that the time needed to navigate by the chosen route increases as the boat moves upstream. The GWC parameter gets the value 50.

GFF - In Wroclaw, the shipping season currently lasts a maximum of 7 months in the spring and summer. The GFF parameter gets the value 20.

TCW - The average length of sections of a waterway without obstacles is only 6 km. The length of the longest collision-free section is 8.7 km. These are very low values. The TCW parameter gets the value 20.

TNT - Nuisance in the form of locks extends the time of travel by water. The TNT parameter gets the value 30.

TNE - In Wroclaw a significantly larger part of the river is positively perceived by waterway users. The TNE parameter gets the value 80.

UCB - Layout of waterway is linear in the predominant east-mid-west direction, which means that the river is not able to support N-S directions as a way of communication. Large service or business centers mostly have no connections with the river. The UCB parameter gets the value 10.

UTH - Public transport interchange nodes are moved away from the river by a minimum of about 350 m. For everyday public transport it is an uncomfortable distance - too large for the need to integrate transport means. The UTH parameter gets the value 25.

UCT - A large part of tourist and recreational attractions is located near water. This makes the Odra river ideal for tourist navigation and attractive for residents in recreational categories. The UCT parameter gets the value 70.

UCW - A large part of the bigger housing estates do not have direct contact with the river, even if they are located nearby. The UCW parameter gets the value 20.

The results of modeling using our fuzzy inference model are presented in the range 0–100%. The model assessment of potential of the waterway in Wroclaw is:

- for public transport PuTR = 38%, it means that the result is between unsufficient and mediocre
- for tourist transport TuTR = 89%, it means that the result is between good and very good.

This evaluation has been compared to the results of expert tests performed in the classic way by the authors [3] and to the current decisions of the city authorities included in the planning documents [1]. The results are convergent, which means that our model correctly evaluates the potential of the waterway.

A similar analysis was carried out for Hamburg which is a city with operating watercourse lines (the function verified in practice). Although water transport covers only part of the city, it works good as a complement to the whole structure of the public transport network. It should be emphasized that a large part of the city's tourist attractions is accessible via the waterway. Similarly to the analysis of Wrocław, the input data for the fuzzy model of inference for the city of Hamburg have been determined:

GCT - 80, GLT - 10, GWC - 20, GFF - 20, TCW - 90, TNT - 10, TNE - 20, UCB - 90, UTH - 80, UCT -80, UCW - 80.

The results obtained from the fuzzy inference model:

- for public transport PuTR = 70%, it means that the result is good
- for tourist transport TuTR = 90%, it means that the result is between good and very good.

The obtained results are consistent with the actual functionality of the Hamburg system of public transport.

5 Conclusions

The proposed fuzzy inference model has been validated on the example of the city of Wroclaw. The obtained results are convergent with the current decision of the city authorities made in 2018 regarding the possibility of using the structure of the water network in Wroclaw and with the results of previous analysis carried out and published by the authors. Our model can be useful in making strategic decisions. It should be emphasized that the analysis using the model is easy to carry out, gives the effect in a short time at a low cost. Quick analysis can be very useful because it happens that the structure of a water network in the city that apparently offers great opportunities to be used for introducing regular water transport lines, in fact does not meet the requirements for such a function. Due to this analysis, we can quickly ascertain it. The research conducted on the example of Wroclaw has shown that the communication functionality of the water network is too small in the context of everyday travel of the city's inhabitants. However, it is suitable for tourist traffic services.

References

1. Dutkiewicz, R.: Studium uwarunkowań i kierunków zagospodarowania przestrzennego Wrocławia. Study of conditions and directions of spatial development, Biuro Rozwoju Wrocławia (2018)
2. Kovačić, M., Zekić, A., Violić, A.: Analysis of cruise tourism on Croatian rivers. "Naše more" Int. J. Marit. Technol. **64**, 27–32 (2017)
3. Lower, A., Lower, M.: Analysis of the possibility of using the Wrocław waterway system for the introduction of a waterbus line. In: 5th International Multidisciplinary Scientific Conference on Social Sciences, Arts (eds.) Urban planning, architecture & design, issue 5.2, Urban studies: planning and development, vol. 5, pp. 37–44. SGEM, Sofia, September 2018
4. Lower, M., Lower, A.: Evaluation of the location of the p&r facilities using fuzzy logic rules. In: Zamojski, W., Mazurkiewicz, J., Sugier, J., Walkowiak, T., Kacprzyk, J. (eds.) Theory and Engineering of Complex Systems and Dependability, pp. 255–264. Springer, Cham (2015)
5. Lower, M., Lower, A.: Determining the criteria for setting input parameters of the fuzzy inference model of P&R car parks locating, pp. 239–248. Springer, Cham (2016)
6. Majima, T., Katuhara, M., Takadama, K.: Analysis on Transport Networks of Railway, Subway and Waterbus in Japan, vol. 56, pp. 99–113, January 1970
7. Takadama, K., Majima, T., Watanabe, D., Katsuhara, M.: Exploring quantitative evaluation criteria for service and potentials of new service in transportation: Analyzing transport networks of railway, subway, and waterbus. In: Yin, H., Tino, P., Corchado, E., Byrne, W., Yao, X. (eds.) Intelligent Data Engineering and Automated Learning - IDEAL 2007, pp. 1122–1130. Springer, Heidelberg (2007)
8. Wojewódzka-Król, K., Rolbiecki, R.: Grounds and opportunities for the development of passenger and cargo shipping on the lower Vistula. Acta Energetica nr **2**, 106–117 (2013)
9. Yu, B., Peng, Z., Wang, K., Kong, L., Cui, Y., Yao, B.: An optimization method for planning the lines and the operational strategies of waterbuses: the case of Zhoushan city. Oper. Res. **15**, 25–49 (2015)

The SCIP Interoperability Tests in Realistic Heterogeneous Environment

Piotr Lubkowski[1]([✉]), Robert Sierzputowski[2], Rafał Polak[2], Dariusz Laskowski[1], and Grzegorz Rozanski[1]

[1] Military University of Technology,
Gen. W. Urbanowicza 2, 00-908 Warsaw, Poland
piotr.lubkowski@wat.edu.pl
[2] Transbit Sp. z o.o., Łukasza Drewny 80, 02–968 Warsaw, Poland

Abstract. The secure communications strategy originating from the NNEC (NATO Network Enabled Capability) concept assumes the use of effective and safe transport mechanisms at various levels of the heterogeneous C2 (Command and Control) NATO systems [1]. SCIP (Secure Communication Interoperability Protocol) was introduced by NATO as the primary standard for secure communication between different network devices through networks with restricted bandwidth [2]. The implementation of the SCIP protocol in various end devices requires the unification of transmission and signalling procedures as well as the methods of coding the speech signal. The main challenge associated with this issue is therefore the interoperability of the available SCIP implementations. Studies concerning the practical implementation of SCIP protocol are much fewer in number than theoretical ones. This paper is therefore focused on interoperability testing of practical SCIP implementation in heterogeneous networks with restricted bandwidth. A set of interoperability tests, including call set – up and limits of voice transmission are proposed as well. Results of these tests are then analysed and compared with ITU (International Telecommunication Union) and NATO recommendations.

Keywords: Interoperability testing · SCIP · Secure voice communication · Voice quality

1 Introduction

The NATO SVS (Secure Voice Strategy) was published in January 2012, introducing the recommendation to use the SCIP protocol as the basis for ensuring secure voice communication services in both NATO and national communication systems [3]. It also contains guidelines for verification of secure communications solutions.

The goal of the NATO SVS is to develop guidelines and recommendations for introducing a secure communication system, which meets the increased requirements for interoperability, flexibility, scalability, mobility, and performance on the modern battlefield. The basic premises of the mentioned strategy stem from the assumptions of the FMN (Federated Mission Networks) architecture and the convergence of communication systems towards the widespread use of the IP protocol [4]. Preference is

W. Zamojski et al. (Eds.): DepCoS-RELCOMEX 2019, AISC 987, pp. 311–320, 2020.
https://doi.org/10.1007/978-3-030-19501-4_31

given to techniques and network technologies meeting the necessary interoperability requirements, whereas in the case of a highly heterogeneous scenario, ensuring secure communication in the end – to – end relationship seems problematic, as various cryptographic devices use different technologies focused on the type of network used. The solution to this problem is the use of SCIP gateways, in order to ensure interoperability for Secure Voice Services between heterogeneous communication systems. Another problem resulting from the multitude of used network technologies is the many different types of terminals and computer applications dedicated to specific networks. In this situation, the solution to providing secure communication in a heterogeneous network environment is the use of SCIP terminals.

The introduction of SCIP technology will ensure interoperability between users regardless of the telecommunications network they are connected to, both in national and in allied operations, and will significantly reduce the number of secure voice terminals used. Technological progress and allied commitments have resulted in several research works on the implementation of the SCIP protocol, as well as testing the interoperability of the resulting solutions in national and allied communication systems. Research related to interoperability with the SCIP implementation focuses on the one hand on theoretical studies and general concepts [5, 6], and on the practical application of test suites in a realistic environment [7–9] from the other hand. Our work has been motivated by the need to test interoperability with the SCIP implementation developed in the course of the INNOTECH – K2/IN2/14/181896/NCBR/12 project, supported by the National Centre for Research and Development. On the basis of recommendations regarding interoperability testing, a number of tests of the newly developed SCIP implementation were carried out in a testbed environment. The established testbed reflects a realistic heterogeneous environment of national and coalition communication systems.

The remaining part of this paper is organized as follows. An overview of the SCIP protocol is described in Sect. 2. Further, in Sect. 3, basic information about interoperability tests and the testbed in which the implementation is performed is given. Section 4 presents the most important results of the tests along with discussion and recommendation for future work. Finally, the paper ends with the conclusions contained in Sect. 5.

2   SCIP Overview

The SCIP is an international standard developed for the provision of secure voice and data communication. SCIP technology is designed to provide interoperable end – to – end connectivity between a variety of communication systems, from military radio communication, traditional telephone communication system through satellite communication channels to VoIP and various types of mobile telephony standards. It is designed to operate at the application layer with minimal dependency on the characteristics of the lower layers. SCIP enables simplex operation in point – to – point or point – to – multipoint modes. It is primarily designed for 2.4 kbps transmission, to ensure communication via narrowband KF/VHF radio. However, if more bandwidth is available, SCIP enables multimedia data transfer at speeds up to 10 Mbps. It is also worth noting that user data encryption via the SCIP channel is implemented with minimal bandwidth requirements, and encryption requires bit transparency for signalling and data. An important advantage of the SCIP protocol is the use of the classical model of VoIP services. Session Initiation Protocol (SIP) and Real – Time Protocol (RTP) are both used by SCIP terminals to establish an IP connection between SCIP terminals. SCIP session includes a crypto – sync message used for establishing a secure mode of voice/data transmission.

The presented implementation of the SIP protocol refers to the voice service application using MELPe (2.4 kbps) or G.729D (7.2 kbps) codecs. It is worth noting that the developed implementation has been, and remains, the subject of a series of interoperability tests as part of NATO exercises aimed at presenting the use of the SCIP protocol in military and other communication systems in a heterogeneous environment. It is also worth noting that modifications of signalling and data transmission procedures introduced in the SCIP specification allow for the provision of multimedia services, including video transmission. Selected results of interoperability tests carried out as part of the above – mentioned exercises together with the developed research methodology are presented in the following sections.

3 Interoperability Testing Methodology

The ability of SCIP to operate in heterogeneous networks requires the developed implementations of the SCIP protocol to be validated in a wide range of network environments. Practical testing is an important element of NATO strategy for SCIP. The testing strategy is focused on ensuring that SCIP implementations can operate with sufficient level of security, reliability, efficiency and interoperability with other SCIP implementations in military and other environments. This includes conformance, interoperability and network and key management tests. In [10] the Minimum Interoperability Profile (MIP) was defined, and a description of the minimum set of characteristics, which the SCIP implementation must fulfil, was given. They are limited to the technical specification, for which the testing methodology is described in [11]. A number of recommendations concerning interoperability testing are given in [12].

Taking these sources into consideration, as well as the experience of authors in the field of testing [13], we designed interoperability testing scenarios and infrastructure, which reflect realistic heterogeneous communication systems (Fig. 1). We assumed

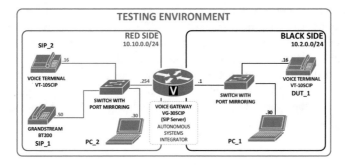

Fig. 1. General reference configuration for interoperability testing of SCIP implementation.

that interoperability tests should ensure interconnectivity of the provided SCIP implementation (DUT – Device Under Tests) across typical IP and military communication networks.

The following input data was included in the test scenario planning process:

(1) System architecture – this group of information describes the test environment in detail. It contains a network diagram with separated autonomous systems subjected to the process of interoperability testing, and constitutes a technical specification, which strictly specifies the network traffic engineering issues. In this information set, in addition to the description of the standard tested with its version, transport protocols and protocols additionally intermediating or directly related to data transmission during testing are defined. Additional information is network services that can be provided during system tests.

(2) Test participants – Providers, Mediators, Distributors, Observers, and Consumers. The most important participants are Providers and Consumers, as they are representatives of autonomous systems between which interoperability is identified. Mediators are an intermediary in data transmission and are responsible for processing data generated by Providers into a representation, which is comprehensible to the Consumers. Data Distributors ensure distribution of transport streams between nodes of autonomous systems of Providers and Consumers. They do not modify the data representation, but they can adapt the distribution to the needs of users. Test observers have access to data exchanged between all network nodes. The observers do not interfere in the way of data representation, exchange method or test result – they can only provide insights into these aspects.

(3) Test variants – tests related strictly to the tested solution are determined and described. Checks are performed to verify whether signalling, code, and the data coding/decoding processes are correct in relation to the SCIP protocol.

(4) Validation criterion – this information is treated as a result for each test variant and on its basis, a positive or negative result of interoperability is found. The following test results are determined: Success, Limited Success, Interoperability Issue and Not Tested. "Success" occurs when all intended goals for information exchange have been achieved and full functionality is achieved – therefore the systems are interoperable. "Limited Success" means that the exchange of information has been partially achieved or the quality/performance of transmission have deteriorated. This result is also assigned when changes in the configuration of software or components of the tested system had to be made during the tests to achieve at least minimal functionality. The "Interoperability Issue" status is assigned to an outcome, in which the exchange of information between the systems in a given test variant, was not possible and the functionality was not achieved due to reasons that must be identified, e.g. non-conformance of standards, or a configuration problem. The status "Not tested" will also include the reason why the tests have not been performed, e.g. the lack of implementation of the appropriate functionality.

Hence, the set of test scenarios used is as follows:

- SIP call flow interoperability, which includes call setup, call release, call setup to a busy line, no route to destination and unlogged user.
- Interoperability of the correct selection and cooperation of VoIP codecs, testing at the least the MELPe and G.729D codecs.
- Evaluation of the SCIP/VoIP connections quality under conditions of simulated impairments.
- Assessment of the correctness of setting up SCIP connections under conditions of simulated IP impairments.

All elements of testing environment have been pre – configured to meet the adopted conditions and test mode. The secure domain (BLACK SIDE) consists of DUT_1, which represents VoIP terminal with proposed SCIP implementation. A PC_1 terminal is representing a monitoring point running Wireshark. The secure domain is connected to a non – secure domain (RED SIDE) over a Voice Gateway (VG – 30 SCIP) which fulfils the role of interoperability point and reflects the heterogeneous backbone network. In the non – secure domain, SIP_1 and SIP_2 terminals are used as voice connection clients. The PC – 2 terminal has the same role as the already mentioned PC – 1. A network emulator NE, which was included between the RED and BLACK networks, was used to simulate packet delays and packet losses (not shown in the diagram).

The reasoning of correctness for the specification of network components in terms of data transmission and reflecting information in the testing environment is based on a statistical estimation of reliability of the software and hardware platform forming the service chain. Products of renowned hardware and software vendors were used in the testing environment. Therefore, it appears reasonable to conclude that the specified system is a correct and highly reliable testing environment.

4   Results of Interoperability Testing

4.1   Interoperability of SIP Call Flow

The purpose of the tests was to verify the proposed implementation in the target operating environment. Tests were focused on confirming whether the calls are properly established and completed according to the requirements given in [12]. First, interoperability tests of connections to various types of terminals were made. The test was carried out for the default configuration of the DUT_1 device, which works with the G.711 telephone codec. In the first series of tests, subscribers were called using the physical number of the terminal (subscriber number), while in the second, they were called using the IP address. In each series, the correctness of the disconnection of the call by the calling subscriber and the called subscriber was also verified. The result of the test performed between DUT_1 and SIP_2 is presented in Fig. 2. As it can be seen, the exchange of messages and control signals takes place between the devices is correct and occurs in accordance with the SIP specification. The stability of the connection establishment and the ability to disconnect the connection via DUT_1 was noted as well. This is extremely important, given that at the same time the VG – 30 gateway converts data between the secure (BLACK) and non – secure (RED) networks. After establishing the connection, a conversation was carried out between the subscribers of the cooperating devices (RTP flow in Fig. 2). A similar effect was obtained in the relation DUT – 1 – SIP_1.

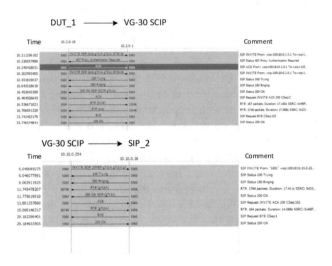

Fig. 2. The flow of signaling messages between DUT_1 (SCIP implementation) and SIP_2.

Subsequently, the tests related to establishing a connection to a busy user, to a user with no destination route and to an unlogged user were carried out. All tests ended with a positive result according to the [12] and the DUT_1 device sent the right information message each time.

4.2   Interoperability of VoIP Codecs Selection and Cooperation

Next the tests of correct selection and cooperation of VoIP codecs were performed. DUT_1 implemented support for the following codecs: G.711 – μlaw – 64k G.711 – Alaw, G.726 – 16k, G.726 – 24k, G.726 – 32k, G.726 – 40k, G.729A, G.729D, G.723, GSM – 06.10, iLBC, iLBC – 13k3, iLBC – 15k2, MELP. During testing in the DUT_1, only one of the specified codecs was used to handle VoIP calls. All connections were supported by Voice Gateway as it can be seen in Fig. 1.

The DUT_1 device was able to establish connections using the G.711 – μLaw – 64k, G.711 – ALaw – 64k and G.729A codecs with SIP_1. In the case of other codecs, the connection did not take place ("line out of service" message), as SIP_1 does not support other codecs. Summing up, it can be concluded that DUT_1 passed the test, as it correctly signals the selected codec and performs the connection using that codec in line with the requirements given in [10, 11].

The next test was related to codec negotiation, and the result is the choice of the codec indicated in the list of supported codecs and common for both cooperating devices. DUT_1 was supporting the G.711 – μlaw – 64k, G.729D, and G.723 codecs while the DUT_2 G.726 – 16k, G.729A and G.723 codecs. DUT_1 negotiated the G.723 codec and passed the test which confirmed compliance with the recommendations given in [11].

4.3 Evaluation of the SCIP/VoIP Connection Quality Under Conditions of Simulated Impairments

The tests were carried out in the configuration shown in Fig. 1. A few connections for selected codecs were made for the DUT_1. Connections were carried out under conditions of changing packet loss (%) and the delay(s) in the testing environment. Test results indicating the quality of SCIP and VoIP connections are based on the following assessment criteria:

– connection/no connection established;
– negotiation/no negotiation of SCIP session parameters;
– distortion/lack of speech signal distortion.

In the case of the last criterion, the characteristic values of a voice connection such as intelligibility, loudness or auditory effort were used. Tests were performed for the wide range of voice codecs supported by DUT_1. An example results of the test of DUT_1 using SCIP (G.729D) codec are shown in table Table 1 ([15]).

Fields marked in red define connections, which allowed to establish a transmission channel, and in the case of SCIP connections, established a cryptographic parameter as well, but were unusable due to very high distortions of voice. The '–' sign means that no connection has been established. The fields marked in grey and dark green indicate connections where the other person was difficult to understand (grey color – high disturbance). The light green color means good hearing conditions and good connection quality. Disturbances in the form of packet losses and delays do not significantly affect the quality of the voice connection.

However, exceeding the critical values, especially in the case of packet losses, makes it impossible to make a voice call. The reason for this is the lack of proper exchange of data packets (high level of packet loss), leading to a lack of intelligibility of the caller or the lack of opportunities to exchange signaling data.

4.4 Assessment of the Correctness of Setting up SCIP Connections Under Conditions of Simulated IP Impairments

The purpose of the last of the conducted interoperability tests was to determine the impact of IP network impairments on the correctness of SCIP session setup. Connections were setup between DUT_1 and SIP_2 under conditions of changing packet loss in the testing environment. Two types of codec were used: MELP and G.726D. For each of the considered codecs, 2 attempts of SCIP connection setup were made. Negotiation of the connection in the SCIP mode is sensitive to network interference, which may lead to a failure of connection setup. Table 2 presents the result of the test for MELP codec while the Table 3 for the DUT_1 using the G.729D codec ([15]).

Table 1. The SCIP connection quality as a function of packet loss ratio and delay.

Packet Loss Ratio (%) / Delay (s)	0	5	10	15	20	25	30
0	+	+	Low distortion	Higher distortion	Limit of intelligibility	Very high distortion	–
0.25	+	+	Low distortion	Higher distortion	Limit of intelligibility	Very high distortion	–
0.5	+	+	Low distortion	Higher distortion	Limit of intelligibility	Very high distortion	–
0.75	+	+	Low distortion	Higher distortion	Limit of intelligibility	Very high distortion	–
1	+	+	Low distortion	Higher distortion	Limit of intelligibility	Very high distortion	–
1.25	+	+	Low distortion	Limit of intelligibility	Very high distortion	–	–
1.5	+	+	Low distortion	Limit of intelligibility	Very high distortion	–	–
1.75	+	Low distortion	Higher distortion	Limit of intelligibility	Very high distortion	–	–
2	+	Low distortion	Higher distortion	Limit of intelligibility	Very high distortion	–	–
2.5	+	Low distortion	Higher distortion	Limit of intelligibility	Very high distortion	–	–
3	+	Low distortion	Higher distortion	Limit of intelligibility	Very high distortion	–	–
4	+	Low distortion	Higher distortion	Limit of intelligibility	Very high distortion	–	–
5	+	Low distortion	Higher distortion	Limit of intelligibility	Very high distortion	–	–
6	+	Low distortion	Disconnection	–	–	–	–
7	–	–	–	–	–	–	–

In the table sign '+' represents a successful attempt, while the '-' indicates an unsuccessful attempt of SCIP connection setup.

It is evident that packet loss exceeds 45%, practically preventing the SCIP connection from being established. It is worth noting, however, that with such a large loss of packets, problems with the implementation of multimedia services, which also include voice services, would generally occur in any system.

Table 2. The SCIP connection setup as a function of packet loss ratio.

Packet loss ratio (%)	0	10	20	30	50	60	70	75
MELP	++	++	++	++	++	++	+ −	− −

Table 3. The SCIP connection setup as a function of packet loss ratio.

Packet loss ratio (%)	0	10	20	30	40	45	50
G.729D	++	++	++	++	++	+ −	− −

5 Conclusions and Recommendations

We presented a result of interoperability tests of developed SCIP implementation which provides a secure voice and data communication capabilities over wired heterogeneous networks. The results of the SIP signaling compliance tests, the quality of the voice connection and the reliability of SCIP signaling procedures remain in line with the ITU and NATO recommendations. Thus, the obtained results clearly confirm the legitimacy of using the SCIP protocol within the scope of providing secure multimedia communication in military and civilian heterogeneous networks. The research environment, together with a set of developed tests, allows testing of SCIP implementations installed in a variety of terminal devices.

Although the tests were conducted in a wired environment, however, the developed SCIP implementation may also operate successfully in a wireless (radio) network.

References

1. NATO C3 Agency: NATO Network Enabled Capability Feasibility Study – Executive Summary, 28 (2005)
2. Collura, J.S.: Secure Communications Interoperability Protocols (SCIP), IEEE Military Communications, 19–1–19–10 (2006)
3. NATO C3B: NATO Secure Voice Strategy. C3B Annex 1 AC/322 – D (2012)0001, 27 (2011)
4. NATO SHAPE CIS: BI – SC Secure C2 Data Strategy, v.1.0, 88 (2010)
5. Griffeth, N., Hao, R., Lee, D., Sinha, R.K.: Interoperability testing of VoIP systems. In: Global Telecommunications Conference, GLOBECOM 2000, vol. 3. pp. 1565–1570 (2000)
6. Hao, R., Lee, D., Sinha, R.K., Griffeth, N.: Integrated system interoperability testing with applications to VoIP. IEEE/ACM Trans. Netw. (TON) **12**(5), 823–836 (2004)
7. Alvermann, J.M., Kurdziel, M.T., Furman, W.N.: The secure communication interoperability protocol (SCIP) over an HF radio channel. MILCOM **2006**, 1–4 (2006)

8. Küyük, R.T., Celebi, H.B., Hökelek, I., Ören, Ö., Yeni, A., Saribudak, A., Kara, F., Vicil, G.M., Uyar, Ü.: Interoperability of secure VoIP terminals. In: Proceedings of First International Black Sea Conference on Communications and Networking (BlackSeaCom), pp. 172–176 (2013)
9. Dilli, O., Koyuncu, M., Akçam, N., Ögüşlü, E.: Secure communication tests carried out with next generation narrow band terminal in satellite and local area networks. In: 6th International Conference on Recent Advances in Space Technologies (RAST), pp. 493–498 (2013)
10. SCIP – 221 Document: Minimum Interoperability Profile
11. NATO C3A: Interoperability and Network test Plan. NC3A Test Documentation SCIP – 620 V.1.1 (2010)
12. I3 forum: Interoperability Test Plan for International Voice services. International Interconnection Forum for Services over IP. Rel. 6. 17 (2014)
13. Lubkowski, P., Laskowski, D., Maslanka, K.: On supporting a reliable performance of monitoring services with a guaranteed quality level in a heterogeneous environment. Theory Eng. Complex Syst. Dependability 365, 275–284 (2015)
14. Lubkowski, P., et al.: Provision of the reliable video surveillance services in heterogeneous networks, safety and reliability: methodology and applications. In: Proceedings of the European Safety and Reliability Conference, ESREL 2014. CRT Press, A Balkema BOOK, pp. 883–888 (2015)
15. Report on the research of the BSWD communication component model, W5/1: INNOTECH – K2/IN2/14/181896 (2013)

Softcomputing Art Style Identification System

Jacek Mazurkiewicz[1]([✉]) [iD] and Aleksandra Cybulska[2]

[1] Faculty of Electronics, Wrocław University of Science and Technology,
ul. Wybrzeże Wyspiańskiego 27, 50-370 Wrocław, Poland
jacek.mazurkiewicz@pwr.edu.pl
[2] DreamLab, ul. Curie-Skłodowskiej 12, 50-381 Wrocław, Poland
209209@student.pwr.edu.pl

Abstract. The paper discuss the possibility to use the softcomputing methods to create effective and useful system for art style identification. The system should operate on the samples of paintings. The assumption is to use only the small parts of pictures with no high resolution. Different types of preprocessing methods are tested to create the input vectors for Convolutional Neural Network (CNN) which is an identification tool. The experiments are done for the significant dataset covering ten most classic art styles of paintings. Different types of CNN topology is discussed. The promising results could be an interesting subject for custodians, art historians or scientists. This may help them not only recognize the style with some certainty but also compare and mark the similarities and differences between styles or artists. The paper can be extended to help them in authenticating and determining the timeline of paintings.

Keywords: CNN · Art style · Softcomputing · Identification

1 Introduction

Humanity has always tried to express themselves throughout creativity. Images have always been the closest due to the simplicity of copying the world. Each person sees it individually as it is understood more psychologically than geometrically, and this gives many ways of expression. Because of them, the styles change throughout history. Through the style of each painting, we can discover a lot about the period in history and the matters which concerned humanity at the time. The investigation of art styles helps cataloguing databases and understanding culture of the past generations. People are drawn to the past just as much as to the future. Art was always an inspiration both for artistic and non-artistic individuals. Even in Middle Ages, the sculptures from ancient Greece and Rome were noticed and appreciated in the old continent. Then, the Renaissance took even more inspiration out of them to create the most magnificent sculptures and paintings of their own. Obviously, no one would have wanted to copy the old masters. Every reputable artist has a call to be extraordinary and create the art style which later would be recognized as theirs. Recently more and more artists use computers to help themselves in the process of creation and so the storing of art became easier by using the modern databases and searching systems. This process is also applicable for the identification of art. Beforehand using only the knowledge of the

© Springer Nature Switzerland AG 2020
W. Zamojski et al. (Eds.): DepCoS-RELCOMEX 2019, AISC 987, pp. 321–330, 2020.
https://doi.org/10.1007/978-3-030-19501-4_32

past, now it can be simplified by machine learning. It has become a very famous research subject in recent years. Considering the variety of methods available in machine learning, the number of usages can be countless. Each request containing enough data can be a suitable to create a helpful tool. Considering the style of art and usage of such methods, the work of custodians may become easier and more precise, depending on network they would use. In this paper we want to examine how neural networks can be applied to recognition of art styles. We believe it could be an interesting subject for custodians, art historians or scientists. This may help them not only recognize the style with some certainty but also compare and mark the similarities and differences between styles or artists. The paper can be extended to help them in authenticating and determining the timeline of paintings.

2   Art Styles Identification

2.1   Problem Statement

Art style is separated by art movements, period, place or group of artists. It shows individual characteristic of art and it is often categorized later in history. A lot of styles are also divided to early, middle and late stadiums which can be interpreted as slow shaping of the used methods or subjects. Recent history had a big increase in styles which made historians avoid the classification as far as it is possible. [5] Art identification is a wide process which can take a lot of time. In the process, the art historian compares the given subject with other examples from the alleged artist. Because of the expert knowledge of the historian, it can be examined through its theme, material and technique and be compared to the artists habits. Then the investigation is taken over by photographing the painting in normal, ultraviolet and infrared light, with different illuminations to reveal irregularities on the surface. Then the layers of the painting are examined. A painting is made up of four layers: support, ground, paint, varnish. The materials used are often characteristic to the period of creation. A lot of materials can be re-created, which is the reason why the expert has to be careful in their research. The age of the wood or canvas on which the painting was made (support) can be carried out by dendrochronology or carbon-14 dating. Then the ground layer is measured in used pigments. A lot of them were used in a certain period then to be replaced. The interesting part of identification is uncovering of underdrawings with infrared reflectography. It can be very helpful because it reveals early sketches or the parts of the concepts that later changed. Many artists have their own manner of painting or drawing and that can be a confirmation of authenticity. [5] The styles themselves are often compared to the other paintings from the period and other pieces by the same artist. Plenty of painters were involved in more than one style so the exact timing can be crucial to determine the style. Furthermore, the style is not only determined by the technical details such as colors and methods of painting. Often the topic of the art is as important. In many periods, certain society was struggling with problems shared and remembered by the paintings. Symbols and allegories are as important. A lot of styles have their certain motives which were duplicated due to their popularity among people. For instance, Romantic paintings have a lot of nature, cliffs and ships and Renaissance focused on Roman and Greek gods as much as symbolic representation of Christian religion. [11] In 2012, *ACM Journal on Computing and Cultural Heritage* published a research by scientists Lior Shamir and Jane Tarakhovsky of Lawrence Technological University in Michigan. They used approximately 1000 paintings of 34 artists and let the algorithm analyze similarities in visual content of the paintings, and without any human guidance. The computer provided a network of similarities between painters that is largely in agreement with the perception of art historians.

2.2   State of Art

In paper written by Adrian Lecoutre, Benjamin Negrevergne and Florian Yger, it is described how such networks work. They write about accuracy they tried to improve by using popular neural networks used for art recognition. They used the whole WikiArt database with class distribution being not equal. Then they used AlexNet and ResNet which are very advanced convolutional neural networks to get the results. The ResNet has 53 convolutional layers which happen to work well with the given data however could overfit due to its size. Other network is smaller however does not receive as good results [7]. Another research done in Israeli university connected algorithms from other research to create a new way to categorize paintings. They used SVM algorithm to pretrain network and to create binary classifiers. Then they compared five classifier algorithms such as SMV, Naïve Bayes and kNN to choose 5NN to classify their images. For the purpose they used Matlab framework VLFEAT. The precision they reached was set around 40% in different methods and tests [1]. Stanford University students in their identification took examples of seven different artists and taught the network to recognize each of them. Their sets were rather small, they had 200 paintings per artist, but comparing to sources it is not easy to find more paintings by the same artist. As the previous research team, they tested their data on Naïve Bayes and SVM algorithms. They reached the problem of number of data and the overfitting. For different approaches, they tested data in different batch sizes, noticing that the more they give, the more problems the network has. In the two-class problem they tested, they reached over 90% accuracy and then, in the 7 classes – over 80% which is impressive considering the diversity between each input [2].

3   Art Style Recognition System

3.1   Dataset

We decided to choose dataset of 1000 paintings from 10 classes. Each class got an equal value of 100 paintings, chosen by hand to increase the diversity in the set. Through that, our dataset did not have more than 10% paintings by the same artist, the motives used in certain period are widely discovered and not focused on only one. We also value the quality of the images. In order to test one of our assumption, we needed paintings with as many details as possible. The dataset used for that purpose was *Wikiart.org* which has a major database divided to different art styles [3]. The classes were named from 0 to 9 in such manner: 1 – High Renaissance 2 – Baroque, 3 –

Romanticism, 4 – Impressionism, 5 – Art Nouveau 6 – Surrealism, 7 – Cubism, 8 – Neoclassicism 9 – Pop Art, 0 – Neoclassicism. Apart from the style identification, we also used the name of the artist in the image to help in gathering the data. Through that, we were able to quickly check if the painting was already in the set and if one artist is not used too many times. The data was then split to train and test sets. The default value was set to 75–25 images per class however it is expected to test other splits in the experiment Section (90–10; 70–30). [9]. The preprocessing is presented in (Fig. 1).

Fig. 1. Preprocessing steps block diagram

3.2   Convolutional Neural Network

Convolution itself is a mathematic operator, which is often used within signal processing to simplify equations. It processes two functions and modifies them. In the matter of neural networks, it processes network's inputs. The image usually is a three-dimensional matrix composed of three sets of width and height of the image, each of them of different color in RGB model [6]. The input is processed by kernel. It is also known as a mask in image processing and it is used to alter the image. It can blur, sharpen the image or be used to detect edges etc. In CNN, it is used as a filter that creates the map of features [4]. Convolutional Neural Network is made up of identical neurons which is not common in regular neural networks. All neurons have equal parameters and weights. It reduces the number of parameters controlled by the network which makes the network more efficient. The connection between nodes is limited to local connection patterns. In other words, inputs of each node connect only with neighboring receptors. In images, it means spatial neighboring. Layers can be assembled as a representation of data, the lower layer, the more abstract the features. If enough layers are connected they can cover the whole image. Weights are shared by all the nodes of each layer. It should prevent learning through nodes of the unpredictable set of local parameters. Filters in convolutional layer work in one set and process input data together. The standard training is done with the backpropagation algorithm, which is a common procedure in neural networks. The goal of backpropagation is to minimalize the cost function, also known as error function. To do that, it goes backward and changes weights of the neurons to fit the model more precisely [8]. Repeating the training is crucial to learn each kernel. One epoch equals one iteration through all the inputs. Meaning, the more epochs, the more times the inputs are passed through the network. The convolutional layer's parameters consist of a set of learnable filters. Every filter has four parameters: size, depth, stride and pooling. During the forward pass, the network convolves each filter across the width and height of the input volume and compute dot products between the entries of the filter and the input at any position. As it convolves the filter over the width and height of the input volume we will produce a 2-dimensional activation map that gives the responses of that filter at every spatial position. The network will learn filters that activate when they see some type of visual feature such as an edge of some orientation or a blotch of some color on the first layer. Apart from perceptron (fully connected), every layer is partially connected, and each layer use many feature maps [10]. This helps creating expanded set of filters (Fig. 2).

In our network we decided to use the available in Keras methods to change the parameters of the network. It refers to: Filter size, Pooling size, Stride size, Number of convolutional layers, Number of hidden layers, Number of epochs, Activation type. To create a correct model, the network requires information what size of input it should expect: Height = 100, Width = 100, Depth = 3. Two different types of layers are in use: convolutional and regular one called Dense. Flatten method was also used to quite literally flatten the output. The convolutional layer's output was three dimensional and to use them as inputs in regular layer they had to be set in one dimension. To do that they were simply turned to a vector that had length equal to multiplication of width, height and depth of the output. Keras offers a lot of activation functions. Ones used in our network were tested to reach the best effects and it appears the ReLU function worked the most efficiently. However, for the last layer, softmax activation was used. It is often used to work with multiple classes and gives probability of belonging to each class which sums to 1. In Table 1 all datasets used in learning are presented. Dataset 3 was checked to determine how the network will behave with bigger images.

Table 1. Datasets used in experiments

Dataset No.	Classes to identify	Number of inputs	Image size	Additional information
1	10	4000	100	Image normalized to 200 pixels and cropped to 4 pieces
2		1000	100	Image normalized to 100 pixels
3		1000	250	Image normalized to 250 pixels

The experiments were mainly focused on adjusting the parameters of the network, especially number of filters, their size, size of stride and pooling. To improve a network that was promising, We were testing different number of dense and convolutional layers, then to check how it will work with more epochs. Epochs however, were the least to change since the trend could be predicted from smaller numbers, too.

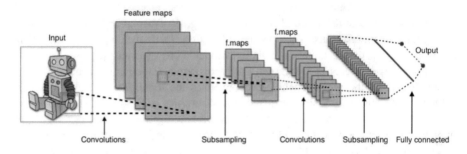

Fig. 2. Typical convolutional neural network structure

The first experiments were done to establish how complex the network should be to receive the best results. First tests (shown in Table 2) were done as classic network examples and then the network was extended to more convolutional and dense layers. Hidden layers were very highly useful to balance the network's results. The outputs were based on less amplitude and were closer to one another. Number of epochs tested were mainly for 100 and 200. Small impact and easy to predict result in smaller numbers made me decide on using only numbers in interval bigger than 50. The examples in one epoch were determined as in dataset in Table 1, however the proportion between the test and train data would vary. It was 25% of test data and 75% of train data as starting value but then we tested more combination to determine which approach is better (Table 3).

Full painting inputs gave better results because of smaller overfit. The same network settings trained by pieces of the painting usually received better results in the matter of percentage but with much bigger overfit. Even network set explicitly for that dataset, the overfit did not disappear. Since it was our main goal to reach the lowest overfit, after a couple of tests it completely disqualified the cropped database. In the figures below, the fluctuations are much bigger in the full paintings network but the angle of overfitting (red) function is much smaller (Fig. 3).

Max pooling is used to down-sample a network representation. Through that, the nodes which were not activated or with values that do not give any meaning to the network are omitted. This is one of the ways to avoid overfitting and not to deal with large networks.

Most of our tests had a max pooling function right after each convolutional layer. To test their importance in image recognition, we decided to learn a network without one or more of these functions.

Table 2. First three examples of tested networks' convolutional layers

Network No.	Conv. layer No.	Number of filters	Filter size	Pooling size	Stride size
1	1	32	2	2	2
	2	64	2	2	2
	3	128	2	2	2
2	1	32	2	2	2
	2	32	2	2	2
	3	64	2	2	2
3	1	32	2	2	2
	2	32	2	2	2
	3	64	2	2	2
	4	128	2	2	2

Table 3. First three experiments details

Experiment No.	Dataset No.	Network No.	Number of epochs	Final percentage [%]	Additional information
1	2	1	200	35	overfit
2	2	2	200	35	overfit
3	2	3	100	36	overfit

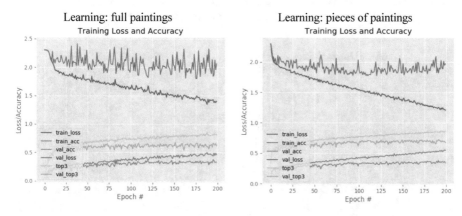

Fig. 3. Network trained by pieces or full paintings: Grey - accuracy of the network, Red - loss function, Blue - training set accuracy, Purple - testing set accuracy, Yellow - training set top3 accuracy, Green - testing set top3 accuracy

We believe the tests without one or two max pooling functions can be promising, since the results did not vary much from the network with max pooling functions after each convolutional layer. However, they are important in image recognition since without them the network is completely damaged (Fig. 4).

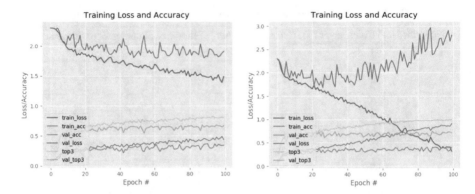

Fig. 4. Network without one MaxPooling function – on the left, Network without three MaxPooling functions – on the right

Dropout is yet another function in Keras which supposedly should decrease overfit. This is also often used in other architectures of neural networks. We discovered it while experimenting and it was very promising, especially when combined with MaxPooling functions. With right dropout values, they may have big impact on the success of the network without big overfit value.

We examined the same network with 0.1 and 0.2 valued dropout. The values mean that accordingly 10% and 20% of the previous layer's results were deleted to make the network more flexible for different values (Fig. 5). The 20% dropout seemed to have better results and smaller overfit angle. Top1 (purple line) values are growing but the Top3 (green line) is possible to get further from its trained value (yellow line).

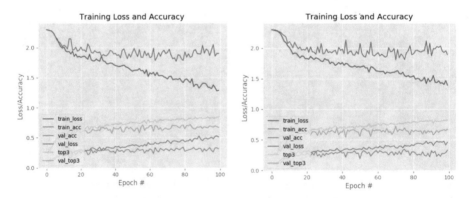

Fig. 5. Network with dropout 0.1 – on the left, Network with dropout 0.2 – on the right

We have also tested our networks for dataset with bigger pictures. The networks used for 100 × 100 samples did not have good results for them, but after application of changes for larger size, the results were similar. Which can imply that networks will work with similar results no matter whether the input sized would grow (Fig. 6).

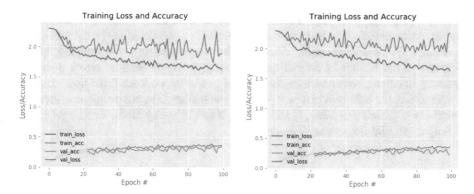

Fig. 6. Network 1 (Table 2) – on the left, Best Network (Table 2), dataset 3 (Table 1) – on the right

4 Conclusions

As mentioned earlier, the biggest problem we were trying to overcome was overfitting. Due to highly detailed images, varying in content, it was most important not to let the network create the model just to recognize the examples from training batch. Throughout the experimentation period we have noticed that hidden layers were for reduction of spikes on the chart, convolutional layers for increasing the recognition. If these were balanced, there was a possibility of a network learned with a good percentage and small overfit. Nonetheless, in the case of too many layers, this may turn the other way. Machine learning is not yet ready to completely replace human being in artistic style recognition. The study in this direction still requires more precision and tests to be able to fully recognize paintings. In our opinion, it is possible, but it might need much more precision. Comparing our first experiments with smaller number of classes, the results were similar and the change in the number of inputs or size of them did not change much in the networks with similar number of layers. Our guess would be, the network is either too simplified for the problem of style identification or the problem is too complicated to be handled without more complex data. In this we mean especially the schemes used in different styles and the age of the painting or the artist information.

We strongly believe further samples would get much better results even without determination of age or artist since the network will have such connections by itself. In our opinion, the research should be broadened not only to teach the network the pixels and their values, but also pretrain them with the whole idea of the style. For example, check how much it is Romanticism by existence of landscapes and nature on the painting or check the Baroque in how dark the painting is.

We suppose even without the determination of the year of the painting, the network would be able to set the weights much better than simply focusing on given colors and gradients.

The promising results could be an interesting subject for custodians, art historians or scientists. This may help them not only recognize the style with some certainty but also compare and mark the similarities and differences between styles or artists. The paper can be extended to help them in authenticating and determining the timeline of paintings. We assume it could be also applied in art schools while studying the differences in the history of art and also to challenge the students to create their own painting in certain style and then check if the neural network confirms their attempts. Furthermore, we believe it can be a great tool to introduce to not people not interested in art. By combining technology and culture, it can be a way to show more people the beauty of art and not to treat it as boring.

References

1. Bar, Y., Levy, N., Wolf, L.: Classification of artistic styles using Binarized features derived from a deep neural network (2015). http://pdfs.semanticscholar.org/2b20/33af5ae4e705b90e 970a586e0431678374b2.pdf. Accessed June 2018
2. Blessing, A., Wen, K.: Using machine learning for identification of art paintings. https://pdfs. semanticscholar.org/1d73/0a452a5c03cc23f90d4fde71c08864f31c35.pdf.   Accessed   May 2018
3. Google Arts and Culture. artsandculture.google.com
4. Hearty, J.: Advanced Machine Learning with Python. Packt Publishing, Birmingham (2016)
5. Hockney, D., Gayford, M.: History of Pictures: From the Cave to the Computer Screen. Thames & Hudson Ltd., London (2016)
6. Krizvsky, A., Skutskever, I., Hinton, G.: ImageNet Classification with Deep Convolutional Neural Networks. https://www.nvidia.cn/content/tesla/pdf/machine-learning/imagenet-classifi-cation-with-deep-convolutional-nn.pdf. Accessed June 2018
7. Lecountre, A., Negrevergne, B., Yger, F.: Recognizing art style automatically in painting with deep learning, France (2017). http://www.lamsade.dauphine.fr/~bnegrevergne/ webpage/documents/2017_rasta.pdf. Accessed Mar 2018
8. Nielsen, M.: Neural Network and Deep Learning. Determination Press (2015). http:// neuralnetworksanddeeplearning.com/. Accessed Apr 2018
9. Pedregosa, F., Varoquaux, G., Gramfort, A., Michel, V., Thirion, B., Grisel, O., Blondel, M., Prettenhofer, P., Weiss, R., Dubourg, V., Vanderplas, J., Passos, A., Cournapeau, D., Brucher, M., Perrot, M., Duchesnay, E.: Scikit-learn: machine learning in Python 2011. http://scikit-learn.org/. Accessed Apr 2018
10. Singh, V.: Convolutional Neural Network for Image Classification (2017). www. completegate.com/2017022864/blog/deep-machine-learning-images-lenet-alexnet-cnn. Accessed May 2018
11. Zaki, F.: Identify This Art (2015). http://www.identifythisart.com/. Accessed Apr 2018

Intelligent Agent for Weather Parameters Prediction

Jacek Mazurkiewicz[1]([✉]) [iD], Tomasz Walkowiak[1] [iD],
Jarosław Sugier[1] [iD], Przemysław Śliwiński[1] [iD], and Krzysztof Helt[2]

[1] Faculty of Electronics, Wrocław University of Science and Technology,
ul. Wybrzeże Wyspiańskiego 27, 50-370 Wrocław, Poland
{jacek.mazurkiewicz, tomasz.walkowiak, jaroslaw.sugier,
przemyslaw.sliwinski}@pwr.edu.pl
[2] Teleste sp. z o.o, ul. Szybowcowa 31, 54-130 Wrocław, Poland
krzysztof.helt@teleste.com

Abstract. The paper shows how the typical and not sophisticated topology of the neural network trained by easily implemented gradient method can fulfil the practical needs of the intelligent agent to be useful for weather parameters prediction. If we are able to accumulate the significant set of weather events recording temperature, atmospheric pressure, wind speed, etc. we have the real input for correct prediction in the future. The size of the training vectors can be limited as well as the number of the training epochs. Better results of prediction we can expect when we use the combination of weather events for the training vectors creation. It is possible to create the type of intelligent agent to predict the value of the weather parameters with acceptable low-level error at different climate zones. This way the idea of the weather Complex Event Processing systems seems to be sensible where Event Processing Agents (EPAs) can typical sensors to test the values of the weather parameters as well as intelligent tools created based on big data sets stored year by year.

Keywords: MLP · Weather forecast · Intelligent agent · Processing system

1 Introduction

Weather forecasting is nowadays relying mainly on computer-based models which take into consideration multiple atmospheric factors [2]. However there is still needed human input to choose which prediction model is best for forecasting, to recognize weather impacting patterns, and to know if models are performed better or worse. Big data collected and used for weather forecasting results require extensive computational needs. There are two main forecast approaches. The first one is based on the big collection of the single weather parameter, accumulated year by year temperature for example - as a source of information for the same parameter forecasting in next time period. The simplicity of sensor structure is the main advantage of this approach. On the other hand it is and quite easy to make the unified necessary data preprocessing. Opposite solution tries to use the combination of the different factors which can describe the weather as the complex input for the single parameter forecasting. This

W. Zamojski et al. (Eds.): DepCoS-RELCOMEX 2019, AISC 987, pp. 331–340, 2020.
https://doi.org/10.1007/978-3-030-19501-4_33

approach seems to be more sophisticated and potentially can describe the state of weather more precisely. So based on more valuable data the results of forecasting can be closer to the reality. Of course the problem we can notice is how to combine the set of factors into sensible input vector for the forecasting. It means the parameters normalization process and the weights to define the importance of each factor. In this paper we try to build the intelligent agent – the artificial neural network to forecast the temperature based on the accumulated data from the previous years. We try to tune the length of the forecast – measured by the number of days ahead - as well as we compare the quality of the forecast if we use only the previously stored temperature data and if we built the training vectors based on the set of different factors stored year by year. The tests are made for three cities located in completely various climatic zones. The forecast precision is verified by the real data also already recorded.

Section 2 presents the idea of the intelligent agent construction. Section 3 briefly describes the possible approaches to weather forecast problem. Section 4 shows how the artificial neural network can be used as weather forecasting tool. Finally Sect. 5 is the case study report.

2 Intelligent Agent Idea

Artificial Intelligence seemed to be natural choice for interpreting data which is provided in "human" way [10, 13, 14]. Weather forecasting is almost always related to phenomenon occurring in a specific region. Weather forecast models aggregate the events: weather factors – temperature, wind, pressure, precipitation measured in the pointed place on the Earth. It is obvious that the models build for persistence or for climatology prediction cannot provide any reasonable results to other place in the world. The forecasting by analogies is the only approach which can take advantage from different locations. The analog method uses data to find analogous events in the past. It can take into account data based on events from other, similar location to enhance predicting weather state when no analogies can be found. It is crucial to thoroughly adjust another location model.

Valuable results for other forecasting models can be achieved by averaging models build from surrounding locations. These locations cannot be far away and cannot include extreme locations. Often weather model is very similar for whole region, where multiple data collecting stations are present. They can complement each other data and provide better, and more complete results.

However sometimes there are weather stations which location is on top of the hill or mountain and data collected there contain more raw temperature, pressure or wind speed measurements. Whole process of finding locations with analogous weather models or finding locations which are close enough to provide good results can be called weather events processing. That means that resulting model for location of interest is actually build as the union of similar models, adapted to provided best results.

From the functional and practical point of view we try to describe the climate models as Complex Event Processing (CEP) system. It is a network of some basic components which can communicate with themselves by sending event messages. The primary source of *raw weather events* (i.e. the basic events which enter the system) are external *event producers* which can be as simple as real-world weather sensors or as complex as other information systems capable of communicating their output in the form of events. The essential part of the processing is done within an *Event Processing Network* (EPN) which is a composition of *Event Processing Agents* (EPAs). Each agent is capable of reading events on its inputs, analyzing them according to some specific processing scheme and generating *derived events* on its outputs. The derived events can be forwarded to other agents within the network for further processing or can be sent to *event consumers* as the result of system operation.

The weather raw events used by us are taken from HadCRUT3 records from the Intergovernmental Panel on Climate Change (IPCC). These data are designated by the World Meteorological Organization (WMO) for use in climate monitoring process. They contain monthly sets of weather events values for more than 3000 land stations. National Climatic Data Center from Asheville, NC, USA reports collecting global summary of the day data (GSOD). Data which creates the Integrated Surface Data (ISD) daily summaries is obtained from USAF Climatology Center. Beginning from 1929, over 9000 land and sea stations are monitored. Daily reports from GSOD contain multiple collectible data types which can serve as perfect input to big data processing [6, 7]. Data are given in Imperial units so we covert them to Metric system.

3 Approach to Weather Data Analysis

The classic weather forecasting began near 650 BC, when the Babylonians started to predict weather from cloud patterns. Even early civilizations used reoccurring events to monitor changes in weather and seasons. There are five main weather forecasting methods [15].

Persistence - This is the most simple and primitive method of weather forecasting. The persistence method takes as main input information that the conditions will not change during whole time of the forecast. This method predicts weather correctly for regions where weather does not change so often like sunny and dry regions of California in the USA.

Trends - Basing on speed and directions of weather changes simple trend model can be calculated. Atmospheric fronts, location of clouds, precipitation and high and low pressure centers are main factors of this method. This model is accurate if we generate the weather forecast for only next hours, especially for detecting storms or upcoming weather significant changes.

Climatology - This method bases on statistics and uses long term data accumulated over many years of observations. It is first method that can take into account wider spectrum of changes occurred in particular region and thus can provide more accurate local predictions. The extremes ignoring and the averaged output are the main disadvantage of it. It does not react on climate changes, so it can provide incorrect output.

Analog - Analog method is more complicated than previous ones. It involves combination data describing current weather and analogous conditions in the past. This method assumes that actual set of weather factors is same as found in archival data.

Numerical Weather Prediction – It is the most common nowadays method of forecasting. In 1920 the first attempt to predict weather this way is noticed but due to insufficient computer power realistic results were not provided until 1950. This method uses mathematical models of the atmosphere to calculate possible forecast. It can be used for both - short and long term weather predictions [12]. Due to chaotic nature of atmosphere simulations is almost impossible to solve multiple equations considering them without any error. This is the reason why this approach provides valid forecasts for up to one week only.

4   Neural Network as Forecasting Device

4.1   State of Art

Artificial Intelligence was successfully used in meteorology before. In [5] authors try to use four different Multilayer Perceptrons for one day ahead prediction of single parameter – temperature [8]. Model was built for city of Kermanshah located in west Iran. Datasets are based on ten-years set of meteorological data. They were used to describe the weather of each season. The approach provided promising results of temperature forecasting with average difference between predicted and expected temperature less than 2 °C. Similar work is presented in [9]. Authors used Multilayer Perceptron, Elman Recurrent Neural Network, Radial Basis Function Network and Hopfield model [10, 13, 14] to forecast temperature, wind speed and humidity. All predictions are for city in southern Saskatchewan, Canada. The researchers prepared four models for each season and they produced one day look ahead forecast. Most accurate forecast were provided by combination of multiple neural networks. Andrew Culclasure in [3] presents another application of using Artificial Intelligence for weather forecasting. Three different neural networks were created and trained with data collected in Stetesboro in Georgia, USA. Researcher uses small scale and imperfect datasets for creating forecast for 15-min, 1-h, 3-h, 6-h, 12-h and 24-h ahead. Despite using neural network for weather forecasting, it has been also used to nonlinear meteo-image processing [4, 11]. It has been proven by authors that Artificial Intelligence can be used for object recognition and feature extraction. This way the weather forecasting can also be done [10, 13, 14].

4.2   Neural Network Features

The weather changes prediction using neural network can be compared to two forecasting methods: *Climatology* - neural network can predict weather conditions according to many years of data collection [1]. The result of forecast is an average state of temperature, pressure, wind speed or other parameters taking into account the part of the year. Such forecasting is done within climatology model. *Analog* - without consideration the part of the year, forecasting using neural network is similar to forecasting using analogies. Based on [5], [9] and theoretical research Multilayer Perceptron (MLP) with configurable topology is the forecasting tool [8]. Application created as a result of this work allowed to create networks. Sample configurations are presented in Table 1.

Table 1. MLP forecasting tool configurations

Configuration	Input parameters	Input neurons	Output parameter	Output neurons
Seven days forecast based on temperature only	Temperature	7	Temperature	1
Seven days forecast based on temperature and number of the day of the year – season is known	Temperature, day of the year	8	Temperature	1
Three days forecast based on temperature, mean atmospheric pressure and wind speed for the day	Temperature, atmospheric pressure, mean wind speed	3	Temperature	1

We assumed the input temperature from previous two days, pressure and wind speed from one previous day should provide one day temperature forecast. The algorithm needs to keep history of three days of measurements. Two days history of events create the neural network input, and the last day events are used for validation or testing if provided answer is correct.

The Multilayer Perceptron forecasting tool is trained by backpropagation [8, 10]. The neural network is not being trained until convergence. It is impossible to find optimal solution to create model which perfectly corresponds to weather conditions in any situation. The network topology is calculated as follows. If the input vector described – for example – the temperature of seven days and one look ahead temperature prediction MLP has seven input neurons and one output neuron.

Three places in the world: Opole, Helsinki and Mexico City are chosen to benchmark neural network performance and effectiveness. The choice is driven by the weather fluctuation within one year in each place. Opole is a city in southern Poland on the Odra River. It can be characterized by moderate winters and summers. Neural network model build for this city achieves error rate around 0.01 after 5 epochs of learning (Fig. 1).

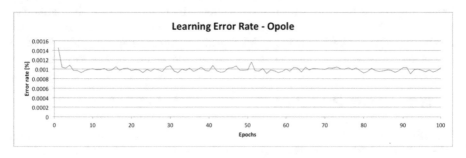

Fig. 1. MLP forecasting tool error rate - prediction for Opole, Poland

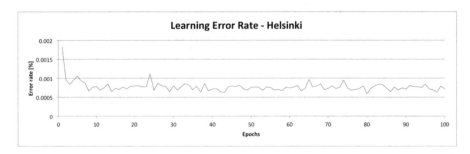

Fig. 2. MLP forecasting tool error rate - prediction for Helsinki, Finland

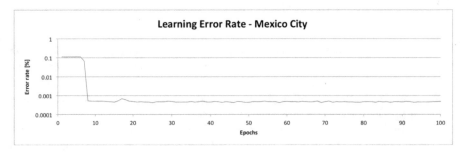

Fig. 3. MLP forecasting tool error rate - prediction for Mexico City, Mexico

Helsinki is the capital and largest city of Finland located in southern Finland, on the shore of the Gulf of Finland, an arm of the Baltic Sea. This city has a humid continental climate and it has lower minimal temperature than Opole. Neural network has less stable but lower error rate (Fig. 2).

Mexico City is capital of Mexico. It has a subtropical highland climate, and due to high altitude of 2.240 m over sea level the temperature is rarely below 3 °C and above 30 °C. Due to almost constant temperature along the year network has very low error rate (Fig. 3).

Neural network learning procedure shows significant correlation: weather model created for the specific city with real climate describing the place on the Earth. The constant error rate is available after 10 epochs of training. If the city has more constant weather conditions like Mexico City MLP achieves lower error rates (Fig. 3). More fluctuating temperature in Helsinki and Opole provide slightly higher error rates.

5 Temperature Prediction

The aim of the experiment is to create weather forecast for three chosen cities based on different atmospheric parameters. Weather prediction was targeted to whole year 2013. MLP tool is loaded by data from 1973 to 2012. Year 2013 was used for testing forecasting possibilities. First forecast is based on seven days temperature input.

Additionally MLP has information about the part of the year by providing the number of the day of the year as one of the network inputs. Forecasting results for Opole are presented in Fig. 4. Average error is estimated to 2.06 °C, and standard deviation is equal to 1.59 °C. The worst predictions are around 200th day of the year, in July. It seems the neural network has problems with temperature extremes detection. Better results are visible for months with temperature closer to yearly average.

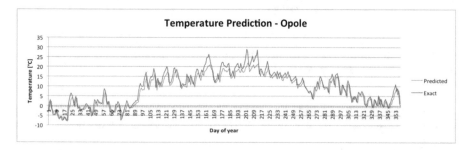

Fig. 4. MLP forecasting tool - seven days forecast based on temperature and number of the day of the year - for Opole, Poland

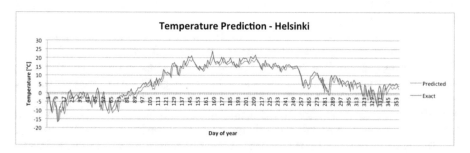

Fig. 5. MLP forecasting tool - seven days forecast based on temperature and number of the day of the year - for Helsinki, Finland

Similar experiment has been performed for Helsinki in Finland. Results are presented in Fig. 5. Results for this city does not correspond with the results for Opole. MLP better predicts in summer - weather conditions were moderate. Network is not able to correctly detect temperature drop in the end of February. The weather in February 2013 stands significantly out from the average, so neural network taught by big set of quite typical temperature events probably forecast wrong temperature.

Best results are for Mexico City (Fig. 6). Average difference between predicted and observed temperature is equal to 1.49 °C and standard deviation equal to 1.21 °C. Such great results are because of nearly constant temperature in Mexico City. Training procedure is much faster - no more than five training epochs are needed. Only one significantly wrong temperature forecast is provided. Again - the unusual weather conditions are the source of this mistake.

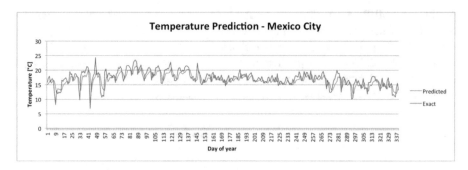

Fig. 6. MLP forecasting tool - seven days forecast based on temperature and number of the day of the year - for Mexico City, Mexico

Opole is chosen for further experiments. Configuration with three days and twenty days temperature based forecast is fixed. The results are presented in (Figs. 7 and 8) respectively.

The average difference between predicted and exact value for equal to 3.37 °C for three days forecast, 4.14 °C for twenty days forecast. The probable reason for such MLP behavior is not correctly selected number of input vectors used during learning procedure. The answers of the network are above the real values. In previous experiments they are below.

The last experiment is the three days temperature forecast, where the input vectors combining temperature, mean atmospheric pressure and wind speed for the day. Results are presented in (Fig. 9). The predicted value of temperature is again constantly above the real values and follows the exact value thorough all days of the year with the same stable error equal to 2.02 °C and standard deviation is equal to 1.32 °C. It means the wider package of weather events including not only data about the temperature allows to describe the possible fluctuations of single parameter in more detailed way.

The dynamic and unexpected changes of the temperature are much better pointed in the forecast. It is important and interesting especially because the place of experiments – Opole is characterized by moderate climate, but with temperature changes in one year period. It seems that MLP forecasting tool could be interesting alternative approach for the classic methods.

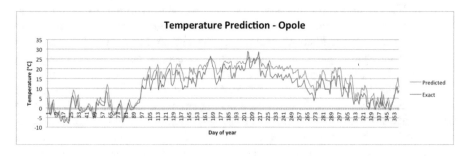

Fig. 7. MLP forecasting tool - three days forecast based on temperature and number of the day of the year - for Opole, Poland

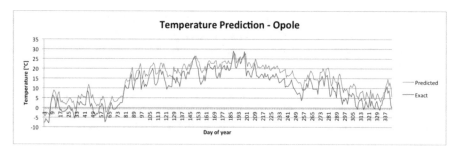

Fig. 8. MLP forecasting tool - twenty days forecast based on temperature and number of the day of the year - for Opole, Poland

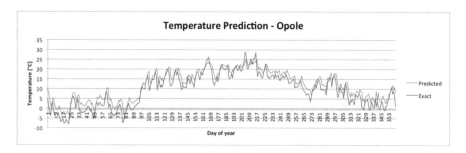

Fig. 9. MLP forecasting tool – three days temperature forecast based on temperature, atmospheric pressure, wind speed and number of the day of the year - for Opole, Poland

6 Conclusions

The results of Artificial Intelligence usage for the temperature forecasting are very promising. It is possible to create the type of intelligent agent to predict the value of the weather parameters with acceptable low-level error. This prediction could be correct for even one year period taking into account four seasons and exchange of weather conditions caused by them. The answers of this agent could be found as valid in different locations on the Earth – for different climate zones. It means the climate model can be successfully created by the structure of the artificial neural network and large enough set of historical data to describe the weather conditions for the pointed localization. The paper shows the typical and not sophisticated topology of the neural network trained by easily implemented gradient method can fulfil the practical needs of the intelligent agent to be useful for weather parameters prediction. If we are able to accumulate the significant set of weather events recording temperature, atmospheric pressure, wind speed, etc. we have the real input for correct prediction in the future. The size of the training vectors can be limited as well as the number of the training epochs. It sounds really pretty – it means the training procedure could be not time consuming process. Better results of prediction we can expect when we use the combination of weather events for the training vectors creation. It means we find the set of weather sensors correlated by time and localization as the best source of data for intelligent agent.

This way the idea of the weather Complex Event Processing systems seems to be sensible where *Event Processing Agents* (EPAs) can typical sensors to test the values of the weather parameters as well as intelligent tools created based on big data sets stored year by year.

Acknowledgements. This work was supported by the Polish National Centre for Research and Development (NCBR) within the Innovative Economy Operational Programme grant No. POIR.01.01.01-00-0235/17 as a part of the European Regional Development Fund (ERDF).

References

1. Bonarini, A., Masulli, F., Pasi, G.: Advances in Soft Computing, Soft Computing Applications. Springer (2003)
2. Bell, I., Wilson, J.: Visualising the Atmosphere in Motion. Bureau of Meteorology Training Centre in Melbourne (1995)
3. Culclasure, A.: Using Neural Networks to Provide Local Weather Forecasts. Electronic Theses & Dissertations. Paper 32 (2013)
4. Francis, M.: Future telescope array drives development of exabyte processing. http://arstechnica.com/science/2012/04/future-telescope-array-drives-development-of-exabyte-processing/
5. Hayati, M., Mohebi, Z.: Temperature forecasting based on neural network approach. World Appl. Sci. J. **2**(6), 613–620 (2007)
6. Hoffer, D.: What does big data look like? visualization is key for humans. http://www.wired.com/2014/01/big-data-look-like-visualization-key-humans/
7. Katsov, I.: In-stream big data processing. http://highlyscalable.wordpress.com/2013/08/20/in-stream-big-data-processing/
8. Kung, S.Y.: Digital Neural Networks. Prentice-Hall, Englewood Cliffs (1993)
9. Maqsood, I., Riaz Khan, M., Abraham, A.: An ensemble of neural networks for weather forecasting. Neural Comput. Appli. **13**, 112–122 (2004)
10. Pratihar D.K.: Soft Computing. Science Press (2009)
11. de Ridder, D., Duin, R., Egmont-Petersen, M., van Vliet, L., Verbeek, P.: Nonlinear image processing using artificial neural networks (2003)
12. Sandu, D.: Without stream processing, there's no big data and no internet of things. http://venturebeat.com/2014/03/19/without-stream-processing-theres-no-big-data-and-no-internet-of-things/
13. Sivanandam, S.N., Deepa, S.N.: Principles of Soft Computing. Wiley, Hoboken (2011)
14. Srivastava, A.K.: Soft Computing. Narosa PH (2008)
15. Guideline for developing an ozone forecasting program. U.S. Environmental Protection Agency, Office of Air Quality Planning and Standards (1999)

Mathematical Modeling of the Hot Steam-Water Mixture Flow in an Injection Well

Nail Musakaev[1,2]($\boxtimes$), Stanislav Borodin[1], Sergey Rodionov[1,3], and Evgeniy Schesnyak[3]

[1] Tyumen Branch of Khristianovich Institute of Theoretical and Applied Mechanics of the Siberian Branch of the Russian Academy of Sciences, Taymirskaya Street 74, Tyumen, Russia
musakaev@ikz.ru
[2] Industrial University of Tyumen, Volodarskogo Street 38, Tyumen, Russia
[3] RUDN University, Miklukho-Maklaya Street 6, Moscow, Russia

Abstract. In this paper, for the problem of pumping heat-transfer agent into an oil reservoir, a mathematical model of the processes occurring during the movement of a hot steam-water mixture in an injection well has been proposed. This model takes into account phase transitions occurring in a two-phase mixture "water-steam", and external heat exchange of the well with surrounding rock (including permafrost). Algorithm has been constructed that allows calculating for different time moments of the well operation: the heat-transfer agent parameters, the temperature distribution in the surrounding rock and the thawing radius of permafrost along the depth of the injection well. According to the proposed algorithm, a computer code was developed and numerical experiments were carried out to find the parameters of the downward flow of a hot steam-water mixture in an injection well. By calculation it is shown that using of heat-insulated pipes leads to a lesser heat loses of a heat-transfer agent along the depth of an injection well. Also, using of heat-isolating materials on the outer surface of the well lifting column allows to increase the depth and quantity of steam (compared with non-heat-insulated lifting column). It is shown that with an increase in the heat insulating layer, the thawing radius of permafrost decreases, which is caused by a decrease in heat transfer from the well product to surrounding rocks.

Keywords: Mathematical model · Algorithm · Injection well · Steam-water mixture · Permafrost · Thawing zone

1 Introduction

The last decades have been characterized by an increase in hydrocarbon consumption, especially in developing countries [1]. However, most researchers believe that only 20–30% of the residual world oil reserves can be extracted using traditional methods of oil field development [2, 3]. Therefore, various methods are being developed and applied in the world to increase the oil recovery factor. Also the oil extraction from

© Springer Nature Switzerland AG 2020
W. Zamojski et al. (Eds.): DepCoS-RELCOMEX 2019, AISC 987, pp. 341–348, 2020.
https://doi.org/10.1007/978-3-030-19501-4_34

fields with difficultly recoverable hydrocarbon reserves is also being actively studied. Such deposits include deposits of highly viscous and bituminous oils (heavy oils). The main method used today for the development of such deposits is the injection of heat-transfer agent (hot water or steam) into an oil-saturated reservoir [3, 4]. This method currently accounts for up to 90% of all oil produced from heavy oil fields.

The temperature of a reservoir and saturating fluids increases, when a heat-transfer agent is injected into the reservoir [4, 5]. An increase in temperature leads to a rather sharp decrease in the viscosity of heavy oil, increasing its mobility [5, 6]. Accordingly, the rate of filtration flow increases and the oil recovery rate increases. As a result, the development of fields with heavy oils becomes economically justified.

The production of a heat-transfer agent requires a large amount of energy. In addition, effective thermal action on an oil reservoir requires a significant amount of the heat-transfer agent at sufficiently high rates of the gas-liquid mixture injection into the reservoir. When using steam as a heat-transfer agent, it is necessary that it has a high degree of dryness (the mass fraction of steam in the two-phase "water-steam" mixture) at the entrance to a reservoir. All this indicates the need to optimize this thermal method of developing oil fields. Therefore, we need calculations of the injected heat-transfer agent parameters (temperature, steam content, etc.) throughout its movement from the well head to a porous reservoir [4]. Such calculations should be performed using a special computer code developed on the basis of a mathematical model, which takes into account various aspects of the studied processes.

Theoretical study of the hot steam-water mixture downward flow in a vertical channel has been carried out. The constructed mathematical model takes into account vapor condensation as the two-phase mixture advances to the well bottom and external heat exchange with surrounding rock taking into account the thawing of permafrost.

2   Problem Statement and Basic Equations

Consider the problem of the heat-transfer agent (steam-water mixture) movement in the injection vertical well from the well head to the well bottom. The z axis will be directed vertically downwards, the origin of coordinates coincides with the well head. We will consider the well operation mode stationary. The well passes through permafrost. The parameters of the heat-transfer agent at the well head (at the outlet of a heater or steam generator) are known, the mass flow rate m is constant:

$$z = 0: \quad p = p_{or}, \quad T = T_{or}, \quad k_v = k_{or}. \tag{1}$$

Hereinafter, the subscripts l or v are assigned to the parameters related to the liquid water or steam, respectively; subscript or is attributed to the well head parameters; p and T are the two-phase flow pressure and temperature; k_v is the mass fraction of steam in steam-water mixture (steam dryness).

In mathematical modeling we accept the following assumptions: the temperature for each section of a channel is the same for both phases; phase transitions occur in the equilibrium mode; the flow in a wellbore steady and inertialess; thermal conductivity in the axial direction of a well is negligible compared to convective heat transfer;

the rocks surrounding a well are uniform and isotropic; when the front of phase transitions moves in frozen rocks, there is no mass transfer of liquid fluids [4, 7]. We will neglect the influence of the seasonal change in surface temperature on the near-surface area of the ground.

The system of basic equations describing a two-phase flow in a vertical channel consists of the equations of conservation of masses (2), impulses (3), and energy (4). This system in one-dimensional case with the accepted assumptions has the form [8–10]:

$$m_l + m_v = m = \text{const} \tag{2}$$

$$\frac{dp}{dz} = -F_w + (\rho_l(1-\alpha) + \rho_v \alpha) g \tag{3}$$

$$mc\frac{dT}{dz} = \frac{m_v}{\rho_v^0}\frac{dp}{dz} - L_{lv}\frac{dm_v}{dz} - Q_w \tag{4}$$

$$F_w = \frac{\lambda_w}{4r_w}\left(\rho_l\frac{(1-\varphi)^2}{1-\alpha} + \rho_v\frac{\varphi^2}{\alpha}\right)W^2, \quad \lambda_w = 0,067\left[\frac{158}{\text{Re}} + \frac{\varepsilon}{r_w}\right]^{0,2},$$

$$mc = m_l c_l + m_v c_v, \quad m_l = \rho_l(1-\varphi)\,S\,W, \quad m_v = \rho_v\,\varphi\,S\,W, \quad S = \pi r_w^2, \quad \rho_l = \text{const},$$
$$p = Z_v \rho_v R_v\,T, \quad Q_w = 2\pi\,r_w q_w,$$

$$k_v = \frac{m_v}{m}, \quad \varphi = \left(1 + \frac{\rho_v\,m_l}{\rho_l\,m_v}\right)^{-1}, \quad W = \frac{1}{S}\left(\frac{m_l}{\rho_l} + \frac{m_v}{\rho_v}\right),$$

$$\alpha = \begin{cases} 0.833\,\varphi, & \varphi \le 0,9 \\ \left[0,833\,\varphi + 0.167\left(1 + \frac{\rho_l(1-\varphi)}{\rho_v\,\varphi}\right)^{-1}\right]\varphi. & \varphi > 0,9 \end{cases}$$

where m_i and $\rho_i(i = l, v)$ are the mass flow and true density of the i-th phase; $c_i(i = l, v)$ is the specific heat capacity at a constant pressure of the i-th phase; λ_w is the coefficient of friction between the flow of a steam-water mixture and the inner surface of the well lifting column; Re is the Reynolds number; ε is the roughness size; φ is the volumetric flow steam content; R_v is the gas constant for water vapor; L_{lv} is the specific heat of vaporization; q_w is the intensity of heat removal per unit area [7]; α if the volumetric steam content [11]; Z_v is the gas supercompressibility factor for steam [12]. In Eq. (3), the terms that are associated with the inertial effects were neglected.

The temperature of the two-phase flow in a well casing is significantly influenced by the temperature on the external well surface $T_c(z)$, the value of which, in turn, depends on the temperature distribution in the rocks surrounding the well. The problem of heat distribution in the surrounding permafrost, taking into account possible thawing, can be solved on the basis of the following system of equations [7]:

$$t > 0, \quad r_c < r \le r_c + \theta : \quad \frac{\partial T_{th}}{\partial t} = \chi_{th} \frac{1}{r} \frac{\partial}{\partial r} \left(r \frac{\partial T_{th}}{\partial r} \right), \tag{5}$$

$$t > 0, \quad r_c + \theta < r < \infty : \quad \frac{\partial T_{fr}}{\partial t} = \chi_{fr} \frac{1}{r} \frac{\partial}{\partial r} \left(r \frac{\partial T_{fr}}{\partial r} \right), \tag{6}$$

$$t > 0, \quad r = r_c : \quad -\lambda_{th} \frac{\partial T_{th}}{\partial r} = \beta (T_w - T_c), \tag{7}$$

$$\lambda_{fr} \frac{\partial T_{fr}}{\partial r} \bigg|_{r=r_c+\theta+0} - \lambda_{th} \frac{\partial T_{th}}{\partial r} \bigg|_{r=r_c+\theta-0} = \rho w L_i \frac{d\theta}{dt}, \tag{8}$$

$$t = 0, \quad r_c \le r < \infty : \quad T_{fr} = T_{geo}, \tag{9}$$

$$t > 0, \quad r \to \infty : \quad T_{fr} = T_{geo}, \tag{10}$$

where T_{fr} and λ_{fr} are the temperature and thermal conductivity coefficient of permafrost; T_{th} and λ_{th} are the temperature and thermal conductivity coefficient of thawed rocks; w is the ice content; ρ is the density of the soil; L_i is the specific heat of ice melting; θ is the thawed zone length; T_{geo} is the geothermal temperature.

The problem of heat propagation at the rocks surrounding a well, located below the permafrost zone, will be solved on the basis of the following system of equations:

$$t > 0, \quad r_c < r < \infty : \quad \frac{\partial T_{ext}}{\partial t} = \chi_{ext} \frac{1}{r} \frac{\partial}{\partial r} \left(r \frac{\partial T_{ext}}{\partial r} \right), \tag{11}$$

$$t = 0, \quad r_c \le r < \infty : \quad T_{ext} = T_{geo}, \tag{12}$$

$$t > 0, \quad r = r_c : \quad -\lambda_{ext} \frac{\partial T_{ext}}{\partial r} = \beta (T_w - T_c), \tag{13}$$

$$t > 0, \quad r \to \infty : \quad T_{ext} = T_{geo}, \tag{14}$$

where T_{ext}, λ_{ext} and χ_{ext} are the temperature, coefficient of thermal conductivity and temperature-conductivity of the surrounding rocks, which are located below the permafrost zone.

3   Solving Scheme and Calculation Results

Figure 1 presents the algorithm for calculating the parameters of the downward two-phase flow in an injection well, taking into account the thermal interaction of the well product with surrounding rock (including permafrost).

According to the above algorithm, the problem of finding the parameters of a heat transfer agent in an injection well, taking into account the thermal interaction with surrounding rock is divided in two parts. In the first part, a steady downward flow of a hot steam-water mixture in the well column is considered, taking into account heat losses to surrounding rock. In the second part, the unsteady propagation of heat from the well in the rock is calculated taking into account possible thawing of permafrost.

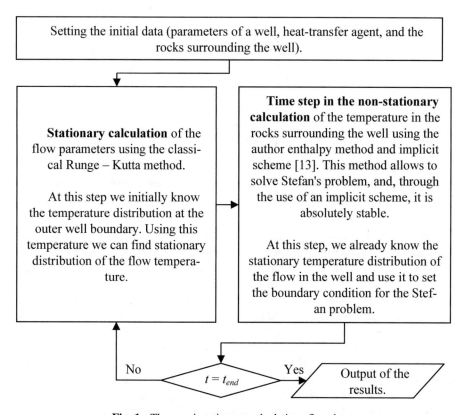

Fig. 1. The quasi-stationary calculations flowchart.

According to the proposed algorithm, a computer code was developed and numerical experiments were carried out to find the downstream two-phase flow parameters in an injection well. Figure 2 shows the change with depth of the mass concentration of steam, temperature and pressure. During the calculations, the following values of the parameters were used: $m = 0,45$ kg/s; $p_{or} = 9,6$ MPa; $k_{or} = 0,9$; $r_w = 0,031$ m; $r_1 = 0,036$ m; $r_2 = 0,076$ m; $r_3 = 0,084$ m; $r_c = 0,125$ m; $p_{cr} = 21,8$ MPa; $T_{cr} = 647$ K; $\lambda_l = 0,244$ W/(m·K); $\lambda_{met} = 45$ W/(m·K); $\lambda_b = 1,1$ W/(m·K); $c_l = 4200$ J/(kg·K); $R_v = 461$ J/(kg·K); $\mu_l = 1,2\cdot10^{-5}$ Pa·s; $\varepsilon = 0,00001$ m; $g = 9,8$ m/s^2; the heat-transfer agent temperature at the well head corresponds to the equilibrium temperature of the steam-water phase transition for the well head pressure;

the initial temperature of the rock at the well head is 268 K; depth of permafrost is 200 m; well depth is 900 m. There is no heat insulation on the lifting column outer surface; there is water in the annular space. The values of c_v, λ_v, μ_v were determined by interpolating the tabular data for water steam [4].

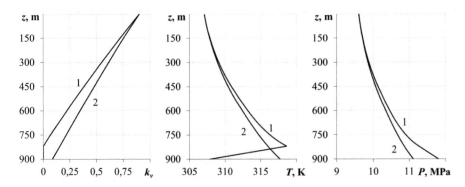

Fig. 2. The distributions of the steam mass concentration, temperature and pressure at the well height. Line 1 corresponds to the initial time moment, and the line 2 – five days after the start of the hot steam-water mixture injection.

Figure 2 shows that when the hot steam-water mixture moves from the injection well head to bottom the mass concentration of steam decreases, due to its condensation. With the parameters taken in the calculations at the initial time moment, all the steam condensed before reaching the well bottom. 5 days after the start of the hot water-steam mixture injection a gradual increase in the temperature of the two-phase mixture at the well bottom occurs due to the heating of the surrounding rocks, and in this case the steam is delivered to a reservoir. Also after 5 days a decrease in the density of the two-phase mixture occurs and a decrease in the pressure gradient.

To realize a higher rate of displacement of oil from a reservoir, it is necessary to reduce heat loss in the injection well bore. One of the ways to reduce heat transfer from the injection well product to surrounding rock is using the heat insulating material on the outer surface of the well lifting column. Figure 3 shows the change with the well depth of the temperature and steam mass concentration at the initial time moment at different thicknesses of a heat insulating material. The coefficient of thermal conductivity of the heat-insulating material is assumed to be 0.03 W/(m·K). From the data presented in Fig. 3, it can be seen that the use of insulated pipes leads to a decrease in the temperature of the steam-water mixture in the well, due to the smaller amount of condensed steam. Thus, the use of insulating material allows to increase the depth and amount of steam, which leads to an increase in the amount of thermal energy supplied to a reservoir. From Fig. 4, it can be seen that with the growth of the heat insulating layer, the radius of the thawed zone decreases, which is caused by a decrease in heat transfer from the well product to surrounding rock.

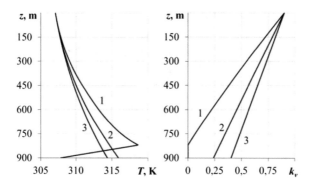

Fig. 3. The distributions of temperature and steam mass concentration at the initial time moment. Lines 1, 2 and 3 correspond to the thickness of the insulation material 0, 2 and 4 mm.

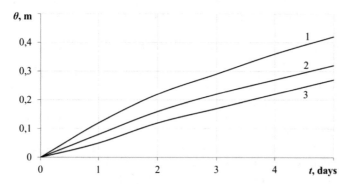

Fig. 4. The evolution over time of the thawed zone thickness. Lines 1, 2 and 3 correspond to the insulation material thickness 0, 2 and 4 mm.

4 Conclusion

Based on the equations of mechanics of multiphase media, the mathematical model is proposed in the form of a system of differential equations describing the downward flow of a hot steam-water mixture in an injection well. This model takes into account condensation of steam as the steam-water mixture moves to the well bottom and external heat exchange with the rock surrounding the well, taking into account the permafrost thaw zone formed.

The main idea of the computational algorithm is that, knowing the initial temperature of the outer well boundary, using the fourth-order Runge–Kutta method we solve the quasistationary problem of finding flow parameters. Then, using the calculated flow temperature, a time step is taken in the nonstationary calculation of the temperature distribution around the well. Non-stationary calculation is performed by the author's enthalpy method using an implicit scheme [13]. The whole procedure is repeated cyclically until the end of the calculation time.

It is shown that over time after the start of the heat-transfer agent injection, due to the heating of the surrounding rocks, a gradual rise in the steam amount at the bottom of the well occurs. When this occurs, the density of the two-phase mixture decreases and the pressure gradient correspondingly decreases. It is shown that thermal insulation leads to a significant increase in the mass concentration of steam delivered into a reservoir, as well as to a decrease in the thawing zone around the well, which will have a positive effect on the reliability of the well design.

Acknowledgement. The research was supported by the Russian Science Foundation (project number 18-19-00049).

References

1. Makogon, Y., Holditch, S., Makogon, T.: Natural gas-hydrates - a potential energy source for the 21st century. J. Pet. Sci. Eng. **56**, 14–31 (2007). https://doi.org/10.1016/j.petrol.2005.10.009
2. Thomas, S.: Enhanced oil recovery – an overview. Oil Gas Sci. Technol. Rev. IFP **63**(1), 9–19 (2008). https://doi.org/10.2516/ogst:2007060
3. Rodionov, S., Pyatkov, A., Kosyakov, V.: Influence of fractures orientation on two-phase flow and oil recovery during stationary and non-stationary waterflooding of oil reservoirs. AIP Conf. Proc. **2027**, 030044 (2018). https://doi.org/10.1063/1.5065138
4. Burger, J., Sourieau, P., Combarnous, M.: Thermal Methods of Oil Recovery. Editions Technip, Paris (1985)
5. Willhite, G.: Over-all heat transfer coefficients in steam and hot water injection wells. J. Pet. Technol. **19**(5), 607–615 (1967). https://doi.org/10.2118/1449-PA
6. Shagapov, V., Yumagulova, Y., Gizzatullina, A.: High-viscosity oil filtration in the pool under thermal action. J. Eng. Phys. Thermophys. **91**(2), 300–309 (2018). https://doi.org/10.1007/s10891-018-1749-4
7. Musakaev, N., Borodin, S.: Mathematical model of the two-phase flow in a vertical well with an electric centrifugal pump located in the permafrost region. Heat Mass Transf. **52**(5), 981–991 (2016). https://doi.org/10.1007/s00231-015-1614-3
8. Nigmatulin, R.: Dynamics of Multiphase Media. Hemisphere Publishing Corporation, New York (1991)
9. Shagapov, V., Musakaev, N., Khabeev, N., Bailey, S.: Mathematical modelling of two-phase flow in a vertical well considering paraffin deposits and external heat exchange. Int. J. Heat Mass Transf. **47**(4), 843–851 (2004). https://doi.org/10.1016/j.ijheatmasstransfer.2003.06.006
10. Bondarev, E., Rozhin, I., Argunova, K.: Modeling the formation of hydrates in gas wells in their thermal interaction with rocks. J. Eng. Phys. Thermophys. **87**(4), 900–907 (2014). https://doi.org/10.1007/s10891-014-1087-0
11. Chisholm, D.: Two-phase flow in pipelines and heat exchangers. Longman Higher Education, London (1983)
12. Shagapov, V., Urazov, R., Musakaev, N.: Dynamics of formation and dissociation of gas hydrates in pipelines at the various modes of gas transportation. Heat Mass Transf. **48**(9), 1589–1600 (2012). https://doi.org/10.1007/s00231-012-1000-3
13. Borodin, S.L.: Numerical solution of the Stefan's problem. Tyumen State Univ. Herald Phys. Math. Model. Oil Gas Energy **3**, 164–175 (2015). https://doi.org/10.21684/2411-7978-2015-1-3-164-175

Concept of Preventive Maintenance in the Operation of Mining Transportation Machines

Dinara Myrzabekova[1], Mikhail Dudkin[1], Marek Młyńczak[2(✉)],
Alfiya Muzdybayeva[1], and Murat Muzdybayev[1]

[1] East Kazakhstan State University, Protozanov Str., 69,
070004 Ust-Kamenogorsk, Kazakhstan
[2] Faculty of Mechanical Engineering,
Wrocław University of Science and Technology,
Wyb. Wyspiańskiego 27, 50-370 Wrocław, Poland
`marek.mlynczak@pwr.edu.pl`

Abstract. The origin of the problem arises from numerous limitations existing in the operation of wheeled mining machines including: loaders, drilling vehicles, storage vehicles, etc. It should be mentioned mainly operational and environmental limitations and requirements, such as: ensuring the efficiency and continuity of the mining process, spoil disposal, closed operation system with limited human and technical resources, difficult environmental conditions (high temperature and humidity, high dustiness or muddiness, limited space). In those difficult conditions, the maintenance according to periodic strategy seems to be inadequate as work and degradation processes are variable. The more appropriate approach is to operate according to the state, wherein monitoring of diagnostic parameters would allow setting a reasonable service time. The example of a bolt joint is one of many elements of these machines subject to ageing, for which preventive maintenance according to the state is the most appropriate. Paper describes most important failures occurring in operation and propose the use of modern information systems to gather, transmit and archive diagnostic parameters. It has been assumed that data will be sent periodically, but with high intensity, e.g. every day after the end of the work shift in the machine parking space, via wireless (e.g. RFID, Bluetooth) while passing the machine through a specific gate. The data acquired for each machine as a function of time and work time will allow determining the trend line of the change of the tested parameter and predict the appropriate time of maintenance.

Keywords: Maintenance · LCC · Mining machines

1 Introduction and Objectives

In a certain group of mining machines, accelerated wear of the pivoted joints was observed, causing the pins to slip out and immobilization of the machines. The phenomenon appeared irregularly, depending on the current state of the environment and the intensity of use. Excessive wear was mainly caused by bad environmental

© Springer Nature Switzerland AG 2020
W. Zamojski et al. (Eds.): DepCoS-RELCOMEX 2019, AISC 987, pp. 349–357, 2020.
https://doi.org/10.1007/978-3-030-19501-4_35

conditions such as high dustiness, humidity, temperature and the occurrence of mud. Unfortunately, the way of repair was a simple removal of effects no causes. The reaction to these damages was securing the bolt from being pulled out of the socket by welding a rod, sheet or piece of pipe. It was a makeshift solution, and the existing clearance between the bolt and the socket was had grown even more intense.

The main objective of the study is: description of the problem, gathering failure data directed on introducing preventive maintenance and propose a concept of informatics based management of these preventive maintenances.

2   Research Method

In the operation of mining machines, periodic maintenance strategy does not seem to be sufficient, since the wear rate of many elements depends mainly on the variation of environmental conditions in time and intensity [6]. The timing of the maintenance may not coincide with the moment of the critical state of the connector, what usually leads to sudden machine stop due to damage between services. A more appropriate approach here is the maintenance according to the actual condition, in which the diagnostic parameter responsible for degradation is observed and gives information about the state of the element [3, 5, 10].

Monitoring of this parameter will allow for establishing a rational maintenance time. An example is a bolted connection of the frame, one of many machine elements subject to wear, for which preventive maintenance according to the state is most appropriate [1]. The project involves the use of modern information systems for the generation, transmission and archiving of diagnostic symptoms. It is assumed that the data on the size of the gaps will be transmitted periodically, but with high intensity, for example, every day after the end of the work shift at the parking space of the car over the wireless connection (for example, RFID, Bluetooth) while the car passes through a specific gate. The data acquired for each machine as a function of time and operating time will allow determining the trend line of change of the tested parameter and predicting the likely moment of reaching the limit value.

3   Failure Analysis of Loading and Shuttle Mining Machines

As an example of machines working in difficult conditions, the loading and shuttle car is considered. The operational system and process is located in the underground mine of East Kazakhstan. For a group of machines data of downtime and repair is acquired and processed.

Two types of machines are taken into account: the Underground Mining Loader Caterpillar R1300G (see Fig. 1a) and the Underground Mining Shuttle Car Sandvik EJC417 (see Fig. 1b).

The analysis concerns the downtime of the main functional units of machines due to failures, which led to the loss of ore production. The average downtime per year in hours for main units for Caterpillar R1300G is shown in Table 1.

The total downtime of one loader is estimated as 2400 hours per year, what gives availability of the machine on the level of 0,72 only.

Fig. 1. (a) Underground Mining Loader Caterpillar R1300G. (b) Underground Mining Shuttle Car Sandvik EJC417.

Table 1. Ranking of functional units of underground mining loader Caterpillar R1300G due to its downtime

Machine unit	Downtime [h]	Rate of downtime [%]
Transmission system	649	27
Hydraulic system	624	26
Frame with joints	314	13
Engine	301	12
Bucket	208	9
Electrical system	181	7
Braking system	87	4
Bucket arm	60	2

To rank the systems and units for their reliability, a Pareto diagram is proposed and shows the least unreliable systems (the dominant influence is about 80% of the total time to failure). It includes four systems: transmission, hydraulics, frame with hitch assemblies and engine with its auxiliary systems. Figure 2 shows the longest machine downtime due to failures of transmission and hydraulics elements. Together, they caused almost half of all machine failures. However, according to the criterion of work safety, the most critical is the frame with hitch assemblies having a ratio of the downtime of 13% of the total machine downtime.

In the specified group of structural elements of the loader, the failure of the hinge assemblies in the frame occurs most frequently. The cause of the failure is the wear of the seat of the hinge pin in the eye socket. To keep the hinge from spontaneous dismantling until the scheduled repair work by the shift mechanics, it was decided during an operation to install the element (lever thrust out of the bar) that holds the bolt from leaving the hinge socket. The bar was fixed with one edge to the body by welding the bar pressing the hinge pin, preventing it from moving upwards (Fig. 3).

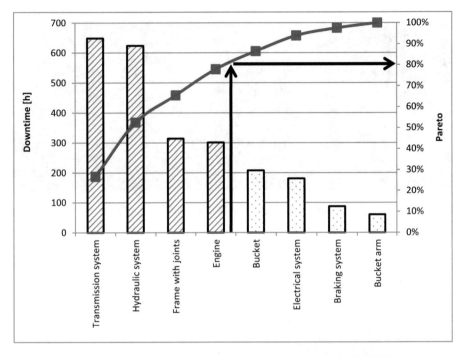

Fig. 2. Downtime ranking of functional units of Caterpillar R1300G.

The correct way of repair should be installing a new bolt with a higher diameter. The action of increasing the diameter of the eyelet by swivel head is shown in Fig. 4.

Similar data analysis is done for another significant machine in the mining operation which is underground mining shuttle Car Sandvik EJC417. Table 2 shows the downtime statistics for Sandvik EJC417.

Ranked units of shuttle Car Sandvik EJC417 in the Pareto diagram is shown in Fig. 5. In total, the group of the least unreliable systems (80% of the total of downtime) covers three systems: transmission, frame with hitch assemblies and engine with its auxiliary systems. From Fig. 5 it can be seen that the longest machine downtime is due to the failure of transmission elements. It should be noted that the frame and hitch assemblies, which, by the criterion of safety work, are the most critical (rank second) is responsible for about 22% of the total downtime.

The total downtime of one loader is estimated as 2200 hours per year, what gives slightly higher availability of the machine on the level of 0,75.

Similarly to the loader, also in case of shuttle car, so-called weak element is a "frame coupling". It is observed intense wearing process leading to escaping of the bolt and stopping the machine. The cause of failure is abrasive-corrosive wear of the hinge pin seat in the eyelet socket (Fig. 6).

Table 2. Ranking of functional units of underground mining shuttle Car Sandvik EJC417 due to its downtime

Machine unit	Downtime [h]	Rate of downtime [%]
Transmission system	752	34
Frame with joints	489	22
Engine	387	17
Hydraulic system	268	12
Electrical system	176	8
Body	88	4
Braking system	62	3
Chassis	1	>0

Fig. 3. A temporary solution preventing the spontaneous dismounting of the hinge pin when the hinge of the swivel mechanism is worn-out.

4   "Fast Truck" for Solution of Significant Technical Problems

The instantaneous and primitive solution of the securing the bolt in the place is installation the addition, extra element holding the bolt in the hinge [4]. Thus, statistics on machine downtime and the results of their analysis confirm the relevance of the chosen research topic, as well as the practical significance of its results.

Fig. 4. Correct way of repair of the eyelet of the frame of the loader Caterpillar R1300G.

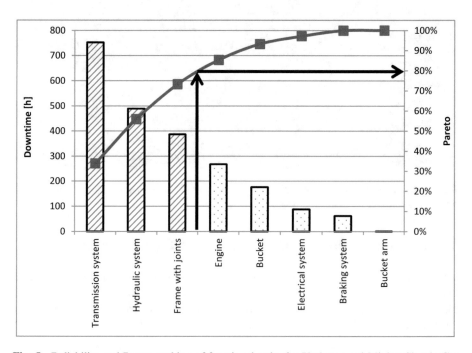

Fig. 5. Reliability and Pareto ranking of functional units for Underground Mining Shuttle Car Sandvik EJC417.

Fig. 6. Worn-out of the hinge of Sandvik EJC417

Solutions aimed at improving the reliability and testability of the hinge mechanism. The most appropriate solution to reduce wear on the hinge assembly seems to be creating a protective structure. It is quite possible to use all-metal caps made by casting or forging. The cover must have sufficient strength because it should not be damaged or destroyed by falling debris (pieces) of the mined material (ore). Placing the cover is supposed to be on top - above the nest of the top eyelet of the hinge assembly. To protect against the ingress of dust and moisture should provide an element sealing the inner space of the cover. An alternative method of sealing can be considered as applying to the flange surface of an automotive sealant. Which of the most rational ways can be found out experimentally [8]. Methods of fastening the cover to the hinge housing (on the eye) include mounting to the mounting bolts or studs with nuts. The bolted version is more technologically advanced and structurally simple [4]. Anyway, the above remedies are possible to introduce during corrective maintenance or planned maintenance. The approach based on maintenance 4.0 should help in the reduction of unplanned stops in operation [2, 7, 10].

5   Concept of Maintenance 4.0 in Underground Machines Fleet Operation

Discussed system of underground mining machines is closed system limited in space and subjected to heavy conditions with the recurrent scheme of actions. It allows for introducing maintenance strategy using operational data to prevent catastrophic failures stopping the machine or even the entire production system [7, 10]. Employment of wireless data transmission eliminates man and shortens the time of accessing information about the current state of the technical system [9].

Making modification of the hinge design improving its isolation from aggressive environment it is suggested to include special sensors monitoring a state of the hinge. They should ensure continuous monitoring of the technical condition of the hinge by at least two parameters: the state of the hinge according to the radial wear and the radial clearance (slack) of the finger in the upper hinge socket. The variants of principles used to select the proper sensor require anyway additional analysis. However, one can preliminarily consider the concept of displacing the axis of the finger relatively to the axis of the eye. The intensity of the displacement of the axis of the finger will be the initial information for predicting events of significant wear of the hinge, when the freedom of its axial displacement appears. Then, another sensor is needed to control the axial displacement of the finger in the hinge eye. At the same time, this sensor is an indicator of the appearance of an additional degree of freedom of the finger in the axial direction, which causes a critical displacement of the finger from the eyelet socket and the subsequent spontaneous disassembly of the hinge. When the second sensor is triggered, the repair should be performed as soon as possible.

To accommodate the sensors should allocate the necessary space in the proposed cover above the hinge. Taking into account the fact that the sensors can be made on the modern microelectronics, the necessary space inside the cover for their placement will be not critical parameter.

The flow of information and actions is presented in Fig. 7. Signals from sensors are stored in the internal databank and could be continuously or periodically read throughout the work shift. Data transmission should be preferably wireless and after verification and processing sent to Maintenance Decision Module which generates command of maintenance. Then, Maintenance Operation Module, on the basis of information from each machine and given maintenance strategy, specifies the instant of maintenance, scope of work, crew and orders spare parts. After completing all required resources preventive maintenance takes place.

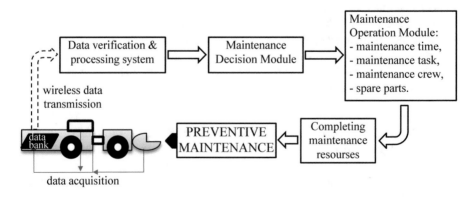

Fig. 7. Scheme of preventive maintenance based on maintenance 4.0 concept.

6   Results and Summary

Paper describes the early stage of the project, oriented on introducing a concept of Maintenance 4.0. It assumes monitoring of technical object, onboard data acquisition and periodic transmission to the databank. Decision-related to maintenance module will set up a maintenance task. It is unmanned process or with limited access of human what may speed-up the process and prevent from human errors. Especially in heavy environmental conditions influencing very much mining staff and a state of technical systems is an important factor. Intensive wear-out phenomena are usually observed in mining operations where temperature, humidity and dust cause fast degradation of machine elements, so that durability of very expensive machines usually doesn't exceed 5 years.

The concept of machine operating system based on the approach of Maintenance 4.0 is presented. As the system reduces human participation in the information chain it may result in increasing a safety level and effectiveness. The advantage of the project is expected also in the development and construction of a monitoring and information system for acquisition and processing data on current degradation processes as well as working out maintenance strategy for group of machines. This system provides informative models of preventive maintenance, the purpose of which is mainly to increase the technical availability of machines and mining system.

References

1. Andrzejczak, K., Młyńczak, M., Selech, J.: IT system for supporting cost-reliability analysis of fleet vehicles. J. Konbin **46**, 87–98 (2018). https://doi.org/10.2478/jok-2018-0025
2. Chesworth, D.: Industry 4.0 Techniques as a Maintenance Strategy (A Review Paper). Glyndwr University (2018)
3. Kobbacy, K.A.H., Murthy, D.N.P. (eds.): Complex System Maintenance Handbook. Springer, London (2008)
4. Dhillon, B.S.: Engineering Maintainability: How to Design for Reliability and Easy Maintenance. Elsevier Science & Technology Books, Amsterdam (1999)
5. Dhillon, B.S.: Maintainability, Maintenance, and Reliability for Engineers. CRC Press. Taylor & Francis Group, Boca Raton (2006)
6. Doudkin, M.V., Pichugin, S.Y., Fadeev, S.N.: The analysis of road machine working elements parameters. World Appl. Sci. J. **23**(2), 151–158 (2013). https://doi.org/10.5829/idosi.wasj.2013. 23(02), pp. 13061
7. Li, Z., et al.: Intelligent predictive maintenance for fault diagnosis and prognosis in machine centers: Industry 4.0 scenario. Adv. Manuf. **5**, 377–387 (2017)
8. Gresham, R.M., Totten, G.E.: Lubrication and Maintenance of Industrial Machinery. Best Practices and Reliability. CRC Press. Taylor & Francis Group, LLC, Boca Raton (2009)
9. Mobley, R.K. (ed.): Maintenance Engineering Handbook. McGraw-Hill, New York (2008)
10. Ravnå, R., Schjølberg, P.: Industry 4.0 and Maintenance. Norsk Forening for Vedlikehold (NFV) (2016)

Capabilities of ARCore and ARKit Platforms for AR/VR Applications

Paweł Nowacki and Marek Woda(✉) [ID]

Department of Computer Engineering, Wroclaw University of Technology,
Janiszewskiego 11-17, 50-372 Wrocław, Poland
nowackipawel@yahoo.com, marek.woda@pwr.edu.pl

Abstract. In this paper ARCore and ARkit capabilities were scrutinized and compared. Authors established comparison criteria for both platforms, developed test applications and ran comparison tests. Obtained results can be a help in choosing the right framework to speed up prototyping and development of modern AR/VR applications. This work consists of a comprehensive comparison of these new frameworks in the following respects: general performance (CPU/memory use), mapping of planes on various surface types, influence of light and movement on mapping quality etc.

Keywords: Comparison · ARKit · ARCore · Augmented reality · Virtual reality

1 Introduction

With the development of computer technology, new methods of user interaction with the computer and data presentation appear. Such methods include, among others, Augmented Reality (AR) and virtual reality (VR). Both methods of data presentation differ mainly in the ratio of computer-generated images to the real world. In a nutshell, AR [5] is when the real world is enriched with data generated by the computer, i.e. there are connections between the real and virtual world. VR [9] occurs when computer generated data completely obscures the real world.

Until recently, these methods required efficient computers and specialized equipment (GPUs) [12], however, thanks to the rapid development of mobile technology, millions of users have gained to access to powerful mobile devices [10] capable to deal with AR and VR in almost real-time. In 2017, Apple and Google - two giants of the mobile industry introduced two competitive application programming interfaces supporting the creation of augmented reality applications for mobile devices: *ARKit* (September 19, 2017) and *ARCore* (stable release March 1st, 2018), giving users of iOS and Android devices new opportunities to create immersive applications and games.

According to a forecast [11], VR and AR markets are expected increase almost tenfold to reach the size over 209 billion USD in 2021, from a mere 27 billion in 2018. Many programmers are looking for methods and tools to simplify a complex process of building realistic AR and VR applications. There are several frameworks available that facilitate rapid prototyping and development of AR/VR apps. Many portals [1, 6, 7, 13]

© Springer Nature Switzerland AG 2020
W. Zamojski et al. (Eds.): DepCoS-RELCOMEX 2019, AISC 987, pp. 358–370, 2020.
https://doi.org/10.1007/978-3-030-19501-4_36

recommend different solutions, amongst the most popular ones are: *ARCore, ARKit, ARToolkit, Kudan,* MAXST Wikitude. Besides the fact that *ARCore* and *ARKit* are free of charge, offer plethora of features which were previously only available only in commercial (paid) version of competitive SDKs. ARCore and ARKit though new ones, constantly evolving, they already caught attention of market (Fig. 1).

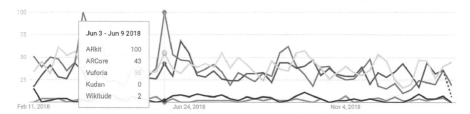

Fig. 1. Worldwide interest (Numbers represent search interest relative to the highest point on the chart for the given region and time. A value of 100 is the peak popularity for the term. A value of 50 means that the term is half as popular. A score of 0 means there was not enough data for this term.) – popular AR/VR frameworks (source trends.google.com)

Even though there are several publications available [2, 8, 9, 14, 15] related to use of newcomers on AR/VR market, it is hard to draw any conclusion which one a worthy competitor is, in comparison to existing commercial solutions, not saying about differences between themselves. Authors in this paper wanted to address the question how good (or bad) new frameworks are, and what are the differences in between them.

2   Comparison Criteria and Test Methodology

To be able to compare capabilities of both frameworks there were developed two applications (iOS and Android) that were taking advantage of given features. Both apps provided following functionalities: detection flat surfaces, imaging of detected flat surfaces/special points, positioning any number of 3D objects on stage, choosing scene lighting based on the actual lighting conditions, support for shadows cast generation by virtual objects, measure the distance between 2 points, saving information about detected planes (and special points) to a file, measurement and display of frames per second during operation, measurement and save information on app startup time. Thanks to the above-mentioned features, it was possible to create a base for tests of both frameworks, and finally presentation of strengths and weaknesses. Apps were tested on several devices (Table 1).

During the evaluation of both frameworks following comparison criteria were established:

- General performance
  - Time to load models, initialize, run application and camera
  - CPU load and memory usage during use of app (with 100 and w/o models)
- "Understanding" the surroundings

Table 1. Test devices – comparison sheet

Device	Processor	Cores	RAM [GB]
ARKit			
iPhone X	Apple A11 Bionic	6	3
iPhone 8	Apple A11 Bionic	6	2
iPhone 7plus	Apple 10 Fusion	4	2
iPhone 7plus	Apple 9	2	2
iPad Pro (2017)	Apple A10X Fusion	6	4
iPad (5th Gen)	Apple A9	2	2
ARCore			
Google Pixel 2XL	Qualcomm Snapdragon 835	8	4
Google Pixel	Qualcomm Snapdragon 821	4	4
Google Nexus 6P	Qualcomm Snapdragon 810	8	4
Google Nexus 5X	Qualcomm Snapdragon 808	6	2
Samsung Galaxy S9	Samsung Exynos 9810	8	4
Samsung Galaxy S8+	Samsung Exynos 8895	8	4
Samsung Galaxy S7	Samsung Exynos 8890	8	4

One of the key elements of augmented reality is the "understanding" by the application of the surrounding space. One of the components is surface detection. Well-mapped surfaces allow for realistic placement of virtual objects in real space. The quality of the surface detection itself translates to the quality of the entire application. If the accuracy is too low virtual models will hang in the air, move, or appear in places where it would be impossible in the real world. Such phenomena spoil the general impression and the comfort of using the application. In addition to accuracy, it is also important time that the application needs to detect surfaces, because too long detection time would weaken the functionality of such applications and again comfort of use. Both frameworks offer detection of flat horizontal surfaces (floors, countertops) and vertical surfaces (walls).

– Percentage of plane detected in relation to the total plane area
– Number of incorrectly detected planes (including false positives)
– Time needed to detect and map a plane [14]
– Accuracy of the scale selection[1]

• Work in various lighting conditions
  – Scene display (quality) based on the actual lighting conditions
  – Impact of brightness and color of light on virtual objects [daylight, incandescent light (100 W, 60 W bulb), 3 W LED bulb]

---

[1] Measurement of a distance between 2 points on a mapped plane vs. real distance.

- Work in unpropitious conditions
  - Impact of a shape on plane mapping [flat surface, single-colored, devoid of pattern and texture (wall, uniform top), flat surface with a pattern (wooden table top), surface with unevenness (grass, uneven scrub), shiny surface (water)]
  - Influence of low-light
  - Work in motion – during rapid movement of a device (number of detected characteristic points, time to detect a plane)

3   Tests Results

Measurements of CPU load and memory usage were performed for two application states. The first state is the state at startup, when no plane has been detected yet and there are no objects on the stage. The second state is when the application has already detected the plane and 100 test objects have been placed on the stage. *Android Profiler* tool, which is part of Android Studio since version 3.0, was used to measure the CPU load and use the memory of the ARCore test application. The test parameters for the ARKit application were measured using *Instruments* tool that is part of the Xcode programming environment.

The purpose of first test was to examine the impact of test applications on device resources. For this purpose, the CPU load (Fig. 2) and memory usage (Fig. 3) were measured. The results were checked for applications that do not have any models on the stage and applications with the stage on which there are 100 test objects.

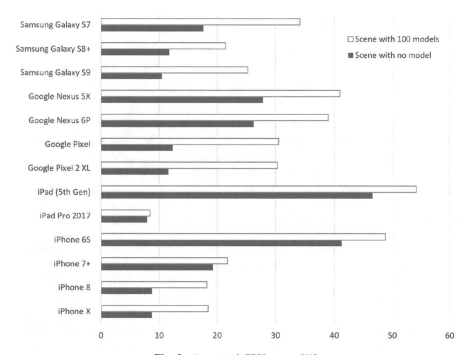

Fig. 2. Avg. total CPU usage [%]

For Android devices, memory usage was almost 3× higher than for iOS. Differences can arise from many factors, such as library requirements, nature of OS (its memory management). Keep in mind that increased memory usage for Android devices should not be a problem, as most of the tested devices had more memory available than iOS devices.

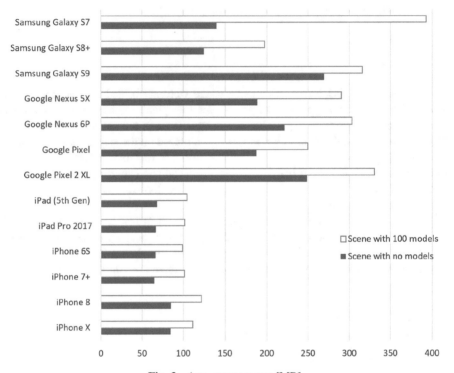

Fig. 3. Avg. memory use [MB]

Plane Detection. The tests were carried out in conditions enabling the best performance of both platforms (daylight, planes with a different surface pattern). It should be mentioned that the tests tried to achieve a reasonable compromise between the accuracy of the measurements and the time of surface mapping, because this reflects the best real use of both platforms. Therefore, during the surface mapping, each try didn't last longer than 25 s. The measurements were carried out using the best available devices supporting the discussed platforms (Google Pixel 2XL and iPhone 8+).

Accuracy of Plane Detection. The measured parameter was the percentage coverage of the test plane by the mapped surface expressed as a percentage. 30 measurements were taken for each platform based on an attempt to map a test surface measuring 90 cm × 60 cm. The tests showed the tendency of the ARKit platform to cover a smaller area of the tested plane. In addition, taking the standard deviation value, the results obtained for the Apple framework were more diverse than in the case of ARCore. Greater diversity means that the result of the operation is more difficult to predict, sometimes the application will be able to cover the surface to a satisfactory degree (over 90%) and sometimes not. The Google Framework was characterized by a much greater repeatability, 2/3 of all measurements were in the range of 93.54–96.60%, which is a good result (Table 2).

Table 2. Accuracy of plane detection

Platform	Average coverage [%]	Standard deviation
ARCore	94,833	2,731
ARKit	90,712	4,725

Percentage of Incorrectly Detected Planes. The exact measurement of the surface area of the detected plane is an important factor in determining the accuracy of the detected plane. To determine this parameter, a script was prepared calculating the area of each detected plane based on a model consisting of a triangle grid, which is the basis of each plane. In addition, to make the location of a given plane in space, screenshots were made showing the detected plane plotted on the actual object. An additional advantage of the screenshot is the ability to determine the extent to which the virtual flat surface coincides with the real one. Thus, the examined aspect is the amount of the actual area covered by the virtual expressed in percent, as well as the amount of virtual surface detected, which does not coincide with the real object (false positive).

The test consisted in approaching the device with cameras at various angles on a mapped object, in this case a table, to cover as much as possible, the whole surface of the table top. Next, a screen shot was made with device devices aimed perpendicular to the table top to easily calculate the percentage coverage. The screen capture itself was combined with a function that calculated the surface area of the virtual plane, which was then saved to a file for further analysis. The collected data was processed, and then based on them were calculated percentage coverage values according to the formula:

$$S_{pk} = \frac{100\,P_a}{P_t}$$

where S_{pk} is the amount of coverage of the surface to be tested by the detected area expressed in %, P_a is the area of the correctly detected plane, and P_t is the surface area of the test plane - in this case the table.

The size of incorrectly determined planes (false positives) was calculated based on the formula:

$$S_{fp} = \frac{100\,P_b}{P_a + P_b}$$

where S_{fp} is the part of the detected surface that has been incorrectly determined expressed in %, P_a is the area of the correctly detected plane, and P_b is the area of the incorrectly determined surface (the one that does not coincide with the real object).

To calculate the surface area of the abovementioned planes, the ImageJ tool was used based on a screenshot. ARCore performed noticeably worse, in more than half of measurements, detected plane exceeded the real object surface. During tests, Apple devices only sporadically drawn incorrectly detected surfaces (5 times exceeding 3.2% of the surface) (Table 3).

Table 3. Percentage of incorrectly determined plane area

Platform	Incorrectly determined planes [%]	Standard deviation
ARCore	5,518	5,970
ARKit	1,468	2,809

Time to Detect the First Plane. ARKit was faster to than the ARCore in this test. Apple framework achieved much more repeatable results, which also confirms the superiority of ARKit in this test (Table 4).

Table 4. Time to detect the first plane

Platform	Average time [s]	Standard deviation
ARCore	5,604	1,314
ARKit	4,809	1,101

Time to Map the Surface. Another test was to check how much time it takes to cover the 90 × 60 cm test plane (in at least 80%) (Table 5).

Table 5. Time to map the surface

Platform	Average time [s]	Standard deviation
ARCore	12,978	2,965
ARKit	13,665	2,211

ARCore despite the worse result in the previous test, now is ahead of ARkit. Although the detection of the first plane in the case of Android takes more time, then for larger areas better results in terms of quality coverage are achieved. It was observed that plane coverage elements were greater than in the case of ARKit, which translates into shorter mapping times.

Table 6. Influence of surface type on mapping

Surface (Fig. 5)	Mapped?		Time [s]	
	ARCore	ARKit	ARCore	ARKit
A. Single-color wall	**No**	*Partially*	-	39,249
B. Furniture board (vertical)	Yes	Yes	**46,765**	5,529
C. Upholstery fabric (green)	Yes	Yes	5,337	4,447
D. Poster board	**No**	**No**	-	-
E. Upholstery fabric (black)	Yes	Yes	8,295	10,571
F. White canvas	Yes	Yes	6,203	4,857
G. Wooden table top	Yes	Yes	4,924	5,057
H. Worktop	Yes	Yes	4,879	5,173

Influence of Surface Type on Mapping Quality. Tests were divided into two parts. In the first part, the applications were tested against various surfaces inside a room. The second part consisted in checking the possibility of detecting planes on various type of terrain outside (very good lighting conditions). Google Pixel 2XL and iPhone X devices were used for tests. ARCore demonstrated performed significantly worse during the mapping of vertical planes. This was manifested by a much longer detection time of the plane and a smaller number of detected characteristic points (case with a vertical furniture board). In addition, ARKit in case of surfaces whose pattern has low contrast (e.g. a uniform wall, white canvas) increases the contrast to detect more specific points.

Table 7. Influence of terrain type on mapping

Surface (Fig. 5)	Mapped?		Time [s]	
	ARCore	ARKit	ARCore	ARKit
A. Low grass ($\sim$5 cm)	Yes	Yes	4,791	5,006
B. Med grass ($\sim$15 cm)	Yes	Yes	5,038	5,576
C. High grass ($\sim$40 cm)	Yes	*Partially*	5,292	6,686
D. High scrub ($\sim$70 cm)	Yes	**No**	18,620	-
E. Thick bushes ($\sim$50 cm)	Yes	Yes	6,201	5,276
F. Sheet of water	**No**	**No**	-	-
G. Concrete	Yes	Yes	5,832	7,273

The worst result was obtained during technical paper mapping, where the number of detected points was so low that none of the platforms was able to identify the plane. Partial mapping of a single-color wall by the Apple framework was possible only after the device was moved closer (approx. 20 cm) to the tested surface (Fig. 4).

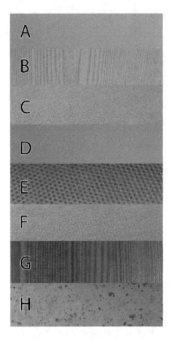

Fig. 4. Internal test surface (see Table 6)   **Fig. 5.** External test surface (see Table 7)

Outdoor tests indicate the advantage of Google tools. ARCore was able to detect plane where ARKit wasn't (high scrub) or covered a much larger area (high grass). Surface mapping using the ARCore took less time in all cases except thick bushes. In addition, faster mapping of large areas by Android system was observed (with each movement, increase of covered plane was larger), and greater range (mapped space was larger than in case of ARKit). A complete surprise was the detection of the surface in high scrub by ARCore. Surprisingly, the mapped plane was equal to ground level (SIC!), and not the plane defined by the upper part of the bushes, as it was the case with thick bushes. Sheet of water mapping in both cases was disappointing, but ARCore was able to detect much more characteristic points.

Work in Low Light Conditions. The parameters examined were the detection time of the first plane and the number of specific points detected. The results presented were averaged based on 10 attempts made for each setting. The devices used were Google Pixel 2XL and iPhone 8+. Together with the reduction of lighting power, the detection time of the planes increased. This can be seen in both ARCore and ARKit. The performance of ARKit is significantly reduced in lower light conditions. This is manifested by slower or completely unsuccessful surface detection. In situations where ARCore was still able to detect the plane, ARKit failed. Better results obtained by ARCore may be related to the fact that the application increases the image contrast, thanks to which it is much better at dealing with surface detection in low light conditions (Table 8).

Table 8. Influence of light conditions on mapping

Light Source (Fig. 5)	Mapped?		Time [s]	
	ARCore	ARKit	ARCore	ARKit
Daylight (high noon)	Yes	Yes	4,965	5,637
Incandescent (100 W bulb)	Yes	Yes	6,564	8,849
Incandescent (60 W bulb)	Yes	**No**	27,620	–
Incandescent (3 W LED bulb)	Yes	**No**	–	–

Work in Motion. Since the position of the device in space is calculated in real time, rapid sudden movements of the device in many directions may cause the application to run out of balance. An additional factor is blurred image caused by rapid movements. The tests (repeated 50 times) assumed making attempts to map a surface, with three exemplary models, during random, sudden movements (however repeatable) of devices. Both devices were connected, and tests were made in simultaneously. ARCore performed worse, losing 14 times (vs. only 5 in case of ARKit) the location of mapped plane, however both were able to recover the detected plane just after 2 s after the movement stopped. Accidental covering of camera lens (ARCore) results in losing the location of detected plane, and in the case of ARKit, a plane remains visible on the screen, but it is a bit shifted. Reopening lens results in gaining focus much faster on devices with ARCore (>0,5 s.) whereas in ARKit focus time is about 5 s.

It should be mentioned that it is very difficult to cause abnormal imaging. The movements were so violent that it deviates significantly from the normal use cases. The purpose of the test was however to determine which platform better handle rapid movements. ARKit can react faster to unwanted disturbances thanks to having possibility of handling twice as many frames per second (iOS and camera specific). In addition, ARKit largely relies on the device's accelerometer data, so that even after covering camera lens completely, the models and planes remain on the screen (Table 9).

Table 9. Comparison results ARCore vs. ARKit

Field of comparison	ARCore	ARKit
Detection of first plane		*Faster than ARCore*
Mapping a plane	*Faster to map larger surfaces*	
Plane detection	**Tendency to incorrectly detect planes by going beyond the boundaries of mapped surfaces**	
Estimation of lighting conditions		Slightly higher sensitivity to changes in lighting
Memory usage[a] [MB]	Greater (187-392)	Smaller (84–121)

(continued)

Table 9. (*continued*)

Field of comparison	ARCore	ARKit
Frames per second	Limited to 30	*Always 60 FPS*
Performance	**Performance drops on older devices (Google Nexus 6P and Google Nexus 5X)**	Smooth operation on all tested devices
Mapping of uneven/complex planes	Better suitable for outside mapping (good in mapping of complex planes like high grass)	Better suitable for inside mapping (vertical planes, not overly complex)
Work in low light conditions	*More resistant to interference due to insufficient lighting*	
Work in motion		*More resistant to interference due to fast, variable movements of the device*
Documentation for application in UNITY	*Fully documented*	Partially documented
Working examples	A few, simple ones	*Many, sophisticated examples documenting application of framework*
Shadow casting	Requires own implementation	*Built-in*
Software requirements	Android 7 and ARCore libs installed on a device	iOS 11 or higher
Other	Sharing of characteristic points for multiplayer games	In iPhone X (and newer) - *TrueDepth* feature allowing face-mapping

[a]Test application size ARCore – 36,85 MB, ARKit 52.8 MB.

4 Conclusions

It should be noted that some of the advantages of ARKit arise from the fact that it runs on more powerful devices. For example, at this point it is not possible to run the ARCore application in 60 frames per second mode due to hardware limitations. The Apple framework thanks to this feature can respond better to rapid camera movements. Another example is faster application startup, or faster detection of the first plane, which is probably caused by faster and better optimized devices.

Both platforms differ in the approach to delivering the required library to devices on which AR applications run. In the case of ARCore it is necessary to install an additional library, which can be downloaded from the Google Play store, while ARKit is delivered together with the iOS version 11 or higher. These approaches are conditioned by the current state of the mobile devices market. Most Android devices do not have the latest version OS installed, so delivering ARCore together with subsequent versions of the operating system would significantly limit the development of this platform, because it would reach a small group of users. In the case of ARKit, most devices have the latest version of the operating system, so delivering subsequent library updates together with the system seems to be a sensible approach.

Both platforms have their strengths and weaknesses, but it is difficult to find technology that would be objectively better. Both platforms are better suited to the situation. In the final analysis, the choice of technology should be conditioned by the current state of the market and your own preferences.

Because the technologies discussed are relatively new (both presented in the second half of 2017), there are many areas where improvements can be done. Both are new enough to sometimes give the impression of unfinished. It should be borne in mind that platforms are actively developed, by increasing functionality and fixing issues, so the impression of an unfinished product can quickly pass away. While making this comparison, Google announced support for shared characteristic points enabling multiplayer interaction [4] on one stage, then Apple a month later announced similar functionality in new ARKit2 [3]. Such behavior indicates high competitiveness of platforms and prompt reaction to appearance of competitive features.

In authors opinion following are areas for improvement.

- estimation of lighting color on the stage
- estimation of the angle from which shadows fall to improve the realism of the objects displayed on the stage
- possibility of mapping the face based only on image from a camera
- improved plane detection accuracy, elimination of occasional fault lines on a surface
- (ARKit only) low light performance conditions - allowing longer exposure time by e.g. contrast increase or temporarily limiting camera frame rate
- (ARCore only) – better plane detection on less diverse and/or vertical surfaces

References

1. 6 Best Augmented Reality SDKs and Frameworks, Themindstudios. https://themindstudios.com/blog/5-best-augmented-reality-sdks-and-frameworks/. Accessed 01 Feb 2019
2. ARCore Resources https://developers.google.com/ar/discover/. Accessed 28 Feb 2019
3. ARKit 2. https://www.apple.com/newsroom/2018/06/apple-unveils-arkit-2/. Accessed 01 Feb 2019
4. Gosalia, A.: Experience augmented reality together with new updates to ARCore. https://www.blog.google/products/arcore/experience-augmented-reality-together-new-updates-arcore/
5. Azuma, R., Baillot, Y., Behringer, R., Feiner, S., Julier, S., MacIntyre, B.: Recent advances in augmented reality. Naval Research Lab Washington DC (2001)
6. Best AR SDK kits. https://thinkmobiles.com/blog/best-ar-sdk-review/. Accessed 01 Feb 2019
7. Best AR SDK for development for iOS and Android in 2019. ThinkMobiles. https://thinkmobiles.com/blog/best-ar-sdk-review/. Accessed 01 Feb 2019
8. Bergquist, R., Stenbeck, N.: Using Augmented Reality to Measure Vertical Surfaces. Bachelor thesis, Linköping University (2018)

9. Duan, G., Han, M., Zhao, W., Dong, T., Xu, T.: Augmented reality technology and its game application research. In: 2018 3rd International Conference on Automation, Mechanical Control and Computational Engineering (AMCCE 2018). Atlantis Press, May 2018

10. Halpern, M., Zhu, Y., Reddi, V.J.: Mobile CPU's rise to power: quantifying the impact of generational mobile CPU design trends on performance, energy, and user satisfaction. In: 2016 IEEE International Symposium on High Performance Computer Architecture (HPCA), pp. 64–76. IEEE, March 2016

11. Forecast AR and VR market size worldwide (2016-1022). https://www.statista.com/statistics/591181/global-augmented-virtual-reality-market-size/. Accessed 01 Feb 2019

12. Willemsen, P., Jaros, W., McGregor, C., Downs, E., Berndt, M., Passofaro, A.: Memory task performance across augmented and virtual reality. In: 2018 IEEE Conference on Virtual Reality and 3D User Interfaces (VR), pp. 723–724. IEEE, March 2018

13. Six Top Tools to Build Augmented Reality Mobile Apps InfoQ. https://www.infoq.com/articles/augmented-reality-best-skds. Accessed 01 Feb 2019

14. Wang, W.: Plane detection. In: Beginning ARKit for iPhone and iPad, pp. 261–297. Apress, Berkeley (2018)

15. Zhang, W., Han, B., Hui, P.: Latency mobile augmented reality with flexible tracking. In: Proceedings of the 24th Annual International Conference on Mobile Computing and Networking, pp. 829–831. ACM, October 2018

Log Based Analysis of Software Application Operation

Daniel Obrębski and Janusz Sosnowski[✉]

Institute of Computer Science, Warsaw University of Technology,
Warsaw, Poland
j.sosnowski@ii.pw.edu.pl

Abstract. Event logs provide the capability to gain insight into system operation under the real workload. They have been widely used to detect anomalies, evaluate dependability including security, service time analysis, etc. The paper outlines log analyses schemes described in the literature and presents a new approach which takes into account specificity of applications embedded into system environment. It takes into account a wide scope of logs, defines and extracts their features helpful in assessing application operation in the considered lifetime span. The presented methodology is illustrated with results related to a complex commercial system used as a case study.

Keywords: Event logging · Error detection · Dependability evaluation

1 Introduction

Software applications become more sophisticated and quite often they are modified or upgraded to cover new functionalities and environment changes. During runtime software applications generate various logs, which provide useful traces of their behavior [1, 2, 4, 21]. The number of generated and reported log entries increases rapidly with the operation time resulting in large files. They comprise diverse textual messages, their interpretation needs some effort or expert experience.

There are many publications devoted to event log analysis. Most of them are targeted at identification of system anomalies or errors ([1, 8, 14, 15] and references therein). For this purpose various algorithms and tools have been developed and illustrated with some results mainly for open source projects. They trace mostly critical log reports. Unfortunately, they are limited to a single type of logs (system fault logs) and neglect detailed analysis of information content within variety of logs, which is especially important in the case of application logs created by developers.

During monitoring of system operation, it is reasonable to take into account all available event logs and extract appropriate information to characterize the working profile of the application, identify resource usage, environment interactions, efficiency problems, errors, anomalies, etc. Hence, an important issue is to discover representative log attributes and properties. The main contribution of this paper is a wide scope view on available logs and derivation of features which can be helpful in characterizing application/system operation. These features are found by log message text mining and extracting various statistical properties of registered log entries in time and other

© Springer Nature Switzerland AG 2020
W. Zamojski et al. (Eds.): DepCoS-RELCOMEX 2019, AISC 987, pp. 371–382, 2020.
https://doi.org/10.1007/978-3-030-19501-4_37

perspectives (e.g. usage profile). The developed log analysis methodology and tools relate to a case study of a commercial system implemented in a distributed environment and involving many users from different countries.

Section 2 presents the background and related work. System and application log features are outlined in Sect. 3. Section 4 presents research goals and relevant original analysis schemes. In Sect. 5 the usefulness of this approach is illustrated with experimental results related to a commercial system. Section 6 concludes the paper.

2 Background and Related Work

The runtime information comprised in software logs is useful in anomaly detection fault diagnosis, program verification and performance evaluation [8, 10, 21]. For this purpose various data mining algorithms are used, which are supported with log collection and log parsing schemes, e.g. [7, 9]. Often, they are targeted at specific logs with relatively simple structures, e.g. console logs [18], code performance logs [5]. In [11] some analysis has been performed with regular expressions which follow changing templates of logs due to application upgrades and modifications. Special tools (e.g. [6, 13, 16, 20]) are useful to visualize log information for assessing computer operation. Another group of publications provides diverse algorithms for log classification ([12] and references therein). Some faults can be identified by finding specific keywords in log messages (e.g. fatal. failed, error, unavailable, unsuccessful). Deriving time and spatial correlations of log entries is helpful in this process.

It is quite useful to identify log similarities and creating log classes (categories) to decrease the space of analysis. For this purpose various comparison metrics are used. In [2, 11] the authors take into account not only Levenstein distance between words in compared log messages but also their positions. Some other modification to this metric are also proposed. In the process of log categorization we can also identify variable and stable phrases [12] to make the log entries more compact by replacing variables with a universal symbol *. In [21] fault correlation with messages is calculated by creating $m \times n$ matrix, where m denotes the number of different words in a sample logs, n represents the fault type, a_{ji} element of the matrix denotes the probability of the i-th word belonging to the j-th catalog of faults. Unfortunately, in many cases we do not have representative catalog of messages related to known fault types. Quite often new fault types may appear in a longer time span.

Derived statistics on logs give a better view on their significance and analysis problem. In [21] we have statistics according to fault types (disc, input/output, memory, network, processor, database, web, file). Interesting statistics of logs according to severity levels (information, warning, panic, critical, error) are presented in [16]. They relate to 5 open source projects: most logs were classified as info category 70–94%, error logs contributed 5–28%, warning logs contributed 0.3–5.5%, panic and critical logs were in the range of a fractional percentage. The message type distribution related to coarse grained classification: sshd, kernel, syslog-ng and RPC was in the range of a single percent to over 50% and differed significantly on projects.

In [10] authors discuss the problem of assuring high log quality and give guidelines on logging schemes. They base on some analysis of texts used in logging statements (performed at the code level). Within the analyzed 17 projects 70–80% logging statements comprised textual logging descriptions. The distribution of the length of the logging descriptions showed dominating logs with up to 10 tokens (90%). It is much lower than in the case of commercial projects we analyzed. In [10] the authors distinguish 3 group categories of logs: program operation (completed, current, next), error message (exception, value-check) and semantic description (correlated variable, function, branch). Our experience with wide spectrum of logs shows that this categorization is not sufficient. Logging practices are also reported in [2, 3, 7, 15], they mostly relate to open source projects. Some logging improvements are presented in [10, 20]. In an industrial project [4] logging instructions embedded into the code contributed about 3% of code lines (they were created with 25 procedures). Similar logging point frequencies have been reported for open source projects in [19].

Assessing system operation we should investigate not only errors but also many aspects of the application performance, usage profiles, impact of performed upgrades, environment interactions, etc. To the best knowledge of the authors, all these issues except errors and service times are neglected in the literature. Complex commercial systems comprise diverse log files targeted at various issues. The reported log entries may differ in structure and textual messages. They are generated by various log mechanisms embedded into program codes. Analyzing a wide scope of logs extends the space of studying many application aspects and behavioral features. In contrast to our comprehensive approach, most publications are restricted to selected problems in relevance to a single log type.

3   System and Application Logs

In general, log entries comprise the registration timestamp and a text message, often enhanced with other attributes, e.g. unique ID, severity level, event generation source. Each log entry is correlated with a logging mechanism (e.g. logging point in the code and the type of used procedure) specific for the used software platform (e.g. *log4j*). The density of logging depends upon developers and assumed practices in the company. Typically, it is in the range of a few percent of the project line codes (compare Sect. 2). In the runtime logging instructions can be activated many times (producing many log entries) or not activated due to system inconsistency with the performed operational profile, lack of a targeted error detection, etc.

The goals of logging can be attributed to state dump, execution tracing, event reporting, performance and application dependent features. State dump entries report specific data values with a limited or none textual descriptions. Execution tracing entries notify initialization or termination of a function call, branch execution, etc. Event reporting entries comprise usually some textual descriptions of the detected situation (e.g. security threat, error, warning). Performance and application dependent entries relate to resource usage, service time, user activities, etc.

The goal of our analysis was to derive features describing application operation including users and environment behavior. In the case of application logs we have a lot of specific information directly referred to its operation, user activity profiles, performance issues, etc. In practice, appropriate separate log files are used with the structure adapted to the comprised information. In Java we can use *Log4j* library to create logs which comprise 3 elements: *logger* - responsible for generating logs; *appender* – pointing logging sites, log file names and file rotation policy; *layout* - describing visual aspects of log entries. The configuration file specifies details.

The analyzed system is a typical e-commerce application (ECA) serving users and clients (internet shop). The user is responsible for a company selling various goods, a client can send buying requests served by the users. Within ECA (managed by its owner) various businesses can be created. ECA is hosted in different countries. ECA and correlated micro services (for integration with external services) are handled by many machines including database and backup, machines for monitoring, synchro-nization and collecting logs (e.g. Nagios). ATM router with firewalls filters and dis-tributes internet traffic to appropriate machines. Apache HTTP servers are responsible for network traffic and handling internet pages. ECA is based on Java, Java script and SQL files with Docker architecture (about 6 millions of LOCs). Problems reported in Jira were at the level 0.04 per month/1 Kloc, confirming high quality. The number of users was almost 4 millions (only 30% active, with 2 millions of orders/year).

In the considered analysis we take into account application and HTTP logs, which comprise relatively interesting information. In particular, we consider business level log *aol.log*, technical level log *server.log*, session level log *session.log*, statistical and access logs (*staistics.log, access.log*). The *aol.log* comprises exceptions, logs added by programmers with *Log4j* tool with different severity levels (info, warn, error), in addition it can included information on long executed SQL queries. In the include messages we can have class names related to the event, specification of the users (ID, country). For exceptions we have more data, e.g. the whole stack trace which can be quite big text comprising many parameters, such as: http protocol, session ID, server name, remote host or port addresses, detailed data describing requests. Text messages comprise natural language words (NLs), acronyms, numbers, concatenation of various words or symbols, etc. *Aol.log* allows to diagnose business problems of the application. We can identify classes or methods with frequently reported exceptions as well as correlated servers. Typical error log entry comprises several sections: time stamp and general information on exception, attributes of the request and the server exception, headers of the request, information on HTTP session, details of the request and exception stack trace (in particular names of classes preceded with word at), etc.

The *server log* reports technical problems, e.g., related to connections with queues, database, external services. It also includes information on exceeding connection times and other anomalies. This is useful to detect problems with external services, inte-gration issues. *Info* level entries describe application operation, e.g. details of creating services. Below we give an example of an error entry:

Ex. 1: *2017-08-13 02:13:11,793 ERROR [org.apache.activemq.ActiveMQSession] (default-threads - 13) error dispatching message: javax.ejb.EJBTransactionRolled-backException: Could not send Message.*

It specifies a problem with queue (ActiveMQ) and no capability of sending a message using the specified queue JMS.

Information on created and deleted sessions is included in *session.log*. Typically, it gives the information on creation time of the session, IP address responsible for this, origin country and session ID. Some statistics on page loading times are included in *statist.log*. In addition, it gives information on used devices (e.g. smartphone, workstation), operating systems, internet browser or engine (including their versions and periods of using them). This is useful to correlate reported problems with used software or devices, as the one showing data and time of loading a page:

Ex. 2: *32017-08-13 07:32:35,906 INFO pl.xxxxx.aol.war.common. statistics. StatisticsActions [hu, user: kata46]: Page load time(ms): 6184 Before unload time: 1502609549374 Unload time:1502609550610 DOM ready time: 1502609554273 Load time: 1502609555558 Referrer:* https://www.xyz.hu/all-tna *User host: 46.130.180.74 User agent: Mozilla/5.0 (Windows NT10.0; WOW64; rv:54.0) Gecko/20100101 Firefox/54.0*

It gives also user details (including country), page loading time (6184 ms), accessed page (https:www…) from IP address 46.130…), used Firefox browser version 54.0 under Windows 10.

Logs generated by Apache HTTP servers (separate for each country covered by the analyzed application) are stored in *access.log* files. In particular, they allow to trace information flow to users (in bytes), visited pages, responses sent to users, frequencies of generated requests to the system. For an illustration we give an example entry:

Ex. 3: *194.143.135.100 - - [15/Aug/2017:13:23:02 +0000] "GET/user/apanov HTTP/1.1" 301 240 "-" "Mozilla/5.0 (Macintosh; Intel Mac OS X 10_11_6) AppleWebKit/603.3.8 (KHTML, like Gecko) Version/10.1.2 Safari/603.3.8" "-" 0* www. xyz.ua

It describes request Get to page /user/apanov from specified IP address (194….) on specified time, correct response 301, sent 240 bytes, the used browser Safari on device MAC OS X (version 10.11.6).

Depending upon the log file we have some knowledge on the structure of involved entries. We can identify specific positions describing correlated information (e.g. time stamp, address IP, user name). The text included within a position may comprise several words, hence in the analysis we operate on tokens related to single words.

4   Deriving Log Features

Searching for characteristic log features we have developed three classes of statistical analysis: general numerical event entry statistics, lexical analysis of event entries, and special purpose statistics (application dependent). For this purpose we have developed appropriate algorithms and scripts supported with relevant tools.

General statistics include distribution in time of the number of event entries in the log, distribution of their lengths (in bytes or words). For some logs (e.g. *access.log*) these statistics can be projected on countries, users, etc. These statistics can be presented in fine or coarse grained time perspectives and correlated with other logs (e.g. user activities). To identify time periods with suspected behavior we generate box plots showing minimal, maximal, average, 1st and 3rd quartile values. More detailed statistics (e.g. percentile distribution) can be also included for deeper analysis.

Lexical analysis of event entries simplifies extracting semantic meaning of registered reports and in consequence deriving most interesting words or phrases. For this purpose we have developed special algorithms using Hadoop *map* and *reduce* schemes (due to large volume of analyzed logs). Having included log files into Hadoop system files the adapted *map* algorithm filters out unimportant data (specified by regular expressions). The extracted words are correlated with natural language (NL) thesaurus (based on [https://github.com/dwyl/english-words]), dictionary of positive and negative words (using [https://www.cs.uic.edu/~liub/FBS/ sentiment-analysis.html#datasets]). In this process we identify also non-NL words, e.g. IP addresses, numerical parameter values, class names, etc.

Beyond single word analysis we look also for word phrases composed of n succeeding words (n-grams). In this case we take a current word from the created dictionary for the analyzed log and look for another words preceding this word. This process is extended for 3, 4, etc. grams, The results of map algorithm for each log are used by the *reduce* algorithm, which counts appropriate statistics, word or phrase frequency, cumulative plots showing the number of used unique words in function of time (log files), etc. Some specific words or phrases (e.g. negative) may appear mostly in a limited time period. Identification of such periods are important traces. Similarly, repetitions of some phrase within a short period of time (e.g. access unavailable) can also be considered as a suspicious log feature. Many words can be attributed to specific names describing classes, operating systems, web browsers, users, countries, servers, etc. They allow us to check deeper correlations of event entries as well as deriving cross-sectional statistics to get a better view on the system operation.

Depending upon the usage scope of the application we can formulate features characterizing various aspects of its behavior. This is correlated with information contents included in available logs. Hence, searching for such features we have developed a special program *LogAT*. This program provides lexical analysis of event reports and various numerical statistics targeted at predefined features (general and special). Basing on semantic and syntactic analysis of logs it is adapted to structural properties of considered logs. Fortunately, the log structure is well correlated with different information types (e.g. dates, performance issues, usage profiles), which facilitates creating appropriate analytical scripts. For an illustration we give a list of basic analysis schemes related to the considered application: statistics of transmitted data, distribution of server responses to the users, statistics of page loading times, distribution of exceptions, distribution of visited pages and correlated IP addresses (based on HTTP and *access.log*), operation system and browser usage. An important issue is to find cross correlations between different objects, selecting statistical perspectives (hourly, daily, monthly), ranges (e.g. most frequent or unique features), cross-section views, e.g. exceptions by operating system, user site, methods.

The developed analysis tool is tuned to logs generated with *Log4j*, which assures some uniform log structure with fixed positions and tags attributed to specified data comprised in logs, e.g. time and date stamps; IP; characteristic, important or unique features. Taking into account this property we decided to develop appropriate scripts specified in *awk* language and using standard Unix tools such as: *Sed* - used to replace specified text, *Grep* - used for searching in the pointed text specified patterns (directly or with regular expression), using switches (-r) and (-z) we can perform recurrent searches and in compressed files. In these scripts we specify the analyzed log files, searched information, performed aggregation (e.g. average value), and final result presentation file (e.g. sorted with decreasing values). Here, we should notice that the log entry position may be inconsistent with tag number in *awk*, (some data within the log may cover several tokens). As an input we specify appropriate set of log files to be analyzed. For an illustration we present a script for finding most frequently visited pages from *access.log*. We have this information on event log field position 5 corresponding to token 7 (compare example 3 in Sect. 3):

find path -name *access.log.$i*.gz -type f | xargs gunzip -c | awk '{print $7}' | sort -r | uniq -c | sort -n -k1 > path + most_often_visited_sites.log

This script extracts the searched information and response codes. The collected data is sorted, grouped and stored in the output file. Another script is used to investigate distribution of exceptions, taking into account the number of their occurrences:

find "path_to_logs" -name "server.log*" -type f | xargs zgrep -oh "\w*Exception" | sort -r | uniq -c | sort -r

For the given set of logs are selected patterns **Exception*, the results are sorted and grouped according to the number of occurrences. More complex is the script for analyzing exceptions in function of methods from *server.log*:

for i in 05 06 07 08 09; do find "path_to_file" -name "server.log.2017-$i-*" -type f | xargs pcregrep -ohM "\w*Exception\s + at.*name.packet(\.\w*)*" | awk 'ORS = NR %2?" ":"\n" ' |sed 's/at//g'| sort -r | uniq -c | sort -r > ~/output_path _$i.log; done

It generates statistics for subsequent months (05–09, i.e. May–September), in Java after throwing an exception we get information of its place (specified after keyword at). Within the selected entries we delete unimportant text (using *awk*) which after sorting are included in the output stream.

Creating analysis scripts we take into account not only the knowledge of their structures but also results of lexical analysis which point interesting semantical aspects. This process has been consulted with developers due to diversity of log files and relevant log entries.

5   Experimental Results

The presented log analysis methodology has been used to evaluate the e-commerce system ECA (Sect. 3). Here, we present some illustrative results. An important issue is to find the semantical scope of logs. For this purpose we use the developed lexical analysis, which identifies the used words in logs (unique words), their usage frequency and keywords. In Fig. 1a we give the distribution in time of unique words in *aol.log*. Each day a new log file is generated and some new words are introduced. Similar plots have been generated for other logs (they saturate more quickly). Within one day we usually get 15–20% of unique words, 5 day and 60 day logs cover 49–66% and 72%–89% unique words, respectively. Depending upon log type unique words tend to level 2000–4500 - the used dictionary is relatively small. We can neglect some stop and other unimportant words, so the analysis of the used dictionary can be done manually to select keywords or word classes of some specific significance.

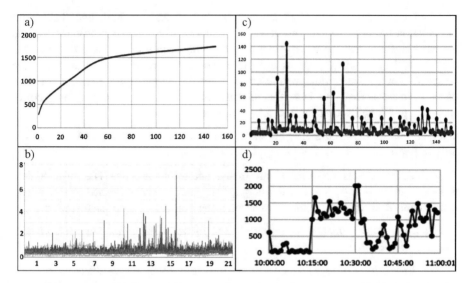

Fig. 1. Statistics of ECA system: (a) cumulated number of unique words in *server.log* for 150 days, (b) distribution of page load times in seconds (21 weeks), (c) distribution of exceptions for 150 days (in thousands), (d) distribution of exceptions on 130th day (one hour period).

Another important feature is frequency of word appearance and sentimental meaning, i.e. positive and negative words (bad words in [17]). Words directly correlated with potential problems, e.g. error, exception contributed very small percentage. Negative words in *aol.log*, *server.log*, *stataistics.log* and *session.log* constituted 79%, 99%, 2% and 52%, respectively. The most frequent negative words were: aborted, bad, disabled, error, expired, failed, failure, false, incorrect, invalid, invalidate, leak, limit, rejected, refused, unavailable, undefined, unexpected, unknown, unsafe, warned, warning, wrong.

Word phrases (n-grams) may give some additional view on information contents. Selecting most frequent word pairs (exceeding 500 appearances) for *aol.log*, *server.log*, *stataistics.log* and *session.log* we obtained 708, 390, 13, 467 pairs, respectively. The session log has good log structuring and low number of unique information (as opposed to other logs). Some examples of word pairs are: connect timeout, unreliable session, update timestamp, upgrade insecure, illegal state, time-out, expired session, session create. Derived phrases to some extent point possible information in considered log entries. In particular, for *aol.log* they relate to requests, technical problems (connect timeout, unreliable session), logging actions, library usage, thread creation, validating requests, some characteristics of data base contents, user profiles, etc. *Server.log* provides us with time between requests, incorrect states within application (illegal state, session), information on queues, service usage, integration with external services, etc. *Statistics.log* provides general statistics, execution time of various operations (load time, unload time, ready time), visited pages (order details, order payment, self-registration, shopping list), logging and authorization actions. *Session.log* provides information on countries with created sessions, creation, canceling, expiration of sessions, user profiles, session duration, etc.

Quite important is the analysis of exceptions. In particular, exception categories occurrence distribution, exception distribution in time projected on methods, users, countries. Extracting exceptions reported in *server.log* and *aol.log*, we have got the following distribution: *SocketException* - 32%, *EJBTransactionRolledbackException* - 24%, *ClassNotFoundException* - 12%, *SOAPFaultException* - 8%, *WebServiceException* - 8%, *NullPointerException* - 5%, *SocketTimeoutException* - 4%, *RuntimeClassNotFoundException* - 3%, *IndexOutOfBoundsException* - 3%, *NotSerializableException* - 1%. *SocketException* denotes some internet problems, similarly *SocketTimeoutException* also characterizes network performance.

The analyzed application uses Webservice or SOAP, and EJB components in read and writes to the database (java EE framework). *NullPointerException* is generated by erroneous application code (5% cases, e.g. referring to non-existent object), *IndexOutOfBoundException* and *NotSerializable* exceptions relate to implementation deficiencies within methods. Distribution of exceptions over related methods showed that some of them existed till some time and then disappeared (after method correction). On the other hand some exceptions appeared with newly included methods or classes. *InvalidStateException* related to inappropriate activity of users. Some exceptions have appeared continuously, they did not annoy users, so no effort has been done to eliminate them. Most exceptions related to network infrastructure features.

Time distribution of the number of exceptions showed some fluctuations (Fig. 1c), some spikes can be suspicious and needed further drilling with higher resolution (Fig. 1d). Hence, it is also reasonable to generate box plots to get better view on average operation characteristics and outliers (maximal values). The most frequent exception generated by the analyzed application was *NullPointerException* (however distributed over different methods). Some excessive numbers of exceptions in specific classes were reported to developers for refactoring. We have also calculated the ratio of exceptions within logs, typically it was on average 0.5–1.5‰ per day. For a few days it raised to 4–6‰ and within these days some problems appeared. We also give exception distribution for subsequent 5 months M1–M5 (calculated assuming day granularity), subsequent numbers in the brackets relate to parameter values (in thousands): maximal, quartiles Q3, Q2, Q1, minimal: M1 - (145, 13, 7, 2.5, 1), M2 - (58, 10, 8, 6, 2, 1), M3 - (116, 9, 7, 3, 2.5), M4 - (25.5, 10, 7, 6, 3), M5 - (42, 11, 7, 6.5, 2.5).

Here, we can observe significant maximal values, however they are rather sporadic due to relatively low Q3 quartile. Having detected days with higher values we can trace logs with higher resolution. Similar analysis can be performed for all events. The total number of registered logs within a single sever was on average in the range of 3000–5000 events per hour (up to 100000 per day). Monthly analysis showed Q3 quartile fluctuation in the range 80000–125000 events per day, maximal values were in the range of 120000–200000. One of the critical days was 130th day, during this day some anomalous high frequency events appeared within 10 and 11 h. This is illustrated in the detailed plot (Fig. 1d) with minute resolution. Similar plot with hourly resolution shows a single peak at 4500 level (with neighboring values at 500 events per day). This problem related to the session database access limitation.

Analyzing IP addresses we get information on user origin, possible DDoS attacks, frequency of requests and their distribution in time, distribution of unique addresses in time (e.g. monthly perspective), a box plot of unique addresses correlated with subsequent months (time periods). The maximal number of unique addresses is below 2.5 millions, it confirms low number of new clients (mostly are old clients).

Quite useful is the analysis of HTTP responses, i.e.: information codes (1xx), success codes (2xx), redirection codes (3xx), errors of client application (4xx) and HTTP server errors (5xxx). The distribution of codes was as follows: 200 (72%), 304 (11%), 301 (10%), 302 (6%), 404 (1%), and the remaining ones with a negligible percent 404, 410, 503, 409, 400, 409, 400, 403, 206, 500, 502, 429. The application is almost all the time available (low number of 503 code, typically below 0.1%), 27% requests are redirected to another page, only 1% were not found (probably the users had obsolete links). In each month most responses were 200, 302, 304 or 301. However, on one month the percentage of 503 codes was 0.25%, more than for other months. In most months responses 200 constituted 80%, in a subsequent one 66%. This resulted from changes of 200 into 303 (it increased from 3% to 18,7%). Analyzing distribution of errors (5xx) over countries we observed two countries with significantly higher monthly frequency (3rd quartile: 17 and 60), as compared to 0 or 1 for some other countries. This resulted from configuration and administration problems.

Analyzing pages visited by users we can trace and identify their behavior or find critical paths (application code correlated with frequently visited pages needed optimization). The main page was the mostly visited one, 15% of users changed default language (English), over 20% of users did not decide to log, internet bots generated only 1% of the whole traffic. An important parameter is the page loading time. The developed script allows us to derive its distribution in time (e.g. with daily granularity – Fig. 1c) related to different countries. The application was quite stable – box plots with monthly resolutions showed small monthly differences and moderate maximal values. The first and third quartile for the considered months ranges were 610–605 ms and 620–645 ms, respectively, sporadically maximal values appeared mostly at the level not crossing 660 ms. This confirms quite good system performance.

Tracing user operational profiles we have also studied operating system usage and upgrades in time. Distribution of OS usage was as follows: MS Windows 67% (dominating Windows 7 and 10), Linux 26% (including Android 4.4.2–7.0), Mac OS 3%, others 4%. This confirmed the need of assuring compatibility with quite big span of OS versions during upgrades of the application.

The presented study revealed a large space of interesting information hidden in event entries. This resulted in diverse statistics and relevant features. In a large extent they are application specific. Nevertheless, in other systems we can also expect many of them and generate similar operation profile reports. The gained experience can be generalized, in particular the lexical analysis is universal and important in the case of various logs comprising complex text messages related to different application contributors. The preliminary analysis can be consulted with developers to adapt and refine feature selection scripts. A systematic wide scope log analysis creates an application specific knowledge database, which can provide a feedback to developers.

6   Conclusion

The survey of a wide scope computer logs shows a large number of generated entries comprising rich data. These data submitted to appropriate processing schemes provide useful information and many statistics for runtime evaluation of system operation. Data exploration should be targeted at tracing characteristic features (in diverse perspectives) indicating normal or anomalous system behavior, e.g.: failures, imperfections, deficiencies, security threats, non-satisfactory performance metrics. Well structured logs simplify analysis algorithms (scripts). An additional semantic analysis based on text mining processes is helpful in exploring log entry messages (finding keywords, negative words or phrases, etc.). Despite voluminous logs, this analysis is quite effective due to limited cardinality of unique words and n-grams.

Some detected negative features are not critical while they occur sporadically. Usually, many software applications include a possibility of tuning their operation with some parameters. Finding correlations between log features and application configurations is useful in system optimization. Further research can combine event log analysis with monitoring performance counters and bug reports. Here, we can base on our experience reported in [12] and references therein.

References

1. Aue, J.: Log analysis from A to Z: a literature survey. MSc thesis, Delft University (2016)
2. Chen, C.H., Singh, N., Yajnik, D.: Log analytics for dependable enterprise telephony. In: IEEE 9th European Dependable Computing Conference, pp. 94–101 (2012)
3. Chen, B., Jiang, Z.: Characterizing logging practices in Java-based open source software projects. Empirical Softw. Eng. 22(1), 330–374 (2017)
4. Cinque, M., Cotroneo, D., Della Corte, R., Pecchia, A.: Assessing direct monitoring techniques to analyze failures of critical industrial systems. In: 25th IEEE International Symposium on Software Reliability Engineering, pp. 212–222 (2014)
5. Ding, R., Zhou, H., Lou, J-G., Zhang, H., Lin, Q., Fu, Q., Zhang, D.: Log2: a cost-aware logging mechanism for performance diagnosis. In: USENIX ATC Conference, pp. 139–150 (2015)
6. Fu, X., Ren, R., Zhan, J., Zhou, W., Jia, Z., Lu, G.: LogMaster: mining event correlations in logs of large-scale cluster systems. In: 31st IEEE Symposium on Reliable Distributed Systems, pp. 71–80 (2012)
7. Fu, Q., Zhu, J., Hu, W., Lou, J., Ding, R., Lin, Q., Zhang, D., Xie, T.: Where do developers log? An empirical study on logging practices in industry. In: Companion Proceedings of the 36th International Conference on Software Engineering, ICSE 2014, pp. 24–33 (2014)
8. He, S., Zhu, J., He, P., Lyu, M.R.: Experience report: system log analysis for anomaly detection. In: Proceedings of International Symposium on Software Reliability Engineering, pp. 207–218 (2016)
9. He, P., Zhu, J., He, S., Li, J., Lyu, M.R.: Towards automated log parsing for large-scale log data analysis. IEEE Trans. Dependable Secure Comput. 15(16), 931–944 (2017)
10. He, P., Chen Z., He, Z., Lyu, M.L.: Characterizing the natural language descriptions in software logging statements. In: 33rd IEEE/ACM ASE International Conference, pp. 178–189 (2018)

11. Jain, S., Singh, I., Chandra, A., Zhang, Z.L., Bronevetsky, G.: Extracting the textual and temporal structure of supercomputing logs. In: 16th IEEE International Conference on High Performance Computing, pp. 254–263 (2009)
12. Kubacki, M., Sosnowski, J.: Holistic processing and exploring event logs. In: 9th International Workshop, SERENE 2017, LNCS, vol. 10479, pp. 184–200. Springer (2017)
13. Nagaraj, K., Killian, C., Neville, J.: Structured comparative analysis of systems logs to diagnose performance problems. In: 9th USENIX Conference on Networked Systems Design and Implementation, pp. 26–31 (2012)
14. Pecchia, A., Russo, S.: Detection of software failures through event logs: an experimental study. In: 23rd International Symposium on Software Reliability Engineering, pp. 31–40 (2012)
15. Pecchia, A., Cinque, M., Carroza,G., Cotroneo, D.: Industry practices and event logging: assessment of a critical software development process. In: 37th IEEE International Conference on Software Engineering, pp. 169–178 (2015)
16. Salman, M., Welch, B., Tront, J., Raymon, D., Marchany, R.: Designing PhelkStat: big analytics for system event logs. In: HICSS Symposium on Cybersecurity Big Data, Analytics, pp. 1–7 (2017)
17. Stearley, J., Oliner, A.J.: Bad words: finding faults in Spirit's Syslogs. In: 8th IEEE International Symposium on Cluster Computing and the Grid, pp. 765–770 (2008)
18. Xu, W., Huang, L., Fox, A., Patterson, D., Jordan, M.: Detecting large-scale system problems by mining console logs. In: 22nd ACM SOSP Conference, pp. 117–132 (2009)
19. Yuan, D., Park, S., Zhou, Y.: Characterizing logging practices in open-source software. In: 34th International Conference on Software Engineering, pp. 102–112 (2012)
20. Zhu, J., He, P., Fu, Q., Zhang, H., Lyu, R., Zhang, D.: Learning to log: Helping developers make informed logging decisions. In: 37th International Conference on Software Engineering, pp. 415–424 (2015)
21. Zou, D.Q., Qin, H., Jin, H.: UiLog: improving log-based fault diagnosis by log analysis. J. Comput. Sci. Technol. **31**(5), 1038–1052 (2016)

The Impact of Strong Electromagnetic Pulses on the Operation Process of Electronic Equipment and Systems Used in Intelligent Buildings

Jacek Paś⬤, Adam Rosiński$^{(\boxtimes)}$ ⬤, Jarosław Łukasiak⬤, and Marek Szulim⬤

Faculty of Electronics, Military University of Technology, Gen. Witolda Urbanowicza 2, 00-908 Warsaw, Poland
{jacek.pas,adam.rosinski}@wat.edu.pl

Abstract. Electronic equipment and systems are operated in intelligent buildings in specific environmental conditions. When operated in these environmental conditions, they should most often stay in a state of fitness. This is the basic operation process state, also guaranteed by the application of redundancy of the technical solution. All electronic equipment and systems operated in this should exhibit electromagnetic compatibility – both externally and internally. The appearance of strong electromagnetic field pulses, generated by an interference source, can lead to a catastrophic damage to electronic elements, devices, subsystems and systems, which are used to control the living functions in an intelligent building. This elaboration contains basic considerations in the field of a reliability-operational analysis of electronic equipment and systems, which can be operated in an environment exposed to the impact of strong electromagnetic pulses. A radiation source of strong electromagnetic pulses uses directional and omni-directional antennas, which allow to appropriately direct the energy in order to incapacitate the aforementioned technical objects. A relationship graph was suggested for specified antenna characteristics, which made it possible to formulate a set of Kolmogarov-Chapman equations. This enabled the calculation of the probability values of staying in the states of: full fitness, safety hazard and safety unreliability.

Keywords: Antenna · Propagation · Model · Electromagnetic field

1 Introduction

Electronic devices and systems function in different, often extreme, operational condition. One of the fundamental prerequisites for the correct operation of electronic equipment and systems is ensuring electromagnetic compatibility. Due to the high saturation of intelligent buildings with electronic systems responsible for all environmental functions (e.g. lighting, heating, security, etc.) implemented within a given facility, it is very important to determine the resistance and vulnerability to electromagnetic interference generated in an intentional or unintentional manner. Generating

© Springer Nature Switzerland AG 2020
W. Zamojski et al. (Eds.): DepCoS-RELCOMEX 2019, AISC 987, pp. 383–392, 2020.
https://doi.org/10.1007/978-3-030-19501-4_38

strong external electromagnetic interference in an intentional manner is aimed at partial or complete incapacitation of the aforementioned equipment. Electronic equipment and systems may be in the states of: full fitness, safety hazard and safety unreliability [8, 9, 12, 20]. Incapacitation of electronic equipment and systems is a function of the following parameters:

- limited space LS. The coverage of strong electromagnetic pulses depends on wave propagation, terrain (so-called radiolocation range), environmental conditions, technical parameters of the radiation source - e.g. power, spectrum of generated signals of varying frequency, transmitter antenna gain, radiated wave length λ, etc. The limited space of wave propagation depends also on the building materials – partitions used to construct facilities and buildings - electromagnetic wave attenuation;
- time T, strong electromagnetic pulse radiation sources have a limited power and after travelling a certain distance resulting from the conditions of wave propagations in different environment are significantly attenuated. The hazard of exposure to a strong electromagnetic pulse occurs for a given time interval Δt and over a limited, large-area terrain ΔS;
- impact planes IP, the use of strong electromagnetic pulse sources means a range of very high incapacitating frequencies with a varying amplitude. These signals freely propagate over a free space with a waveguide impedance of Z_f, however, they do not propagate deep into the ground. They are strongly attenuated, in contrast to very long waves [1, 4, 6, 11, 17];
- secondary radiation SR resulting from the induction of the electromotive force in electrically conductive elements, which are in the way of electromagnetic waves propagating within these buildings. Interfering voltages and current with parasitic inductance intercept only a part of the strong electromagnetic pulse fundamental signal. The interference can penetrate an ICT, power, control or other systems;
- electromagnetic wave reflections EMWR from metal surfaces (e.g. roofing, metal parts of building facades – railings, etc.), which are in the wave of electromagnetic wave propagation - Fig. 1.

The division of buildings with electronic equipment and systems subject to protection in the course of the operation process shall depend on the manner of using the sources of microwaves with strong electromagnetic pulses. There are two methods of sources of strong electromagnetic pulses impacting buildings shown in Fig. 2, which are (a) with the use of a directional antenna and (b) with the use of an omni-directional antenna [2, 9, 16, 23, 24].

At present, research is being conducted on the impact of strong electromagnetic pulses on electronic systems, but only selected research teams share their results [1, 4, 7, 11, 17]. The above mentioned studies present the results of research (spectrum, time courses), e.g. induced currents, voltages in elements, electronic circuits without discussing the issues of the impact of these phenomena on the operation process. The article presents a new approach to these studies taking into account the impact of these impulses (e.g. parameters - duration, antenna characteristics, type of buildings) on the process of system operation.

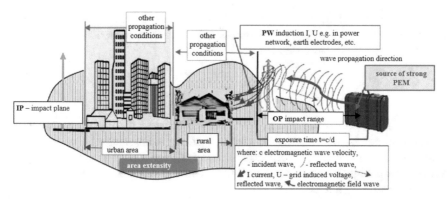

Fig. 1. The impact of strong electromagnetic pulses on selected buildings subject to protection [own study].

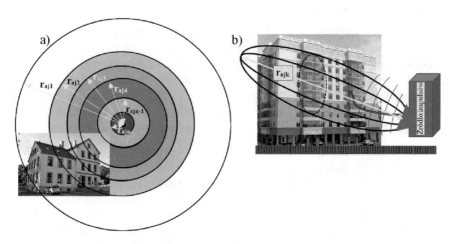

Fig. 2. Classification of buildings exposed to strong electromagnetic pulses – pictorial drawing, (a) using a pulse source of omni-directional character, (b) using a pulse source of directional character, where: r_{aj1} – pulse absence border – no hazard, r_{aj2} – lack of pulse impact border, r_{aj3} – pulse impact border (catastrophic damage of digital circuits), r_{aj4} – pulse impact border (catastrophic damage of analogue circuits), …, r_{ajn-1} – pulse impact border (catastrophic damage of electronic circuits, equipment and systems), r_{ajn} – pulse impact border - catastrophic damage of electronic and electrical circuits, equipment and systems – n.

2 Reliability-Operational Modelling of Electronic Equipment and Systems Used in an Intelligent Building, Taking into Account the Impact of Strong Electromagnetic Pulses

The reliability-operational modelling for intelligent buildings in terms of their exposure to strong electromagnetic pulses shall also take into account the total impact (use) η efficiency of the source [3, 10, 15, 18]. The total efficiency η can be defined as a ratio of the general cubic volume of an intelligent building in [m³], which is exposed to strong electromagnetic pulses to given radiation source antenna characteristic in a 3D space, expressed in [m³] – relationship (1).

$$\eta = \frac{V_{ko}}{A_{A\dot{z}}} \qquad (1)$$

where: V_{ko} - cubic volume of a building [m³] exposed to the impact of strong electromagnetic pulses, $A_{A\dot{z}}$ - characteristics of a strong electromagnetic pulse radiation source antenna within a 3D space expressed in [m³] – Fig. 2. The use efficiency of a strong electromagnetic pulse radiation source within a 3D space shall fall within the limits expressed by a relationship (2):

$$\eta = \ <1,0> \qquad (2)$$

Two extreme cases can be distinguished:

1. $\eta = 1$ - cubic volume of a building under protection [m³] exposed to the impact of strong electromagnetic pulses is fully covered by $A_{A\dot{z}}$ - characteristics of a strong electromagnetic pulse radiation source antenna within a 3D space expressed in [m³],
2. $\eta = 0$ - cubic volume of a building [m³] exposed to the impact of strong electromagnetic pulses, of which the coverage by $A_{A\dot{z}}$ - characteristics of a strong electromagnetic pulse radiation source antenna within a 3D space expressed in [m³] is equal to zero – no impact, Fig. 2 - r_{aj2} – lack of pulse impact border.

The total use efficiency of a strong electromagnetic pulse radiation source η within a 3D space is a function of many technical parameters of this device - a generator (e.g. pulse power P_i of source, interference signal λ wavelength, directional antenna gain G, interference signal duration t_i, etc.). We can introduce the concept of partial efficiency - η_c it is responsible for catastrophic damage to individual elements (e.g. digital circuits, transistors, diodes, etc.), devices (laptop, radiostation, etc.), electronic and electrical systems located within an intelligent building. Partial fitness η_c can then be defined as a ratio of a cubic volume separated from a given, overall building space V_{kwo} [m³] with electronic vulnerable equipment exposed to strong electromagnetic pulse impact to a 3D space with a strong electromagnetic pulse radiation source antenna characteristics [m³] $A_{A\dot{z}}$ – expression 3 [16, 23, 24].

$$\eta_c = \frac{V_{kwo}}{A_{A\dot{z}}} \qquad (3)$$

where: V_{kwo} [m³] - cubic volume separated from the general building space V_{kwo} [m³] with vulnerable electronic equipment exposed to the impact of strong electromagnetic pulses, $A_{A\dot{z}}$ - strong electromagnetic pulse source antenna characteristics within a 3D space expressed in [m³].

Based on the definition of the total η and partial efficiency η_c, we can classify the building in terms of their exposure to the impact of strong electromagnetic pulses as follows:

- BE$_j$ - strong electromagnetic pulse(s) able to cover the entire cubic volume (regardless of the height, width and length of the building, and its cubic volume) and cause catastrophic damage to electronic and electrical elements, circuits, devices and systems [5, 7, 13, 21];
- BE$_j$1 - strong electromagnetic pulse(s) able to cover only a limited cubic volume of a building and cause catastrophic damage to the aforementioned technical elements, which are least resistant to the impact of a electromagnetic field with strong pulses. An impact zone range can be determined based on the resistance of the weakest link – element(s) of electronic devices located within this zone. This can be expressed as an electromagnetic field power [J, W/m^2], which is sufficient to generate the aforementioned damage – Fig. 3;

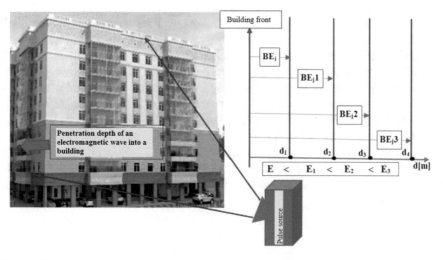

Fig. 3. Classification of facilities exposed to an electromagnetic field, where: BE$_j$, ..., BE$_j$3 zones within a facility with a possibility of a catastrophic damage in the event of using high-power pulses E, E$_1$, E$_2$ i E$_3$, respectively

- BE$_j$2 - strong electromagnetic pulse(s) able to cover only a limited cubic volume of a building and cause catastrophic damage to the aforementioned technical elements, which exhibit medium resistance to the impact of a electromagnetic field with strong pulses;
- BE$_j$3 - strong electromagnetic pulse(s) is (are) able to cover only a limited cubic volume of a building and cause catastrophic damage to electronic circuits, equipment and systems, which are most resistant to the impact of a strong pulse electromagnetic field [4, 10, 15, 19].

The impact range of strong electromagnetic pulse can be defined as the maximum distance – (interference signal source – intelligent building), from which the electronic equipment (systems) operated in the aforementioned buildings with assumed probability for the pertaining propagation conditions will be incapacitated – rendered harmless. From the point of view of reliability and operation, taking into account the impact of strong electromagnetic pulses, electronic equipment and systems used in intelligent buildings can be in the following states of:

- full safety (fitness),
- safety hazard (partial fitness),
- safety unreliability (unfitness).

Elaborations regarding the reliability-operational analysis of electronic equipment and systems used in intelligent buildings are numerous and concern various technical aspects [4, 14, 18]. They include various reliability structure [25–27], redundancy and the influence of electromagnetic interference - including compatibility. However, available publications do not contain an assessment of strong electromagnetic pulse impact and the impact of an interference source antenna characteristics. Specific state transitions undergoing over a considered time interval are adopted for the developed models. In order to map the operational process of electronic equipment and systems, the authors of this elaboration used a directed graph, with its vertices being the reliability-operational states and the arc representing transitions between the states. Considerations of the behaviour of electronic equipment and systems during the impact of strong electromagnetic pulses need to take into account the presence of safety hazard states. These states appear in the course of operation of the aforementioned buildings in the event of a partial coverage of an intelligent building by a strong electromagnetic pulse, and the total use efficiency of the source is $\eta < 1$. The electronic system operation process graph shown in Fig. 4 takes into account the use efficiency η of an electromagnetic field source taking into account the characteristics of the antenna (directional, omni-directional).

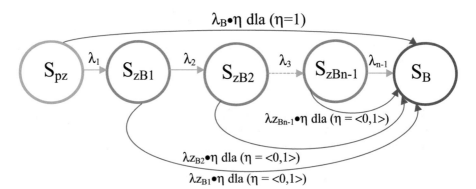

Fig. 4. Relationship occurring within electronic equipment and systems operated in an intelligent building, taking into account the safety hazard states and the possibilities of a transition from state S_{PZ} and states S_{ZB1}, S_{ZB2}, ..., S_{ZBn-1} to state S_B, taking into account the use efficiency η of a strong electromagnetic field source, where: green lines mean "normal" wear process within electronic equipment and systems, red lines mean transitions under the impact of a strong electromagnetic field source.

A model of electronic equipment and system operation is an ordered triple of the form:

$$M = \langle SB, RE, FR \rangle \qquad (4)$$

where:

$$SB = \{S_{PZ}, S_{ZB1}, S_{ZB2}, \ldots, S_{ZBn-1}, S_B\} \qquad (5)$$

SB is a set of operational states of electronic equipment and systems. Individual states are interpreted as follows:

- S_{PZ} – state of full fitness,
- S_{ZB1} – state of safety hazard 1,
- S_{ZB2} – state of safety hazard 2,
- ...,
- S_{ZBn-1} – state of safety hazard n − 1,
- S_B – state of safety unreliability.

The second element RE of the ordered M triple is a set of pairs with elements interpreted as follows:

- (S_{PZ}, S_{ZB1}) informs about the possibility of a transition from state S_{PZ} to state S_{ZB1} impacted by a "normal" wear process occurring in electronic equipment and systems,
- (S_{zB1}, S_B) informs about the possibility of a transition from state S_{zB1} to state S_B impacted by a strong electromagnetic field source,
- ...,
- (S_{zBn-1}, S_B) informs about the possibility of a transition from state S_{zBn-1} to state S_B impacted by a strong electromagnetic field source,
- (S_{pz}, S_B) informs about the possibility of a transition from state S_{zB2} to state S_B impacted by a strong electromagnetic field source.

Each element from the set RE is assigned a number from a set with a transition intensity interpretation. In particular:

- $\lambda(S_{pz}, S_{zB1}) \equiv \lambda_1$ is interpreted as a transition from a state of full fitness S_{PZ} to a state of safety hazards S_{zB1}, impacted by a "normal" wear process of equipment or systems,
- $\lambda(S_{zB1}, S_B) \equiv \lambda_{zB1} \cdot \gamma$ for $(\eta = <0.1>)$ is interpreted as a transition from a state of safety hazard S_{zB1} to a state of safety unreliability S_B, impacted by a strong electromagnetic field source,
- ...,
- $\lambda(S_{zB2}, S_B) \equiv \lambda_{zB2} \cdot \gamma$ for $(\eta = <0.1>)$ is interpreted as a transition from a state of safety hazard S_{zB2} to a state of safety unreliability S_B, impacted by a strong electromagnetic field source,

- $\lambda(S_{zBn-1}, S_B) \equiv \lambda_{n-1}$ is interpreted as a transition from a state of safety hazard S_{zBn-1} to a state of safety unreliability S_B, impacted by a "normal" wear process of equipment or systems,
- ...,
- $\lambda(S_{zBn-1}, S_B) \equiv \lambda_{zBn-1} \cdot \gamma$ for ($\eta = <0.1>$) is interpreted as a transition from a state of safety hazard S_{zBn-1} to a state of safety unreliability S_B, impacted by a strong electromagnetic field source.

In order to determine the probabilities of an electronic device and system staying in individual states of interest to us, the model shown in Fig. 4 shall be described with the following Kolmogorov-Chapman equations:

$$\begin{aligned}
R_0'(t) &= -\lambda_1 \cdot R_0(t) - \lambda_B \cdot \eta \cdot R_0(t) \\
Q_{ZB1}'(t) &= \lambda_1 \cdot R_0(t) - \lambda_2 \cdot Q_{ZB1}(t) - \lambda_{ZB1} \cdot \eta \cdot Q_{ZB1}(t) \\
Q_{ZB2}'(t) &= \lambda_2 \cdot Q_{ZB1}(t) - \lambda_3 \cdot Q_{ZB2}(t) - \lambda_{ZB2} \cdot \eta \cdot Q_{ZB2}(t) \\
\cdots \\
Q_{ZBn-1}'(t) &= \lambda_{n-2} \cdot Q_{ZBn-2}(t) - \lambda_{n-1} \cdot Q_{ZBn-1}(t) - \lambda_{ZBn-1} \cdot \eta \cdot Q_{ZBn-1}(t) \\
Q_B'(t) &= \lambda_B \cdot \eta \cdot R_0(t) + \lambda_{n-1} \cdot Q_{ZBn-1}(t) + \lambda_{ZB1} \cdot \eta \cdot Q_{ZB1}(t) + \\
&\quad + \lambda_{ZB2} \cdot \eta \cdot Q_{ZB2}(t) + \ldots + \lambda_{ZBn-1} \cdot \eta \cdot Q_{ZBn-1}(t)
\end{aligned} \tag{6}$$

Assuming the baseline conditions:

$$\begin{aligned}
R_0(0) &= 1 \\
Q_{ZB1}(0) &= Q_{ZB2}(0) = \ldots = Q_{ZBn-1}(0) = Q_B(0) = 0
\end{aligned} \tag{7}$$

and using the Laplace transform for the set of Eqs. (6), we get a set of linear equations (in a symbolic approach). Next, by using the reverse transformations, we get the values of the probabilities for the electric devices and systems staying in distinguished states. In equation no. 7 the theoretical values of the initial states of electronic systems were theoretically accepted. All electronic systems in buildings in the "nominal" operation process without the presence of strong electromagnetic pulses are in a fully operational state, S_{PZ} (Fig. 4), i.e. the probability of the system being in this state is $R(t) = 1$. The article presents the initial studies of the impact of strong electromagnetic pulses on the operation of electronic systems. The operation of these systems during the impact of strong pulses should take into account the parameters: among others interfering generator, materials from which building partitions were made - means all elements located on the propagation path of the electromagnetic wave.

3   Conclusion

The research paper discusses issues associated with the process of operating electronic equipment and systems, which are used in an intelligent building for controlling environmental parameters [8, 12, 22]. Electronic equipment and systems operated in a building should meet the external and internal electromagnetic compatibility requirements, due to resistance, strength and vulnerability. Satisfying these requirements does not mean protecting the used buildings against the impact of a strong source of electromagnetic field generated intentionally to incapacitate the aforementioned equipment and systems. The process of strong electromagnetic field source impact on an intelligent building depends on the technical parameters of an interference source end device – i.e., an antenna. Due to the vast area (cubic volume) of an intelligent building, it is possible to use two available types of interference signal antennas – a directional and omni-directional sectoral antenna. Using these antennas enables covering selected zones within an intelligent building in order to induce catastrophic damage to the operated electronic equipment and systems. Electromagnetic interference reaches these technical objects through direct, indirect and induced impact (power and signalling cables). A graph of relationships within a considered electronic device or system, taking into account the γ impact factor for the strong electromagnetic pulse generated by an antenna was developed. The graph was described using a set of Kolmogorov-Chapman equations. This can be used as a base to determine relationships that would allow to calculate the probabilities of electronic devices and systems being in the states of: full fitness, safety hazard and safety unreliability. The strong electromagnetic impact factor γ was taken into account during the modelling of an electronic device and system operation process.

Acknowledgments. The article was edited due to the implementation of a project financed by The National Centre for Research and Development, concerning research and development works for defense and national security no. DOB-1-3/1/PS/2014 "Methods and systems for protection against HPE-M pulses".

References

1. Kuchta, M., Paś, J.: Electromagnetic terrorism—threats in buildings. Biuletyn WAT **LXIV** (2), pp. 135–147 (2015)
2. Burdzik, R., Konieczny, Ł., Figlus, T.: Concept of on-board comfort vibration monitoring system for vehicles. In: Mikulski, J. (ed.) Activities of Transport Telematics, pp. 418–425. Springer, Heidelberg (2013)
3. Paś, J.: Shock a disposable time in electronic security systems. J. KONBiN **2**(38), 5–31 (2016)
4. Chernikih, E.V., Didenko, A.N., Gorbachev, K.V.: High Power Microwave pulses generation from Vircator with inductive storage, Materiały konferencyjne Międzynarodowej Konferencji EUROEM, Electronic Environments and Consequences, Bordeaux (1994)
5. Charoy, A.: Interference in electronic devices. WNT, Warsaw (1999)
6. Dziula, P., Paś, J.: The impact of electromagnetic interferences on transport security system of certain reliability structure. In: 12th International Conference on Marine Navigation and Safety of Sea Transportation TransNav 2017, Gdynia, Poland, pp. 185–191 (2017)
7. Chen, S., Ho, T., Mao, B.: Maintenance schedule optimisation for a railway power supply system. Int. J. Prod. Res. **51**(16), 4896–4910 (2013)
8. Paś, J., Rosinski, A.: Selected issues regarding the reliability-operational assessment of electronic transport systems with regard to electromagnetic interference. Eksploatacja i Niezawodnosc – Maintenance and Reliability **19**(3), 375–381 (2017)
9. Paś, J.: Operation of electronic transportation systems. Publishing House University of Technology and Humanities, Radom (2015)

10. Paś, J., Rosiński, A., Wiśnios, M., Majda-Zdancewicz, E., Łukasiak, J.: Electronic security systems. In: Introduction to the Laboratory. Military University of Technology, Warsaw (2018)
11. Dras, M., Kałuski, M., Szafrańska, M.: HPM pulses – disturbances and systems interaction – basic issues. Przegląd elektrotechniczny 11, 11–14 (2015)
12. Dyduch, J., Paś, J., Rosiński, A.: The Basic of the Exploitation of Transport Electronic Systems. Publishing House of Radom University of Technology, Radom (2011)
13. Jin, T.: Reliability Engineering and Service. Wiley (2019)
14. Lheurette, E. (ed.): Metamaterials and Wave Control. ISTE and Wiley (2013)
15. Loeffler, C., Spears, E.: Uninterruptible power supply system. In: Hwaiyu Geng, P.E. (eds.) Data Center Handbook, pp. 495–521. Wiley (2015)
16. Ogunsola, A., Mariscotti, A.: Electromagnetic compatibility in railways. In: Analysis and Management. Springer (2013)
17. Przesmycki, R., Wnuk, M.: Susceptibility of IT devices to HPM pulse. Int. J. Saf. Secur. Eng. 8(2), 223–233 (2018)
18. Reddig, K., Dikunow, B., Krzykowska, K.: Proposal of big data route selection methods for autonomous vehicles. Internet Technol. Lett. 1(36), 1–6 (2018)
19. Rosiński, A.: Modelling the Maintenance Process of Transport Telematics Systems. Publishing House Warsaw University of Technology, Warsaw (2015)
20. Siergiejczyk, M., Krzykowska, K., Rosiński, A.: Reliability assessment of integrated airport surface surveillance system. In: Zamojski, W., Mazurkiewicz, J., Sugier, J., Walkowiak, T., Kacprzyk, J. (eds.) Proceedings of the Tenth International Conference on Dependability and Complex Systems DepCoS-RELCOMEX, pp. 435–443. Springer (2015)
21. Siergiejczyk, M., Paś, J., Rosiński, A.: Issue of reliability–exploitation evaluation of electronic transport systems used in the railway environment with consideration of electromagnetic interference. IET Intel. Transport Syst. 10(9), 587–593 (2016)
22. Siergiejczyk, M., Rosiński, A., Paś, J.: Analysis of unintended electromagnetic fields generated by safety system control panels. Diagnostyka 17(3), 35–40 (2016)
23. Siergiejczyk, M., Paś, J., Dudek, E.: Reliability analysis of aerodrome's electronic security systems taking into account electromagnetic interferences. In: Safety and Reliability – Theory and Applications, London, pp. 2285–2292, 27th European Safety and Reliability Conference ESREL 2017, PORTORAŽ, Słowenia (2017)
24. Kuchta, M., Dukata, A., Paś, J., Kubacki, R.: Analysis of the propagation of the electromagnetic field to the internal structures of the selected building structure. In: Selected Problems of Diagnosing and Using Devices and Systems, pp. 125–140. WAT (2015)
25. Caban, D., Walkowiak, T.: Dependability analysis of hierarchically composed system-of-systems. In: Zamojski, W., Mazurkiewicz, J., Sugier, J., Walkowiak, T., Kacprzyk, J. (eds.) Proceedings of the Thirteenth International Conference on Dependability and Complex Systems DepCoS-RELCOMEX, pp. 113–120. Springer (2019)
26. Stawowy, M., Kasprzyk, Z.: Identifying and simulation of status of an ICT system using rough sets. In: Zamojski, W., Mazurkiewicz, J., Sugier, J., Walkowiak, T., Kacprzyk, J. (eds.) Theory and Engineering of Complex Systems and Dependability, DepCoS-RELCOMEX 2015, pp. 477–487. Springer (2015)
27. Stawowy, M.: Comparison of uncertainty models of impact of teleinformation devices reliability on information quality. In: Nowakowski, T., Młyńczak, M., Jodejko-Pietruczuk, A., Werbińska-Wojciechowska, S. (eds.) Safety and Reliability: Methodology and Applications - Proceedings of the European Safety and Reliability Conference ESREL 2014, pp. 2329–2333. CRC Press/Balkema, London (2015)

Modelling the Safety Levels of ICT Equipment Exposed to Strong Electromagnetic Pulses

Jacek Paś⬤, Adam Rosiński(✉)⬤, Marek Szulim⬤,
and Jarosław Łukasiak⬤

Faculty of Electronics, Military University of Technology,
Gen. Witolda Urbanowicza 2, 00-908 Warsaw, Poland
{jacek.pas,adam.rosinski}@wat.edu.pl

Abstract. The reliability-operational analysis of ICT equipment (i.e. used in transport) shows that they function in varying, often extreme conditions. In the course of operation, they should be in the state of fitness. It depends not only on the reliability of ICT subsystems but is also impacted on rational operational management. When conducting operational analyses, it is also important to take into account the vulnerability and resistance of ICT equipment to strong electromagnetic pulses. Therefore, we need to, i.e., determine the functional safety levels of ICT equipment exposed to the impact of strong electromagnetic pulses. This will enable rational application of solutions (e.g. design, organizational), which would lead to increased probability of equipment being in a state of fitness.

This elaboration contains considerations in the field of reliability-operational analysis of an ICT device, which is operated in an environment exposed to strong electromagnetic pulses. Next, it proposes a relationship graph for the device in question, which makes it possible to develop a system of Kolmogarov-Chapman equations. Another step may be the determination of relationships that would allow to calculate the probability of a device being in the states of: full fitness, safety hazard and safety unreliability.

Keywords: Operation · Modelling · Strong electromagnetic pulses

1 Introduction

The conducted analysis of the operation process of selected ICT equipment (e.g. used in communications, transport traffic control systems) confirms that they are operated in varying environmental conditions. In the course of operation, they should be in the state of fitness. It depends not only on the reliability of ICT subsystems forming them, but also on the optimal management of the operational process [10, 22, 26]. An important aspect in the functioning of these devices is also the degree of vulnerability and resistance to the impact of strong electromagnetic pulses [4, 17, 21]. Therefore, the selection of proper indicators of the strong electromagnetic pulse impact on the functionality of an ICT device requires a broader view than just an EMC analysis [29].

ICT devices operate within a specified electromagnetic environment, which depends on numerous external factors. A natural electromagnetic environment shaped

© Springer Nature Switzerland AG 2020
W. Zamojski et al. (Eds.): DepCoS-RELCOMEX 2019, AISC 987, pp. 393–401, 2020.
https://doi.org/10.1007/978-3-030-19501-4_39

by phenomena that occur on Earth is most often seriously distorted within transport areas [11] (railway, in particular [1, 5, 19]). It is caused by the application of a large number of devices, which are the source of electromagnetic fields radiating in intended or unintended ways [20, 30]. Every electrical or electronic device is powered by electricity, and therefore, generates its own electromagnetic field, which is associated with its operation [7, 18].

The fact that ICT devices are characterized by diverse sensitivity to a strong electromagnetic pulse (with the same parameters) is also an important issue [24]. An ICT device has different sensitivity to a strong electromagnetic pulse. This results from, among others, the used design solutions and the position relative to the source of electromagnetic interference [6, 8, 16].

Solutions increasing the safety level are applied in order to increase the resistance of ICT devices to impacting strong electromagnetic pulses. Such an approach is reasonable, but it should be noted that not all of them are effective and feasible (i.e. due to technical or economic reasons) [13–15, 23].

From the point of view of reliability and operation, taking into account the impact of strong electromagnetic pulses, ICT devices can be in the following states of: full safety (fitness), safety hazard (partial fitness), safety unreliability (unfitness).

There are numerous elaborations on the reliability-operational analysis of ICT devices [3, 12]. They take into account, i.e., reliability structures (serial, parallel, serial-parallel) and the impact of electromagnetic interference. However, these publications do not contain comprehensive solutions in the field of evaluation and protection against the impact of strong electromagnetic pulses on ICT equipment.

ICT equipment is faced with many requirements. The most important include: miniaturization, limiting electricity consumption, appropriate functionality [9, 25, 31] and reliability [28], resistance to vibrations [2], as well as the quality of information [27, 32]. Satisfying these requirements leads to a decreased difference between the level of useful signals and the level of interference generated by interference sources. Therefore, the issue of determining the safety level for ICT devices exposed to the impact of strong electromagnetic pulses is important. The considerations in this regard are presented in this paper.

2   Reliability-Operational Modelling of an ICT Device, Taking into Account the Impact of Strong Electromagnetic Pulses

Modelling of an ICT equipment operation process, which takes into account the impact of strong electromagnetic pulses, involves the presentation of their behaviour in various environmental conditions. It is usually assumed that a developed model shows changes of the states, undergoing within a considered time interval. The number of distinguished reliability-operational states of an ICT device is a finite set. It depends on the objectives and the adopted modelling accuracy. In order to map the operational process of an ICT device, the authors of this elaboration used a directed graph, with its vertices being the reliability-operational states and the arc representing transitions between the states.

In the simplest case, an ICT device can be in one of the two following states: state of full fitness (S_{PZ}), state of safety unreliability (S_B).

The graphical interpretation of the aforementioned situation is as follows (Fig. 1).

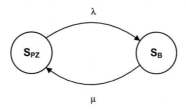

Fig. 1. Relationships within a device. Marking in the Fig.: λ – damage intensity, μ – repair intensity.

Let us adopt a factor Γ determining the impact of a strong electromagnetic pulse on the analysed ICT device. Its value fall within the following range:

$$\Gamma \in \langle 0, 1 \rangle \tag{1}$$

We assume that:

- $\Gamma = 0$ for the lack of strong electromagnetic pulse impact (the applied solutions completely eliminate the impact of a strong electromagnetic pulse on an ICT device),
- $\Gamma = 1$ for the impact of a strong electromagnetic pulse (no solutions aimed at mitigating the effects of the impact of a strong electromagnetic pulse on an ICT device).

The strong electromagnetic pulse impact factor Γ regarding the analysed ICT device shall be used when determining the safety levels. Therefore, the device relationship graph shown in Fig. 1 takes the form as in Fig. 2.

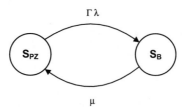

Fig. 2. Relationships within a device, taking into account the Γ strong electromagnetic pulse impact factor.

Then, the probability of a device being in a state of full fitness will be described by the following relationship:

$$K_{g1} = P_{PZ} = \frac{\mu}{\mu + \Gamma \cdot \lambda} \tag{2}$$

The graphs of relations in the device presented in Figs. 1 and 2 take into account the possibility of repair. In real research on the impact of strong electromagnetic pulses on devices (especially mobile ones), there is usually no way to take corrective action. For this reason, the repair process is not taken into account in further considerations.

The system relationship graph shown in Fig. 2, which takes into account the strong electromagnetic pulse impact factor, does not reflect the possibilities of a device moving from a state of full fitness to states other than safety unreliability. In order to reflect the actual situations regarding the use of ICT devices, taking into account the impact of strong electromagnetic pulses, further modelling including the transitions to states of partial fitness was conducted. Therefore, the device relationship graph shown in Fig. 2 will take the form as in Fig. 3.

The specificity of the developed graph of relations results from the inclusion of partial states of fitness, the number of which is related to the applied methods of protection against the impact of strong electromagnetic pulses. This approach will allow to compare different types of technical solutions (mechanical, electromechanical, electronic, construction) or organizational measures (operation procedures) to increase the level of device security against electromagnetic interference. Thanks to this, it will be able to decide on a specific solution to increase the level of object security against the impact of strong electromagnetic pulses. Also the values of intensity of transitions between the distinguished states and the Γ coefficient are determined, among others, on the basis of device immunity tests on HPM impulse [5, 16, 23].

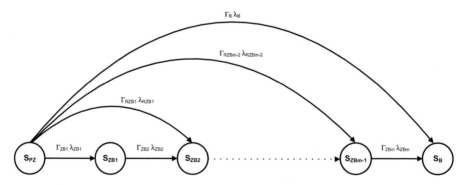

Fig. 3. Relationships within a device, taking into account the states of safety hazard and the possibility of moving from the S_{PZ} state to S_{ZB} states and the HPM pulse impact factor.

Considering the above, an operational model of a device is an ordered triple of the forms:

$$M = \langle SB, RE, FR \rangle \tag{3}$$

where:

$$SB = \{S_{PZ}, S_{ZB1}, S_{ZB2}, \ldots, S_{ZBm-1}, S_B\} \tag{4}$$

SB is a set of operational states of a device and is interpreted as follows: S_{PZ} – state of full fitness, S_{ZB1} – state of safety hazard 1, S_{ZB2} – state of safety hazard 2, ..., S_{ZBm-1} – state of safety hazard m − 1, S_B – state of safety unreliability.

The states belonging to the **SB** set shall be interpreted as the states of: full fitness, safety hazard 1, safety hazard 2, ..., safety hazard m − 1 and safety unreliability, and then we can continue deliberations regarding the model, taking into account the impact of strong electromagnetic pulses.

The second element **RE** of the ordered M triple is a set of pairs with elements interpreted as follows:

- (S_{PZ}, S_{ZB1}) informs about the possibility to move from state S_{PZ} to state S_{ZB1},
- (S_{PZ}, S_{ZB2}) informs about the possibility to move from state S_{PZ} to state S_{ZB2},
- ...,
- (S_{PZ}, S_{ZBm-1}) informs about the possibility to move from state S_{PZ} to state S_{ZBm-1},
- (S_{PZ}, S_B) informs about the possibility to move from state S_{PZ} to state S_B,
- (S_{ZB1}, S_{ZB2}) informs about the possibility to move from state S_{ZB1} to state S_{ZB2},
- (S_{ZB2}, S_{ZB3}) informs about the possibility to move from state S_{ZB2} to state S_{ZB3},
- ...,
- (S_{ZBm-1}, S_B) informs about the possibility to move from state S_{ZBm-1} to state S_B.

Thus:

$$RE = \left\{ \begin{array}{l} (S_{PZ}, S_{ZB1}), (S_{PZ}, S_{ZB2}), \ldots, (S_{PZ}, S_{ZBm-1}), (S_{PZ}, S_B), \\ (S_{ZB1}, S_{ZB2}), (S_{ZB2}, S_{ZB3}), \ldots, (S_{ZBm-1}, S_B) \end{array} \right\} \tag{5}$$

so

$$RE \subset S \times S \tag{6}$$

Let us assume that element **FR** is a set of function, each of which is determined on the set **RE** and adopts a value from a set of positive real numbers, i.e., R^+. In particular, this function λ has the form:

$$\lambda : RE \longrightarrow R^+ \tag{7}$$

Therefore, each element from the set **RE** is assigned a number from the set R^+ with a transition intensity interpretation. In particular:

- $\lambda(S_{PZ}, S_{ZB1}) \equiv \lambda_{ZB1}$ is interpreted as the intensities of transition from a state of full fitness S_{PZ} to a state of safety hazard S_{ZB1},

- $\lambda(S_{PZ}, S_{ZB2}) \equiv \lambda_{RZB1}$ is interpreted as the intensities of transition from a state of full fitness S_{PZ} to a state of safety hazard S_{ZB2},
- $\lambda(S_{PZ}, S_{ZB3}) \equiv \lambda_{RZB2}$ is interpreted as the intensities of transition from a state of full fitness S_{PZ} to a state of safety hazard S_{ZB3},
- ...,
- $\lambda(S_{PZ}, S_{ZBm-1}) \equiv \lambda_{RZBm-2}$ is interpreted as the intensities of transition from a state of full fitness S_{PZ} to a state of safety hazard S_{ZBm-1},
- $\lambda(S_{PZ}, S_B) \equiv \lambda_B$ is interpreted as the intensities of transition from a state of full fitness S_{PZ} to a state of safety unreliability S_B,
- $\lambda(S_{ZB1}, S_{ZB2}) \equiv \lambda_{ZB2}$ is interpreted as the intensities of transition from a state of safety hazard S_{ZB1} to a state of safety hazard S_{ZB2},
- $\lambda(S_{ZB2}, S_{ZB3}) \equiv \lambda_{ZB3}$ is interpreted as the intensities of transition from a state of safety hazard S_{ZB2} to a state of safety hazard S_{ZB3},
- ...,
- $\lambda(S_{ZBm-1}, S_B) \equiv \lambda_{ZBm}$ is interpreted as the intensities of transition from a state of safety hazard S_{ZBm-1} to a state of safety unreliability S_B.

In order to determine the probabilities of an ICT device staying in individual states of interest to us, the model shown in Fig. 3 shall be described with the following Kolmogorov-Chapman equations:

$$R_0'(t) = -\Gamma_B \cdot \lambda_B \cdot R_0(t) - \Gamma_{ZB1} \cdot \lambda_{ZB1} \cdot R_0(t) - \Gamma_{RZB1} \cdot \lambda_{RZB1} \cdot R_0(t)$$
$$-\Gamma_{RZB2} \cdot \lambda_{RZB2} \cdot R_0(t) - \ldots - \Gamma_{RZBm-2} \cdot \lambda_{RZBm-2} \cdot R_0(t)$$
$$Q_{ZB1}'(t) = \Gamma_{ZB1} \cdot \lambda_{ZB1} \cdot R_0(t) - \Gamma_{ZB2} \cdot \lambda_{ZB2} \cdot Q_{ZB1}(t)$$
$$Q_{ZB2}'(t) = \Gamma_{ZB2} \cdot \lambda_{ZB2} \cdot Q_{ZB1}(t) - \Gamma_{ZB3} \cdot \lambda_{ZB3} \cdot Q_{ZB2}(t) + \Gamma_{RZB1} \cdot \lambda_{RZB1} \cdot R_0(t)$$
$$Q_{ZB3}'(t) = \Gamma_{ZB3} \cdot \lambda_{ZB3} \cdot Q_{ZB2}(t) - \Gamma_{ZB4} \cdot \lambda_{ZB4} \cdot Q_{ZB3}(t) + \Gamma_{RZB2} \cdot \lambda_{RZB2} \cdot R_0(t)$$
$$\ldots$$
$$Q_{ZBm-1}'(t) = \Gamma_{ZBm-1} \cdot \lambda_{ZBm-1} \cdot Q_{ZBm-2}(t) - \Gamma_{ZBm} \cdot \lambda_{ZBm} \cdot Q_{ZBm-1}(t) + \Gamma_{RZBm-2} \cdot \lambda_{RZBm-2} \cdot R_0(t)$$
$$Q_B'(t) = \Gamma_B \cdot \lambda_B \cdot R_0(t) + \Gamma_{ZBm} \cdot \lambda_{ZBm} \cdot Q_{ZBm-1}(t)$$

$$(8)$$

Assuming baseline conditions:

$$R_0(0) = 1$$
$$Q_{ZB1}(0) = Q_{ZB2}(0) = \ldots = Q_{ZBm-1}(0) = Q_B(0) = 0 \qquad (9)$$

and using the Laplace transform for the set of Eq. (8), we get a set of linear equations (in a symbolic approach). Next, by using the reverse transformation, we get the values of the probabilities for the device staying in distinguished states.

3   Conclusion

The elaboration includes a reliability-operational analysis of an ICT device, which can be impacted by strong electromagnetic pulses. A graph of relationships within a considered device, taking into account the Γ strong electromagnetic pulse impact factor was developed. Next, it was described using a set of Kolmogorov-Chapman equations. This can be used as a base to determine relationships that would allow to calculate the probabilities of an ICT device being in the states of: full fitness, safety hazard and safety unreliability. The authors plan to increase the number of safety hazard states (by entering parallel branches in the proposed relationship graph in the device) and the transitions between them in future studies. This will enable a more thorough mapping of the impact of various types of solutions mitigating the impact of strong electromagnetic pulses on ICT devices.

The strong electromagnetic impact factor Γ was taken into account during the modelling of an ICT device operation process. It enables determining the effectiveness of the implemented solutions (i.e., technical, functional and organizational) aimed at increasing the safety level. This will allow to evaluate and compare various variants of the protection against the impact of strong electromagnetic pulses, and to make a rational choice.

Acknowledgments. The article was edited due to the implementation of a project financed by The National Centre for Research and Development, concerning research and development works for defense and national security no. DOB-1-3/1/PS/2014 "Methods and systems for protection against HPE-M pulses".

References

1. Badyor, M.P., In'kov, Yu.M.: Electromagnetic compatibility of a traction power supply system and infrastructure elements in areas with high traffic. Russ. Electr. Eng. **85**(8), 488–492 (2014)
2. Burdzik, R., Konieczny, Ł., Figlus, T.: Concept of on-board comfort vibration monitoring system for vehicles. In: Mikulski, J. (ed.) Activities of Transport Telematics, pp. 418–425. Springer, Heidelberg (2013)
3. Caban, D., Walkowiak, T.: Dependability analysis of hierarchically composed system-of-systems. In: Zamojski, W., Mazurkiewicz, J., Sugier, J., Walkowiak, T., Kacprzyk, J. (eds.) Proceedings of the Thirteenth International Conference on Dependability and Complex Systems DepCoS-RELCOMEX, pp. 113–120. Springer (2019)
4. Charoy, A.: Interference in Electronic Devices. WNT, Warsaw (1999)
5. Chen, S., Ho, T., Mao, B.: Maintenance schedule optimisation for a railway power supply system. Int. J. Prod. Res. **51**(16), 4896–4910 (2013)
6. Chmielińska, J., Kuchta, M., Kubacki, R., Dras, M., Wierny, K.: Selected methods of electronic equipment protection against electromagnetic weapon. Przegląd elektrotechniczny **1**, 1–8 (2016)
7. Corsi, S.: Voltage Control and Protection in Electrical Power Systems. Springer, Heidelberg (2015)
8. Dras, M., Kałuski, M., Szafrańska, M.: HPM pulses – disturbances and systems interaction – basic issues. Przegląd elektrotechniczny **11**, 11–14 (2015)
9. Duer, S., Zajkowski, K., Płocha, I., Duer, R.: Training of an artificial neural network in the diagnostic system of a technical object. Neural Comput. Appl. **22**(7), 1581–1590 (2013)
10. Dyduch, J., Paś, J., Rosiński, A.: The basic of the Exploitation of Transport Electronic Systems. Publishing House of Radom University of Technology, Radom (2011)

11. Dziubinski, M., Drozd, A., Adamiec, M., Siemionek, E.: Electromagnetic interference in electrical systems of motor vehicles. In: Scientific Conference on Automotive Vehicles and Combustion Engines (KONMOT 2016). IOP Conference Series-Materials Science and Engineering, vol. 148, pp. 1–11 (2016)

12. Jin, T.: Reliability Engineering and Service. Wiley, Hoboken (2019)

13. Kierzkowski, A., Kisiel, T.: Airport security screeners reliability analysis. In: Proceedings of the IEEE International Conference on Industrial Engineering and Engineering Management IEEM 2015, Singapore, pp. 1158–1163 (2015)

14. Kierzkowski, A., Kisiel, T.: Simulation model of security control system functioning: a case study of the Wroclaw Airport terminal. J. Air Transp. Manag. 64(B), 173–185 (2016)

15. Kornaszewski, M., Chrzan, M., Olczykowski, Z.: Implementation of new solutions of intelligent transport systems in railway transport in Poland. In: Communications in Computer and Information Science, pp. 282–292. Springer (2017)

16. Kuchta, M., Paś, J.: Electromagnetic terrorism—threats in buildings. Biuletyn WAT LXIV (2), 135–147 (2015)

17. Lheurette, E. (ed.): Metamaterials and Wave Control. ISTE and Wiley, Hoboken (2013)

18. Loeffler, C., Spears, E.: Uninterruptible power supply system. In: Hwaiyu Geng, P.E. (eds.), Data Center Handbook, pp. 495–521. Wiley (2015)

19. Ogunsola, A., Mariscotti, A.: Electromagnetic Compatibility in Railways. Analysis and Management. Springer, Heidelbrerg (2013)

20. Ott, H.W.: Electromagnetic Compatibility Engineering. Wiley, Hoboken (2009)

21. Paś, J., Rosinski, A.: Selected issues regarding the reliability-operational assessment of electronic transport systems with regard to electromagnetic interference. Eksploatacja i Niezawodnosc – Maint. Reliab. 19(3), 375–381 (2017)

22. Paś, J.: Operation of Electronic Transportation Systems. Publishing House University of Technology and Humanities, Radom (2015)

23. Paś, J., Rosiński, A., Wiśnios, M., Majda-Zdancewicz, E., Łukasiak, J.: Electronic security systems. Introduction to the laboratory. Military University of Technology, Warsaw (2018)

24. Przesmycki, R., Wnuk, M.: Susceptibility of IT devices to HPM pulse. Int. J. Saf. Secur. Eng. 8(2), 223–233 (2018)

25. Reddig, K., Dikunow, B., Krzykowska, K.: Proposal of big data route selection methods for autonomous vehicles. Internet Technol. Lett. 1(36), 1–6 (2018)

26. Rosiński, A.: Modelling the maintenance process of transport telematics systems. Publishing House Warsaw University of Technology, Warsaw (2015)

27. Siergiejczyk, M., Stawowy, M.: Modelling of uncertainty for continuity quality of power supply. In: Walls, L., Revie, M., Bedford, T. (eds.) Risk, Reliability and Safety: Innovating Theory and Practice: Proceedings of ESREL 2016, pp. 667–671. CRC Press/Balkema, London (2017)

28. Siergiejczyk, M., Krzykowska, K., Rosiński, A.: Reliability assessment of integrated airport surface surveillance system. In: Zamojski, W., Mazurkiewicz, J., Sugier, J., Walkowiak, T., Kacprzyk, J. (eds.) Proceedings of the Tenth International Conference on Dependability and Complex Systems DepCoS-RELCOMEX, pp. 435–443. Springer (2015)

29. Siergiejczyk, M., Paś, J., Rosiński, A.: Issue of reliability–exploitation evaluation of electronic transport systems used in the railway environment with consideration of electromagnetic interference. IET Intell. Transp. Syst. 10(9), 587–593 (2016)

30. Siergiejczyk, M., Rosiński, A., Paś, J.: Analysis of unintended electromagnetic fields generated by safety system control panels. Diagnostyka 17(3), 35–40 (2016)

31. Stasiuk, O.I., Grishchuk, R.V., Goncharova, L.L.: Mathematical differential models and methods for assessing the cybersecurity of intelligent computer networks for control of technological processes of railway power supply. Cybern. Syst. Anal. 54(4), 671–677 (2018)

32. Stawowy, M.: Comparison of uncertainty models of impact of teleinformation devices reliability on information quality. In: Nowakowski, T., Młyńczak, M., Jodejko-Pietruczuk, A., Werbińska-Wojciechowska, S. (eds.), Safety and Reliability: Methodology and Applications - Proceedings of the European Safety and Reliability Conference ESREL 2014, pp. 2329–2333. CRC Press/Balkema, London (2015)

Prioritization of Tasks in the Strategy Evaluation Procedure

Henryk Piech and Grzegorz Grodzki[✉]

Czestochowa University of Technology, Dabrowskiego 69,
Czestochowa, Poland
{henryk.piech,grzegorz.grodzki}@icis.pcz.pl

Abstract. The main goal of the work is to support the marketing strategy using the characteristics created on base of the game theory and uncertain knowledge.

In market strategic games we create presumed players who embody the current or upcoming threats in a given economic sphere. Different classifications are therefore subject to these rules. In addition, criteria and parameters are created for the assessment of the scale or intensity (called features across the work), such as the level of certainty, coverage, credibility, believing function, plausibility, suspicion (mistrustability), possibility, etc. In the case of threats, we usually use their range, making them measurable values. Due to the specificity of each of the threats, their basic parameters will be defined differently. For example, the threat of competition resulting from the reduction of prime costs is obviously dangerous when they fall below our costs, and the opposite may be true for marketing expenditure. Many parameters are associated with each other and partly overlap, which can also be included in the parameter definition and estimation. The subject of the analysis partially presented in this article is to pay attention to the advisability and effectiveness of using measurable components (parameters, criteria) in creating the structure of strategic games, i.e. in the assessment of the scale of payoffs of individual players representing the threats as well as our assets. The consists of introduction, four sections and conclusions. In Sects. 2 and 3 we have propositions of characteristics functions which support marketing decision making and in Sect. 4 we have example exploiting proposed strategic.

Keywords: Strategic games · Uncertain knowledge · Threat classification ·
Payoff estimation parameters

1 Introduction

In general, we treat threat parameters as quantities within specific ranges but on one axis. The introduction of additional dimensions does not change the way they are defined in a significant form. This is how other authors refer to this problem [1]. The assessment of these threats can be defined using approximate calculus [2–5] interval [6], fuzzy [7] and deterministic.

However, the definitions of parameters themselves will be fundamentally different because they will be adapted to the type of threat and the significance of its properties and character. We can assume that strategies $\{A, B, C,...\}$, both ours and our

© Springer Nature Switzerland AG 2020
W. Zamojski et al. (Eds.): DepCoS-RELCOMEX 2019, AISC 987, pp. 402–410, 2020.
https://doi.org/10.1007/978-3-030-19501-4_40

oppositionists', are built on the basis of parameters such as: competition (*p1*), trends (*p2*), costs (*p3*), marketing (*p4*), sales (*p5*), and other (*p6*). These are represented by features when assessing players' payoffs. The formal description of these features (*C(i)*, *i* = *1*, ..., *k*) is a priority task and poses considerable difficulties; it must be adapted to a problem under investigation. However, it is worthwhile to use developed methods and methodologies [8]. Nevertheless, the general approach proposed by various authors is worth modifying in order to assess more accurately and precisely the levels of characteristic function values which in turn influence the amount of players' payoffs for different strategies (Fig. 1).

The share or level of significance of a parameter in the structure of a strategy can be defined binary, where zero means insignificant impact (on weight and payoff for a given strategy) and one: significant, for example: A = {*wp1*, *wp2*, *wp3*, *wp4*, *wp5*, *wp6*} = {1, 0, 1, 1, 1, 0}.

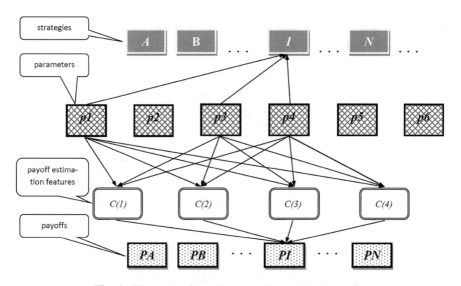

Fig. 1. Hierarchy of creating strategies and their payoffs.

Further considerations suggest some features of parameter evaluation and related formalisms. Let's start with the measurement axis. A one-dimensional structure related to the threat scale is a convenient basis for evaluation of parameters. We can include here ranges of costs, demand, effectiveness of marketing, competitive threats, support for business partners, etc. (example in Fig. 2).

First of all, let's choose a general reference range in which we will use a standardized form of assessment, i.e. we will "move" in range [0, 1] using percentages or fractional values (Fig. 2). Level "0" will correspond to the minimum value of the parameter that is financially, technologically and organizationally possible, and "1" to the maximum value for the current or reasonable foreseeable time period ("0" => min {*Rpi*}, "1" => max{*Rpi*}).

However, we remember that these are discretionary values and it is in our interest to avoid manipulation as early as at this level of analysis.

Rm – the range of parameter values (costs),

$\underline{Rm}, \overline{Rm}$ – the lower and upper limits of the parameter range.

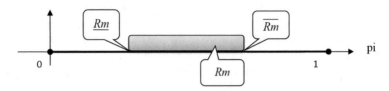

Fig. 2. Range of selected pi parameter values on the threat axis: Rm, where m refers to the strategy of our player.

For the player of our opponent the range is marked as Ro. Using the graphic presentation we have an opportunity to show the zones of the proposed features such as: threaten zone, apprehensive, competition and plausibility zone (Fig. 3).

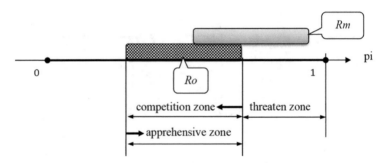

Fig. 3. Zones of estimation of parameter features.

Features relating to their occupation (coverage) are related to the zones described above.

2 Formalisms Concerning Mnemonotechnic Features Constituting the Basis for Estimation of Strategic Parameters and Players' Payoffs

Their following formal definitions are proposed.

2.1 Competition Feature and Supporting Function

$$Cc \sim Rm \cap Ro \text{ and adequate function (Fig. 4)}, \tag{1}$$

$Fc = (Rm \cap Ro)/(Rm \cup Ro),$

$fc = \left(\min\left(\overline{Rm}, \overline{Ro}\right) - \max(\underline{Rm}, \underline{Ro})\right)\big/\left(\max\left(\overline{Rm}, \overline{Ro}\right) - \min(\underline{Rm}, \underline{Ro})\right).$

For negative (*Neg* ∈ {costs, location in relation to the center, competition}) and positive (*Pos* ∈{demand, marketing effectiveness, sales volume, adjustment to trends}) parameters.

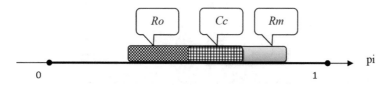

Fig. 4. Feature estimation zones for negative parameters (e.g. costs).

2.2   Apprehensive Feature and Supporting Function

$$Ca \sim (Rm \cup Ro)\backslash Rm \text{ and adequate function (Fig. 5)}, \qquad (2)$$

$fa = (\underline{Rm} - \underline{Ro})/(Rm \cup Ro)$ (for negative parameters) and
$fa = (\underline{Ro} - \underline{Rm})/(Rm \cup Ro)$ (for positive parameters).

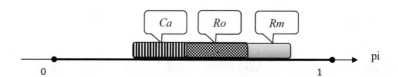

Fig. 5. Feature estimation zones for negative parameters (e.g. costs).

2.3   Threaten Feature and Supporting Function

$$Ct \sim (Rm \cup Ro)\backslash Ro \text{ and adequate function (Fig. 6)}, \qquad (3)$$

$ft = \left(\overline{Rm} - \overline{Ro}\right)\big/\left(1 - \overline{Ro}\right)$ (for negative parameters) and $(Rm \cup Ro)\backslash Rm$ and adequate function $ft = \left(\overline{Ro} - \overline{Rm}\right)\big/\left(1 - \overline{Rm}\right)$ (for positive parameters).

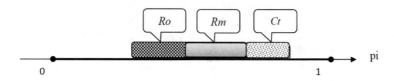

Fig. 6. Feature estimation zones for negative parameters (e.g. costs).

2.4   Plausibility Function

$fp = (\underline{Ro} - \underline{Rm}) + (\overline{Ro} - \overline{Rm})$ (for negative parameters) and

$fp = (\underline{Rm} - \underline{Ro}) + (\overline{Rm} - \overline{Ro})$ (for positive parameters),

$$Fa = (\underline{Ro} - \underline{Rm})/(Rm \cup Ro), Fa = (\underline{Rm} - \underline{Ro})/(Rm \cup Ro). \tag{4}$$

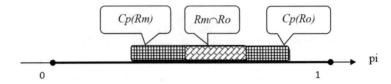

Fig. 7. Feature estimation zones for negative parameters (e.g. costs).

Attention should be paid to polarization of parameters and their features. This is reflected in conversion of signs of the supporting functions. For example, a negative sign of the negative value of the feature "threaten" or "apprehensive" transforms the feature into the group of positives $\{Neg; F < 0\} \rightarrow \{Pos; |F|\}$ and $\{Pos; F < 0\}$ $\rightarrow \{Neg; |F|\}$. After the polarization process we obtain sets of negative and positive features (with positive values of the supporting functions: $\{C-(i), i = 1, ..., n-\}$ and $\{C+(i), i = 1, ..., n+\}$ where $n-$, $n+$ - the numbers of negative and positive features after polarization.

3   Strategy Evaluation Algorithm and Indirect Indicators

In order to comprehensively illustrate polarization effects in relation to particular parameters we will use the $MP(t, j), t = 1, ..., k, j = 1, ..., m$ polarization matrix, where k – the number of features, m – the number of parameters. The example of a polarization matrix with negative (−) and positive (+) features is as follows.

$$MP = \begin{pmatrix} - & + & + & - & - & + \\ + & - & - & + & - & - \\ - & - & + & + & - & + \\ + & - & - & + & + & + \end{pmatrix}, \tag{5}$$

where −, + negative or positive polarization effect.

In order to take into account the affiliation of a parameter to the strategy, we will use the $MI(i, j), i = 1, ..., n, j = 1, ..., m$ inclusion matrix, where n – the number of strategies, m – the number of parameters.

An example of such a matrix looks as follows:

$$MI = \begin{pmatrix} 1 & 0 & 0 & 1 & 0 & 1 \\ 0 & 1 & 0 & 1 & 0 & 0 \\ . & . & . & & . & . \\ 0 & 0 & 1 & 1 & 0 & 0 \end{pmatrix}, \tag{6}$$

Another goal is to develop a payoff [9–14] valuation estimator which is the same for positive and negative components, which will increase (or decrease) as the level of threats increases (or decreases) and will not exceed unity. The Bayesian formula [5] for estimation of the conditional probability, which will be adapted for the structure of features and parameters, will be helpful in the form of:

for $k^{+(-)} = 1$

$$e^{+(-)} = f^{+(-)}(1)$$

for $k^{+(-)} = 2$

$$e^{+(-)} = \frac{f^{+(-)}(1) + f^{+(-)}(2) - f^{+(-)}(1) * f^{+(-)}(2)}{f^{+(-)}(1) + f^{+(-)}(2)}$$

for $k^{+(-)} = 3$

$$e^{+(-)} = \frac{f^{+(-)}(1) + f^{+(-)}(2) + f^{+(-)}(3) - f^{+(-)}(1) * f^{+(-)}(2) - f^{+(-)}(1) * f^{+(-)}(3)}{f^{+(-)}(1) + f^{+(-)}(2) + f^{+(-)}(3)}$$
$$+ \frac{-f^{+(-)}(2) * f^{+(-)}(3) + f^{+(-)}(1) * f^{+(-)}(2)f^{+(-)}(3)}{f^{+(-)}(1) + f^{+(-)}(2) + f^{+(-)}(3)} \tag{7}$$

and next in the counter appear further components which are products of the combination of $cc = \binom{t}{s}$ selected (accordingly to the MP array) of supporting functions, where t – index changing from 1 to $k^{+(-)}$, and s – parameter of the next combination set; $1 \leq s \leq t$, $f^{+(-)}$ – strategic characteristics function for positive (negative) polarization, $e^{+(-)}$ – estimator of payoff for positive (negative) polarization.

For the parameter q, the formal notation of a positive features level will take the form:

$$PP^+(q) = \frac{\displaystyle\sum_{\substack{t=1 \\ MP(t,q)="+"}}^{k+} (-1)^{t-1} \sum_{s=1}^{t-1} \prod_{\substack{komb \\ cc=\binom{t}{s}}} f^+(cc,q)}{\displaystyle\sum_{\substack{t=1 \\ MP(t,q)="+"}}^{k+} f^+(t,q)}, \tag{8}$$

and correspondingly for negative features:

$$PP^-(q) = \cfrac{\displaystyle\sum_{\substack{t=1 \\ MP(t,q)="-"}}^{k-} (-1)^{t-1} \sum_{s=1}^{t-1} \prod_{\substack{komb \\ cc=\binom{t}{s}}} f^-(cc,q)}{\displaystyle\sum_{\substack{t=1 \\ MP(t,q)="-"}}^{k+} f^-(t,q)}, \tag{9}$$

Moving on to strategies, we get an additional index, i.e. strategy code u and the associated matrix of affiliation (importance, materiality) of a parameter to the strategy (*MI*). Ultimately, we have two elements of threats and assets, namely negative and positive impact (10). We treat the factors of these components as independent because the parameters listed as elements of the strategy may fulfill this role.

$$PS^+(u) = \prod_{\substack{q=1 \\ MI(q,u)\neq 0}}^{m} PP^+_{(q,u)}, \quad PS^-(u) = \prod_{\substack{q=1 \\ MI(q,u)\neq 0}}^{m} PP^-_{(q,u)}, \tag{10}$$

Ultimately, the baseline indicator that directly influences the amount of payoff for the given strategy u will be:

$$PB(u) = PS^+(u) - PS^-(u). \tag{11}$$

4   Example Analysis of the Effect of Strategy Selection

An exemplary situation is illustrated for the data presented in Fig. 6. The following parameters relevant to strategy X were selected := {competition, costs, marketing} [15–17]. For each parameter the upper and lower limits of its level range were defined. The values of supportive functions for four mnemotechnic features: competition, apprehensive, threaten and plausibility (*fc, fa, ft, fp*) were calculated according to the methodology presented above.

The application of the algorithm to evaluate the payoff of strategy X is illustrated in Fig. 7.

Table 1. Strategy described by three parameters $X = \{1, 0, 1, 1, 0, 0\}$ = {competition, 0, costs, marketing, 0,0} using interval variables.

competition								
	Rm		Ro		fc	fa	ft	fp
	low	up		up				
	0.37	0.69	0.32	0.77	0.7111	-0.111	0.258	0.0666
negative					0.7111		0.258	0.0666
positive						0.1111		

costs								
	Rm		Ro		fc	fa	ft	fp
	low	up	low	up				
	0.56	0.72	0.43	0.77	0.4705	0.382	-0.178	0.235
negative					0.4705	0.382		0.235
positive							0.178	

marketing								
	Rm		Ro		fc	fa	ft	fp
	low	up	low	up				
	0.66	0.81	0.52	0.75	0.3103	-0.482	-0.315	-0.689
negative					0.3103			
positive						0.482	0.315	0.689

Table 2. Payoff estimation for strategy $X = \{1, 0, 1, 1, 0, 0\}$.

competition	PP+	PP-
negative		0.7723
positive	0.1111	

costs	PP+	PP-
negative		0.6894
positive	0.178	

marketing	PP+	PP-
negative		0.3103
positive	0.5987	

X={1,0,1,1,0,0}	PS+	PS-
positive	0.0118	
negative		0.1652

X={1,0,1,1,0,0}	PB
positive-negative	-0.153

5   Conclusion

The proposed approach is based on an analysis of specific strategic features that have a significant impact on players' payoffs and ultimately on the performance of the market game. Therefore, it is a device allowing to pre-test and predict the development of the situation in real conditions. We can consider an unlimited number of economic parameters and their functional features. The paper proposes four features: competition, apprehensive, threats and plausability and the possibility to assess their scale (fc, fa, ft, fp), separately for each of the selected economic parameters or comprehensively, i.e. in relation to their set. This approach can easily be algorithmized and implemented as part of the system of analysis, planning, threat assessment or organization.

References

1. Piegat, A.: Fuzzy Modeling and Controlling. Akademicka Oficyna Wydawnicza Exit, Warsaw (2003)
2. Pawlak, Z.: Rough sets and fuzzy sets. Fuzzy Sets Syst. **17**(1), 99–102 (1985)
3. Beynon, M., Cosker, D., Marshall, D.: An expert system for multi-criteria decision making using Dempster-Shafer theory. Expert Syst. Appl. **20**, 357–367 (2001)
4. Dempster, A.P.: Upper and lower probabilities induced by a multi-valued mapping. Ann. Math. Stat. **38**, 325–339 (1967)
5. Dempster, A.P.: A generalization of Bayesian inference. J. Roy. Stat. Soc. Ser. B. **30**(2), 208–247 (1968)
6. Jaulin, L., Kieffer, M., Didrit, O., Walter, E.: Applied Interval Analysis. Springer, London (2001)
7. Łachwa, A.: Fuzzy World of Files, Numbers, Relations, Facts, Rules and Decisions. Academic Publishing House Exit, Warsaw (2001)
8. Curiel, I.: Cooperative Game Theory and Applications: Cooperative Games Arising from Combinatorial Optimization Problems, vol. 16. Springer Science & Business Media, Dordrecht (2013)
9. Kałuski, J.: An n-person stochastic game with coalitions. Int. Sci. J. Kibernetika i sistemnyj analiz **3**, 90–100 (2002). National Academy of Science of Ukraine
10. Kałuski, J.: Game theoretical model of the assembly line balancing problem. In: Game Theory and Applications, III, pp. 39–51. Nova Science Publishers, Inc., New York (1997)
11. Osborne, M.J., Rubinstein, A.: A Course in Game Theory. The MIT Press, Cambridge (1994)
12. Owen, G.: Game Theory. PWN, Warsaw (1982)
13. Tyszka, T.: Conflicts and Strategies: Some Applications of Game Theory. WNT, Warsaw (1978)
14. Myerson, R.B.: Game Theory. Harvard University Press, Cambridge (2013)
15. Qin, Z., et al.: E-commerce Strategy, pp. 1–33. Springer, Heidelberg (2014)
16. Ptak, A., Bajdor, P., Lis, T.: The use of social media in European Union Enterprises - comparative study. In: The International Academic Forum (IAFOR), The Asian Conference on the Social Sciences (ACSS 2016), Kobe, Japan, 09–12 June 2016, pp. 299–311 (2016)
17. Skalen, P., Hackley, C.: Marketing-as-practice. Introduction to the special issue. Scand. J. Manage. **27**, 189–195 (2011)

Cost Analysis of Water Pipe Failure

Katarzyna Pietrucha-Urbanik$^{(\boxtimes)}$
and Barbara Tchórzewska-Cieślak$^{(\boxtimes)}$

Faculty of Civil and Environmental Engineering and Architecture,
Rzeszow University of Technology,
Al. Powstańców Warszawy 6, 35-959 Rzeszow, Poland
{kpiet, cbarbara}@prz.edu.pl

Abstract. In recent years, considerable development of water supply system is observed, which, unfortunately, do not protect against the failure occurrence of the water network. The water network constitutes a large part of water company assets. Failures are inseparably related to costs of their removal. Water companies are obliged to deliver potable water according to existing regulations. As to ensure water supply to every recipient at any time, they have to regard the costs of investment of failure network removal. If failure removal costs are not too high, appropriate action should be taken, which will improve safety by reducing the risk of the failure occurrence. Therefore in the paper, the detailed cost analysis was presented, based on real data regarding water network functioning.

Keywords: Water network · Cost of failure removal · Water supply

1 Introduction

Repair and failure removal costs of water supply systems are the heavy burden for the water company. The consequence of failure is lack of supplied water, causing financial losses for the water company. Also, the water recipients are more aware of their rights, which causes that the water supply market must be more adapted to the needs of the consumers, who bear the costs of services regarding water supply and therefore have the right to receive an adequate level of service.

Analysis of the failure removal cost constitutes a trigger to choose such investment projects that will be associated with maintaining reliable operation of water supply to consumers and allow perspective planning of the water supply system modernization [1, 10, 14, 17]. Conducting cost analysis will also adjust the level of reliability to the existing economic conditions [6, 33, 36]. The methodology for determining the costs of repairing the network and water supply fittings was presented in [8, 9]. The assumption of the method presented in [9] is that knowing the failure rates and the unit repair costs for damages for the type and pipe diameters, it is possible to determine, in relation to one year, the stream of repair costs for pipe of diameter d and length L.

The assessment of the consequences of the water pipes failure was presented in, e.g. [25, 28, 31]. Losses associated with the occurrence of undesirable events in water distribution systems can be divided into financial losses of the water producer incurred

© Springer Nature Switzerland AG 2020
W. Zamojski et al. (Eds.): DepCoS-RELCOMEX 2019, AISC 987, pp. 411–424, 2020.
https://doi.org/10.1007/978-3-030-19501-4_41

by water companies directly related to water losses as a result of failures or leaks and water consumers losses. Water producers bear the costs of restoring the subsystem for proper operation, through failure removal, network disinfection and possible compensation for not supplied water [27]. Losses of water consumers concern the possibility of health or life losses often described by means of linguistic variables and assessments of the so-called health risk, domestic and economic difficulties and related costs of buying bottled water.

Losses may also affect industry, as a result of not providing water needed for production [5, 12, 13]. From the point of consumers and water producers view, consequences of the water pipes failure depend on the frequency and duration of failure. The estimation of the duration of failure from the moment a failure to the point of its removal was based on the failure protocols. Waiting time for repair includes the acceptance of the application, call services, selecting equipment, preparing to leave, access to the location of the failure, which depends on the distance and weather conditions. Next, the location of the failure and its identification occur, whether it is a mechanical failure or pitting, on which pipe depending on the material and the size of the leak. Waiting time for repair usually lasts from half to two hours. The proper renewal time includes closure of valves, trenches and preparatory works, and depending on the type of material, the process of welding.

Also important are water losses due to the so-called hidden failures as leaks, during which water gets into the ground and depend on the type and the duration of outflow, as well as pressure in the area of failure [2, 7, 11]. Additional losses are related to the failure of the water supply network, which are connected to the necessity of flushing the pipelines after repairing them. That is why it is so important for the risk-reduction process to take into account the analysis of profits and incurred costs. It is necessary to specify the risk level at which the costs of further reduction are incomparably high [23, 24, 29, 30, 32, 38].

Failures of the water supply system constitute a serious problem in the process of network operation. Pipe failures can concern pipe itself, as well as connections and fittings, as valves, fire hydrants, vent and drain [3]. Failure may be due to ageing, corrosion, pipeline operation conditions or inadequate maintenance. Failure occurrence can be a result of random events, human intervention or emergence of combined factors [20, 39]. Also, external factors as undesirable environmental conditions through soil, flooding, frost, drought or seismic activity can lead to pipeline deterioration, through connection damage, low crack resistance, brittleness. [4]. The reasons for the failure of water supply network can also be wrong concept of the network structure, poorly chosen hydraulic conditions of the network operation, such as too high operating pressure, lack of fittings protecting against hydraulic impact, soil corrosion, temperature changes [15, 19, 21, 26, 37]. Also crucial problems constitute the leaching of microorganisms from the biofilm and the biological stability of water [34, 35].

The cost of water pipe failure removal constitutes a large part of water company expenses. As to minimize potential cost of failure removal it is important to take the decision about providing corrective (reactive) or preventive (proactive) approach, as to perform the most proper maintenance activities. Usually, the option of corrective maintenance is incorporated when the failure rate of given segment of pipe reaches the critical value. In the case of the preventive approach, it can be considered as inspection, overhaul, upkeep or servicing.

The following factors influence the cost of water supply system repair [18]:

- type of pipe material,
- failure size,
- operating conditions,
- place of failure occurrence, the necessity to restore failure place to its initial state, depending on where the pipe is, e.g. under road, footpath or verge,
- whether water company has a repair brigade,
- removing the potential losses associated with the pipe failure resulting from land flooding or landslides.

The aim of the presented analysis is to characterize the water network cost associated with failure occurrence.

2 Materials and Methods

The cost of pipe failure removal can be defined in the following way, as presented in Fig. 1.

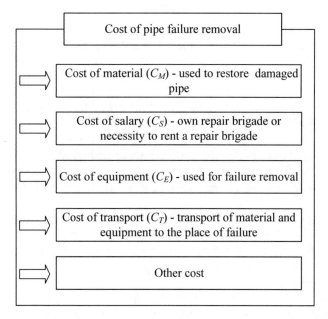

Fig. 1. Costs incurred as a result of water pipe failure.

In detail the aforementioned costs were determined in the following way:

- the cost of material, determined in the division to mains, distributional pipes, and water supply connections,
- the cost of salary, including cards of employees and masters, including rates per working,
- the cost of equipment, including both the price of the work equipment, such as excavators, and equipment used for failure removal, e.g. pumps. They were calculated by multiplying the number of hours of work equipment and the unit price for the machine hours of work equipment,
- the cost of transport, determined on the base of the distance that the vehicles had to pass from the workshop to the place of failure and the unit price per kilometre for each car, taking into account all the vehicles involved in the failure removal.

The cost of removing one failure can be defined as [6, 8, 9, 17, 28]:

$$C_n = C_s + C_t + C_m + C_l + o_{se} \tag{1}$$

where C_s is the cost of salary, C_t is the cost of transport, C_m is the cost of material, C_l is the cost of water losses, o_{se} is the cost of overhead of supply expenses.

The cost of overhead of supply costs o_{se} can be calculated by the formula [6, 8, 9, 17, 28]:

$$o_{sc} = i_{sc} \cdot C_m \tag{2}$$

where i_{sc} is the cost of overhead of water supply cost.

The cost of water losses C_l is determined according to the formula [6, 8, 9, 17, 28]:

$$C_l = Q \cdot t \cdot C_{tp} \tag{3}$$

where Q is the amount of water losses, t is the time of leakage (the duration of the failure), C_w is the tariff price of water.

Other costs associated with failure repairs concern the cost of restoring the failure place to the state before failure, as well as costs of preliminary works including the separation of the failure place, location of water pipe failure, industry supervision, approval and acceptance inspection, marking place in the case of traffic organization change.

Labour costs were calculated on the basis of reports on water supply failures and labour sheets of employees taking part in the removal of these failures. The costs for the course of the vehicles owned by the municipal enterprise associated with failure repairs of water network were calculated on the basis of the failure report attached to the road cards of cars.

Additional costs related to the unreliability of the water supply system are costs of maintenance brigades. As well as water loss costs other than the unreliability of the system, so-called unavoidable water loss occurring on the water network or during rinsing the network.

The annual unreliability of the water storage systems includes the total annual water losses as a result of the water supply disruption to the recipients related to the costs of water production (treatment costs, energy costs, materials, etc.) and gross water sales costs.

In order to determine the annual repair costs for the unit of the i-th length of the water supply system, the cost index of damage repair C_{ci}(in $EUR \cdot km^{-1} \cdot a^{-1}$) was calculated as the product of the failure rate of the water supply network expressed in $km^{-1} \cdot a^{-1}(\lambda)$ and the average cost of liquidation of one i-th failure of water supply network (c_{navgi}) [8, 9]:

$$C_{ci} = \lambda \cdot c_{navgi} \tag{4}$$

Next, the total overall cost of repairs, expressed in $EUR \cdot km^{-1}$, of i-th network with the length l was determined, which was expressed as the product of the cost indicator of damage repair and length l of i-th network [8, 9]:

$$C_{zi} = \frac{C_{ci}}{\sum_{i=1}^{n} l_i} \tag{5}$$

As to estimate the costs of repair of the water network the information obtained from water company was used. The examined city is located in Poland's Subcarpathian province, in south-eastern Poland, it covers a total area of 49 km^2 with 50 thousands inhabitants.

3 Results and Discussion

In Fig. 2 the range of the failure rate for mains, distribution pipes, water-supply connections and the whole network is presented.

The standards of failure rate were proposed in the following research [16, 22]. The standards proposed in [22] stated that failure rates should not exceed the following values, in case of mains less than 0.3 no. of failures/(km$^{-1} \cdot a^{-1}$), for the distributional pipes λ_D should be less than 0.5 no. of failures/(km$^{-1} \cdot a^{-1}$), and for water connections $\lambda_{WC} \leq 1.0$ no. of failures/(km$^{-1} \cdot a^{-1}$). In the work [16] the following proposals of classification of criteria values of failure rates for the whole water network were presented and classified in terms of its reliability, high failure rate concerning low reliability when $\lambda \geq 0.5$ no. of failures/(km$^{-1} \cdot a^{-1}$), high reliability when $\lambda \leq 0.1$ no. of failures/(km$^{-1} \cdot a^{-1}$), and average reliability between mentioned criteria values $0.1 < \lambda < 0.5$ no. of failures/(km$^{-1} \cdot a^{-1}$).

According to presented data in the Fig. 2 mains and water supply connections respectively $\lambda_{WCavg} = 0.573$ no. of failures/(km$^{-1} \cdot a^{-1}$) and $\lambda_{WCavg} = 0.465$ no. of failures/(km$^{-1} \cdot a^{-1}$) are most often damaged and the distribution is the least often damaged ($\lambda_{Davg} = 0.334$ no. of failures/(km$^{-1} \cdot a^{-1}$). The occurrence of more failures in mains and water supply connections is related to the age and type of material from

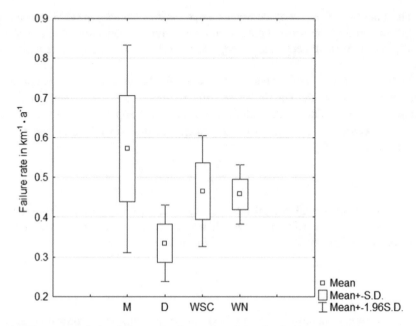

Fig. 2. The range of the failure rate, for: mains (M), distribution pipes (D), water-supply connections (WSC) and the whole network (WN).

which they are made. Over the years, the failure rate of distributional pipes is reduced. This testifies to conducting modernization of the water supply network in the city.

In Figs. 3, 4, 5, 6 and 7 the range of the failure removal cost per one failure associated with material, salary, equipment, transport and other costs including, for example, restore the area to its original state after a failure related to the individual failure in the examined water network, is shown.

Considering the labour costs, the largest median per single failure was recorded for the mains, amounting to EUR 116.02, compared to the average value of 168.70 EUR. In turn, in the distribution pipes the median of labour cost was EUR 80.71 (on average 114.83), while the lowest labour costs were for the failure of water supply connections with the median amounting to 42.88 EUR and an average of 63.29 EUR. In total, for the entire water supply network, the median of labour for single failure amounted to 60.53 EUR and on average of 92.31 EUR.

In the case of transport costs, the highest median was recorded in the case of water main failure, amounting to 24.14 EUR per one failure, on average of 33.5 EUR, compared to other pipes, as distributional pipes and water supply connections, the medians were, respectively, 16.91 EUR (on average 30.49 EUR) and 8.69 EUR (on average 18.01 EUR).

Similarly, the costs incurred for failure repair equipment were distributed, the largest costs of equipment occurred in the case of mains, on average of 176.30 EUR, similar equipment costs occurred on distribution lines, on average 118.80 EUR and for water connections 100.66 EUR.

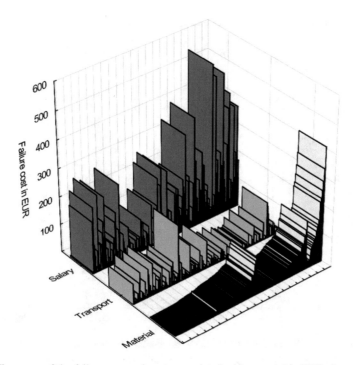

Fig. 3. The range of the failure removal costs associated with material in EUR, for: mains (M), distribution pipes (D), water-supply connections (WSC) and the whole network (WN).

Also, the largest material costs generate failures occurring in water mains. This may be related mainly to large diameters of main pipes, and usually longer sections of replaced or repaired pipes.

Both the sum and average costs of materials used for removing failures in distribution pipes (on average of 39.65 EUR) and water supply connections (on average 23.22 EUR) were significantly smaller than in the case of mains (on average of 149.35 EUR). The median of the cost of water supply materials was 25.80 EUR. The low cost of materials used in the removal of failures is caused by the fact that failures on connections usually have short sections to replace and work on restoring the area to the initial state does not generate relatively significant costs.

The results of the total failure removal costs are shown in Fig. 8.

The average cost of removing a single failure for the mains amounted to 511.76 EUR, for distribution pipes 271.42 EUR, and for water supply connections 159.64 EUR. High costs of removing failures in mains are associated with higher costs of surface restoration after a failure, which often constitute a significant percentage of the total cost of failure removal.

In the Fig. 9 the percentage of individual components in the failure cost removal are summarized.

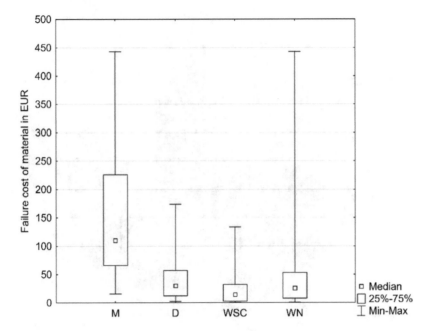

Fig. 4. The range of the failure removal costs associated with material in EUR, for: mains (M), distribution pipes (D), water-supply connections (WSC) and the whole network (WN).

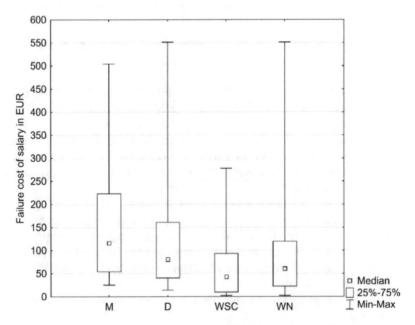

Fig. 5. The range of the failure removal costs associated with salary in EUR, for: mains (M), distribution pipes (D), water-supply connections (WSC) and the whole network (WN).

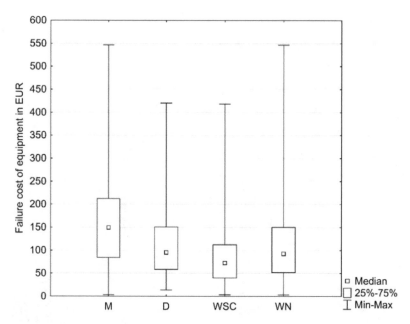

Fig. 6. The range of the failure removal costs associated with equipment in EUR, for: mains (M), distribution pipes (D), water-supply connections (WSC) and the whole network (WN).

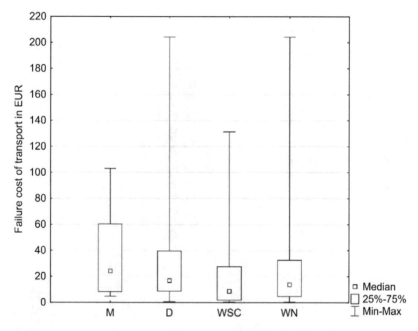

Fig. 7. The range of the failure removal costs associated with transport in EUR, for: mains (M), distribution pipes (D), water-supply connections (WSC) and the whole network (WN).

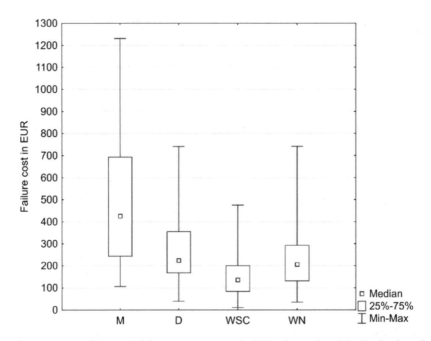

Fig. 8. The range of the total failure removal costs in EUR, for: mains (M), distribution pipes (D), water-supply connections (WSC) and the whole network (WN).

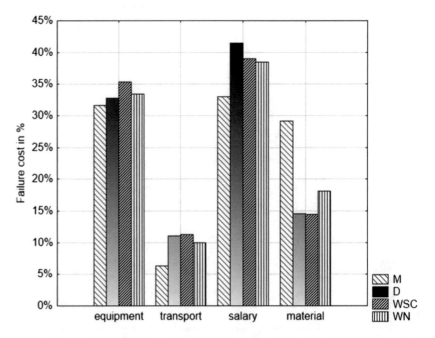

Fig. 9. Percentage of individual components in the failure cost removal in %, for: mains (M), distribution pipes (D), water-supply connections (WSC) and the whole network (WN).

As can be seen, for each types of pipes, labour costs play a significant role in generating failure costs, with the highest share in the case of distribution pipes, as much as 41.48%, smaller for water supply connections (38.96%) and the smallest for mains (32.96%).

Comparing the transport costs related to the failure removing for different types of pipes, it is noted, that the largest share is for water supply connections (11.28%) and distribution pipes (11.08%), while the lowest for mains (6.27%).

As it is to check whether there is a dependence between the costs incurred to remove failures in the form of material, transport and salary costs, the correlation was calculated. In the present case, the correlation between the components of the cost of failure removal is very weak.

Taking into account the percentages of equipment costs, they are at a similar level, for mains 31.58%, distribution pipes 32.83% and water supply connections 35.33%.

The percentage share of material costs is the highest in the case of mains (29.18%), for distribution pipes and water supply connections about 11%. The highest repair costs occurred in mains and amounted to 209.82 EUR $\cdot$ km^{-1} $\cdot$ a^{-1}, lower in distribution pipes (103.14 EUR $\cdot$ km^{-1} $\cdot$ a^{-1}) and in water supply connections (73.43 EUR $\cdot$ km^{-1} $\cdot$ a^{-1}). The total overall cost of repairs amounted to 102.88 EUR $\cdot$ km^{-1}.

4 Conclusions and Perspectives

Water company should increase pipe condition, as to minimize failure occurrence and therefore cost of failure removal. It should be indicated, that costs incurred to failure removal constitute large part of water company expenses, therefore water supply networks should be characterized by high reliability and low operating costs.

Reliable operation of water distribution subsystem depends on minimizing the risk of failure, which can be achieved through the renewal of water network. Proper management of the water supply system must be connected with maintaining its functional and technical properties. After several years of operation, as a result of ageing processes, the failure rate of water network increases. It is up to the managers of the water supply system to choose between removing more failures or perform technical renewal of the pipes through major repairs. It should be mentioned that a significant problem in the operation of water supply systems is the minimization of water losses, and thus ensuring reliability of operation and safety, which should be an expression of good engineering practice of the entire system. Considering the criteria for selecting pipes to be replaced, the water companies exchange network while performing investment works related to the reconstruction of underground infrastructure or the construction of new large-surface facilities or public squares and roads reconstruction. Also the replacement of water pipe occurs during failure, when sections of the network are destroyed to such extent, that they are only qualify for replacement. As to provide preventive approach should implement some inspection programs, also new technology. Decision about the renewal of the waterworks parts should be taken on the basis of economic calculations. Ensuring reliable and safe access to water of appropriate quality often requires high costs for water supply companies. These costs may be related to the minimization of risks related to the possibility of various water accidents occurring in the system, various undesirable events, that may be incidental or may have health-threatening effects or even the lives of water consumers. The quality and safety of supply of tap water to recipients affects certain economic costs, which should be appropriately and transparently considered in relation to the expected benefits. The important element in the process of cost and benefit assessment in a water supply company should be the analysis of acceptance by consumers of measures taken to reduce risk, because they affect the price of water. The acceptance in this case depends to a large extent on the awareness of threats and the expected conditions for ensuring the quality of provided services. The analysis assessment of costs may constitute the possibility as to justify the need of incurring the costs associated with a reduction in risk, and consequently prioritize the various risk control options.

References

1. Boryczko, K., Piegdon, I., Eid, M.: Collective water supply systems risk analysis model by means of RENO software. In: Safety, Reliability and Risk Analysis: Beyond the Horizon. CRC Press, Taylor & Francis Group, Boca Raton, pp. 1987–1992 (2014)
2. Cichon, T., Krolikowska, J.: Efforts to reduce water losses in large water companies. Ekonomia i Srodowisko – Econ. Environ. **60**(1), 161–170 (2017)
3. Debón, A., Carrión, A., Cabrera, E.: Comparing risk of failure models in water supply networks using ROC curves. Reliab. Eng. Syst. Saf. **95**, 43–48 (2010)
4. Gheisi, A., Naser, G.: Simultaneous multi-pipe failure impact on reliability of water distribution systems. Procedia Eng. **89**, 326–332 (2014)
5. Hippe, Z.S., Zamorska, J.: A new approach to application of pattern recognition methods in analytical chemistry - II. Prediction of missing values in water pollution grid using modified KNN-method. Chemia Analityczna **44**, 597–602 (1999)
6. Hotloś, H.: Methods for predicting the costs of water-pipe network repair. Ochrona Środowiska **28**(1), 49–54 (2006)
7. Hotloś, H.: Pressure limitation in water supply system as a factor decreasing damage sensitivity and cost of pipes damage repair. Gas, Water Sanit. Tech. **5**, 180–184 (1999)
8. Hotloś, H.: Quantitative assessment of the effect of some factors on the parameters and operating costs of water-pipe networks. Wrocław University of Technology Publishing House, Wrocław (2007)
9. Hotloś, H., Mielcarzewicz, E.: Failure rate and repair costs of water supply pipelines. Gas, Water Sanit. Tech. **1**, 25–28 (1996)
10. Iwanejko, R., Bajer, J.: Determination of the optimum number of repair units for water distribution systems. Arch. Civ. Eng. **55**, 87–101 (2009)
11. Kowalski, D., Kowalska, B., Kwietniewski, M.: Monitoring of water distribution system effectiveness using fractal geometry. Bull. Polish Acad. Sci. **63**, 155–161 (2015)
12. Kozlowski, E., Kowalska, B., Kowalski, D., Mazurkiewicz, D.: Water demand forecasting by trend and harmonic analysis. Arch. Civ. Mech. Eng. **18**(1), 140–148 (2018)
13. Krolikowska, J., Debowska, B., Krolikowski, A.: An evaluation of potential losses associated with the loss of vacuum sewerage system reliability. In: Conference on Environmental Engineering IV, Lublin, Poland, 03–05 September 2012. Environmental Engineering IV, pp. 51–57 (2013)

14. Krolikowska, J., Krolikowski, A.: Fees for a storm water discharge - needs and possibilities. Rocznik Ochrona Srodowiska **15**, 1143–1152 (2013)
15. Kutyłowska, M.: Prediction of failure frequency of water-pipe network in the selected city. Period Polytechnica Civ. Eng. **61**, 548–553 (2017). https://doi.org/10.3311/ppci.9997
16. Kwietniewski, M.: Failure of water supply and wastewater infrastructure in Poland based on the field tests, In: XXV Scientific-Technical Conference, Międzyzdroje, Poland, 24–27 May, pp. 12–140 (2011)
17. Kwietniewski M., Roman M.: Influence of the degree of reliability of the water supply pipeline on its costs. In: Materials of the International Conference: Issues of Water Supply For Towns and Villages. Publisher PZITS O/Wielkopolski, vol. III, pp. 15–31 (1988)
18. Kwietniewski, M., Roman, M., Trębaczkiewicz-Kłoss, H.: Water and Sewage Systems Reliability. Arkady Publisher, Warszawa (1993)
19. Harbulakova, V.O., Purcz, P., Estokova, A., Luptakova, A., Repka, M.: Using a Statistical Method for the Concrete Deterioration Assessment in Sulphate Environment
20. Pietrucha-Urbanik, K., Tchórzewska-Cieślak, B.: Approaches to failure risk analysis of the water distribution network with regard to the safety of consumers. Water **11**, 1679 (2018)
21. Pozos-Estrada, O., Sanchez-Huerta, A., Brena-Naranjo, J.A., Pedrozo-Acuna, A.: Failure analysis of a water supply pumping pipeline system. Water **8** (2016). https://doi.org/10.3390/w8090395
22. Rak, J.: Bases of Water Supply System Safety. Polish Academy of Science, Lublin (2005)
23. Rak, J.R.: Some aspects of risk management in waterworks. Ochrona Srodowiska **29**, 61–64 (2007)
24. Rak, J.: Selected problems of water supply safety. Environ. Prot. Eng. **35**, 23–28 (2009)
25. Rak, J.R., Tchórzewska-Cieślak, B., Pietrucha-Urbanik, K.: A hazard assessment method for waterworks systems operating in self-government units. Int. J. Environ. Res. Public Health **16**(5), 767 (2019). https://doi.org/10.3390/ijerph16050767
26. Tabesh, M., Soltani, J., Farmani, R., Savic, D.: Assessing pipe failure rate and mechanical reliability of water distribution networks using data-driven modelling. J. Hydroinformatics **11**, 1–17 (2009)
27. Tchorzewska-Cieslak, B.: A fuzzy model for failure risk in water-pipe networks analysis. Ochrona Środowiska **33**, 35–40 (2011)
28. Tchorzewska-Cieslak, B.: Estimating the acceptance of bearing the cost of the risks associated with the management of water supply system. Ochrona Środowiska **29**, 69–72 (2007)
29. Tchórzewska-Cieślak, B.: Matrix method for estimating the risk of failure in the collective water supply system using fuzzy logic. Environ. Prot. Eng. **37**, 111–118 (2011)
30. Tchorzewska-Cieslak, B.: Risk management in water safety plans. Ochrona Środowiska **31**, 57–60 (2009)
31. Tchorzewska-Cieslak, B., Szpak, D.: Proposal of a method for water supply safety analysis and assessment. Ochrona Srodowiska **37**, 43–47 (2015)
32. Tchorzewska-Cieślak, B., Pietrucha-Urbanik, K., Urbanik, M., Rak, J.R.: Approaches for safety analysis of gas-pipeline functionality in terms of failure occurrence: a case study. Energies **11**, 1589 (2018)
33. Wieczysty, A.: Methods of Assessing and Improving the Reliability of Municipal Water Supply Systems. Committee of Environmental Engineering Sciences, Cracow (2001)
34. Zamorska, J.: Biological Stability of Water after the biofiltration process. J. Ecol. Eng. **19**, 234–239 (2018)
35. Zamorska, J., Papciak, D.: Activity of nitrifying biofilm in the process of water treatment in diatomite bed. Environ. Prot. Eng. **34**, 37–52 (2008)

36. Zayed, T., Mohamed, E.: Budget allocation and rehabilitation plans for water systems using simulation approach. Tunn. Undergr. Space Technol. **36**, 34–45 (2013)
37. Zimoch, I.: Pressure control as part of risk management for a water-pipe network in service. Ochrona Środowiska **34**, 57–62 (2012)
38. Zimoch, I., Lobos, E.: Comprehensive interpretation of safety of wide water supply systems. Environ. Prot. Eng. **38**, 107–117 (2012)
39. Zimoch, I., Lobos, E.: Evaluation of health risk caused by chloroform in drinking water. Desalin. Water Treat. **57**, 1027–1033 (2016)

Attack on Students' Passwords, Findings and Recommendations

Przemysław Rodwald$^{(\boxtimes)}$ ⓘ

Polish Naval Academy, ul. Śmidowicza 69, 81-127 Gdynia, Poland
p.rodwald@amw.gdynia.pl

Abstract. Passwords are still the most widespread method of authentication. It is well known and very common for users to create weak passwords. We decided to check the strength of passwords of real systems by cracking MD5 hashes. The results have dismayed us given that 94,94% of passwords were cracked within just a few days. In order to understand the results of cracking better, we asked students about their password conventions, and the strength of selected passwords. We report herein on the most interesting findings as well as their recommendations.

Keywords: Passwords · Cracking passwords · Computer security

1 Introduction

A password, a secret string, remains the widely used method for user authentication in the majority of IT systems. Students and young people aged between 19–23 years old reported by the Pew Research Centre post-millennials [1], are the group which have used them since childhood. On the other hand, they hear about passwords leakages and cracks as a daily basis. This study consists of two main parts. The first one is an attempt to crack the password of one of the real student's system. The second part is a survey about passwords conducted among students. The most interesting findings as well as security recommendations are proposed in the final part.

2 Background

The inherent problem with data security is the human factor. Vulnerability will exist even with the most advanced security mechanisms in place, if there is a human component. Currently, access to restricted systems or data is protected very often by passwords. Therefore, the password is a key element for ensuring security. In this chapter an overview of three password related topics is presented: the commonness of passwords among websites and related pitfalls; methods of password storage and finally password techniques, or hashes and cracking, to be more precise.

W. Zamojski et al. (Eds.): DepCoS-RELCOMEX 2019, AISC 987, pp. 425–434, 2020.
https://doi.org/10.1007/978-3-030-19501-4_42

2.1   The Commonness of Passwords

Passwords with a combination of username (email or nickname) are still the most widespread method of authentication [2]. The commonness of this technique is based on a few factors. Firstly, it is relatively easy to implement: password based authentication mechanisms are widely provided. Secondly, users are accustomed to this: login form with password is almost on every website with user accounts. Thirdly, it is cheap or even free: users do not need any special equipment (hardware token, USB security key, fingerprints reader, etc.). Finally, passwords are convenient and "portable": users can use passwords in any place and on any device. Thus, password-based authentication, especially among websites, will probably remain the predominant security mechanism for the foreseeable future, although some attempts of replacement are under strong development [3]. Rather than using passwords, alternative authentication mechanisms use biometrics (facial recognition or fingerprints), typing patterns, and mobile authentication, to mention just a few.

For years users have had the tendency to choose weak passwords [4]. Furthermore, they reuse passwords across multiple sites (if passwords are leaked from one site, attackers can gain access to other sites effortlessly), share information on social networks (date of birth, maiden name, and pet name is often a password [5]) and inadvertently click on phishing links and provide credentials on fake login forms (phishing techniques are becoming more and more sophisticated).

Large companies are trying to mitigate those bad habits, e.g. by forcing users to use password managers (most popular ones: 1Password, Dashlane, KeePass, LastPass); encouraging to use two-factor authentication (the largest services such as Google, Facebook, Instagram, WhatsApp, Twitter, Apple, Microsoft, Amazon offered 2FA, the White House once had a campaign asking to #TurnOn2FA); proposing global scale solutions (Automatic Strong Passwords provided by Apple [6]).

2.2   Password Storage

The methods of securing passwords stored in IT systems have evolved over the years [7]. Starting from storing passwords as plaintext, through ciphering passwords and finally by the usage of cryptographic hash functions. The last approach is currently the most popular one but the security depends on the hash function used. Usage of fast cryptographic hash functions like MD5 or SHA-1 for storing passwords is not recommended any more, but still many web-based systems are using them [8]. Techniques of strengthening this approach, like salting and multiple iterations, are only short-time mitigation rather than a long-term secure solution. In the current state of art in password cracking, it is strongly recommended to use adaptive password algorithms: computation-hard ones (like bcrypt or PBKDF2) or memory-hard ones (like Argon2 or Balloon). It is much more difficult to attack algorithms from the last group using GPUs or dedicated hardware due to memory requirements.

2.3   Password Cracking

The generic password cracking methodology could be divided into the following steps. Step 1 involves hash extraction. Target hashes must be downloaded, extracted, cleaned, formatted and identified (what hash algorithm is used) depending on the attack. Step 2 involves the calculation of cracking rig capabilities. Depending on the identified hash algorithm and available devices (CPU-based personal PC, GPU-based PS, dedicated rig) the attacker should estimate his capabilities. In the most popular cracking software built-in benchmark command can be used (`john.exe --test` in John the Ripper and `hashcat -b` in Hashcat). Step 3 involves preparing a plan of attack. The attacker is able to adjust the plan of attack to hardware and time constraints based on the benchmark results. Step 4 is the attack itself. Additionally, after successfully cracking a sufficient amount of hashes, the attacker could analyze the results for finding any clues, rules or patterns typical for an analyzing dataset. A more custom attack could be performed based on this knowledge.

Various classes of password cracking methods are reviewed in many papers, for example [9, 10]. Dictionary, brute-force, rule, and mask are the most popular. A *dictionary attack,* known as a *wordlist attack*, uses a precompiled list of words to attempt to match a password. *Rule attack* generates permutations against a given dictionary by modifying, trimming, extending, expanding, reverse, and lower/upper casing, etc. *Brute-force attack*, called *exhaustive key search*, attempts every possible combination of a given character set, up to a certain length. Finally *mask attack* is a form of *brute-force attack* by using placeholders for characters in certain position (i.e. ?u?l?l?l?l?d?d)[1].

3   Cracking Student's Real Passwords

3.1   The Target of the Attack

To check the "quality" of passwords used by students, the author decided to crack real hashes from the system at the Polish Naval Academy called the *Virtual University*[2] (VU). After possessing the required approval from University Authorities, an administrator of this system provided the list of password hashes for the purpose of our research. All hashes are from the database dump of the system dated December 2016 For security reasons. The list contains only hashes, without any additional information (student names, student identifiers, identifiers, etc.). "Users passwords are encrypted by the MD5 algorithms"[3] based on the information provided from the system administrator and system designer (PCG Academia). The target of our attack was the list of 6146 MD5 hashes. Approximately 95% of those passwords belong to 19–24 years old students.

---

[1] A notation used for masks in popular software (Hashcat, JohnTheRipper): ?l - lower-case letters [a-z], ?u - upper-case letters [A-Z], ?d – digits [0-9], ?s – special chars [!"#$%&'()*+,-./:;<=>?@[\]^_`{|}~], ?a – all character set [?l?u?d?s].

[2] It is a component of "Student Information System University" designed by PCG Academia (https://pcgacademia.pl/solutions-for-heis/student-information-system/).

[3] Such a statement (cited from email received from pcgacademia.pl domain) is not fully true, because encryption is a reversible process, while hashing is, by definition, non-reversible.

3.2   Attack

According to the previously mentioned attack methodology step 1, except for the extraction of hashes, encompasses cleaning. The main aim of such cleaning is to remove hash duplicates. The list of 5714 unique hashes was obtained following step 1. Step 2 involves the calculation of cracking capabilities. A dedicated rig with 6 x GPU (MSI GeForce GTX 1080 GAMING X 8G) was used as hardware[4]. The benchmark[5] for MD5 hash algorithm is 100 GH/s[6]. In comparison, a MD5 benchmark for a sample CPU based laptop is 238,1 MH/s[7]; a sample GPU personal computer is 7,7 GH/s. With such computational power, during step 3, the following plan of attack was prepared: 1. *brute-force attack* for mask up to eight chars ?a?a?a?a?a?a?a?a; 2. *mask attack* for the most popular masks coming from the RockYou breach (file with 20560 masks available in hashcat software as file `rockyou-7-2592000_cleaned.hcmask`, but only a mask larger than 8 chars were selected what reduced the set to 6016 masks); 3. *dictionary attack* where three different dictionaries were tested: a. RockYou (14 690 131 words), b. CrackStation (1 212 334 060 words) and c. EvilGhost (10 579 626 523 words); 4. *rule attack* where to the smallest wordlist from the previous attack (Rock-You) Rules Generation option was activated (with parameter 10000). Table 1 provides the statistics, where number of checked words, time of attack and numbers of cracked passwords are presented for all the above mentioned attacks. Time of attack: for *brute-force* and *mask attack* it mainly depends on hardware (GPU), for *dictionary* and *rule attack* it depends on software and the CPU.

As a total result 94,94% of passwords were cracked within just a few days. Some statistics about passwords are similar to other research about password habits. In our attack for example 39% of passwords are 10 chars length, 17% are 8 chars length and 11% are 9 chars length. 43% of passwords apply for the simple mask string digit. A more detailed statistical analysis is available on the dedicated website [11].

4   The Survey and Its Results

The author, concerned about the total number of cracked passwords and the ease of breaking them, decided to check the knowledge of students about password security. The main purpose of this survey was to collect students' opinions on the selected passwords on their strength. At the end of 2018 an anonymous questionnaire among students at Polish Naval Academy was carried out after obtaining the required permission to conduct the survey. The survey consisted of five parts: respondent metric, general questions, difficulty level for passwords categories, particular password strength questions, and one open question. The survey was conducted among both Polish and foreign students, that is why two language versions were prepared. The initial group of questions concerned gender of the responder, faculty, and division.

---

[4] Detailed specification available at https://www.rodwald.pl/blog/1156/hashkiller-1080-spec.

[5] Full benchmark available at https://www.rodwald.pl/blog/1161/hashkiller-1080-benchmark.

[6] GH/s – 10^9 hashes per second.

[7] MH/s – 10^6 hashes per second.

Table 1. Statistics of attacks and its results.

No	Type of attack	Number of checked words	Time of attack [hours]	Number of cracked passwords
1	brute-force	6 704 780 954 517 120	37,2	1733 (30.33%)
2	mask	25 291 710 028 075 184	140,5	3212 (56,21%)
3a	dictionary [RockYou]	14 690 131	0,002	1020 (17.85%)
3b	dictionary [Crackstation]	1 212 334 060	0,08	2984 (52.22%)
3c	dictionary [EvilGhost]	10 579 626 523	0,62	3099 (54.24%)
4	rule	146 901 310 000	47,6	4149 (72.61%)

No personally identifiable information was collected. In the part 1 (general questions), the following questions were asked: Do you use the same password on different websites? How many unique passwords do you have to protect your Internet accounts/profiles? How do you store your passwords? Next, we asked our responders to set the level of difficulty of the password they used (or they would use) for the following categories of Internet services: Financial Services (internet banks, stock exchanges), Social Networks (Facebook, Instagram), Access to email (Gmail, Outlook), Entertainment Services (YouTube, Netflix), E-commerce Services (Allegro, AliExpress). In the third part, the main one, 14 selected passwords[8] were presented and students were asked to setup the difficulty of each one, based on a 5-point scale: Very Weak (grade 1 in figures), Weak (grade 2), Medium (grade 3), Strong, Very Strong (grade 5). Provided passwords were intentionally selected as belonging to one category. A detailed list of passwords is presented in Table 2. In the real questionnaire passwords belonging to one category were not presented one next to the other, the position in the questionnaire is presented as the Order column in Table 2.

At the end of the questionnaire, we asked respondents to provide free-formatted comments on their general thoughts on what a safe password should look like. We were able to get answers from 460 respondents: 410 Polish and 50 foreign students.

The following results present the most interesting findings from part 1 (general questions): 56% of respondents reuse passwords among websites; 52% of students have 2–3 unique passwords, and only 14% have more than 5 unique passwords. 392 respondents (85%) declared that they store passwords in their mind; only 27 students (5.7%) use a special Password Manager like for example KeePass, Last-Pass, 1Password; 96 students (20.9%) remember passwords in the browser; 55 users (11.9%) store passwords on paper. Among the answers, 104 respondents (25%) selected more than one answer. A graphical representation of above the results keeping the proportions of sets is presented on Fig. 1.

---

[8] Passwords vary in both language versions of the questionnaire. For example, the equivalent of password "KasiaNowak" in the Polish version is "SteveJobs" in the English version.

Table 2. The passwords used in the questionnaire.

No	Order	Password	Length	Symbols	Category number	Category description
1	2	1467234690	10	?d	1	only digits
2	11	19860425	8	?d		
3	12	qazwsxed	8	?l	2	keyboard sequence
4	6	Q2W3E4R5	8	?u?d		
5	1	zxcvbn12	8	?l?d		
6	13	KasiaNowak	10	?u?l	3	name and surname (with modifications: LeetSpeak, reverse)
7	3	K@$i@NOw@k	10	?u?l?s		
8	7	KawonAisak	10	?u?l		
9	5	iwonka21	8	?l?d	4	name and digits (with modifications)
10	14	KarOlcia9	9	?l?u?d		
11	4	Aizuz1996!	10	?l?u?d?s		
12	8	RDE1095Ja	9	?l?u?d	5	car plate
13	9	passw0rd	8	?l?d	6	popular password
14	10	Q5!gH$aWd	10	?l?u?d?s	7	random password

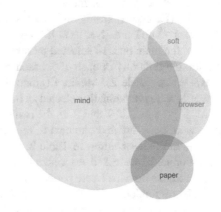

Fig. 1. The Venn diagram showing proportions of answers for the question: How do you store your passwords?

The results of part 2, where students were asked to set a difficulty level for passwords protecting some categories of services, are presented in Fig. 2. It is not surprising that the majority of students set the level of password difficulty for Financial Services Group as Very Strong. But surprising is the fact that the majority of them did not set up the same level for Access to the email Group. Nowadays, an email account is like a cyber-gate to many internet services. In many cases users are able to change the password to other services (Social networks, Entertainment, E-commerce) with access to a compromised email account.

Answers for questions about strength of particular passwords (part 3 of the questionnaire) are most interesting when we compare answers for passwords belonging to the same category. For example, users accurately decided that password "KasiaNowak" (name and surname) is Weak or Very Weak: 423 (92%) what is presented in Fig. 3. However, the same password with LeetSpeak transformation "K@$i@N0w@k" is considered as Strong by most students. What is not true from the crackers point of view, because such a password could be relatively easily cracked with the power of a dictionary attack with common LeetSpeak rules. Similar cracking technique, dictionary attack with simple reverse rule, could be used for an attack on a third password from this group "KawonAisak".

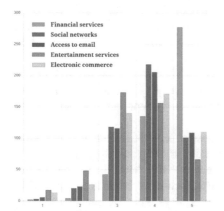

Fig. 2. The numbers of answers for the question about security level of passwords from the group of services (grade scale: 1 - Very Weak, 2 - Weak, 3 - Medium, 4 - Strong, 5 - Very Strong).

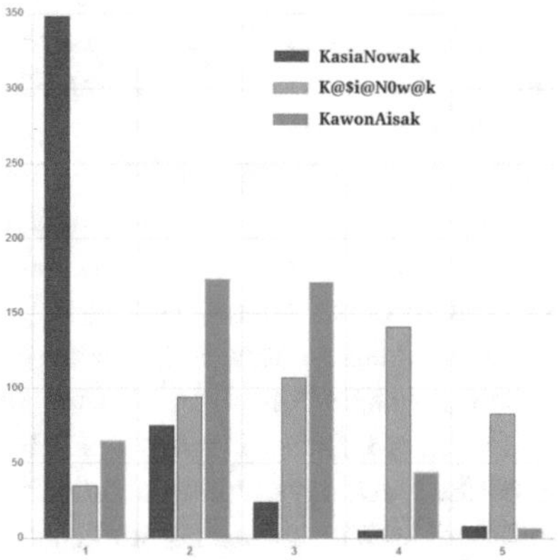

Fig. 3. The numbers of answers for the question about security level of passwords belonging to group 3 (grade scale: 1 - Very Weak, 2 - Weak, 3 - Medium, 4 - Strong, 5 - Very Strong).

For another category of passwords – keyboard pattern passwords (category 3), the answers were correlated with the popularity of the pattern. One of the popular patterns[9] "qazwsxed" as a password was classified as Weak or Medium by 316 (68.7%) students. Passwords "zxcvbn12" and "Q2W3E4R5" were classified by majority of users ($\approx$70%) as Medium or Strong. More details from the survey with correlated results and figures are available on the dedicated website [12].

---

[9] The most popular keyboard pattern is "qwerty".

5   Findings

In recent years, many researchers have investigated password related topics. They focused on: the users' weak password habits [13, 14], the statistical analysis of passwords [15–17], password composition policies [18, 19]. Most of the data based in mentioned articles is obtained by: only survey users or only breaches results. By contrast in this study we firstly measured what students actually do (how they passwords really looks like in presence of weak password policy) and then compared it with what they say.

The ease of cracking passwords is frightening. There are at least three factors behind this. First, people create easy to remember (and easy to guess) passwords. Second, passwords are stored in an un-recommended format, like MD5. Third, the computational power of attackers increases all the time. For example, one of the most up to date GPU Nvidia RTX 2080 FE, dated January 2019, has an efficiency of 50 GH/s for MD5 [20].

88% of students rely mainly on their memory. They have only up to 5 unique passwords. Such an approach makes them vulnerable for loosing access to many systems in case of leakage in one of them. Students, in general, are not aware of the associated risks and repercussions.

Students reuse passwords among many websites. For the direct question about password re-usage "only" 56% of respondents claim that they use the same password on different websites. However, 87% of student say that have up to five unique passwords. Nowadays, young people for sure have much more than 5 accounts/profiles on different websites. Similar studies confirm our findings. For example, in a study of password habits among American consumers [21] 76% 18 to 24-year-old respondents admit to reusing passwords, which is more than any other age group (average is 61%).

More than a half of passwords were cracked in a *dictionary attack*. Users use passwords which could be easily found in available breaches.

An attack also revealed that the average password is between 8 to 10-character long. This is rather an optimistic result, in the absence of a minimum number of characters in the password in the analyzed VU system.

The analyzed VU system as a starting password used user's date of birth. Additionally, the system did not force users to change their initial passwords. Our findings show that students tend to stay on starting passwords (approximately 30%) when a system does not enforce a password change at the first login.

6   Recommendations

The first and foremost recommendation is for computer system designers: the usage of cryptographic hash functions like MD5, SHA-1, SHA-2 family for storing passwords should be definitely stopped [7]. It is strongly recommended to use dedicated, adaptive password algorithms, computation-hard ones or memory-hard ones, like the aforementioned bcrypt, PBKDF2, Argon2, Balloon, etc. Unfortunately, recent research [22, 23] shows that freelancers and developers seldom store passwords securely.

The second recommendation is directed to system administrators. Eight-character length and shorter passwords should be replaced by at least ten-character length passwords, especially in systems where sensitive data are stored (banking systems, e-mail). System administrators should enforce strong password policies, enforce starting password change and monitor users' credentials for compromise.

Students do not understand techniques of password cracking used by hackers which causes a lot of misconception in password strength estimation. One of the most important aspects of the password creation process should be feedback on why a password is weak and what should be done to improve its strength. Currently, the majority of websites deploy password strength meters to provide timely feedback. However, an interesting type of feedback is fear appeals i.e. messages that are intended to increase the perception of a threat and one's ability to cope with it [24]. It is strongly recommended to educate users during the process of password creation/change by providing gentle suggestions on how to improve password strength and the time it would take a brute-force attack to crack a password [25].

Acknowledgments. We have obtained approval for this research from the Rector of the Polish Naval Academy in Gdynia. The author wishes to thank Pawel Wierciszewski, a student, for his help in conducting this survey among students from PNA.

References

1. Dimock, M.: Defining generations: Where Millennials end and post-Millennials begin (2018). http://pewrsr.ch/2GRbL5N. Accessed 13 Jan 2019
2. Bonneau, J., Herley, C., Van Oorschot, P.C., Stajano, F.: Passwords and the evolution of imperfect authentication. Commun. ACM **58**(7), 78–87 (2015)
3. Reilly, M.: Google Has a Plan to Kill Off Passwords (2016). https://www.technologyreview.com/s/601575/. Accessed 13 Jan 2019
4. Bishop, M., Klein, D.V.: Improving system security via proactive password checking. Comput. Secur. **14**(3), 233–249 (1995)
5. Kimmel, J.: What is Your Password? (2015). http://youtu.be/opRMrEfAIiI. Accessed 13 Jan 2019
6. Pugh, C., Abbasian, R., Papadopoulos, H.: Automatic strong passwords and security code AutoFill. In: WWDC 2018 (2018). https://developer.apple.com/videos/play/wwdc2018/204/. Accessed 13 Jan 2019
7. Rodwald, P., Biernacik, B.: Password protection in IT systems. Bull. Mil. Univ. Technol. **67**(3), 73–92 (2018). https://doi.org/10.5604/01.3001.0011.8036
8. Vigilante.pw. https://vigilante.pw. Accessed 13 Jan 2019
9. Rodwald, P.: Choosing a password breaking strategy with imposed time restrictions. Bull. Mil. Univ. Technol. **68**(1) (2019, accepted to print)
10. Picolet, J.: Netmux LLC: Hash Crack: Password Cracking Manual (2017)
11. Password braking PNA (2016). http://www.rodwald.pl/blog/1142/password-braking-pna-2016. Accessed 13 Jan 2019
12. Survey PNA (2018). http://www.rodwald.pl/blog/1136/survey-pna-2018. Accessed 13 Jan 2019
13. Inglesant, P.G., Sasse, M.A.: The true cost of unusable password policies: password use in the wild. ACM (2010). https://doi.org/10.1145/1753326.1753384

14. Shay, R., Komanduri, S., Kelley, P.G., Leon, P.G., Mazurek, M.L., Bauer, L., Christin, N., Cranor, L.F.: Encountering stronger password requirements: user attitudes and behaviors. In: Proceedings of the Sixth Symposium on Usable Privacy and Security, p. 2. ACM (2010). https://doi.org/10.1145/1837110.1837113
15. Dell'Amico, M., Michiardi, P., Roudier, Y.: Password strength: an empirical analysis. In: 2010 Proceedings IEEE INFOCOM, pp. 1–9 (2010)
16. Weir, M., Aggarwal, S., Collins, M., Stern, H.: Testing metrics for password creation policies by attacking large sets of revealed passwords. In: Proceedings of the 17th ACM Conference on Computer and Communications Security, pp. 162–175. ACM (2010)
17. Bonneau, J.: The science of guessing: analyzing an anonymized corpus of 70 million passwords. In: 2012 IEEE Symposium on Security and Privacy, pp. 538–552 (2012). https://doi.org/10.1109/sp.2012.49
18. Komanduri, S., Shay, R., Kelley, P.G., Mazurek, M.L., Bauer, L., Christin, N., Cranor, L.F., Egelman, S.: Of passwords and people: measuring the effect of password-composition policies. In: Proceedings of the SIGCHI Conference on Human Factors in Computing Systems, pp. 2595–2604. ACM (2011)
19. Mayer, P., Kirchner, J., Volkamer, M.: A second look at password composition policies in the wild: comparing samples from 2010 and 2016. In: Thirteenth Symposium on Usable Privacy and Security, pp. 13–28 (2017)
20. Nvidia Gigabyte RTX 2080 TI Hashcat Benchmarks. https://gist.github.com/codeandsec/1c1f2c7bd81abba6fa9736b061944675. Accessed 13 Jan 2019
21. Consumer survey: password habits (2012). http://www.csid.com/wp-content/uploads/2012/09/CS_PasswordSurvey_FullReport_FINAL.pdf. Accessed 13 Jan 2019
22. Naiakshina, A., Danilova, A., Tiefenau, C., Smith, M.: Deception task design in developer password studies: exploring a student sample. In: Fourteenth Symposium on Usable Privacy and Security, pp. 297–313 (2018)
23. Naiakshina, A., Danilova, A., Gerlitz, E., von Zezschwitz, E., Smith, M.: If you want, I can store the encrypted password. A Password-Storage Field Study with Freelance Developers (2019)
24. Vance, A., Eargle, D., Ouimet, K., Straub, D.: Enhancing password security through interactive fear appeals: a web-based field experiment. In: 2013 46th Hawaii International Conference on System Sciences (HICSS), pp. 2988–2997. IEEE (2013)
25. Gamified password change. https://www.edug.pl/password.php?form=game&lang=en. Accessed 13 Jan 2019

Algorithm-Aware Makespan Minimisation for Software Testing Under Uncertainty

Jarosław Rudy[✉]

Department of Computer Engineering, Wrocław University of Science
and Technology, Wybrzeże Wyspiańskiego 27, 50-370 Wrocław, Poland
jaroslaw.rudy@pwr.edu.pl

Abstract. In this paper a modification of the parallel machines scheduling problem with machines capacity for the purpose of running software tests in a cloud environment is considered. The goal function to minimise is the sum of the makespan and the time needed to obtain the schedule. The processing times of tests are unknown, but can be estimated with the use of the history of execution of previous tests. A mathematical model of the problem is presented and two solving methods, adapted Largest Processing Time algorithm and Simulated Annealing metaheuristics are implemented and compared in a computer experiment using real-life data. The results indicate that the Simulated Annealing method is better up to around 14 000 tests despite its longer running time. Results also show that, when paired with a simple method of estimation of software tests duration, both solving methods are fairly robust, providing schedules with similar quality to the ones obtained without estimation.

Keywords: Discrete optimisation · Parallel machines ·
Machine capacity · Processing time estimation · Metaheuristics ·
Software testing

1 Introduction

Software testing is one of the crucial and most costly stages in software development, often consuming 50% to 60% of the entire project costs [7]. Furthermore, in addition to simple and quick unit tests, some IT projects require more complex testing. One example is software used in telecommunication and Internet of Things where the hardware used is also modelled through software for testing purposes. In result, the tests themselves take more time and the entire testing process gets more complicated.

One way to facilitate the testing process it to move it into the cloud, a concept known as Testing-as-a-Service, or TaaS, cloud computing model [13]. Of course, the process of cloud software testing is very complex, but in this paper we focus on specifically on scheduling of independent software tests.

One testing method is acceptance testing. It is performed less often, for example once a week (which works well with modern agile software development

© Springer Nature Switzerland AG 2020
W. Zamojski et al. (Eds.): DepCoS-RELCOMEX 2019, AISC 987, pp. 435–445, 2020.
https://doi.org/10.1007/978-3-030-19501-4_43

methodologies). In this case, the cloud receives entire set of software tests for a given project at once and the goal is to execute them in the shortest time possible. Such minimisation of makespan for a set of software tests can be thought of as a practical application of the problem of scheduling jobs on identical parallel machines form scheduling theory. This problem is denoted as $P||C_{\max}$ in the Graham's notation and has been long known to be NP-hard in general case [5].

However, minimisation of software testing is not a typical $P||C_{\max}$ problem. First, unlike most theoretical scheduling problems like job shop, where a single machine can only process one job at a time, in the cloud a single computer node (machine) might be able to run multiple programs at once. However, the programs might affect each other, so there is a limit on how many jobs can be run on a single computer, called machine capacity. Second, the duration of each job is unknown, which greatly complicates the problem. However, duration of jobs can be estimated if a history of their previous executions times is available.

Finally, most research on scheduling considers the quality of solutions (*i.e.* makespan) and the time in which they were obtained as two separate aspects. However, in the considered problem the solution can be constructed only after the data (software tests to be executed) arrive in the system. That means the time spent obtaining the solution counts towards the final makespan. In result, there is a trade-off: will the makespan obtained with more computation time will be short enough to make up for the additional time spent on obtaining it?

In this paper we tackle the problem of optimising such "algorithm-aware" makespan for software testing with machine capacities, uncertain job processing times using real data and researching the aforementioned trade-off. For simplicity the paper focuses solely on scheduling aspect of software testing process. The rest of the paper is organised as follows. Section 2 contains brief literature overview. Section 3 presents mathematical model of the problem. Section 4 described two solving methods for this problem. Section 5 presents the results of computer experiments. Finally, Sect. 6 contains the conclusions.

2   Literature Overview

Over the decades a great number of papers in the topic of parallel machines scheduling was published. Thus, in this section we will restrict ourselves to a few examples, most relavant to the topic of the paper.

Optimizing software testing can be very different dependent on the assumptions made, including the level of detail. For example, a 2005 paper by Lastovetsky [9] considers a practical study of accelerating software testing in parallel environment for one existing software. The results shows good speedup efficiency and the effectiveness of the Largest Processing Time (LPT) first dispatching rule. The paper also deals with low-level details and the possibility of jobs being dependant on each other. The processing time of a software test is estimated based on its average processing time in the past, but this might be misleading: even small code changes can vastly change the processing time, especially in some IT project.

One of the variants of parallel machines problem is introduction of the so-called machine eligibility constraints. This means that some jobs cannot be executed on some machines. For such a problem Centento and Armacost [3] showed that in realistic case, the LPT algorithm outperforms the Least Flexible Job first rule be Pinedo. Simple dispatching rules were also used Su *et al.* for minimisation of total weighted tardiness [10]. Aside from simple dispatching rules, metaheuristic methods are also commonly employed for this problem. One example is work by Yu *et al.*, who developed a new decoding method for representation of permutation of jobs for Genetic Algorithm and proved its effectiveness, taking into consideration machine eligibility [12].

While interesting and obviously practical, the concept of machine eligibility does not fully conform to the problem we consider in this paper, as each machine can still only process one job at a time. However, in literature there is also concept of machine capacity. For example, Cheng *et al.* in their work [4] considered practical scheduling problem of parallel batch processing machines with non-identical job sizes (SPBN) and proposed a polynomial approximate algorithm and provided worst-case ratios. However, their work assumes equal capacity for each machine. Wang and Chou considered a similar problem in their paper [11], taking into account release dates of jobs. They successfully employed Simulated Annealing and Genetic Algorithm metaheuristics for this problem.

As for scheduling problems for the purpose of software testing, Binder considered scheduling test cases inside a test suite in order to decrease duration and costs of executing software tests [1]. Minimisation was performed using integer programming. In another paper Zhongsheng and Wang employed a Genetic Algorithm for resource scheduling in cloud testing, proving the viability of this approach. Finally, Zheng *et al.* proposed modified ACO algorithm [14], which made use of testing task dependencies, which was shown to achieve better execution speed and load balancing compared to several other approaches.

To summarise this section, it can be seen that machine eligibilities and capacities are touched upon in the literature. Similarly, scheduling theory is applied for software testing in cloud environments. Moreover, metaheuristic methods are successfully employed in both fields. However, to the best of our knowledge, no such paper considers estimation of uncertain processing times of jobs or extension of the makespan goal function by the running time of the solving algorithm.

3 Problem Formulation

In this section we formalise the problem of optimising the aforementioned software testing process as a specific parallel machine problem with machine capacities, uncertain processing times and modified makespan criterion.

Let $\mathcal{J} = \{1, 2, \ldots, n\}$ be a set of n jobs to be processed. For each job $j \in \mathcal{J}$ let p_j denote its true (actual) processing time and let s_j denote its size. Next let $\mathcal{H}_j = (p_j^1, p_j^2, \ldots, p_j^{h_j})$ be the sequence of h_j historical processing times of job j. This means that job j was previously executed h_j times. We also assume that p_j^1 is the processing time of the most recent execution.

Next, let $\mathcal{M} = \{1, 2, \ldots, m\}$ be the set of m machines. All machines have the same speed (the processing time of a job does not depend on the machine chosen), however, they are not fully identical as each machine i has capacity denoted as c_i (not to be confused with C_j).

To define a schedule of jobs we have to specify two vectors (sequences): S and M. $S = \{S_1, S_2, \ldots, S_n\}$ is the vector of starting times of jobs, while $M = \{M_1, M_2, \ldots, M_n\}$ is the vector of jobs-to-machine assignments. Thus, in schedule (S, M) job j starts at time S_j on machine M_j. Moreover, we can define vector of completion times of jobs $C = \{C_1, C_2, \ldots, C_n\}$, where $C_j = S_j + p_j$.

A job can be processed by at most one machine at a time. However, it might be possible for a machine to process more than one job at a time. The condition is that the total size of all jobs currently processed on a machine can never exceed the machine capacity. In other words:

$$\underset{t}{\forall} \underset{i \in \mathcal{M}}{\forall} \sum_{j \in \mathcal{M}_i(t)} s_j \leq c_i, \tag{1}$$

where t is time and $\mathcal{M}_i(t)$ is the set of jobs processed on machine i at time t.

With schedule (S, M) we can define its basic makespan $C_{\max}(S, M)$ as:

$$C_{\max}(S, M) = \max_{j \in \mathcal{J}} C_j = \max_{j \in \mathcal{J}} S_j + p_j. \tag{2}$$

The basic goal is to obtain schedule $(S*, M*)$ that minimises the makespan:

$$(S^*, M^*) = \arg \min_{S, M} C_{\max}(S, M). \tag{3}$$

Schedule (S^*, M^*) is called the optimal schedule.

Under normal conditions, the jobs in parallel machines problem can be started as soon as they arrive in the system (at time 0 in our case). However, that assumes the schedule is known by time 0. Unfortunately, in our case, we can start the algorithm to obtain the schedule only after jobs have arrived in the system and running the algorithm takes some time. Due to this, jobs cannot be started before the algorithm finishes. This overhead ends up increasing the final makespan depending on the running time of the chosen algorithm. We can model this by modifying our goal function:

$$C_{\max}(S, M, A) = \max_{j \in \mathcal{J}} C_j + t(A), \tag{4}$$

where A is the chosen algorithm and $t(A)$ is its running time. We can call this criterion "algorithm-aware makespan". Then the goal is to obtain triple (S^*, M^*, A^*) that minimises said makespan:

$$(S^*, M^*, A^*) = \arg \min_{S, M, A} C_{\max}(S, M, A). \tag{5}$$

Finally, let us note that the considered problem is more difficult than the classic parallel machines scheduling problem. This is because of additional constraints (machines capacity) and because the processing times of jobs are uncertain. Thus, the considered problem is NP-hard as well.

4   Processing Time Estimation and Solving Methods

In this section we describe the method for estimation of processing times of jobs as well as two inexact methods for solving the considered problem: (1) a modified version of a simple constructive algorithm and (2) a local search metaheuristic.

As mentioned earlier, for each job j, a sequence of h_j historical executions of that job is available. This allows us to estimate the true processing time p_j. Many methods of estimation are possible, but we decided to use the moving average of the k most recent executions, also called the k-rec method. Thus, processing time of job j estimated with k-rec method will be denoted e_j^k and defined:

$$
e_j^k = \begin{cases}
\left\lfloor \frac{1}{k} \sum_{i=1}^{k} p_j^i \right\rceil & \text{if } h_j > k \wedge k > 0, \\
\left\lfloor \frac{1}{h_j} \sum_{i=1}^{h_j} p_j^i \right\rceil & \text{if } h_j \leq k \wedge h_j > 0, \\
D & \text{if } h_j = 0 \vee k = 0,
\end{cases}
\tag{6}
$$

where $\lfloor x \rceil$ is the nearest integer to x and D is the default value used when $k = 0$ or there is no execution history available.

The first of the considered solving methods is modified Largest Processing Time, or LPT, which is deterministic constructive algorithm. For the classical $P||C_{\max}$ problem this algorithm was proven to be $\frac{4}{3} - \frac{1}{3m}$ approximate algorithm by Graham [6] and that this bound was tight. For the considered problem the LPT algorithm needs to be modified. The jobs are sorted in descending order according to their estimated processing times e_j^k. Algorithm assigns jobs to machines one by one. A job is always added to a partial schedule created by assigning previous jobs and jobs once assigned cannot be rescheduled. To schedule job j the algorithm scans all machines. For each machine i value $t_{s_j}^i$ is computed, indicating the earliest time when i will have at least s_j free capacity. For machines with $c_i < s_j$ we assume $t_{s_j}^i = \infty$. The job is scheduled on machine with smallest $t_{s_j}^i$. In the case of ties, a machine with the smallest index is chosen.

The running time of modified LPT method is $O(\text{sort}(n) + nm)$, where $\text{sort}(n)$ is the complexity of sorting n numbers. Since e_j are all integers, it is possible to employ counting sort with running time $O(n)$. Common solutions of using quicksort ($O(n^2)$ worst-case) or merge sort ($O(n \log n)$) are also possible, yielding total running time of LPT from $O(nm)$ to $O(n^2 + nm)$. However, in practice the running time is often below 1 s.

The second employed method is Simulated Annealing, or SA, a local-search metaheuristic. It is a non-deterministic method often used for optimisation in various scheduling problems [2]. The following SA implementation was used:

1. The solution is represented as m permutations. The i-th permutation represents jobs assigned to machine i in the order they will be executed.
2. Starting solution is chosen as the best from 1000 random solutions and one LPT solution.

3. Standard exponential cooling scheme was used, with starting temperature equal to difference between best and worst of the 1000 starting solutions. The cooling factor was set to 0.95.
4. Neighbour candidate solution is obtained by random insert move. Insert works by removing random job from random machine and inserting it into random position at random machine. However, the algorithm guarantees that a job is never placed onto machine with capacity less than size of the job. Thus, only feasible solutions are generated this way. In each iteration 20 neighbours are generated and the best is chosen.
5. Standard acceptance function $exp((C_{\max}(N) - C_{\max}(C))/T)$ was used, where $C_{\max}(x)$ is $C_{\max}$ value of solution x, C is current solution, N is the best neighbour solution and T is the current temperature.
6. After 200 iterations without improvement the temperature was set to half of the starting temperature.
7. The algorithm stops after 10 000 iterations.

The running time of the SA method is $O(LPT(n) + icnm)$, where i is the number of iterations, c is the maximal machine capacity and $LPT(n)$ is the running time of the LPT algorithm for n jobs. Thus, the running time of SA method is between $O(mnc)$ and $(n^2 + cmn)$.

5   Computer Experiment

In this section we describe the computer experiment performed to test the effectiveness of both proposed solving methods. Since no optimal solutions are known for the instances we will be using, we will compare the algorithms to each other and to the results obtained when actual processing times of jobs are used instead of their estimates. All experiments were conducted on laptop with Intel® Core™ i7-6700HQ 2.6 GHz CPU and 7.7 GiB of RAM. All reported time values in all experiments are in seconds.

All computer experiments were carried out using 15 instances generated from real-life data of execution of software testing tasks used in [8]. We divided the history into 15 weeks. At the end of each week we took the most recent execution of each software test seen so far. Those tests (along with their processing time and size) make up our job set $\mathcal{J}$. The previous executions of each test make up the $\mathcal{H}_j$ sets. The size of instances varied from 2 555 to 20 072 jobs. The value D from Eq. (6) was set to 50 s, based on the average job time.

It is more difficult to obtain the data concerning the size (capacity requirement) of each job, however, we assumed that for each instance 20% of jobs were randomly set to have size 4. Remaining jobs had size 1. We also assumed 120 machines, out of which 96 (80%) had capacity 4 (i.e. the machines have more cores, being capable of executing multiple software tests without affecting each other), with the remaining machines having capacity 1.

First, a preliminary experiment was carried out to establish the quality of the chosen k-rec estimation method. Two measures of goodness-of-fit were used:

the Pearson correlation coefficient ρ and the coefficient of determination R^2. In both cases we were comparing the actual processing time of a job with the time estimated (predicted) by the k-rec method. This means ρ measured the strength of linear relationship between p_j and e_j^k, while R^2 measured how well e_j^k explains the variation of p_j. Research indicated that the value of $k = 2$ yielded the best results as shown in Table 1. The value of ρ ranges from 0.756 to 0.909 with the average of 0.859, which indicates fairly strong correlation. As for R^2, the minimal value was 0.568, the maximal was 0.824 and the average was 0.734.

Table 1. Results of Pearson correlation coefficient ρ and coefficient of determination R^2 for estimation through k-rec method with $k = 2$

Instance	ρ	R^2	Instance	ρ	R^2	Instance	ρ	R^2
1	0.877	0.768	6	0.886	0.780	11	0.847	0.714
2	0.900	0.800	7	0.892	0.792	12	0.826	0.679
3	0.850	0.718	8	0.882	0.772	13	0.756	0.568
4	0.834	0.681	9	0.853	0.723	14	0.844	0.709
5	0.909	0.824	10	0.852	0.719	15	0.872	0.758

With the value of k fixed, we carried out the main experiment for comparing the proposed solving methods. Let us start with the LPT method. Since LPT is deterministic and is fast even for large n, the algorithm was run only once, obtaining the basic makespan $C_{\max}$ and the running time of the algorithm $t(LPT)$, which are then summed for the total algorithm-aware makespan.

For the SA method we used a different approach. If we run the algorithm only once, we risk unstable results, because of the non-deterministic nature of the SA metaheuristic. Moreover, we assumed 120 machines in our system (computing cloud), so in that case the remaining 119 machines would be idle. Thus, instead we start multiple runs of the SA method on different machines. Thanks to this, machines will not be idle and we can choose the best result, alleviating the non-deterministic nature of the SA method. In our research we ran the SA method 50 times for each instance. As the value of $C_{\max}$ we choose the best (minimal) out of the 50 obtained makespans and as for the running time $t(SA)$ we choose the worst (maximal) out of the 50 running times.

The result of the experiment, including direct comparison between LPT and SA methods, are shown in Table 2. We can see that despite much longer running time, the SA method can provide shorter total algorithm-aware makespans. The difference between the methods can be large, as in the case of instance 2, where the schedule provided by the LPT method was 77% larger than the one provided by the SA method. Generally, the advantage of SA over LPT decreases as the size of the instance increases. One significant exception to that rule is instance 1, for which both methods are practically equal. This might be caused by the fact that the available history was the shortest for that instance, making estimation of processing times more difficult. As for the remaining instances, we see that for instances 2 through 4 the advantage of SA is high (43% on average) and remains considerable for instances 5 through 8 (10% on average). By instance 12 (n around 14 000) the LPT method starts to surpass the SA method, which is caused by the growing number of jobs per instance.

Table 2. Results of running LPT and SA methods using k-rec estimation ($k = 2$)

Instance	LPT			SA			LPT/SA
	C_{max}	t	$C_{max} + t$	C_{max}	t	$C_{max} + t$	
1 ($n = 2\,555$)	1 222	0.03	1 222.0	1 178	31.6	1 209.6	1.010
2 ($n = 2\,746$)	1 394	0.04	1 394.0	753	33.3	786.3	1.773
3 ($n = 4\,839$)	1 580	0.07	1 580.1	1 206	48.0	1 254.0	1.260
4 ($n = 5\,116$)	1 777	0.10	1 777.1	1 359	51.5	1 410.5	1.260
5 ($n = 6\,421$)	1 890	0.18	1 890.2	1 697	62.8	1 759.8	1.074
6 ($n = 7\,062$)	2 095	0.14	2 095.1	1 765	70.3	1 835.3	1.142
7 ($n = 7\,375$)	1 947	0.27	1 947.3	1 687	85.8	1 772.8	1.098
8 ($n = 9\,499$)	2 444	0.26	2 444.3	2 126	125.0	2 251.0	1.086
9 ($n = 9\,927$)	2 588	0.42	2 588.4	2 307	135.1	2 442.1	1.060
10 ($n = 10\,260$)	2 685	0.47	2 685.5	2 432	148.6	2 580.6	1.041
11 ($n = 11\,337$)	2 708	0.59	2 708.6	2 433	191.8	2 624.8	1.032
12 ($n = 13\,910$)	2 916	0.79	2 916.8	2 979	243.0	3 222.0	0.905
13 ($n = 16\,243$)	3 888	0.80	3 888.8	3 511	301.0	3 812.0	1.020
14 ($n = 19\,638$)	3 901	0.77	3 901.8	3 790	441.4	4 231.4	0.922
15 ($n = 20\,072$)	3 972	5.50	3 977.5	3 719	420.3	4 139.3	0.961

One more experiment was carried out, this time using the actual observed processing times of jobs instead of their estimates. The purpose of this experiment was to compare the quality of solutions obtained with estimates to results obtained from the actual data. The results are shown in Table 3.

First, the analysis of the fourth column supports the conclusions drawn from Table 2 – the SA method performs better up to instance 12 (n around 14 000). Let us also note that number of jobs for instances 1 to 12 is from 2 555 to 13 910. This is much larger than common benchmarks for the job shop problem, which rarely get larger than 50 jobs and 20 machines (yielding 1 000 operations in total). We conclude that the SA method is effective for $n < 10\,000$ and unsuitable for $n > 14\,000$. This effect is also not caused by our estimation method.

Next, we analyse the last two columns of Table 3, in order to see how estimation affects the final results compared to using actual observed data. For the LPT method the results are very good – the algorithm-aware makespan is never more than 4.7% worse compared to using observed data. On average this difference is only 2.1%. For the SA method, the results are worse – for

Table 3. Results of applying the LPT and SA using actual observed values (denoted true) and comparison with results obtained using estimation (denoted krec)

Instance	LPT_{true}	SA_{true}	LPT_{true}/SA_{true}	LPT_{krec}/LPT_{true}	SA_{krec}/SA_{true}
1 $(n = 2\,555)$	1 195.0	842.8	1.418	1.023	1.435
2 $(n = 2\,746)$	1 332.0	948.8	1.404	1.047	0.829
3 $(n = 4\,839)$	1 538.1	1 209.4	1.272	1.027	1.037
4 $(n = 5\,116)$	1 761.1	1 379.0	1.277	1.009	1.023
5 $(n = 6\,421)$	1 848.2	1 652.9	1.118	1.023	1.065
6 $(n = 7\,062)$	2 070.2	1 827.9	1.133	1.012	1.004
7 $(n = 7\,375)$	1 881.3	1 734.9	1.084	1.035	1.022
8 $(n = 9\,499)$	2 480.3	2 109.3	1.176	0.985	1.067
9 $(n = 9\,927)$	2 565.4	2 411.1	1.064	1.009	1.013
10 $(n = 10\,260)$	2 635.5	2 470.1	1.067	1.019	1.045
11 $(n = 11\,337)$	2 649.5	2 600.9	1.019	1.022	1.009
12 $(n = 13\,910)$	2 850.7	2 826.7	1.009	1.023	1.140
13 $(n = 16\,243)$	3 766.7	3 670.0	1.026	1.032	1.039
14 $(n = 19\,638)$	3 839.0	4 045.0	0.949	1.016	1.046
15 $(n = 20\,027)$	3 853.0	4 080.8	0.944	1.032	1.014

instances 1 and 12 using the k-rec estimation provided 43.5% and 14.5% worse schedules than using the observed data. The result for instance 1 is partially explained as the instance suffers from short history of past executions of jobs. The fact remains that instance 12 proved the most difficult, given its size. In all other instances the difference was never higher than 6.7% and with average of only 1.6%.

Let us notice that the increase of makespans due to estimation is relatively small (under 5% in 26 out of 30 cases) even though the estimation itself is not perfect (values of ρ and R^2 were 0.859 and 0.734 on the average, respectively). Thus, we can conclude that both solving methods are fairly robust for this problem when used with the k-recent estimation method.

6   Conclusions and Future Work

In this paper we considered modified parallel machines scheduling problem with machines capacity and uncertain processing times of jobs, modelled after practical problem of running software tests. We presented two solving methods for minimising algorithm-aware makespan for the considered problem. With the use of instances derived from real-life data, we showed that the Simulated Annealing metaheuristic can outperform the Largest Processing Time algorithm for instances up to around 14 000 jobs even with its longer running time. Moreover, we showed that the proposed moving average-based estimation method was effective and robust for this problem, increasing the makespan duration by no more than 5% in all but 4 tested cases.

While the results presented in the paper are promising, they could still be improved by using better optimisation methods or decreasing the running time of those methods. Thus, the next natural stage of this research would be to use other metaheuristic methods, especially hybrid and parallel metaheuristics.

References

1. Binder, R.V.: Optimal scheduling for combinatorial software testing and design of experiments. In: 2018 IEEE International Conference on Software Testing, Verification and Validation Workshops (ICSTW), pp. 295–301 (2018). https://doi.org/10.1109/ICSTW.2018.00063
2. Bożejko, W., Pempera, J., Smutnicki, C.: Parallel simulated annealing for the job shop scheduling problem. In: Allen, G., Nabrzyski, J., Seidel, E., van Albada, G.D., Dongarra, J., Sloot, P.M.A. (eds.) Computational Science - ICCS 2009, pp. 631–640. Springer, Heidelberg (2009)
3. Centeno, G., Armacost, R.L.: Minimizing makespan on parallel machines with release time and machine eligibility restrictions. Int. J. Prod. Res. **42**(6), 1243–1256 (2004). https://doi.org/10.1080/00207540310001631584
4. Cheng, B., Yang, S., Hu, X., Chen, B.: Minimizing makespan and total completion time for parallel batch processing machines with non-identical job sizes. Appl. Math. Modell. **36**(7), 3161–3167 (2012). https://doi.org/10.1016/j.apm.2011.09.061
5. Garey, M.R., Johnson, D.S.: "Strong" NP-completeness results: motivation, examples, and implications. J. ACM **25**(3), 499–508 (1978). https://doi.org/10.1145/322077.322090
6. Graham, R.L.: Bounds on multiprocessing timing anomalies. SIAM J. Appl. Math. **17**(2), 416–429 (1969)
7. Kumar, D., Mishra, K.: The impacts of test automation on software's cost, quality and time to market. Procedia Comput. Sci. **79**, 8–15 (2016). https://doi.org/10.1016/j.procs.2016.03.003
8. Lampe, P., Rudy, J.: Models and scheduling algorithms for a software testing system over cloud. In: Zamojski, W., Mazurkiewicz, J., Sugier, J., Walkowiak, T., Kacprzyk, J. (eds.) Contemporary Complex Systems and Their Dependability, pp. 326–337. Springer International Publishing, Cham (2019)
9. Lastovetsky, A.: Parallel testing of distributed software. Inf. Softw. Technol. **47**(10), 657–662 (2005). https://doi.org/10.1016/j.infsof.2004.11.006
10. Su, H., Pinedo, M., Wan, G.: Parallel machine scheduling with eligibility constraints: a composite dispatching rule to minimize total weighted tardiness. Naval Res. Logistics (NRL) **64**(3), 249–267 (2017). https://doi.org/10.1002/nav.21744
11. Wang, H.M., Chou, F.D.: Solving the parallel batch-processing machines with different release times, job sizes, and capacity limits by metaheuristics. Expert Syst. Appl. **37**(2), 1510–1521 (2010). https://doi.org/10.1016/j.eswa.2009.06.070
12. Yu, C., Semeraro, Q., Matta, A.: A genetic algorithm for the hybrid flow shop scheduling with unrelated machines and machine eligibility. Comput. Oper. Res. **100**, 211–229 (2018). https://doi.org/10.1016/j.cor.2018.07.025. http://www.sciencedirect.com/science/article/pii/S030505481830217X

13. Yu, L., Tsai, W.T., Chen, X., Liu, L., Zhao, Y., Tang, L., Zhao, W.: Testing as a service over cloud. In: Proceedings of the 2010 Fifth IEEE International Symposium on Service Oriented System Engineering, SOSE 2010, pp. 181–188. IEEE Computer Society, Washington, DC (2010). https://doi.org/10.1109/SOSE.2010.36
14. Zheng, Y., Cai, L., Huang, S., Lu, J., Liu, P.: Cloud testing scheduling based on improved ACO. In: International Symposium on Computers & Informatics (ISCI 2015), pp. 569–578 (2015). https://doi.org/10.2991/isci-15.2015.76

Identifying Factors Affecting the Activities of Technology Parks

Elena V. Savenkova, Alexander Y. Bystryakov,
Oksana A. Karpenko[✉], and Tatiana K. Blokhina

Economics Department of Peoples' Friendship University of Russia
(RUDN University), Miklukho-Maklaya Street 6, Moscow 117198, Russia
{savenkova_ev, bystryakov_aya, karpenko_oa,
blokhina_tk}@rudn.university

Abstract. Successful development of technology parks supporting innovations is one of key factors of success of this kind of activity. Technology parks create the favorable conditions for development of manufacturing enterprises of the scientific and technical sphere. At the same time science and technology parks are subject to the risks arising both at the time of their creation, and in the course of their functioning. The main objective of the conducted research is identification of the factors influencing activity of science and technology parks and development of the main actions directed to decrease their negative impact. As a method of identification of risks quantile regression is used. This method is not parametrical, and, therefore, does not assume that distribution size belongs to any parametrical methods. Besides, quantile regression belongs to robust methods, they are steady against emissions. The technology park Academpark was chosen as an object of the research. Following the results of the conducted research authors concluded that the tax burden, the number of the resident companies and number of their employees have the greatest impact on the revenue of science and technology park. The authors offered possible ways for decrease of negative influence of the above-stated factors on the total revenue.

Keywords: Innovative infrastructure · Science and technology park ·
Method of quantile regression · Factors affecting technology park

1 Introduction

Nowadays, there are different types of technology parks and other elements of infrastructure supporting innovative activity: techno polices, zones of development of new and high technologies, scientific and technological parks, the innovative and technological centers and the centers of commercialization of technologies, business incubators.

Science and technology parks are the subjects of innovative infrastructure, they create the conditions favorable for the development of manufacturing enterprises of the scientific and technical sphere. These enterprises have the equipped skilled and experimental base and highly qualified personnel. Science and technology parks provide them a modern infrastructure (for example, buildings, telecommunications, etc.).

© Springer Nature Switzerland AG 2020
W. Zamojski et al. (Eds.): DepCoS-RELCOMEX 2019, AISC 987, pp. 446–455, 2020.
https://doi.org/10.1007/978-3-030-19501-4_44

Development of innovative infrastructure and science and technology parks took place in many countries of the world. However, innovative infrastructure and its elements are not universal. There is a big heterogeneity of spatial forms and trajectories of growth [1, 2]. For example, in Great Britain they use a term "scientific park", in France the similar structure is called as a technopolis, in Germany they use a term "technological park", and in the USA "a research park".

Nevertheless all science and technology parks are always connected with regional economic processes, and therefore there are a lot of factors affecting their activity, for example, the insufficient financing from regional and federal authorities, the low level of attraction of financial resources of private investors, deterioration of the general economic situation in the region and in the country, migration of residents to other science and technology parks or special economic zones, decrease in number of potential projects and numbers of residents of science and technology park. Technology parks are not isolated from regional processes of economic development. So, for example, in a research of European Commission it is claimed that they are closely interconnected with the level of development of the regional industry and scientific potential [3].

Technology parks, as a rule, include the structures which are already created at the regional level and contribute to the development of auxiliary innovative infrastructure [4].

Rosenfeld claims that, technological (innovative) parks – represent active networks of the synergetic organizations which cooperate with various companies in specific industries of the industry [5]. Most of the scientists considers that technological (innovative parks) are included in the local innovative systems (LIS) including territorial organizations that enter into uniform network of the political, economic and public relations [6].

Science and technology parks are subject to a set of the risks arising both at the time of their creation, and in the course of their functioning. Today a little attention is paid to a research of the factors influencing activity of science and technology parks. These factors are the sources of risk. As for the Russian science and technology parks there are a lot of factors which can negatively influence financial results of activity of science and technology park, including the small and medium-sized innovative enterprises. The most sensitive factors of such enterprises are:

- high tax burden;
- deterioration in a situation in the market of highly skilled labor;
- legislative and administrative obstacles in the foreign trade;
- termination of activity of residents of science and technology park or their migration to other special economic zones;
- decrease in access to financial (investment, credit, grant) to resources.

The above-stated list is not exhaustive. There are other factors making negative impact on efficiency of activity of science and technology park and on its financial results.

The main purpose of our article is: to determinate the factors influencing activities of residents of technology park, to chose appropriate methods for definition of these factors and to elaborate possible ways for reduction their negative impact.

We took a technology park Academpark as an object for our research. It is situated in Novosibirsk region in Russia. Today Academpark is included into 12 of priority science and technology parks of Russia and it is one of the most effective.

2 Methodology

In 2013 Academpark became the full member of the European network of the business and innovative centers (EBN) uniting about 250 science and technology parks, business incubators and the venture centers worldwide.

In the course of its development Academpark achieved the highest budgetary efficiency of the projects and also attracted resident companies, created a great number of the jobs and gained a large total revenue.

The main directions of technology park allowing management company to work more purposefully are:

- information and telecommunication technologies;
- biotechnologies and biomedicine;
- knowledge-intensive equipment;
- new materials and nanotechnologies;
- laser systems and photonics.

These directions have the strongest scientific basis in the region thanks to activity of institutes of the Novosibirsk scientific center and also there are very solid and prominent innovative regional companies.

Academpark created a perfect environment for development of the companies of electronics, lighting engineering and chemical technologies.

Main objective of Academpark is ensuring the accelerated development of high-tech industries of economy and their transformation into one of the main driving forces of economic growth of the region.

The local purpose of Academpark is creation not less than 20 successful innovative businesses "from start-up" in a year.

In the Novosibirsk region the technology park has the advanced position in the market of technological services. It creates the structure of services, and conditions (complexity) of their granting. Besides, the value of technology park for the region is defined by its role as institute of development and as center of the advanced competences.

There is a special classification of residents that allows to work more purposefully for the benefit of the residents and to achieve major results of technology park. There are five categories of residents:

- the incubated company;
- accelerated company;
- product company;
- service technological company and service research organization;
- strategic partner.

For the end of 2017 Academpark had the following indicators of development:

- number of residents - 211;
- total number of staff of the accredited companies 4897 persons.
- the budgetary efficiency of the project (the amount of all taxes paid by residents) - 1439 million rubles;
- total revenue from sales of residents - 13150 million rubles (Table 1).

Table 1. Information on the residents in the territory of the Academpark

Name of an indicator	2016 Fact	2017 Plan	Fact	2018 Plan
Share of loading of the areas of Science and technology park by residents, %	99,6	90	98,6	90
Export share in total sales of residents, %	9,0	10,5	9,0	12,0
The budgetary efficiency of the project (the amount of all taxes paid by residents), mln rubles	1 302	1 083	1 128	1 300
Number of the resident companies	212	211	209	211
Number of staff of the resident companies, piece	4 875	4 800	4 897	5 050
Total revenue of residents, mln rubles	13 648	13 600	13 150	14 200
Total area of objects of Science and technology park, sq.m	104 443	114 344	119 501	189 000

Table 2. KPI of science and technology park Academpark

Name indicators/years	2010	2011	2012	2013	2014	2015	2016	2017
The number of the companies–residents	30	85	127	156	171	198	212	209
Number of staff of the companies	218	1395	2868	3830	4500	5540	4875	4897
The number of the projects realized by residents	63	169	249	373	480	530	540	510
Total revenue of residents, mln rubles	100	2 937	5 012	6 460	9 623	11951	13648	13150
The amount of all taxes and insurance premiums paid by residents, mln. rubles	100	300	505	650	962	1255	1590	1439
Export share in total sales of residents, %				6,5	7	8	9	9
Share of loading of the areas by the resident companies				90	99,7	99,7	99,6	98,6

Source: https://academpark.com/about/docs/ [10]

According to the results of 2017 three planned targets are not executed: the export share in total sales of residents remained at the level of 2016, the number of the resident companies was reduced by 3 resident and the total revenue of the resident companies decreased almost by 500 million rubles in comparison with 2016. In general, indicators were rather high (Table 2). The export share in revenue made in 2017 and in 2016 9%.

Rapid development of the project of Academpark, successful performance of construction of facilities, positive dynamics of a return and volumes of services demonstrate negligibility of the risks connected with management. The control system provides necessary stability of the project and takes into account interests of all its residents.

However there are risks connected with insufficient attractiveness of Academpark to the anchor companies. For example, the environment promoting small and medium-sized companies is not created now, moreover, in 2017 there was a decrease in number of small and medium-sized companies – potential partners and customers of anchor residents. It is necessary to take into account, risks of decrease in innovative activity of the residents and the technopark environment in general.

While considering the group of the risks exerting the greatest impact on activity of Academpark we formulated a hypothesis: the level of efficiency of technology park depends on the factors determining revenue, among them are:

- number of the resident companies;
- the number of staff of the accredited companies;
- the amount of the taxes paid by residents;
- the number of the projects which are carried out by residents.

In order to assess the impact of the factors influencing total revenue we took model of quantile regression. This method was for the first time formulated in Roger Koenker's work in 1978 and later was modified for panel data in 2004 [7]. We refused the typical model of the ordinary least squares (OLS) because it is necessary to analyze not only an estimated population mean, but also other conditional distributions.

The results received by the OLS method have the asymptotic efficiency and are equal to normally distributed random variables. However, if there are other kinds of distribution, the asymptotic efficiency of OLS method can be significantly lower. So, for example, in work of Gastwirth and Selwyn [8] is shown that the average weighed quantiles 1/3, 1/2 and 2/3 with scales 0.3, 0.4 and 0.7 respectively has asymptotic efficiency about 80% for normal and logistic distributions and also Laplace and Cauchy's distributions. The average value at the same time has the asymptotic efficiency. However, for Laplace's distribution the asymptotic efficiency is less than 50%, and for Cauchy's distribution it is close to zero.

Besides, OLS results are extremely sensitive to abnormal observations that shift the estimates of these emissions. While creating various statistical procedures every time we proceed from some set of conditions. Generally these conditions concern independence of observations and their identical distribution, the assumptions of the nature of distribution of sample units. In practice the discussed condition are the results of approximation and inevitable idealization. Therefore, these conditions are usually not satisfied, and there are fears connected with correctness of the conclusions drawn by the chosen statistical procedure [9].

The main advantage of quantile regression is that this method is not parametrical, and, therefore, does not assume that distribution of the studied size belongs to any parametrical family. Besides, quantile regression belongs to robust methods, that is it is steady against emissions in observed sizes.

The zero hypothesis of our test consists in equality of influence of various factors on sales of science and technology park Academpark in different quantiles. Several hypotheses were tested in quantiles 0,25; 0,5; 0,75.

Table 3. The quantile assessment, observations of 2010–2017 are used (T = 8) Dependent variable: sales, tau = 0,25. Asymptotic standard errors are considered as independent and equally distributed

	Coefficient	St. errors	t-statistics	P-value
const	−541,135	250,594	−2,159	0,1196
professionals	1,09819	0,244510	4,491	0,0206
firms	0,537088	5,89848	0,09106	0,9332
taxes	7,39068	0,439736	16,81	0,0005
projects	−5,61033	2,53250	−2,215	0,1135

Mean dependent VAR	8041,500	Standard deviation	5012,187
Sum of modules of errors	1772,156	Sum of the sq. residuals	1090358
Likelihood estimation	−53,50554	Criterion Akaike	117,0111
Criterion Schwartz	117,4083	Criterion Hennan-Quinn	114,3321

We carried out quantile assessment for a quantile 0,25. The sales depend on the number of employees (professionals), the companies (firms), taxes (taxes) and the number of projects (projects) (Table 3). The structure of the table includes values of the corresponding coefficient, a standard error and p-value for each of the considered quantiles. Interpretation of the results is as follows: the lower p-value of the factor indicates that it has more influence on revenue. Apparently, as we can see all variables exert impact on the size of revenue of science and technology park differently. So, for example, in a quantile 0,25 high level of dependence of sales from the number of employees-0,537088, at p-value 0,0206, on the same quantile are observed with high dependences on taxation level. The number of projects has an opposite impact on sales. The equation of quantile regression is:

$$sales = -541 + 1,10*professional + 0,537*firms + 7,39*taxes - 5,61*projects$$
$$\quad (251) \quad (0,245) \quad\quad (5,90) \quad\quad (0,440) \quad (2,53)$$
$$T=8, sum\ of\ abs.\ residuals = 1,77e+03 \tag{1}$$
$$(in\ brackets\ standard\ errors\ are\ specified)$$

Let's carry out the analysis for a quantile 0,5 (medians).

Table 4. The quantile assessment, observations of 2010–2017 are used (T = 8) Dependent variable: sales, tau = 0,5. Asymptotic standard errors are considered as independent and equally distributed

	Coefficient	St. errors	t-statistics	P-value
const	−1389,03	346,068	−4,014	0,0278
professional	−0,12905	0,337666	−0,3822	0,7278
firms	18,9058	8,14573	2,321	0,1030
taxes	5,60677	0,607270	9,233	0,0027
projects	6,17951	3,49735	1,767	0,1754

Mean dependent VAR	8041,500	Standard deviation	5012,187
Sum of modules of errors	1321,619	Sum of the sq. residuals	690566,0
Likelihood estimation	−54,40255	Criterion Akaike	118,8051
Criterion Schwartz	119,2023	Crit. Hennan-Quinn	116,1261

For a quantile 0,5 (medians) the most significant variable is a taxation level, its coefficient is equal 5,606777 at the level of value 0,0027 (Table 4). Influence of this factor depends on revenue. The more is the size of revenue the more are the paid taxes. The number of employees has a negative impact on sales. The equation of quantile regression is:

$$sales = -1,39e+03-0,129*professional+18,9*firms+5,61*taxes + 6,18*projects$$
$$\quad (346) \qquad (0,338) \qquad\qquad (8,15) \qquad (0,607) \qquad (3,50)$$
$$T = 8, sum\ of\ abs.\ residuals = 1,32e+03$$
$$(in\ brackets\ standard\ errors\ are\ specified).$$

(2)

Let's carry out assessment for the last quantile 0,75.

Table 5. The quantile assessment, observations of 2010–2017 are used (T = 8) Dependent variable: sales, tau = 0,75. Asymptotic standard errors are considered as independent and equally distributed

	Coefficient	St. errors	t-statistics	P-value
const	−1458,58	275,618	−5,292	0,0132
professional	−0,0843173	0,268927	−0,3135	0,7744
firms	27,1513	6,48750	4,185	0,0249
taxes	5,04369	0,483647	10,43	0,0019
projects	4,09603	2,78539	1,471	0,2378

Mean dependent VAR	8041,500	St. deviation	5012,187
Sum of modules of errors	1572,899	Sum of the sq. residuals	952890,3
Likelihood estimation	−52,55133	Criterion Akaike	115,1027
Criterion Schwartz	115,4999	Criterion Hennan-Quinn	112,4237

In this quantile we can note as key two variables: number of the companies and level of the taxation. So the coefficient of number of the companies is equal 27,1513 at p – value 0,0249, and the level of the taxation has coefficient 5,01369 at p-value 0,0019 (Table 5). The number of employees also has a negative impact on sales. The equation of quantile regression is:

$$sales=-1,46e+03-0,0843*professional +27,2*firms + 5,04*taxes+ 4,10*projects$$
$$(276) \qquad (0,269) \qquad\qquad (6,49) \qquad\quad (0,484) \qquad (2,79)$$
$$T=8, \text{ sum of abs. residuals} = 1,57e+03$$
$$(in\ brackets\ standard\ errors\ are\ specified)$$

$$(3)$$

The indicators of t-statistics tell us about the statistical importance of the above-stated variables. As a result we can reject a zero hypothesis and draw a conclusion that various factors exert unequal impact on total revenue of science and technology park. The taxation level, the number of the companies and number of their employees have the greatest impact on sales. These factors also represent the main sources of risks for science and technology park which influence total revenue of Academpark.

So, during the conducted research the major factors influencing sales of the science and technology park were revealed by quantile regression. There are a taxation level, a number of the enterprises, a number of employees. On each quantile they exert different impact. For a quantile 0,25 the sales depends on the number of employees (professionals), taxes (taxes) and the number of projects (projects), for a quantile 0.5 it depends on a taxation level. In quantile 0.75 there are two main variables: number of the companies and level of the taxation. Thus the indicators influencing on sales can be different for every quantiles Quantile regression allows us to establish the influence of these indicators for various levels. This method allows to determine factors influencing revenue and to elaborate ways of a possible decrease of their negative impact at all levels and quantiles.

The method of quantile regression may be used in the analysis of factors affecting the activity of other science and technology parks. This tool allows to identify risk factors of science and technology park and to develop the ways for decrease in their negative impact on its activity.

3 Conclusion

Definition of the factors exerting the greatest impact on the revenue of science and technology park creates the platform for risk management of science and technology park. It assumes realization of the following events directed to decrease in risks. They are connected with a reduction in taxes, increase in number of residents and revenue:

- the organization of process of attraction of the financial resources necessary for implementation of projects of the resident enterprises, or for development of their innovative activity. It may be done on the principles of public-private partnership. The commercialization of the technologies developed by the resident companies is carried out with use of public funds for capital expenditure;
- use of the mechanism of "venture financing", granting short-term "cheap" borrowed funds to the resident companies;
- development of the program of technological support directed to decrease in costs, terms and improvement of quality of production of prototypes, pilot batches of the knowledge-intensive products, working off of technologies of their mass production;
- creation of mechanisms of cooperation of residents of Academpark and Scientific Research Institute Siberian Branch of the Russian Academy of Science with large industrial partners. Creation of the centers of competences on the basis of Academpark in different scientific and technological directions has to become the main tool of this mechanism;
- carrying out custom research and development for the solution of technological barriers of industrial partners;
- acceleration of the small innovative companies of Academpark, this will diminish risks of the activity of the Academpark.
- expansion of consulting services and technological, engineering. At the same time it is necessary to pay attention to creation of conditions for gaining access to unique services. It may promote the development of small and medium enterprises;
- creation of a granting procedure for the incubated companies of services for the technological companies on the principles of "contract manufacturing". It will provide additional funding to their expansion;
- creation of steady mutual exchange of information about the advanced achievements of science and technology with the leading technology development centers of the world.

Implementation of the above-stated actions will allow to reduce negative impact of various factors on activity of science and technology parks and will promote their growth.

Acknowledgements. The publication was prepared with the support of the "RUDN University program 5-100".

References

1. Castells, M., Hall, P.: Technopole of the World: Making of the 21st Century Industrial Complexes. Routledge, London (1994)
2. Braunerhjelm, P., Feldman, M.: Cluster Gensis: Technology-Based Industrial Development. Oxford University Press, New York (2006)
3. European Commission. Research Intensive Clusters and Science Park (2007)
4. Mytelka, L.: Forum Keynote address (2001)
5. Rosenfeld, S.: Smart systems: A guide to cluster strategies in less favored regions. Regional Technology Strategy (2002)
6. Cooke, P., Heidenreich, M., Braczyk, H.: Innovation System. System: Role of Governance in a Globalized World. Routledge, London and New York (2004)
7. Koenker, R.: Quantile regression for longitudinal data. J. Multivar. Anal. **91**(1), 74–89 (2004)
8. Gastwirth, J.L., Selwyn, M.: The Robustness properties of two tests for serial correlation. J. Am. Stat. Assoc. **75**, 138–141 (1980)
9. Zeng, S., Xie, X., Tam, C.: Evaluating innovation capabilities for science parks: a system model. Technol. Econ. Dev. Econ. **16**(3), 397–413 (2010)
10. Academpark homepage. https://academpark.com/about/docs/. Accessed 28 Dec 2018

Security Analysis of Information Transmission in Intelligent Transport Telecommunications Network

Mirosław Siergiejczyk[(✉)]

Faculty of Transport, Warsaw University of Technology, Warsaw, Poland
msi@wt.pw.edu.pl

Abstract. Selected elements impacting the reliability and security of telecommunications networks supporting transport services were presented in the article. It also discusses distinguished methods and mechanisms, which enable ensuring the required reliability and availability level of telecommunications networks in the failure-free and emergency modes of operation in the aspect of telecommunication security. The paper also addresses the impact of the operating and maintenance method on the security of telecommunications networks, as well as interconnections between telecommunications system security with the manners of information security management by the personnel of organization/decision-making departments.

Keywords: Security · Transport · Telecommunications · Reliability · Availability

1 Introduction

Telecommunications networks design for transport differ from the public ones. Telecom networks in transport are a tool supporting transport services, while public networks are commercial and provide telecommunications services for citizens. Both of the aforementioned network types generally utilize the same technical solutions, but often differ in terms of the architecture, as well as the provided services.

The European Union has already begun conceptual work on the possibility of using broadband integrated radio networks in "critical missions", i.e. crisis situations covering public safety and aid in the event of disasters and accidents, the (PPDR- Public Protection and Disaster Relief) and in sectors within the critical infrastructure, including transport. Addressing these issues is evidenced by the establishment of a CCBG (Critical Communication Broadband Group) Working Group in 2012, with the task to develop next-generation network standards for dedicated private networks of special purpose, such as railway, power, public transport, military communications, crisis communications, etc. The aforementioned ones and other similar applications will gain even more importance since we are now entering an era of unmanned trains and autonomous road vehicles [1, 10, 13].

The transport systems, including intelligent transport, contain specific requirements in the field of communications.

© Springer Nature Switzerland AG 2020
W. Zamojski et al. (Eds.): DepCoS-RELCOMEX 2019, AISC 987, pp. 456–465, 2020.
https://doi.org/10.1007/978-3-030-19501-4_45

ITSs (Intelligent Transport Systems) are defined as systems designed to improve transport activities through the reduction of operating costs, the increase of the level of security, and optimisation of using the existing road infrastructure by moving vehicles. Based on the assumption, these systems should use telecommunications and IT technologies, and also the automation and measurement devices, which in conjunction with advanced control methods, affect the road transport improvement [12, 15, 18].

The ITS has an impact on the improvement of travelling conditions in the multimodal range – dealing with public and private means of road, sea and air transport. The use of ITS is an affordable and easier method for improving the transport conditions than the communication infrastructure expansion in its current form [15, 18]. Each subsystem of ITS has specific requirements for communication channels, which must be chosen adequately to the needs of a given subsystem, its topology, users, and taking into account the costs of the construction and operation of the system.

2 Requirements for Telecommunications Networks in Transport

Telecommunications networks used in transport should, most of all, ensure a widely understood safe provision of the data transmission service. The currently utilized telecom networks consists of hardware/software systems, which is why data transmission security depends on both the hardware, as well as the software [1].

According to one of the definitions, telecommunication security is understood as a set of methods and mechanisms, the application of which ensures a high level of system availability and reliability through matching an appropriate architecture, i.e., specifying, i.a., the redundancy of individual elements.

The security of telecom networks can be divided into [6, 9, 12, 15]:

– IT security, i.e., protection against: hackers, crackers, viruses, trojans, worms, etc.
– ICT security, i.e., self-repair networks, firewalls, passwords, biometrics, tunnelling, encryption, dedicated protocols, etc.

Telecommunication security is also understood as a set of methods and mechanisms, the application of which ensures a required level of availability and service provision continuity through selecting an appropriate structure of a system or network topology or through relevant radio coverage. Whereas communication security involves security measures preventing unauthorized persons to acquire useful information through entering into possession or getting familiar with the transmitted messages.

An important parameter within a telecommunications system, which proves its correct operation is the quality of services (QoS), which involves a specified probability of a false connection, data transmission (transfer) delay, limiter jitter (change of delay within assumed limits) and a specified error rate.

3   The Analysis of the Availability and Reliability of a Telecommunications Network in Intelligent Urban Transport

The task of each telecom network is to transmit information within a specified time and a certain error rate. A telecommunications network is a system, which must exhibit high reliability and ensure a high level of security of the transmitted data. Reliable access to telecommunications services is a largely important issue for an Administrator of transport infrastructure, since it can have a direct impact on the safety of transported persons and goods, as well as traffic flow. The reliability of a telecom network can be breached through:

- Physical interruption of a connection, i.e., broken cable, defective operation of a teletransmission system or the loss of a digital channel.
- Damage or overloading of network equipment.
- Defective operation of a legal or illegal operator.
- Intentional network operation disruption.

The following subsystems can be distinguished within the functional structure of an intelligent urban transport system:

- vehicle monitoring and surveillance system, including video surveillance
- measurement of road traffic, noise, pollution and meteorological parameters,
- road traffic control system, including the control of pedestrian and vehicle traffic, as well as traffic prioritization,
- vehicle stream traffic strategic management system, including traffic planning, directing onto alternative routes and urban logistics,
- road traffic safety management system, including the detection and management of road incidents,
- parking management and access control system,
- driver information system, including road, media and vehicle information.

An example of a role of the vehicle monitoring and surveillance system is providing the operators in a Management Centre with a video preview of the current situation within the monitored areas of intelligent transport (Fig. 1). System functionality can be presented in the following areas:

- Providing Traffic Control;
- Managing Incidents;
- Providing Road Maintenance.

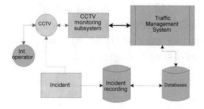

Fig. 1. Monitoring and surveillance subsystem information exchange diagram.

All data recorded by devices of individual subsystems are sent to the Management Centre. They can be stored in a data warehouse, processed by servers, used in software algorithms or received and analysed by the operators. They can also be presented on a web information platform. Based on the received analysis results, the operators or the software adapt the activities to the needs of a situation (e.g. road incident) and send the data to software or hardware forming the subsystems, which they are thus managed.

A teletransmission network is the backbone of each ITS. Ensuring its operational continuity and efficiency of each of its components is crucial for the functioning of the entire ITS. Failure of at least one node can lead to the unavailability of the services, such as parking or meteorological information, and worse, be the cause for interruptions in the operation of the traffic light system. Hence the need for efficient management of a teletransmission networks and its continuous monitoring. Irregularities resulting from intentional activities or accidental damage, simple information on node availability, traffic statistics, data regarding the machines operating within a network – these are just some of the many information an administrator should receive. Of course, proper network management, predicting hazards and their elimination already at the design stage significantly decreases the number of undesirable incidents in the course of operation. However, it should undoubtedly be said that the proper functioning of ITSs requires appropriate methods and tools supporting the monitoring and management of a teletransmission network. It is recommended to apply ring topology in order to guarantee transmission reliability and service continuity in intelligent transport systems.

Ring topology simplifies the construction of a teletransmission network and eliminates the need to use additional intermediate points. It should be noted that the Ethernet standard is the most often used in bus or star topologies, but in order to use it in a ring network, additional measures aimed at preventing the formation of loops should be taken. Any loops within a network based on an Ethernet solution result in the duplication of packages circulating indefinitely, which will quickly lead to decreasing the network efficiency and bandwidth. A solution utilized within ITSs is also a protective link (Fig. 2).

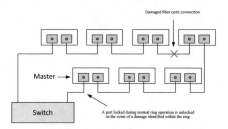

Fig. 2. Protective link application diagram.

In order to protect the rings and ensure reliability of connections, one master switch acting as "Master" appears in each of the rings of the network. This switch blocks one of its ports when the ring is operating normally. In this way, one of the links in the ring (protection link) is inactive, which prevents the formation of loops. The other switches work as slave, "slave", and in the event of damage send information to the master switch, which unlocks the previously blocked port and maintains data transmission, using the inactive link so far.

Urban intelligent transport systems also use local transmission nodes. Such a node includes devices located, for example, at a given crossroad, both the ones used for collecting data (induction loops, video detection cameras, etc.) as well as presenting data (traffic lights, variable content boards, etc.), as well as the ones found in cabinets, such as copper and fibre-optic distribution frames, power supplies or converters. Single-mode fibre-optic cables are most commonly used for connections between local teletransmission nodes. However, due to the fact that within individual crossroad, the devices can be several hundred metres from local node cabinets, there is a need to use metal cables (e.g. ICT twisted pair). Apart from cable connection, it is required to use a wireless solution at some locations, where it is impossible to couple a fibre-optic cable. The PAPN (Private Access Point Name) service, purchased from a telecommunications provider is the most common solution. PAPN utilizes the infrastructure of this provider to create a virtual private network, making it possible for the devices to transmit from any point within the coverage of the radio base station. The equipment within one PAPN cannot communicate with each other or through the Internet, which ensures a high level of security. In order for the devices to be able to communicate with the Management Centre, it is necessary to register a SIM card for a given PAPN at the provider and enter the PAPN name into the GSM modem configuration.

An important issue in terms of ensuring information security is the point of contact with the Internet. The point of contact of an internal network with the Internet is an infrastructural component of increased criticality – it isolates the IT resources of an intelligent transport system from a public network, with a practically non-existent confidence level. This point, which filters the entire traffic between the network infrastructure of the Management Centre and the Internet can be secured through duplicated multi-functional firewalls, namely, UTM (Unified Threat Management) devices, operating in an active-active cluster system. All nodes of a cluster are active in such a system and in the event of damage to one of them, the traffic is distributed between the remaining nodes, without any interruption in Internet access. The integration of numerous solutions within one UTM device guarantees the functionalities of, i.e. a firewall, anti-virus and anti-spam software, content filter and router. It also enables the creation of encrypted VPN (Virtual Private Network) tunnels [4, 17], which allow remote access based on appropriate certificates, intrusion protection, application control and preventing data loss. Connection to the Internet was achieved through coupling a selected port of this module with an Internet-provider device.

The rules are updated on a regular basic, according to the needs (e.g. granting access to temporary collaborators, third parties, etc.). A firewall shall not prevent grouping the rules, which is why they have to be considered from the point of view of logical interfaces, as source addresses and the application of the analysed packages. As a result, specific VLAN interfaces must be defined within such a device [4, 6, 17].

It is extremely important to develop a strategy ensuring the maintenance of a necessary security level and preparing functional plans for the system in the event of particular threats. Such scenarios are referred to as Disaster Recovery strategies (the post-failure infrastructure recovery) and they include processes and procedures associated with resuming or maintaining the technical infrastructure critical to a given organization after a natural or man-induced disaster.

It is extremely important to develop a strategy ensuring the maintenance of a necessary security level and preparing functional plans for the system in the event of particular threats. Such scenarios are referred to as Disaster Recovery strategies (the post-failure infrastructure recovery) and they include processes and procedures associated with resuming or maintaining the technical infrastructure critical to a given organization after a natural or man-induced disaster. The providers of ICT networks offering services for transport should determine the Disaster Recovery strategy for their network, which will be the basis for implementing this functionality. The first step is to thoroughly specify the following issues and requirements [6, 12]:

- failure definition;
- target restoration time;
- level of services, which are priorities after restoration (connection types, value-added services);
- restoration method (manual intervention, remote reprogramming, personnel location).

In fact, Disaster Recovery planning is part of a larger process of planning operational continuity and should involve defining the procedures for resuming application, data, equipment and communication. Three basic stages within the activities regarding catastrophic events are distinguished: preparation stage (prior to a failure or disaster); repair commencement stage (starts with diagnosing a failure or disaster and taking early actions aimed at restoring system efficiency) and repair stage (starts several days or weeks after a failure or disaster).

The network engineering stage should involve developing certain scenarios, where individual system elements are subject to a failure or destruction, e.g., as a result of a fire or natural disaster. Such event scenarios enable the determination of elements critical to the functionality of the entire system, and therefore, enable selecting an appropriate protection measure. A common method, which allows to increase network reliability, security and availability is redundancy; meaning the surplus of devices or the use of additional elements. It refers both to information stored in the registers, as well as hardware elements, which can be duplicated in different ways, e.g., n + 1, 1 + 1, 1:n. Redundancy may involve copying all or just particularly important data. Redundancy may cover [13, 14, 16]:

- an entire system;
- individual subsystems (e.g. teletransmission, commutation, backbone network);
- individual elements included in a system (e.g. management system, commutation node, etc.);
- individual components of system elements (e.g. processor card of communication nodes, interfaces).

It is obvious that the greater the redundancy the more reliable the system, which means a shorter system downtime during a year. However, together with higher redundancy, the system maintenance costs also increase, and the effects of delays resulting from switching between redundant subsystems, and even elements, must also be taken into account.

4 Monitoring an Intelligent Transport Telecommunications Network

Telecom network security also depends on the conducted maintenance activities, which include:

- network administration;
- monitoring the operation of network elements, automatic detection of network threats and congestion;
- teletransmission traffic management;
- resource management;
- service management;

A teletransmission network of an urban ITS is characterized by hardware and software diversity, combining the solution of various manufacturers. Owing to that fact, its management process is complex and multi-faceted, and requires the integration of numerous tools. It requires an administrator to be skilled in numerous programs and systems, as well as familiar with the syntax of commands of a text interface. An undoubted advantage of managing a network with dedicated tools is them exhibiting all the functionalities necessary for this purpose. Supporting them through the use of a seemingly simple, yet practical text interface makes an existing solution become fully complex and sufficiently effective. In this regard, managing a network within an urban ITS shall be assessed as efficient and fully equipped with mechanisms ensuring its constant operation and making it an environment secured against unauthorized access.

Given the actual network management standards, managing a teletransmission network of an urban ITS is executed in a sufficiently effective manner. The sole drawback is the dispersion of management caused by the need to integrate numerous minor dedicated systems, each of which has own limitations and requirements. This entails the time-intensity of the entire process. A solution to this problem is a concept of a network with centralized management, which eliminates the difficulties associated with processes such as:

- introducing frequent changes to the network structure and state,
- supporting network configuration with high-level languages,
- ensuring a sufficient control of tools used for diagnosing and detecting problems within the network.

Attempts to simplify the management of teletransmission networks fail frequently due to difficulties in modifying their fundamental infrastructure. The rigidity of this infrastructure presents few opportunities for innovations or improvements because network devices are proprietary solutions or ones reserved by the manufacturers. In terms of management, the teletransmission network of the intelligent urban transport system usually consists of fixed and permanent standards and devices that require manual administration. Modifications or development of a network in terms of new opportunities and applications require reconfiguration, which may prove to be expensive and time-consuming.

The development of IT technologies, such as virtualization of environments and use of computing clouds forces teletransmission networks to keep up with the progress in increasing the rate and complexity of data processing. Old systems and methods of network management contain significant restraints and methods, which will not be compatible in the new conditions. Network must adapt, becoming more flexible and automated. The concept of virtualization can also be applied in relation to them. A new approach to network management, programmable computer networks SND (Software-Defined Networking) supports separating the "data domain" from the "control domain", making the switches comprising the data domain simple devices transferring packages and leaving a logically centralized program called a controlled, which supervises the behaviour of the entire network. Virtualization, underlying SDN, strips network equipment off network traffic control functions and enables operation regardless of the fundamental infrastructure.

SDN networks are in a certain way modelled upon server virtualization methods and introduce an abstract layer, separating logics and network configuration from physical links and hardware. In this regard, SDN offers software control over both physical, as well as virtual devices, which can dynamically respond to changes within the network environment, utilizing a programmable protocol controlling the flow of packages. Currently, the most popular stack to develop SDN is the OpenFlow open standard, enabling remote control over routing boards. OpenFlow is used to ensure communication between a controller and a single switch, in order to inform the controller on traffic rates and informing the switch on their redirecting methods. The essence of SDN is developing a universal language for programming network switches. API (Application Programming Interface) is the channel used to transfer instructions. A logical diagram of SDN network devices is shown in Fig. 3.

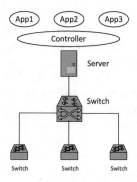

Fig. 3. Logical diagram of SDN network devices (source: own study)

The concept of implementing the management of an urban ITS teletransmission system through the use of a programmable computer network adopted the following assumptions:

- SDN network management is supposed to integrate all physical and virtual network elements,
- control over the network devices is to be implemented through software-based, and not hardware-based, solutions,
- network rules and policies shall be implemented in a high-level programming language (e.g. C++),
- controller-network switches communication is to be through the OpenFlow protocol.

At the same time, such a management method should satisfy the following functional requirements:

- enable a centralized approach towards network configuration and administration,
- facilitate and streamline the introduction of changes, such as adding an extra fibre-optic ring,
- reduce the complex management structure,
- remove restrictions regarding software licences and proprietary solutions of manufacturers,
- be aimed at cooperation with an administrator,
- optimize the utilization of network resources.

5   Conclusion

The security of telecommunications networks supporting transport services is very important because it is associated with or supports the safety of transported persons and goods, as well as effective management of transport companies. It can be concluded that telecom networks are a tool to create better and more effective transport services, hence, must be optimized for individual transport sectors and types. A common feature of the networks is that they are comprised of a wired, as well as a wireless part, which makes the issue of the security of transferred information more complex than in the case of specialized public networks, adding to the fact that the effects of an incorrectly transmitted information can be even tragic.

Another issue, which was not addressed in this article and yet is extremely important, is developing a strategy ensuring the maintenance of an appropriate security level, as well as preparing system functional plans in the case of specific threats. Such scenarios are referred to as Disaster Recovery strategies (the post-failure infrastructure recovery) and they include processes and procedures associated with resuming or maintaining the technical infrastructure critical to a given organization after a natural or man-induced disaster. Providers of ICT networks offering services for transport should work out a Disaster Recovery strategy for their networks, which would be the basis for implementing this functionality.

References

1. Anderson, D.J., Brown, T.J., Carter, C.M.: System of systems operational availability modeling, Sandia National Laboratories (2013)
2. Caban, D., Walkowiak, T.: Dependability analysis of hierarchically composed system-of-systems. In: Zamojski, W., Mazurkiewicz, J., Sugier, J., Walkowiak, T., Kacprzyk, J. (eds.) Proceedings of the Thirteenth International Conference on Dependability and Complex Systems DepCoS-RELCOMEX. Springer (2019)
3. Chen, S., Ho, T., Mao, B.: Maintenance schedule optimisation for a railway power supply system. Int. J. Prod. Res. **51**(16), 4896–4910 (2013)
4. Fry, Ch., Nystrom, M.: Monitoring and Network Security. Helion Publishing House, Gliwice (2010)
5. Jin, T.: Reliability Engineering and Service. Wiley, New York (2019)
6. Karpiński, M.: Information security. Publishing House PAK (2012)
7. Kowalewski, M., Kowalewski, J.: Policy of information security in practice. Library IT Professional (2014)
8. Kowalewski, J., Kowalewski, M.: Cyberterrorism as a particular threat to the national security. Telecommunications and Information Technologies, No. 1–2 (2014)
9. Liderman, K.: Information Security. PWN Publishing House, Warsaw (2012)
10. Qiu, S., Sallak, M., Schön, W., Cherfi-Boulanger, Z.: Epistemic parametric uncertainties in availability assessment of a Railway Signalling System using Monte Carlo simulation – security. In: Steenbergen, et al. (eds.) Reliability and Risk Analysis: Beyond the Horizon. Taylor & Francis Group, London (2014)
11. Paś, J., Rosinski, A.: Selected issues regarding the reliability-operational assessment of electronic transport systems with regard to electromagnetic interference. Maintenance Reliab. **19**(3), 375–381 (2017)
12. Rosiński, A.: Modelling the Maintenance Process of Transport Telematics Systems. Publishing House Warsaw University of Technology, Warsaw (2015)
13. Siergiejczyk, M., Gago, S.: Selected issues associated with the reliability and security of information transmission within a GSM-R system. Problemy Kolejnictwa–Zeszyt 162. Published by Railway Institute, Warsaw (2014)
14. Siergiejczyk, M., Stawowy, M.: Modelling of uncertainty for continuity quality of power supply. In: Walls, L., Revie, M., Bedford, T. (eds.) Risk, Reliability and Safety: Innovating Theory and Practice: Proceedings of ESREL 2016. CRC Press/Balkema, London (2017)
15. Siergiejczyk, M., Wawrzyński, W.: Problem of information security in ITSs. Logistics No. 4 (2014)
16. Siergiejczyk, M., Paś, J., Rosiński, A.: Issue of reliability–exploitation evaluation of electronic transport systems used in the railway environment with consideration of electromagnetic interference. IET Intel. Transport Syst. **10**(9), 587–593 (2016)
17. Stallings, W.: Cryptography and security of computer networks. Concepts and methods of secure communication. Edition V. Helion Publishing House, Gliwice (2012)
18. Williams, B.: Intelligent Transport Systems Standards. Artech House, Inc., London (2008)

Semi-Markov Reliability Model of Internal Electrical Collection Grid of On-Shore Wind Farm

Robert Adam Sobolewski[✉]

Faculty of Electrical Engineering, Bialystok University of Technology,
Wiejska 45D, 15-351 Bialystok, Poland
r.sobolewski@pb.edu.pl

Abstract. Fulfilment of reliability requirements for 'electrical components' of wind farm internal collection grid with combination of good wind resources assure the satisfactory performance of a farm. Concerning on-shore farm the feeder section faults lead to the loss of energy served from either all downstream turbines or all turbines in the farm. The failures of protection systems in terms of failures to trip and false tripping increase the incidence and extent the losses of energy served in greater degree as compared to other designs. Achievement of satisfactory performance (in terms of accepted amount of energy not served) in relation to reliability issues, requires the fulfilment of the reliability requirements. They should be obtained relying on quantitative reliability models and take into account: reliability parameters of the components, the topology of collection grid, quality of renewal action and false tripping of protection systems. In the paper the original approach to reliability modeling of on-shore wind farm internal collection grid is presented in details. It relies on semi-Markov model and takes into account the faults and failures of components, the combinations of the components faults and failures that lead to particular number of generators on outage, and the quality of renewal action and false tripping of protective relays. The reliability measure is expected energy not served by farm within specific period of time. The example of reliability model application is presented in details as well.

Keywords: Wind farm · Reliability model · Semi-Markov model

1 Introduction

More and more wind farms (WFs) are being integrated with distribution and transmission networks in many countries. Since the penetration level of WFs increases, more reliable operation, control and components should be one of the major concerns about WFs planning and operation. Among others, such attention affects 'electrical components' of WFs, such as: generators, transformers, cables, buses, protection systems (couples of one or more protective relays and circuit breaker, that should detect and clear the faults occurred in the components). Fulfilment of the reliability requirements for 'electrical components' with combination of good wind resources assure the satisfactory performance of the WFs. The internal collection grids can be arranged in

© Springer Nature Switzerland AG 2020
W. Zamojski et al. (Eds.): DepCoS-RELCOMEX 2019, AISC 987, pp. 466–477, 2020.
https://doi.org/10.1007/978-3-030-19501-4_46

different ways depending on the size (number and rated power of the turbines) and location (on-shore or off-shore) of a WF. The most common arrangement of on-shore locations is radial one. Wind turbine generators (WTG) are connected in series to one feeder. The number of feeders interconnected with collection bus of a farm can be bigger than one. The main advantage of this design is the total array cable length is smaller as compared to other arrangements (ring, star and so on). Also, it is simple to control and it provides the possibility to taper the cables capacity as a distance from the collection bus increases. The major disadvantage with this design is quite poor reliability provided. First of all, feeder section faults will lead to the loss of energy served from either all downstream turbines or all turbines included in the farm. Secondly, failures of protection systems (PSs) in terms of failures to trip and false tripping, will increase the incidence and extent the losses of energy served in greater degree as compared to other designs. For example, WTG fault requires operation of remote back-up PS located in the feeder while PS of turbine is on outage. In consequence, all turbines interconnected with feeder will be off. Moreover, some of the PS failures can be revealed during either planned maintenance (periodically) or after occurrence of the fault in the component protected by the system.

Concerning PS the different techniques can be used depending on the size of WF, the systems location (grid, WTGs or collection grid) and their role. As refers a grid the two following techniques are quite common, i.e. shaped directional operating characteristics and adaptive distance characteristics that consider the conditions of the WF connected to the grid. The relays that apply the directional and distance characteristics are installed at the point of common coupling (PCC), and if fault occurred in the WF internal grid and the fault is not cleared by PS of the farm – the entire farm is disconnected. For WTGs and section feeders and buses the more conventional techniques of PS are commonly used, i.e. over/under voltage, over/under frequency, instantaneous phase/neutral over-current for generators and LV/MV transformers phase/ground faults, and inverse time phase over-current for generator overload [1]. They can sufficiently protect the feeders, generators and transformers against the damage caused by internal faults. However, the selectivity requirement of wind farm PS can be not fulfilled if they are not able to distinguish a section feeder fault, a collector bus fault, inter-tie fault, and a grid fault. In addition, they cannot distinguish whether it is a connected feeder fault or an adjacent feeder fault. Thus, the section of WF on outage can be wider.

The amount of energy not served within a specific period of time because of the faults and failures of wind farm 'electrical components' depends on: the number of WTGs on outage (one, all in a feeder section, or all in the whole WF), the number and duration of the components' outage and wind conditions. Moreover, the concern about the disconnection of a whole WF is an issue of power grid operator. Ideally, when a fault occurs, the size of the outage zone should be as small as possible.

Achievement of satisfactory WF performance (in terms of accepted amount of energy not served) in relation to reliability issues, requires the fulfilment of the reliability requirements. These requirements should be obtained relying on quantitative reliability models and take into account: reliability parameters of the components, a topology of internal collection grid, quality of components renewal action and false tripping of PS.

The relevant literature offers some approaches, which were used for WF reliability representation in models that describe the characteristics of WTGs and WF power and energy output [2–5]. In [2] the method of component outage leading to WF outage is applied. The method is based on Reliability Block Diagram technique, where all components are connected in series. In [3] a logical diagram construction method is applied with a sequential grouping of the model components in series or in parallel. In [4] a step-by-step procedure is developed for calculation of WF reliability using combinatorial algorithm. All these models feature a simplified representation of the impact of 'electrical components' (especially protection systems) on WF reliability.

In the paper the original approach to reliability modeling of on-shore WF internal collection grid is presented in details. The reliability modeling relies on semi-Markov model and takes into account the faults and failures of components, the combinations of the components faults and failures that lead to particular number of WTGs on outage, the quality of renewal action and the false tripping of protective relays. The reliability measure is expected energy not served by WF within specific period of time. The approach is useful for on-shore WF of radial collection grid topology taking into account the selectivity requirement of PS of the feeders and step-up transformers/generators.

As an example, the model is applied to reliability analysis of internal collection grid of on-shore WF that consists of two feeder sections, eight WTGs and two PS of the feeders, and eight PS of step-up transformers/generators.

2 Wind Farm Internal Collection Grid Arrangement and Protection Systems Operation

Figure 1 shows the internal collection grid of on-shore WF studied in the paper. Two power feeder sections are connected to a collector bus B01, which is interconnected with the grid 'External Grid' through main transformer T00, substation bus B00 and inter-tie line L00. Eight wind turbine generators WTG1, ..., WTG8 are connected to the feeders trough step-up transformers T1, ..., T8, respectively. The first feeder section consists of: cable L01 that links B01 and terminal B1, and cables L12, L23 and L34 that link terminals B2, B3 and B4 of WTG1, ..., WTG4. The second feeder section consists of: cable L05 that links B05 and terminal B5, and cables L56, L67 and L78 that link terminals B6, B7 and B8 of WTG5, ..., WTG8. The transformer T00 is equipped with the two protection systems PS00a and PS00b, whereas protection systems PS01 and PS05 are installed on each feeder in the cables L01 and L05 respectively. Each wind turbine generator (WTG1, ..., WTG8) is equipped with the individual protection system (PS1, ..., PS8, respectively). Let assume that all PS included in the WF grid consist of a overcurrent relay and circuit breaker.

The main role (primary protection function) of PS00a/PS00b, PS01/PS05 and PS1/.../PS8 is to clear the faults that occurred in transformer T00 (or collection bus B00), feeder and step-up transformer/generator, respectively. Moreover, PS00a/PS00b comprise the remote back-up protection function while the fault occurred in feeder and its PS is down. The consequence is disconnection of the whole WF instead of one feeder and interruption of generating the power by the farm. The bigger number of the feeders in a WF is the more serious consequences can be expected. On the other hand, PS01/PS02 provides a remote back-up protection function while the fault occurred in step-up transformer or generator, and their PS is down. The consequence is discon-nection the feeder instead of one WTG and interruption of generating the power by the rest of turbines interconnected with the feeder. The bigger number of turbines in question is the more serious consequences can be expected. The roles mentioned above (both primary and back-up protection functions) are most common for overcurrent relays that should detect and clear a line to line and line to ground faults.

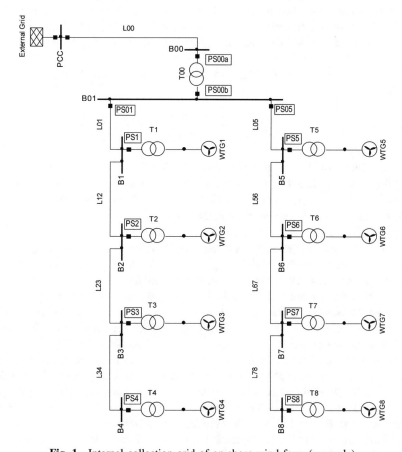

Fig. 1. Internal collection grid of on-shore wind farm (example)

In addition, an incorrect PS operation can arise because of its false tripping. It refers to PS of the feeder while the fault occurred in step-up transformer or generator and their PS is reliable. The PS of the feeder operates instead of available PS comprising a primary protection of the individual component.

3　Reliability Model

The expected energy not served (EENS) by wind farm because of faults/failures in its components is calculated relying on formula

$$\text{EENS} = \sum_g \sum_h M_{gh}(t) \cdot K_{gh} \tag{1}$$

where: $M_{gh}(t)$ – occurrence number of hth combination of reliability states of the components belonging to gth section feeder of the WF within period of time t, K_{gh} – unity mean value of energy not served because of occurrence of hth combination of reliability states of components belonging to gth section feeder, and hth combination of reliability states is one that leads to consequences in terms of energy not served by the farm.

Calculation of $M_{gh}(t)$ can be performed based on semi-Markov model and general formula [5]

$$M_{gh}(t) = \frac{\pi_{gh}}{\sum_{i=0}^{S_g} \pi_{gi} \cdot \tau_{gi}} t \tag{2}$$

where S_g total number of combinations of reliability states of the components belonging to gth section feeder (equivalent to the number of states in semi-Markov model), $h, i \in (0, \ldots, S_g)$, π_{gh} and π_{gi} stationary probabilities of the embedded Markov chain, τ_{gi} mean value of waiting time in the state.

The model refers to one section feeder of the WF collection grid depicted in Fig. 1. The reliability states of the model are described in Table 1. Since the composition of the model (number of states, their descriptions and transition probabilities) is the same for each feeder section the index g in analytical formulas below in Sect. 3 is omitted. One particular component mentioned in Table 1 aggregates all such components that exist in feeder section of the farm in question. For example component 'step-up transformer/generator' consolidates all step-up transformers and generators that are included in the feeder section (e.g. T1…T4 and WTG1…WTG4 in Fig. 1).

The transition diagram of one feeder section of wind farm internal collection grid is depicted in Fig. 2. The states are colored as follows: (i) in blue – the 'up' states of all components in question and each WTG is able to deliver power to external grid, (ii) in green – one or more PS are in 'down' state but each WTG is able to deliver power to external grid and (iii) in yellow – the combination of 'up' and 'down' states of components and one or more WTGs are unable to deliver power to external grid because of faults in components. Each branch in this diagram is labeled by: number of the states, and transition probabilities.

In the reliability model the parameters and characteristics of a semi-Markov process are interpreted as the reliability characteristics and parameters of the internal collection grid of wind farm.

Table 1. The states of semi-Markov reliability model assigned to combinations of reliability states of wind farm components and their consequence in terms of number of WTGs to be off

State	Component				Comments/consequences
	Step-up transformers/ generators	Cables/terminals of the feeder	PS of the step-up transformers/ generators	PS of the feeder	
E0	Up	Up	Up	Up	No consequences
E1	Up	Down	Up	Up	Clearing the fault by PS of the feeder/1 WTG off
E2	Up	Up	Down	Up	No consequences
E3	Up	Up	Up	Down	No consequences
E4	Down	Up	Up	Up	Clearing the fault by PS of the step-up transformer/1 WTG off
E5	Down	Up	Up	Up	False tripping of PS of the feeder/4 WTGs off
E6	Down	Up	Down	Up	Clearing the fault by PS of the feeder/4 WTGs off
E7	Up	Down	Down	Up	Clearing the fault by PS of the feeder/4 WTGs off
E8	Up	Up	Down	Down	No consequences
E9	Down	Up	Up	Down	Clearing the fault by PS of the step-up transformer/1 WTG off
E10	Up	Down	Up	Down	Clearing the fault by PS of the wind farm/8 WTGs off
E11	Down	Up	Down	Down	Clearing the fault by PS of the wind farm/8 WTGs off
E12	Up	Down	Down	Down	Clearing the fault by PS of the wind farm/8 WTGs off

Time to fault of the kth internal collection grid component is nonnegative random variable η_k, $k = 1$ (step-up transformer/generator), $k = 2$ (cable/terminal of the feeder) with exponential distribution

$$F_{Ck}(t) = P(\eta_k \leq t) = 1 - e^{-\lambda_k \cdot t}, \, t \geq 0, \qquad (3)$$

where λ_k is a kth aggregated component fault intensity and $\lambda_k = \sum_m \lambda_{km}$, λ_{km}, is fault intensity of mth feeder section component to be included into kth component (if $k = 1$ then $m = 8$ and $\lambda_{11} = \lambda_{T1}, \ldots, \lambda_{14} = \lambda_{T4}, \lambda_{15} = \lambda_{WTG1}, \ldots, \lambda_{18} = \lambda_{WTG4}$, and if $k = 2$ then $m = 8$ and $\lambda_{21} = \lambda_{L01}, \ldots, \lambda_{24} = \lambda_{L34}, \lambda_{25} = \lambda_{B1}, \ldots, \lambda_{28} = \lambda_{B4}$, see Fig. 1).

Time to failure of the lth PS is nonnegative random variable ρ_l, $l = 1$ (PS of step-up transformer/generator), $l = 2$ (PS of cable/terminal of the feeder) with exponential distribution

$$F_{Pl}(t) = P(\rho_l \leq t) = 1 - e^{-\theta_l \cdot t}, \, t \geq 0 \qquad (4)$$

where θ_l is a failure rate of lth PS and $\theta_l = \sum_n \theta_{ln}$, θ_{ln} is failure rate of nth PS to be included into lth PS (if $l = 1$ then $n = 4$ and $\theta_{11} = \theta_{PS1}, \ldots, \theta_{14} = \theta_{PS4}$, if $l = 2$ then $n = 1$ and $\theta_{21} = \theta_{PS01}$, see Fig. 1).

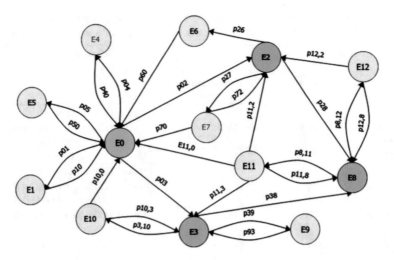

Fig. 2. Transition diagram for the reliability model of one feeder section of wind farm internal collection grid

Probability of PS renewal is p_{Rl} ($l = 1, 2$) and probability of false tripping of PS of cable/terminal of the feeder is q_F.

The random variable T_i (waiting time in state Ei, $i = 0, 2, 3, 8$) and their mean values τ_i are as follows:

$$T_0 = \min(\eta_1, \eta_2, \rho_1, \rho_2) \text{ and } \tau_0 = \frac{1}{\lambda_1 + \lambda_2 + \theta_1 + \theta_2}, \text{ for } i = 0, \tag{5}$$

$$T_2 = \min(\eta_1, \eta_2, \rho_2) \text{ and } \tau_2 = \frac{1}{\lambda_1 + \lambda_2 + \theta_2}, \text{ for } i = 2, \tag{6}$$

$$T_3 = \min(\eta_1, \eta_2, \rho_1) \text{ and } \tau_3 = \frac{1}{\lambda_1 + \lambda_2 + \theta_1}, \text{ for } i = 3, \tag{7}$$

$$T_8 = \min(\eta_1, \eta_2) \text{ and } \tau_8 = \frac{1}{\lambda_1 + \lambda_2}, \text{ for } i = 8. \tag{8}$$

The duration of the other states ($i = 1, 4, \ldots, 7, 9, \ldots, 12$) has the distribution with following CDF

$$F_i(t) = \begin{cases} 0 & \text{for } t \leq 1 \\ 1 & \text{for } t > 1 \end{cases}. \tag{9}$$

The transition probabilities ($p_{ij} > 0, \sum_{j=0}^{12} p_{ij} = 1, i, j = 0, 1, \ldots, 12$) are as follows:

$$p_{01} = \frac{\lambda_2}{\lambda_1 + \lambda_2 + \theta_1 + \theta_2}, \quad p_{02} = \frac{\theta_1}{\lambda_1 + \lambda_2 + \theta_1 + \theta_2}, \quad p_{03} = \frac{\theta_2}{\lambda_1 + \lambda_2 + \theta_1 + \theta_2}, \tag{10}$$

$$p_{04} = (1 - q_F) \cdot \frac{\lambda_1}{\lambda_1 + \lambda_2 + \theta_1 + \theta_2}, \quad p_{05} = q_F \cdot \frac{\lambda_1}{\lambda_1 + \lambda_2 + \theta_1 + \theta_2}, \quad p_{10} = 1, \tag{11}$$

$$p_{26} = \frac{\lambda_1}{\lambda_1 + \lambda_2 + \theta_2}, \quad p_{27} = \frac{\lambda_2}{\lambda_1 + \lambda_2 + \theta_2}, \quad p_{28} = \frac{\theta_2}{\lambda_1 + \lambda_2 + \theta_2}, \tag{12}$$

$$p_{38} = \frac{\theta_1}{\lambda_1 + \lambda_2 + \theta_1}, \quad p_{39} = \frac{\lambda_1}{\lambda_1 + \lambda_2 + \theta_1}, \quad p_{3,10} = \frac{\lambda_2}{\lambda_1 + \lambda_2 + \theta_1}, \tag{13}$$

$$p_{40} = 1, \quad p_{50} = 1, \quad p_{60} = 1, \quad p_{70} = p_{R1}, \tag{14}$$

$$p_{73} = 1 - p_{R1}, \quad p_{8,11} = \frac{\lambda_1}{\lambda_1 + \lambda_2}, \quad p_{8,12} = \frac{\lambda_2}{\lambda_1 + \lambda_2}, \tag{15}$$

$$p_{93} = 1, \quad p_{10,0} = p_{R2}, \quad p_{10,3} = 1 - p_{R2}, \tag{16}$$

$$p_{11,0} = p_{R1} \cdot p_{R2}, \quad p_{11,2} = (1 - p_{R1}) \cdot p_{R2}, \quad p_{11,3} = p_{R1} \cdot (1 - p_{R2}), \tag{17}$$

$$p_{11,8} = (1 - p_{R1}) \cdot (1 - p_{R2}), \quad p_{12,2} = p_{R2}, \quad p_{12,8} = 1 - p_{R2}. \tag{18}$$

The unique stationary distribution of the embedded Markov chain satisfies system of equations $\sum_{i=0}^{12} \pi_i \cdot p_{ij} = \pi_j$, $\sum_{i=0}^{12} \pi_i = 1$, $j = 0, 1, \ldots, 12$, where p_{ij} are expressed by the formulas (10–18).

The occurrence number of hth combination of reliability states of the components within period of time t is obtained from the formula

$$M_h(t) = \frac{\pi_h}{\pi_0 \cdot \tau_0 + \pi_2 \cdot \tau_2 + \pi_3 \cdot \tau_3 + \pi_8 \cdot \tau_8} \cdot t. \tag{19}$$

According to Table 1 the combinations of reliability states of the components, that lead to consequences in terms of energy not served by the farm are following $h = 1, 4, \ldots, 7, 9, \ldots, 12$ (see the states in Fig. 2).

Value of K_{gh} should be calculated relying on probabilistic or statistic model taking into account: the historical data of wind farm output power (or wind resources at the wind farm site), number of turbines on outage because of h combination of reliability state occurrence and maintenance duration of components on outage. The approach of obtaining K_{gh} is not considered in detail in the paper.

If the number of feeder sections in internal collection grid of wind farm is higher than one ($g > 1$) and they all are identical ones (the same number of turbines, the same reliability parameters) the only one reliability model is needed. The results of calculations should be multiplied by number of feeder sections. Otherwise, each feeder section needs individual reliability model and the results of calculations relying on these models should be summed up.

4 Case Study

The approach is implemented on the internal collection grid of the on-shore wind farm, depicted in Fig. 1. Let assume that both section feeders are the same in terms of: interconnections arrangement, components number and their reliability parameters. Thus, it is satisfactory to carry out the reliability analysis of one feeder section and map the results into whole wind farm. Let assume the first section feeder (on the left side in Fig. 1) is considered relying on the model described in details in Sect. 3. The reliability parameters of the components are as follows:

$$\lambda_{T1} = \lambda_{T2} = \lambda_{T3} = \lambda_{T4} = \lambda_{WTG1} = \lambda_{WTG2} = \lambda_{WTG3} = \lambda_{WTG4} = 0.005[1/a], \quad (20)$$

$$\lambda_{L01} = \lambda_{L12} = \lambda_{L23} = \lambda_{L34} = 0.007[1/a], \quad (21)$$

$$\lambda_{B1} = \lambda_{B2} = \lambda_{B3} = \lambda_{B4} = 0.003[1/a], \quad (22)$$

$$\theta_{PS1} = \theta_{PS2} = \theta_{PS3} = \theta_{PS4} = 0.01[1/a], \quad (23)$$

$$\theta_{PS01} = 0.01[1/a], \quad q_F = 0.01, \quad p_{R1} = p_{R2} = 0.95, \quad t = 1\,\text{year} \quad (24)$$

The parameters (21–24) have been used to obtaining the parameters of semi-Markov model. Moreover, let assume that unity mean value of energy not served because of occurrence of the combinations of reliability states of components are following

$$K_{1h} = \begin{cases} 1\,\text{MWh}, & \text{for states}: \quad E4, E9\,(h = 4,9) \\ 4\,\text{MWh}, & \text{for states}: \quad E1, E5, E6, E7\,(h = 1,5,6,7) \\ 8\,\text{MWh}, & \text{for states}: \quad E10, E11, E12\,(h = 10,11,12) \end{cases} \quad (25)$$

Let assume that reliability requirements affect internal collection grid components and can be investigated relying on EENS as a function of the semi-Markov model parameters λ_1, λ_2, θ_1 and θ_2, with two parameters that take the constant values (calculated relying on (20–24)) and two remaining – the ranges from min up to max value. Thus, three variants of parameters' values are investigated (see Table 2).

Table 2. The parameters of semi-Markov reliability model

Number of fig.	λ_1 [1/a]	λ_2 [1/a]	θ_1 [1/a]	θ_2 [1/a]
3	0.04	0.01	0.001...0.1	0.001...0.1
4	0.001...0.1	0.01	0.001...0.1	0.01
5	0.04	0.001...0.1	0.04	0.001...0.1

The results of reliability modeling relying on semi-Markov model and the parameters provided in (20–25) and in Table 2 are shown in Figs. 3, 4 and 5. Figure 3 depicts the EENS against failure rates of both step-up transformers/generators PS θ_1 and PS of the feeder θ_2. As shown, if λ_1 is four times higher than λ_2 the decrease in EENS may be achieved while reducing both of the failure rates, but it is more effective in terms of θ_1. If θ_1 declines from 0.1 to 0.001 ($\theta_2 = 0.1$), the EENS decreases by 57% (0.286 MWh/a $\rightarrow$ 0.123 MWh/a), whereas reducing θ_2 from 0.1 to 0.001 ($\theta_1 = 0.1$) cause the decrease of EENS by 42% (0.286 $\rightarrow$ 0.165). Reducing both failure rates θ_1 and θ_2 from 0.1 to 0.001 results in limitation of EENS by 70% (0.286 $\rightarrow$ 0.087).

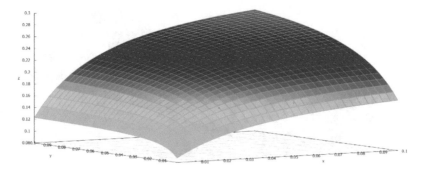

Fig. 3. EENS (z axis) against failure rates of PS θ_1 (x axis) and θ_2 (y axis)

Figure 4 depicts the EENS against fault intensity of step-up transformers/generators λ_1 and failure rate of PS θ_1. As shown, if λ_1 is small (e.g. 0.001), the impact of reducing PS failure rate from 0.1 to 0.001 can be neglected (5% only, 0.065 MWh/a $\rightarrow$ 0.062 MWh/a), whereas if $\lambda_1 = 0.1$ such reducing causes decrease of EENS by 49% (0.323 $\rightarrow$ 0.166). Limitation of both parameters λ_1 and θ_1 from 0.1 to 0.001 results in reduction of EENS by 81% (0.323 $\rightarrow$ 0.062).

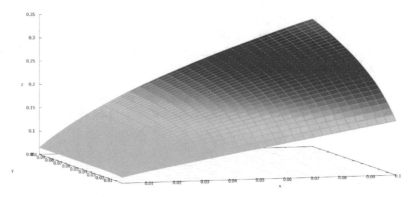

Fig. 4. EENS (z axis) against fault intensity λ_1 (x axis) and failure rate θ_1 (y axis)

Figure 5 depicts the EENS against fault intensity of cables/terminals of the feeder λ_2 and failure rate of PS of the feeder θ_2. As shown, the decrease in EENS may be achieved while reducing both of the parameters, but it is more effective in terms of λ_2. If λ_2 declines from 0.1 to 0.001 ($\theta_2 = 0.1$), the EENS decreases by 74% (0.703 MWh/a $\rightarrow$ 0.183), whereas reducing θ_2 from 0.1 to 0.001 ($\lambda_2 = 0.1$) cause the decrease of EENS by 33% (0.703 $\rightarrow$ 0.473). Reducing both parameters λ_2 and θ_2 from 0.1 to 0.001 results in limitation of EENS by 85% (0.703 $\rightarrow$ 0.108).

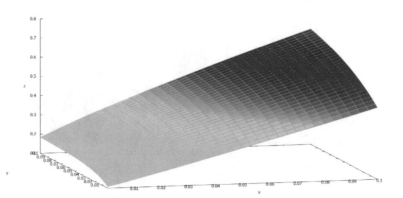

Fig. 5. EENS (z axis) against fault intensity λ_2 (x axis) and failure rate θ_2 (y axis)

5   Conclusions

In this contribution, the semi-Markov model for analyzing an 'electrical components' reliability impact on the EENS by a wind farm is presented. The approach enables to take into account the faults and failures of components, the combinations of the components faults and failures that lead to particular number of WTGs on outage, the quality of components renewal action, and the false tripping of feeder protective relays. The results of the analysis show that the reliability parameters λ_1, λ_2, θ_1 and θ_2 influence the EENS in different way. For example, if θ_1 and θ_2 are of low level the significant decrease in λ_1 and λ_2 respectively, influence the EENS slightly. The higher θ_1 and θ_2 the more significant impact of λ_1 and λ_2 respectively. Moreover, significant reduction of both parameters λ_1 and θ_1 (λ_2 and θ_2) from 0.1 1/a into 0.001 1/a results in depletion of EENS by more than 80%.

Acknowledgment. This work has been partially prepared under the project S/WE/3/2018 at BUT and financially supported by Ministry of Science and Higher Education, Poland.

References

1. Zheng, T.Y., Cha, S.T., Crossley, P.A., Kang, Y.C.: Protection algorithm for a wind turbine generator based on positive- and negative sequence fault components. In: 2011 The International Conference on Advanced Power System Automation and Protection, APAP 2011, pp. 1115–1120. IEEE (2011)
2. Bahirat, H.J., Mork, B.A., Hoidalen, H.K.: Comparison of wind farm topologies for offshore applications. In: IEEE Power and Energy Society General Meeting (2012)
3. Segura-Heras, I., Escriva-Escriva, G., Alcazar-Ortega, M.: Wind farm electrical power production model for load flow analysis. Renewable Energy 36, 1008–1013 (2011)
4. Ali, M., Matevosyan, J., Milanovic, J.V.: Probabilistic assessment of wind farm annual energy production. Electr. Power Syst. Res. 89, 70–79 (2012)
5. Grabski, F.: Semi-Markowskie modele niezawodności i eksploatacji. Instytut Badań Systemowych PAN, Warszawa (2002). (in Polish)

Cracking the DES Cipher with Cost-Optimized FPGA Devices

Jarosław Sugier$^{(\boxtimes)}$ (iD)

Faculty of Electronics, Wrocław University of Science and Technology,
Janiszewskiego Street 11/17, 50-372 Wrocław, Poland
jaroslaw.sugier@pwr.edu.pl

Abstract. This paper examines efficiency of hardware realizations of DES cracking engines implemented in contemporary low-cost Spartan-7 devices from Xilinx, Inc. The engines are designed for the known plaintext attack scheme and find the secret cipher key through brute-force exhaustive search of the entire key space. In order to comprehensively evaluate potential of the selected FPGA family in this task three architectures of DES decoders were tested: the standard iterative organization, the fully unrolled i.e. purely combinational one and the fully unrolled pipelined version. Various sizes of individual decipher units based on these three architectures led to evaluation of optimal ratios of unit speeds vs. their number which fit in one chip. The results are compared with other known hardware platforms and illustrate progress in the cipher cracking systems which was made possible by improvements in the new FPGA technologies.

Keywords: DES cracking · FPGA efficiency · Spartan-7 device ·
Iterative architecture · Pipelining

1 Introduction

Data Encryption Standard (DES), the first standardized and widely used symmetric cipher for computer data, is now obsolete as a cipher method and replaced by its modern successors – first of all by AES – but a task of its cracking is still popular as a benchmark in testing computational efficiency of various dedicated high-end processing systems. Due to relatively short length of the DES key (only 56 bits) the cipher can be broken simply by searching the whole key space and trying every of the 2^{56} combinations. With dedicated massive parallel systems this can be completed in a reasonable time and the speed of key search (expressed e.g. in number of keys tested per nanosecond) can serve as a measure of overall system computational efficiency.

In this paper we evaluate productivity of Spartan-7 FPGA devices form Xilinx, Inc. is a task of building such DES crackers. In order to exhaustively test capabilities of the programmable array for this purpose, the DES decoding units – whose heavy parallel operation is essential in high-speed search – were implemented in various architectures: the smallest standard iterative one, the fully unrolled one and the largest fully unrolled pipelined one. This allowed to illustrate how the automatic implementation tools can handle small size vs. large number of concurrently operating decoding units and what

© Springer Nature Switzerland AG 2020
W. Zamojski et al. (Eds.): DepCoS-RELCOMEX 2019, AISC 987, pp. 478–487, 2020.
https://doi.org/10.1007/978-3-030-19501-4_47

optimal proportion lead to the highest overall processing speed per single FPGA chip utilized to its maximum.

Contents of the paper is organized as follows. As a motivation and state-of-the-art of the problem the next section presents the DES cracking challenges which were proposed twenty years ago to testify the cryptographic strength of the DES cipher and subsequently became a convenient, cross-platform benchmark suitable for evaluation of computational efficiency of any CPU-, GPU- or ASIC-based parallel data processing system, distributed or concentrated. The third section describes original designs of the DES crackers proposed in this paper in three variants and the fourth one – evaluation of the practical results achieved after their implementation in selected Spartan-7 devices.

2   The DES Cipher vs. Progress in Computational Power

2.1   Standardization

The Data Encryption Standard ([8]) was the first worldwide-popular symmetric block cipher used for data protection in computer systems and networks. Originated from an internal IBM project *Lucifer* led by Horst Feistel, in 1974 it was submitted to U.S. National Bureau of Standards in response to the second request calling for a government-approved protection method of unclassified data. After modifications introduced by the U.S. National Security Agency in 1977 it was accepted as a Federal Information Processing Standard (FIPS) for the United States.

Involvement of the NSA in cipher development gave rise to numerous suspicions. Even though the original design called for at least 64-bit key, the final version decided to use only 56 bits which from the beginning was criticized as a weakening of the cryptographic strength already for computer power of that time and insecure for expected future developments. The NSA also imposed, without any explanation, different contents of the substitution boxes which allegedly could provide secret backdoor known only to the agency. It wasn't until 1990 and the first public papers on differential cryptography from Biham and Shamir [1] when it turned out that the modified S-boxes were actually much more resistant to differential attacks – which suggests that this technique was known to NSA already in 1970s.

2.2   DES Challenges

Theoretical deliberations that it is possible to build a dedicated supercomputer capable of testing all 2^{56} key combinations in a reasonable time and thus of decrypting any DES message were discussed already just after publication of the standard. Estimated multi-million dollar cost of such a project made it unfeasible and unrealistic at that time but over the years the situation was changing due to constant – and then very rapid – progress in computational power of digital systems. In the middle of 1990s speculations about state agencies or large business companies being cable to finance construction of such a system became quite realistic, although never officially confirmed in any particular case.

To end these speculations in January 1997 RSA Security, Inc. announced an open contest – a DES Challenge – with an official prize of 10,000$ for anyone who can deliver correct plaintext for a published encrypted message. The winner turned out to be the DESCHALL project [2] which used distributed power of computers connected to Internet whose owners voluntarily enrolled after an announcement published in the Usenet community. The key was found in 96 days with the scan reaching approx. quarter of the whole 56-bit space which allows to estimate the effective average search speed at approx. 2.2 K/ns [Keys per ns]. Total number of computers connected to the central host varied significantly over time: starting from 1000, after two months and popularization of the project it reached about 10,000, peaking at 14,000 just before the solution was found.

The contest stirred lots of research activities and next editions followed (see Table 1):

- DES Challenge II-1 (January 1998) was won in less than 41 days by a non-profit organization distributed.net which also used volunteer computers connected via Internet ([3]). Average search speed was approx. 18.1 K/ns.
- DES Challenge II-2 (July 1998) was won for the first time by a dedicated hardware "Deep Crack" built by Electronic Frontier Foundation [4]. A system consisting of 24 boards, each with 64 custom-designed ASIC chips, each scanning 0.06 keys per ns, offered total search speed of 92.2 K/ns. The solution was found rather luckily in just 56 h, although it would take a little over 9 days to browse the whole key space with that rate.
- DES Challenge III (January 1999) for the first time was solved in less than a day. A joint effort of EFF's "Deep Crack" and about 100,000 computers from all over the world coordinated again by distributed.net broke the code in 22 h and 15 min, reaching the final topmost speed of 250 K/ns.

These results proved indisputably that DES can no longer be considered secure and gave extra boost to intensify development of its successor – the AES standard.

2.3   Cracking DES and Massive Parallel Systems

The solutions developed for DES challenges converged with research interest in generic massive parallel computational systems and new emerging fields of their applications. In 2006 an academic initiative at the University of Kiel built COPACOBANA machine [6] consisting of 120 medium-range Xilinx Spartan-3 1000 FPGA devices. Operating at 100 MHz and holding 4 decrypting units, each device was processing DES keys at a rate of 400 K/μs and the whole machine – 48 K/ns, which was equivalent to simultaneous operation of approx. 22 thousands of Pentium 4 processors. The spin-off company SciEngines GmbH ([5, 9]) continued the development with designs built with newer FPGA devices like Virtex-4 SX35 and, currently, Spartan–6 LX150, aiming at more complex problems which massive parallelism can help to explore. Of other more recent FPGA designs, the website crack.sh still offers a commercial service of breaking any DES ciphertext in a guaranteed maximum of about 26 h using a custom-built machine with Virtex-6 LX240T chips. Total speed of 768 K/ns comes from 48 such devices, each containing 40 DES cores operating at 400 MHz.

Table 1. A brief summary of DES challenge winners and efficiency of other DES crackers

	Year	Speed [K/ns]	Notes
DESCHALL (DES Challenge I)	1997	2.2	Distributed computing (parallel operation of 1 k ÷ 14 k computers for 41 days)
distributed.net (DES Challenge II-1)	1998	18	Distributed computing
"Deep Crack" (DES Challenge II-2)	1998	92	1536 custom ASICs; total cost ~250 k$
COPACOBANA	2007	48	120 FPGA chips (Xilinx Spartan-3 1000); 0.4 K/ns per chip, 600 W power consumption, cost approx. 10 k$
crack.sh	2016	768	48 FGPA chips (Xilinx Virtex-6 LX240T); 16 K/ns per chip, commercial service
GPU 1080Ti (Hashcat)	2017	24	1 x NVIDIA 1080Ti GPU; 350 W, ~1 k$
GPU 2080 (Hashcat)	2018	31	1 x NVIDIA 2080 GPU; 215 W, ~1 k$
8x GPU 1080Ti (Hashcat)	2018	185	8 x NVIDIA 1080Ti GPUs; 3.1 kW, ~30 k$
CPU Intel i7 (Hashcat)	2017	0.046	i7-7700K CPU @ 4.20 GHz (4 cores); 91 W, ~350$
CPU AMD Ryzen7 (Hashcat)	2017	0.165	1800X CPU @ 4.10 GHz (8 cores); 95 W, ~500$

Another computational technique which made its way into massive parallel computing in general and into cipher breaking in particular are contemporary specialized Graphics Processing Units (GPU). Their internal organization heavily uses SIMD data processing modules which, with availability of generic API interfaces to common programming languages, can implement any parallel algorithm not necessarily related to computer graphic. The second part of Table 1 illustrates efficiency of probably the fastest and the most popular "password recovery" software – open-source, multi-platform Hashcat program (https://hashcat.net) – in the task of DES key search which is included in its suite of benchmark tests since 2016. The selected exemplary GPU hardware setups include two prosumer single-card PC configurations and a dedicated multi-card professional machine built especially for massive parallel computations – like cryptocurrency mining, to name one field of application (recently very common) besides cryptography. The last two entries of Table 1 are provided just to illustrate relative potential of performance desktop CPUs which – as can be seen – as DES crackers can offer much more inferior levels of computational speed.

3 Designing a DES Cracker

3.1 Cipher Decrypters

Efficient deciphering units are essential in a cracker which must speedily browse all key combinations. DES decoding ([8]) is – analogously but reversely to encoding – organized in a sequence of $n_r = 16$ rounds, $r = 15...0$, with every round operating on a distinct round key K_r. Fortunately for decryption, in DES the keys can be equally efficiently computed from the external key K in the descending order (K_{15}, K_{14}, ... K_0) as in the ascending one (K_0, K_1, ... K_{15}). In many contemporary ciphers (AES including) keys are computed sequentially using additional internal state and reversing their order is not straightforward.

DES, like any round-based cipher, can be implemented in hardware in different architectures which explore various speed vs. hardware size trade-offs ([7, 10]). One standard approach is to follow software-like mode of data processing: with one cipher round implemented in hardware the input data (ciphertext) goes through it n_r times like in the software loop repeating the same set of instructions. This concept is presented in the left part of Fig. 1 and will be denoted as architecture "X1" (1 round instance). With each round computed in one clock cycle the output is available after n_r cycles and the cracking circuit spends $n_r + 1$ cycles on each key value (the extra one is required for decision whether the proper key is found).

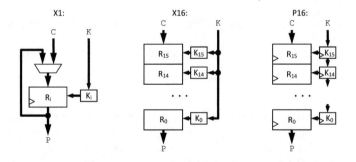

Fig. 1. Three architectures of DES decrypter evaluated in this paper: iterative (X1), combinational (X16) and pipelined (P16).

In search for higher speeds a technique of loop unrolling can be applied: with a cascade of k rounds instantiated in hardware, $k > 1$, the total number of cycles required to compute the output is reduced to n_r/k, up to the point when all n_r rounds are instantiated and the ciphertext goes through them in just one clock cycle – which will be the second architecture investigated here. This case will be denoted as "X16" (16 round instances) and is depicted in the middle of Fig. 1. In this organization the decrypter operates as a combinational circuit: each ciphertext is decoded in one clock cycle, albeit much longer one than that of the X1 case.

Finally, having 16 instances of rounds in hardware one can achieve their parallel operation with pipelining: by adding registers after each round the data propagates through one in each clock cycle but each round can work on a different data – to the effect of 16 different ciphertexts being decoded simultaneously in the cascade. This case is presented in the right part of Fig. 1 and will be denoted as "P16". The pipelined decrypter again needs n_r cycles for completion of each decoding but if such a long latency is not a problem – as is in the case of cipher cracking – the increase in overall system throughput is a fundamental advantage. Once again it should be noted that the pipelined key generation is possible thanks to simplicity of DES key schedule: each next key can be computed directly from the current one $(K_{r-1} = f(K_r))$ and the pipelined stream of the 16 keys can go along with the pipelined stream of 16 decoded ciphertexts without any extra logic or memory resources, as Fig. 1 illustrates.

3.2   The Cracker Architecture

Organization of the complete cracker system is presented in Fig. 2. An array of N decrypters (with N as large as possible for maximum overall speed of key search) is fed by a control unit with a batch of tested keys and the generated plaintexts are compared with the reference input pattern. The control unit includes an "increment by N" counter which goes through the entire 2^{56} space and generates the base value which is then incremented by "+1", "+2", ... "+ $N-1$" at each decrypter. A new batch of keys is generated every 17 clock ticks in case of the X1 units or every clock tick for X16 and P16 ones.

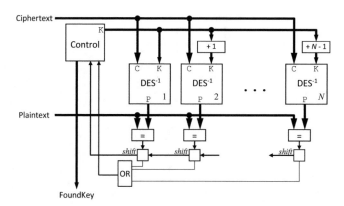

Fig. 2. Organization of N decrypters in the cracker. The largest Spartan-7 device could hold from 40 pipelined (P16) to 256 iterative (X1) decrypters.

Additional functionality was implemented for determining the "FoundKey" value when any of the comparators generates '1' on its output. A straightforward solution would be to apply an $N{:}1$ multiplexer which would pass the selected key to the output but such a circuitry – e.g. a MUX 256:1 with all paths 56-bit wide– would itself occupy substantial amount of programmable FPGA resources. A better alternative is to compute the key by determining the number of the correct decrypter k (0 ... $N-1$) and adding it to the current batch base. The k is found after the OR gate detects '1' at the output of any comparator: the search stops, the values are loaded to an N-bit shift register and its contents is sliced into the control unit. This operation as well as the regular generation of the keys during the search is controlled by a simple FSM. The N-bit shift register, whatever large N would be, is simple in implementation and incurs minimal routing compared to a $N{:}1$ multiplexer.

4   Implementation Results

4.1   Size and Speed of Implemented Designs

The complete cracking engines built in three variants, i.e. with the three types of decrypter architectures, were implemented with an increasing value of the N parameter, i.e. the number of parallel decoding modules, in Xilinx 7S75 Spartan-7 device ([11, 12]). After reaching the limit of the XC7S75 chip one supplementary N case was added reporting the largest design which fit the XC7S100 array. The results generated in the Xilinx Vivado suite are listed in Table 2.

Table 2. Size and speed parameters of the implemented DES crackers

N	T_{clk} [ns]	Speed [K/ns]	Logic delay	Levels of logic	Slices		LUTs	
X1								
1	3,35	0,02	56%	8	92	1%	274	1%
16	4,73	0,20	44%	10	974	6%	3402	7%
32	5,07	0,37	40%	10	1851	12%	6714	14%
64	5,57	0,68	49%	14	3263	20%	12280	26%
128	6,49	1,16	58%	20	7092	44%	24751	52%
234	7,91	1,74	51%	22	15277	95%	46610	97%
256[a]	8,15	1,85	59%	28	15485	97%	49427	77%
X16								
1	32,4	0,03	21%	42	351	2%	1179	2%
16	42,6	0,38	15%	43	4291	27%	16179	34%
42	43,3	0,97	21%	57	12984	81%	47253	98%
54[a]	43,1	1,25	21%	57	15921	100%	60387	94%
P16								
1	3,32	0,30	57%	8	584	4%	1584	3%
16	4,73	3,39	43%	9	5986	37%	20796	43%
32	5,45	5,88	40%	10	14481	91%	46301	96%
40[a]	5,37	7,45	40%	9	15843	99%	57948	91%

[a]implemented in 7S100 device (all others in 7S75)

The first two columns report absolute speed of the crackers as the minimum clock period T_{clk} (estimated by the implementation tools for the completed, routed design) and resulting speed of key search of the whole device. The next two parameters describe the longest propagation path which determined T_{clk}: percentage of its delay that came from logic elements (i.e. not routing) and their total number in the path. The last two columns identify size of the crackers expressed in numbers of occupied slices and LUT generators, given as absolute numbers and percentages of device capacities.

4.2   Comparison of the Three Decrypter Architectures

Parameters included in Table 2 allow for general evaluation of design efficiencies but, putting aside their raw cracking speeds, in this point we will analyse productivity of FPGA implementations for the three variants of decrypter architectures. Figure 3 presents more concealed characteristics of the designs: a number of logic elements constituting the longest propagation path and effectiveness of array utilization which is computed from the basic parameters as the cracking speed divided by the total amount of used LUTs.

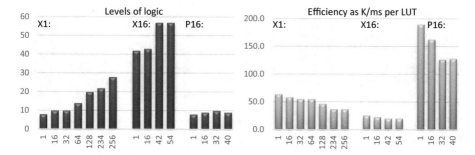

Fig. 3. Internal characteristics of cracker implementations: number of logic elements in the longest combinational path (left) and synthetic cracking efficiency expressed as the key search speed per one LUT element (right).

The $X1|_{N=1}$ case was the smallest and the easiest one for the tools and can serve as the reference implementation. Its longest combinational path spans one cipher round and includes 8 logic elements. An analogous situation is in the P16 designs but in the X16 cases such a path goes across all 16 rounds so here the increase in the number of logic elements is justified and expected. What the left part of Fig. 3 reveals is an increase in this parameter with rising N for X1 designs which indicates problems in their implementation, especially in routing. In contrast, no such increase can be seen in P16 crackers: even in the largest $N = 40$ case this parameter remains under 10, despite the fact that in pure LUT or slice size the design is as large as $X1|_{N=256}$. The number of deciphering units N (over 50 in X1 crackers) makes here the difference or, from another perspective, pipelining significantly helps in P16 implementations.

Contrary to these conclusions, in synthetic efficiency computed as cracking speed of K/ms per one LUT of size (right part of Fig. 3) the fall with increasing N parameter is the largest for P16 designs while moderate for X1 and X16 ones. Nevertheless, the P16 level of raw cracking speed is so high that even in the largest designs ($N = 32$ or 40) the absolute values of this parameter remain clearly the highest. The X16 variants do not offer any advantage and can serve only as a comparative base for evaluation of the two other options.

4.3   Scaling of Size and Speed with Increasing Number of Decrypters

For any kind of decrypter architecture it can be assumed that with rising N parameter there should be a proportional increase both in size of the cracker and in its speed. Taking the $N = 1$ case as the reference, actual parameters of $N > 1$ designs can be compared against expected values computed as "multiply by N" – and such a comparison is presented in Fig. 4.

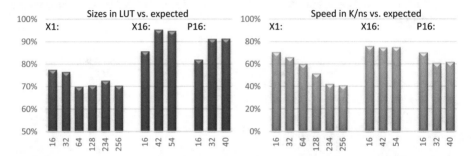

Fig. 4. Actual sizes (left) and speeds (right) of the crackers as percentages of expected values estimated from respective $*|_{N=1}$ cases.

It is known that in FPGA implementations larger designs offer better room for optimization ([10]) and this is again confirmed in the left part of the figure: sizes of the X1 cases are up to 30% smaller than expected while in the unrolled X16 and P16 organizations the reductions is somewhat slighter, by 5–18%. No evident trend related to increasing N can be recognised.

Unfortunately, the optimization is not seen in speed (right part of Fig. 4) as all designs operated at least 25% slower than the extrapolation of their $*|_{N=1}$ cases would suggest. The worst situation again was in X1 crackers: in their largest configurations the speed was reduced to only 40% of what could be achieved if the speed of the $N = 1$ case was preserved, with visible continuous falling tendency. The reductions for P16 designs are also remarkable (down to 60%) but are smaller and without such a trend.

5   Conclusions

In this paper we have thoroughly investigated designing a DES cracker with different architectures of decrypter unit. Implementations in Spartan-7 FPGA devices provided practical data for their evaluation.

The results confirmed that in the case of this – relatively simple as for today's standards – cipher the purely iterative decoding modules (X1) are so small that their very high number which fit one device makes implementation of the whole cracker system difficult and ineffective. The fully unrolled pipelined decrypters offer very high throughput per unit and, due to larger size, their total number in the chip is smaller which eases implementation of the complete cracker and leads to optimal results.

The best solution – the P16 cracker with 40 pipelined decrypters in a 7S100 device – has a theoretical maximum speed of key search estimated by the implementation tools as 7.45 keys per ns, which is approx. 150 times faster than specially designed ASICs used in the EFF "Deep Crack" machine from 1998 (the winner of the DES Challenge II-2), and 18 times faster than Spartn-3 devices used to build COPACOBANA machine in 2007. If 120 such chips were used in a construction analogous to that of COPA-COBANA, the resulting total cracking speed would be 894 K/ns which surpasses any practical solution available today, either GPU or FPGA-based.

References

1. Biham, E., Shamir, A.: Differential cryptanalysis of DES-like cryptosystems. J. Cryptol. **4** (1), 3–72 (1991)
2. Curtin, M., Dolske, J.: A brute force search of DES keyspace. In: 8th Usenix Symposium, pp. 26–29 (1998)
3. distributed.net: Overview of the DES contests. https://www.distributed.net/DES. Accessed Mar 2019
4. Electronic Frontier Foundation: Cracking DES. Secrets of Encryption Research, Wiretap Politics, and Chip Design. O'Reilly Media (1998)
5. Guneysu, T., Paar, C., Pfeiffer, G., Schimmler, M.: Enhancing COPACOBANA for advanced applications in cryptography and cryptanalysis. In: 2008 International Conference on Field Programmable Logic and Applications, pp. 675–678. IEEE (2008)
6. Kumar, S., Paar, C., Pelzl, J., Pfeiffer, G., Schimmler, M.: Breaking ciphers with COPACOBANA – a cost-optimized parallel code breaker. In: Goubin, L., Matsui, M. (eds.) Cryptographic Hardware and Embedded Systems - CHES 2006. Lecture Notes in Computer Science, vol. 4249. Springer, Berlin, Heidelberg (2006)
7. Mazur, F.: FPGA hardware implementation of a cipher algorithm for cryptanalysis. Dissertation for BSc degree. Wrocław University of Science and Technology (2018). (in Polish)
8. National Institute of Standards and Technology: FIPS PUB 46-3 Data Encryption Standard (DES) (1999)
9. SciEngines   Hardware.   https://www.sciengines.com/technology-platform/sciengines-hardware. Accessed Mar 2019
10. Sugier, J.: Spartan FPGA devices in implementations of AES, BLAKE and KECCAK cryptographic functions. In: Zamojski, W., Mazurkiewicz, J., Sugier, J., Walkowiak, T., Kacprzyk, J. (eds.) Contemporary Complex Systems and Their Dependability, Proceedings of the Thirteenth International Conference on Dependability and Complex Systems DepCoS-RELCOMEX. AISC, vol. 761, pp. 461–470 (2019)
11. Xilinx, Inc.: 7 Series FPGAs Configurable Logic Block. UG474.PDF. www.xilinx.com. Accessed Mar 2019
12. Xilinx, Inc.: 7 Series FPGAs Data Sheet: Overview. DS180.PDF. www.xilinx.com. Accessed Mar 2019

Maintaining Railway Operational Requirements in the Context of the GSM-R User Interface Capacity

Marek Sumiła(✉)

Railway Institute in Warsaw, 04275 Warsaw, Poland
msumila@ikolej.pl

Abstract. The main purpose of the article is to answer the question about meeting the operational requirements of the railway through the GSM-R cellular communication system. The first part presents the basic features of the radio interface (subscriber access) to GSM-R. Next has been carried out an analysis of selected railway interoperability requirements of the CCS TSI. In the main part, the theory of telecommunications traffic was used to calculate the probability of call blocking in the area of generic cell in the network. The article ends with summaries and conclusions from the analysis.

Keywords: GSM-R · System maintenance ·
Of traffic telecommunication theory

1 Introduction

Rail transport is a key means of transport in the transport of goods and people. Despite the long history of existence, it still works based on the same principles. Modern rail transport ensures the possibility of moving faster and more comfortable. In order to improve these standards, the European Union adopted a resolution contained in Decision 1692/96/EC [3] and further Council Directive 96/48/EC of 23 July 1996 [4] on the interoperability of the trans-European high-speed rail system and the Directive 2001/16/EC [5] on the interoperability of the trans-European conventional rail system[1]. These documents indicated guidelines for unifying the standards of national railways in the European Union, in such a way that trains could easily cross the borders of EU countries without the need to replace locomotives, train drivers or railway communication systems. This is how the idea of interoperability of the rail system in the European Union was born, the executive acts of which are EU decisions on Technical Specifications for Interoperability (TSI). The specification on the control of train TSI CCS [2] describes the basic assumptions and requirements for the train control system at the infrastructure layer for interoperable trains. The implementation of this objective is to be achieved through the implementation of the ERTMS European Railway Traffic Management System, which consists of two basic components:

---

[1] Both directives were amended.

W. Zamojski et al. (Eds.): DepCoS-RELCOMEX 2019, AISC 987, pp. 488–497, 2020.
https://doi.org/10.1007/978-3-030-19501-4_48

European Train Control System (ETCS) and digital railway radio communication system GSM-R (*GSM for Railway*). The ETCS system is used to control train traffic through electronic and IT solutions unifying and improving train management. The GSM-R system is a system enabling digital voice and data radio communication via a dedicated railway network. Like in other ICT systems [15, 18] GSM-R is used for voice communication and data transmission for train control system ETCS. The state of development of these networks can be followed in UIC publications and on the basis of reports from railway infrastructure managers in individual countries.

Apart from issues related to technological problems with the old technology network presented in works [19, 20] Poland is obliged to implement GSM-R to ensure the interoperability of its rail system. The requirement imposed by the European Union discussed in more detail in part 3. Nowadays, Poland is at the front of a wide implementation of the GSM-R network (about 14,000 km of road track).

The titled capacity of the user interface in GSM-R network is defined as a availability of the radio carrier resources (channels) for the users in the single cell area and becomes significant in light of the current Technical Specifications form 2016 [2] where the legislator introduced requirements that affect on the availability of the communication service in existing GSM-R networks in Europe and on the planned GSM-R network in Poland. The problem lies in the limited number of possible simultaneous connections in the GSM-R cell area. The typical methods of calculating cell load and blocking factor which use of statistical methods based on the relation between the area density and the number of subscribers [21]. The methods were adjusted to the specifics of public cellular networks and cannot be applied for GSM-R because the subscribers use the network for the different purpose and there are additional the requirements comes from the TSI [2]. For these reasons, the Author of the work decided to investigate and describe exposed problem.

In the next section will briefly describe the GSM-R system and its technical capabilities. Section 3 discusses the current requirements and needs of TSI for train control with use GSM-R technology. The main part of the article will analyze the availability of resources in the area of one cell of the network with analysis of telecommunications traffic at different levels and the likelihood of call blocking in the network. There is only few papers discussing this issue in the context of GSM-R [14, 17].

2 Technical Capabilities of GSM-R

The GSM-R system concept was created in the first half of the nineties of the last century. The control over the development of the standard is exercised by UIC as part of the EIRENE (*European Integrated Radio Enhanced Network*) project[2], which aims to determine the technical and functional requirements of GSM-R. The indicated division in the development of the GSM-R standard has led to the creation of two work subgroups dealing with the development of technical system specifications and the development of functional specifications. The current status of these specifications is

---

[2] https://uic.org/gsm-r (last accessed 08.01.2019).

described in the EIRENE SRS [10] and FRS documents [11]. Detailed technical specifications for the GSM-R system were developed as part of the MORANE (Mobile Radio for Railways Networks in Europe) project operating in 1996–1999[3].

Basically, GSM-R is a system operating in the following separate frequency band 876–880 MHz (uplink) and 921–925 MHz (downlink). In addition, Directive 2009/114/EC [6], it is possible to extend the baseband by a further 3 MHz frequency in the bands[4] 873–876 MHz (uplink) and 918–921 MHz (downlink).

Technically GSM-R works in the TDMA/FDMA technique in 19 channels/carriers every 200 kHz[5]. It has a 4 MHz band below the band for cellular public radio-communication systems. Eight time slots are multiplexed within each such channel. One of these slots is used for the transmission of signalling. Each call can be handled in FR (Full Rate) mode by one time slot with the Circuit Switched Data (CSD) on one physical channel/carrier.

The area of work of cellular systems is divided into sub-areas, so-called cells each served by a separate Base Transceiver Station (BTS) working with one or more carrier frequencies[6]. This network structure allows multiple use of the same carrier to support a larger number of subscribers. In cellular architecture it is important that neighbouring areas do not work on the same carrier, as it could lead to co-channel interference. The typical range of a single cell is not large, but in the case of rural areas it can be expanded to a size of about 30–35 km.

3   TSI Requirements for Radio Communications

An earlier decision of the European Commission 2012/88/EU [1][7] stressed the importance of the radio communications system for the safety and interoperability of rail transport. The lack of this availability of radio communications services has operational and commercial consequences, including economic and commercial effects (e.g. decrease in the performance of the railway line, increased delays and stoppages on the line) and effects affecting safety, in particular the making of Railway Emergency Call (REC). For these reasons, the current specification of the CCS TSI [2] and the documents referred to by them [8–11] define the technical conditions for communication in the field of data transmission in the ETCS system and voice communication of railway employees. Application is for:

– voice communication: transmission in full time slot (Full Rate), CSD, pre-emption for connections with higher priority.

---

[3] https://trimis.ec.europa.eu/project/mobile-radio-railway-networks-europe#tab-partners (last accessed 08.01.2019).

[4] The decision to use the above bands for GSM-R purpose was left to the Member States.

[5] The main GSM-R band (UIC band) occupies ARFCN channels numbered 955–973.

[6] A single carrier (channel) transmitter is called TRX.

[7] Currently, the decision has been repealed by the Commission Regulation (EU) 2016/919, nevertheless it is worth to mention the above provision for the significance of the issue.

- data transmission for the needs of the ETCS (level 2 or 3) system: at least two EDOR (ETCS Data Only Radio) devices on the vehicle, Full Rate, CSD or GPRS/EGPRS, continued connection during drive.

A simple analysis of the above requirements allows to state that the average train can occupy two channels in the area of one cell for the needs of the ETCS system and a minimum of two additional channels for the needs of voice communication. One of these channels will be used for communication by the driver and the other for the needs of the train, e.g. the train manager.

In view of the above requirements, it is necessary to ask is there a threat of non-compliance with the operational requirements of the railway as a result of the limitation of the GSM-R network resources?

4 Analysis of the Availability of GSM-R Network Resources

The use of elements of the theory of telecommunications traffic allows to predict how serious is the problem of the availability of resources in the GSM-R network is. The basic parameter describing this concept is the intensity of telecommunications traffic. Traffic intensity A is defined as the ratio of the average occupancy time of network resources t_i to the period of time of the observation T. The classic formula describing this definition is as follows

$$A = \frac{1}{T} \sum_{i=1}^{M} t_i \tag{1}$$

where M is the number of arrival calls.

The guidelines of the Polish GSM-R network operator (PKP PLK S.A.) indicated in OPZ reference number 6060/ICZ4/08860/03374/15/P[8] were used to calculate this parameter. For the calculation of traffic assumed the existence of point-to-point connections and group-broadcasting and alarm. It was assumed that the average number of P2P calls will be 2 and on average they will last about 1 min. For group calls, the average connection time will be 2 min, broadcasting time 30 s, and alarm time 1 min.

On this basis, it was calculated that the average traffic volume will be 0.6 Erl. Due to the fact that the movement may be outgoing, incoming and internal for a given cell, the calculated value should be approximately tripled. Therefore, for further calculations, it should be assumed that the estimated traffic intensity is equal about 1.8 Erl.

4.1 Mathematical Apparatus

The theory of telecommunication traffic includes statistical behavior of network subscribers. Based on many years of observations, it was found that only a small percentage of its users use the network resources at the same time, which is why it is allowed to use a much larger number of users from limited resources. In this case, if the

---

[8] http://www.mapadotacji.gov.pl/projekt/7281390 (last accessed 07.01.2019).

subscribers' activity increases dramatically, the network will not be able to process all calls. Such a system is called a loss system if the network blocks the connection. A queuing system may also be used in the case when calls that cannot be handled immediately are queued. The probability of congestion in the network has been described by the theory of mass service and Markov processes with discrete space of states [12, 13]. Attempts to include the discussed theory in the applications of cellular networks can be found, for example, in the book [21].

For simplicity, the Erlang B Model will be used for further analysis, which well maps the network with the stream of requests described in the Poisson stream, which allows loss of connections in the absence of available network resources, i.e. free channels in the cell area. This model is described by the following formula

$$P(A, N) = \frac{\frac{A^N}{N!}}{\sum_{i=0}^{N} \frac{A^i}{i!}} \qquad (2)$$

where:

P - the probability of blocking
A - average traffic [Erl]
N - number of channels in a given cell

This formula was obtained after assuming that the channel assignment requests have a Poisson distribution, and the subscriber's time of subscribing is exponential. In the Erlang B model, it is also assumed that the number of subscribers is infinitely large and the number of channels is limited. In this model, it is also assumed that channel allocation requests are non-memorable.

The presented model can be treated as a certain simplification, because the GSM-R system has eMLPP feature (enhanced Multi-Level Precedence and Pre-emption) wherein the connections can be queued or preempted [7]. In this case, the Erlang C Model would be used for calculation the probability of falling calls into the queue.

$$P(A, N) = \frac{\frac{A^N N}{N!(N-A)}}{\sum_{i=0}^{N-1} \frac{A^i}{i!} + \frac{A^N N}{N!(N-A)}} \qquad (3)$$

The principle is also to expropriate cell resources for data transmission in the ETCS system. In this case, the vehicle using the radio to the ETCS causes occupancy of the timeslots while staying in the cell area. This means that an active ETCS radio reduces the number of available timeslots for other users in the area.

4.2 Case Studies

The technical interoperability requirements presented in point 3 have an impact on the traffic in the GSM-R network. The operational availability of the network for the needs of railway communication will be considered in limiting to one cell of this network for a route or a small station.

Case 1

There is no train in the area of the cell, voice connections are made by employees in the area of the railway station. The area is covered by a single TRX signal (7 timeslots).

$$P_{ErlB}(A,N) = P_{ErlB}(1.8; \ 7) = 0.02\% \tag{4}$$

Case 2

There is no train in the area of the cell, telephone connections are made by employees in the area of the railway station. Dialed calls can be queued. The area is covered by a single TRX signal (7 timeslots).

$$P_{ErlC}(A,N) = P_{ErlC}(1.8; \ 7) = 0.03\% \tag{5}$$

Case 3

In the cell area there is a moving train with an active one EDOR radio. Telephone calls are made by employees of the train and in the area covered by a single TRX signal (7 timeslots). The probability of blocking is

$$P_{ErlB}(A,N) = P_{ErlB}(1.8; \ 7-1) = 0.78\% \tag{6}$$

The probability of entering a call in the queue is

$$P_{ErlC}(A,N) = P_{ErlC}(1.8; \ 7-1) = 1.11\% \tag{7}$$

Case 4

In the cell area there is a moving train with two active EDOR radios. Voice calls are made by employees of the train and in the area covered by a single TRX signal (7 timeslots). The probabilities of occurrence of the discussed events are

$$P_{ErlB}(A,N) = P_{ErlB}(1.8; \ 7-2) = 2.63\% \tag{8}$$

$$P_{ErlC}(A,N) = P_{ErlC}(1.8; \ 7-2) = 4.05\% \tag{9}$$

Case 5

There are two active trains in the cell area (4 EDOR). Voice calls are made by employees of trains and in the area covered by a single TRX signal (7 timeslots). NOTE: the occurrence of two trains can mean trains going in different directions.

$$P_{ErlB}(A,N) = P_{ErlB}(1.8; \ 7-4) = 18.03\% \tag{10}$$

$$P_{ErlC}(A,N) = P_{ErlC}(1.8; \ 7-4) = 35.47\% \tag{11}$$

Case 6

There are three trains in the area of the cell (6 EDORs). Voice calls are made by employees of trains and in the area covered by a single TRX signal (7 timeslots).

$$P_{ErlB}(A, N) = P_{ErlB}(1.8;\ 7 - 6) = 64.29\% \tag{12}$$

$$P_{ErlC}(A, N) = P_{ErlC}(1.8;\ 7 - 6) > 100.00\% \tag{13}$$

Based on the above calculations, it can be concluded that for more than three trains, voice connections in the area covered by a single TRX signal will be almost impossible to do.

Case 7

There are one or more trains in the cell area. Voice connections are made by employees of trains and in the area covered by the double TRX carrier (14 timeslots) (Table 1).

Table 1. Analysis of probability of call blocking for 2 TRX calculated by using Erlang B and Erlang C Models.

The number of trains with 2 EDORs devices	$P_{ErlB}(1.8;\ 14 - EDOR)$	$P_{ErlC}(1.8;\ 14 - EDOR)$
1	0.00%	0.00%
2	0.00%	0.00%
3	0.05%	0.06%
4	0.78%	1.11%
5	7.50%	12.85%
6	36.65%	85.26%
7	100.00%	100.00%

To sum up, in the analysed cases it should be noted that the appearance of a train with the ETCS radio-active system results in abrupt limitation of network resources for remaining radio connections. Being in the cell area of three or more trains blocks the GSM-R network for a single TRX. The situation may be critical in the areas of railway junctions, where trains run in two directions and the time interval between them is small. The solution to this problem is to assume that in the area covered by ETCS, the number of carriers supporting the cell area is greater than one.

4.3   Availability of Services in Relation to the Cell Area

The previous considerations were conducted due to the theoretical availability of resources in a single cell of the network. In the GSM-R network, the volume of traffic in the cell is affected by the geographical size of the cell, the type of area (railway line/railway station), type of railway line, size of the railway station, additional maneuvering traffic. Due to the large number of factors affecting the volume of telecommunications traffic in the area served by BTS, it is possible to determine the probability of blocking factor in the network based on the volume of telecommunications traffic generated by subscribers and vehicles.

On the basis of calculations, the occurrence of call blocking was performed depending on traffic. For the sake of clarity, it was assumed that the movement was graded on a scale of 1 to 10 Erl and the number of TRX in BTS equal to 3 (Fig. 1).

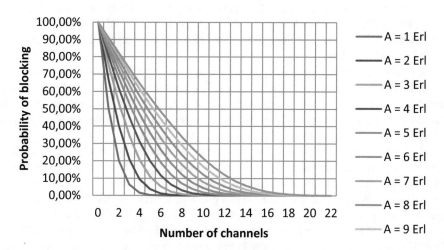

Fig. 1. Charts of probability of call blocking in the function of intensity traffic for different number of channels

On the basis of the presented graph, it is possible to estimate the probability of occurrence of difficulties in establishing a connection in a given area with an increase in the number of connections and their duration. In public networks, the limited availability of the network is acceptable in certain periods during the year, as it is assumed to be temporary. In the case of a communication network working for the needs of railway work, the nature of traffic (its fluctuation) is essentially constant. This means that limited access to radio network resources is not transient. The second element that distinguishes the phenomenon of call blocking in the public network from the railway network is the scope of impact on train traffic and service processes of railway work.

5 Conclusion

The issue presented in the article concerns the availability of communication services with a limited number of radio channels in the GSM-R cell area. The obligation to ensure continuous ETCS data transmission for the train was drawn in the part 3 of the article. The Technical Specification for Interoperability also implies the necessity of establishing two parallel CSD connections for each the train. This means that the network resources in the area of the GSM-R cell are reduced in a fractional way with each train appear. In this sense, the classic statistical methods of calculating cellular network availability fail because they depend directly on the subscribers' activity, but here we are focused on the number of trains in the area and additionally subscribers activity. The comparison the train connection and subscriber connections are not the same. This is incompatible for the author because each train reduce the resources in the cell area in a different way than the subscriber.

In the Sect. 4.2 were presented different cases of changes traffic rate according to specific railroad situation in the GSM-R cell area. Theoretically, 6 trains in the area significantly limits the availability to the network for other subscribers. It is worth remembering, however, that the cell size can be large (see Sect. 2), and the lines are usually double-tracked.

Section 4.3 was calculate the probability of blocking connections in the cell area for different traffic rate and numbers of radio channels available. Such analysis is helpful for planning radio resources in a given area of the network. In this case increasing the number of TRX is not always possible due to the GSM-R system limitation and principles of cellular radio network planning.

To summarize, the presented problem can be serious for the areas of large stations where we have a lot of trains and subscribers calls (p2p calls, broadcast calls, group calls and emergency calls). It creates new challenges in operation and control the train in such areas. For Poland, it is necessary to put a lot of attention and investigate trains' traffic in each railway area during the process of planning GSM-R network.

Article co-financed from the European Funds as part of the task POIR 04.01.01.00-0005/17. Subactivity 4.1.1 Operational Programme "Inteligentny Rozwój Strategiczne programy badawcze dla gospodarki", titled "Standaryzacja wybranych interfejsów komputerowych urządzeń i systemów sterowania ruchem kolejowym (srk)".

References

1. 2012/88/EU: Commission Decision of 25 January 2012 on the technical specification for interoperability relating to the control-command and signalling subsystems of the trans-European rail system (notified under document C(2012) 172)
2. Commission Regulation (EU) 2016/919 of 27 May 2016 on the technical specification for interoperability relating to the 'control-command and signalling' subsystems of the rail system in the European Union
3. Decision No 1692/96/EC of the European Parliament and of the Council of 23 July 1996 on Community guidelines for the development of the trans-European transport network
4. Council Directive 96/48/EC of 23 July 1996 on the interoperability of the trans-European high-speed rail system
5. Directive 2001/16/EC of the European Parliament and of the Council of 19 March 2001 on the interoperability of the trans-European conventional rail system
6. Directive 2009/114/EC of the European Parliament and of the Council of 16 September 2009 amending Council Directive 87/372/EEC on the frequency bands to be reserved for the coordinated introduction of public pan-European cellular digital land-based mobile communications in the Community (Text with EEA relevance)

7. ETSI EN 122 067 (V3.0.1) Digital cellular telecommunications system (Phase 2+) (GSM); Universal Mobile Telecommunications System (UMTS); Enhanced Multi-Level Precedence and Pre-emption service (eMLPP) - Stage 1 (3G TS 22.067 version 3.0.1 Release 1999)
8. ETSI EN 301 515 (V2.3.0) Global System for Mobile communication (GSM); Requirements for GSM operation on railways
9. ETSI TS 102 281 (V3.0.0) Railways Telecommunications (RT); Global System for Mobile communications (GSM); Detailed requirements for GSM operation on Railways
10. EIRENE System Requirements Specification. European Integrated Railway Radio Enhanced Network. GSM-R Operators Group. UIC CODE 951. Version 16.0.0. Paris (2015)
11. EIRENE Functional Requirements Specification. European Integrated Railway Radio Enhanced Network. GSM-R Functional Group. UIC CODE 950. Version 8.0.0, Paris (2015)
12. Freeman, R.L.: Fundamentals of Telecommunications, p. 57. Wiley, New York (2005)
13. Zeng, G.: Two common properties of the Erlang-B function, Erlang-C function, and Engset blocking function. Math. Comput. Model. 37(12), 1287–1296 (2003)
14. Lindström, G.: Is GSM-R the limiting factor for the ERTMS system capacity? Master thesis. Division of Transport and Logistics KTH Railway Group (2012)
15. Losurdo, F., Dileo, I., Siergiejczyk, M., Krzykowska, K., Krzykowski, M.: Innovation in the ICT infrastructure as a key factor in enhancing road safety. A multi-sectoral approach. In: Proceedings of the 25th International Conference on Systems Engineering, ICSEng 2017, pp. 157–162, January 2017
16. ITU-T E.490 Traffic measurement and evaluation – General survey
17. Report Ex-Post Evaluation. Operational Requirements of Railway Radio Communication Systems. ERA (2014)
18. Siergiejczyk, M., Krzykowska, K., Rosinski, A., Grieco, L.A.: Reliability and viewpoints of selected ITS system. In: Proceedings of the 25th International Conference on Systems Engineering, ICSEng 2017, pp. 141–146, January 2017
19. Sumiła, M.: Risk analysis of interference railway GSM-R system in Polish conditions. In: Proceedings of the Eleventh International Conference on Dependability and Complex Systems DepCoS-RELCOMEX, Brunów, Poland, 27 June–1 July 2016. Advanced in Intelligent Systems and Computing, vol. 470. Springer, Heidelberg (2016)
20. Sumiła, M., Miszkiewicz, A.: Analysis of the problem of interference of the public network operators to GSM-R. In: Jerzy, M. (ed.) Tools of Transport Telematics, pp. 76–82. Springer, Heidelberg (2015)
21. Wesołowski, K.: Systemy radiokomunikacji ruchomej. WKŁ, Warszawa (2006)

Benchmarking Comparison of Swish vs. Other Activation Functions on CIFAR-10 Imageset

Tomasz Szandała[(✉)]

Wrocław University of Science and Technology, Wrocław, Poland
Tomasz.Szandala@pwr.edu.pl

Abstract. The choice of the most appropriate activation functions for artificial neural networks has a significant effect on the training time and task performance. Nowadays the most widely-used activation function is the Rectified Linear Unit (ReLU). Despite its "dying ReLU problem" and many attempts to replace it with something better, it is still considered as default choice to begin with creation of network. Two years ago a new, promising function has been described formulated by Google Brain Team. The proposed function - named Swish - was obtained using a combination of exhaustive and reinforcement learning-based search. According to the authors, simply replacing ReLUs with Swish units improves top-1 classification accuracy on ImageNet by 0.9% for Mobile NASNet-A and 0.6% for Inception-ResNet-v2. This paper describes an experiment on CIFAR-10 image set where Swish appears not to outperform ReLU.

Keywords: Activation function · Relu · Leaky relu · Swish · Deep neural networks

1 Introduction

The choice of an activation functions in Deep Neural Networks has a significant impact on the training dynamics and task performance [1]. Currently, the most prosperous and commonly used activation function is the Rectified Linear Unit (ReLU), which is defined as $f(x) = max(0,x)$ [2, 3]. Although various alternatives to ReLU have been proposed, that omit "dying ReLU" problem or assures better accuracy for network [4–6]. Dying ReLU is a situation when all weights of concrete node after training always return negative value for any possible input value. Therefore this node is considered dead, because no matter what error is being backpropagated it will always be neglected due to zero-value-factor of derivative from given neuron.

However none of so far formulated alternatives have managed to replace ReLU due to inconsistent gains in network's performance. In 2017 Google Brain Team has proposed a new activation function, named Swish [7], which is simply x times *sigmoid(x)*. According to the authors, the Swish activation yields significantly better accuracy than rectified linear unit or any other function in this family, at the cost of small training time recession [7, 8]. These are very impressive results from a rather simple looking function.

© Springer Nature Switzerland AG 2020
W. Zamojski et al. (Eds.): DepCoS-RELCOMEX 2019, AISC 987, pp. 498–505, 2020.
https://doi.org/10.1007/978-3-030-19501-4_49

This paper consists of short introduction to Swish function and it will describe a practical comparison of Swish performance in contrast to few other popular activation functions. Each of the most popular activation functions (ReLU, Leaky ReLU, sigmoid and tanh) will be used in the same model of neural network to classify images from CIFAR-10 dataset. Compared criteria would be:

- accuracy of classification,
- speed of training,
- time performance of network during runtime.

Hopefully, this will help the reader decide whether or not to use the Swish, instead ReLU in his own models.

2 Swish Function

Ramachandran et al. from Google brain team announced Swish activation function as an alternative to ReLU in 2017 [7]. The team has employed automatic search to find high-performing novel activation functions. Their search space contained compositions of elementary unary and binary functions such as max, min, sin, tanh, or exp. They found many functions violating properties deemed as useful, such as non-monotonic activation functions or functions violating the gradient-preserving property of ReLU. In the end, their most successful function, which they call Swish, violates both of these conditions.

Despite all ReLUs disadvantages it still plays an important role in deep learning studies [8]. Nonetheless experiments show that this new activation function over performs ReLU for deeper networks [9]. This function can be also known as sigmoid weighted linear unit or SiLU. Swish is defined as x times multiplied *sigmoid* of x function (Eq. 1) (Fig. 1).

$$f(x) = x * \sigma(x) = x * \frac{1}{1 + e^{-x}} \tag{1}$$

In literature exists an enhanced version of Swish, that contains additional parameter *beta* [8] (Eq. 2). A non-zero real value that is being adjusted during training and should improve overall network's accuracy.

$$f(x) = \beta x * \sigma(\beta x) \tag{2}$$

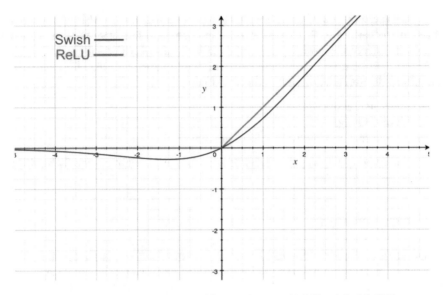

Fig. 1. Chart presenting shape difference between ReLU and Swish [11]

If β gets closer to 0 it resembles linear function - blue line, if β gets closer to infinity, then the function looks like ReLU - yellow line (Fig. 2).

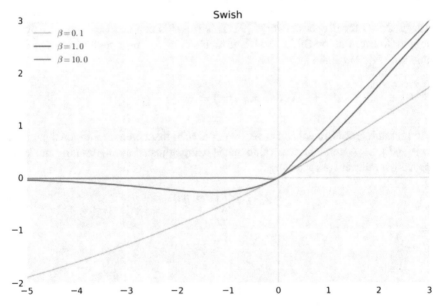

Fig. 2. Plot of behavior of Swish depending on beta parameter value [11]

3   Other Functions

Swish is being said to replace the most popular nowadays Rectified Linear Unit and its variations. The base ReLU is just positive value of *x*. The most well known ReLU mutation is Leaky ReLU. This version instead of constant *0* on negative part, has z small hyperparameter α equal to *0.01* (Eq. 3). This prevents Leaky ReLU nodes from dying [5], because derivative value is always (except *0* itself) different from zero, therefore: always prone to training.

$$f(x) = max(\alpha * x, x) \tag{3}$$

Since Swish is an enhancement of trivial sigmoid [10] (Eq. 4) this one also should be taken into consideration in the research, in order to see how much the *x* multiplier changes. Solely sigmoid as activation function comes with multiple drawbacks like vanishing gradient problem, non-zero centrism and overall weak accuracy in deep neural networks.

$$f(x) = \sigma(x) = \frac{1}{1 + e^{-x}} \tag{4}$$

There is also a tanh function [11], currently fallen out of grace in most cases, due to, similar as sigmoid, vanishing gradient problem (Eq. 5). It used to be an improved sigmoid (also an 's'-shaped function), that is zero-centered version. While it performs much better than sigmoid, it still comes far behind ReLu in case of accuracy and computation cost.

$$f(x) = tanh(x) = \frac{e^x - e^{-x}}{e^x + e^{-x}} \tag{5}$$

4   Experiment

Experiment consists of training the same network model using different activation functions. All nodes have the same activation function each time we test given formula, except the final layer which always goes with softmax. Network has been implemented in Python 3, using keras and tensorflow modules. All operations were performed on single GPU Nvidia GeForce GTX 860 M. The experiment was carried out based on dataset CIFAR-10 [12].

The CIFAR-10 dataset consists of 60000 32 × 32 colour images in 10 classes, with 6000 images per class. There are 50000 training images and 10000 test images. The final accuracy is rather far from perfect, but goal is to compare performance of different activations in the same environment, therefore for experiment a simple network [13] with 2 convolution layers. Complete network schema can be seen on Fig. 3. In order to mitigate randomness there are 3 series of each function usage and the final results are average values for given function.

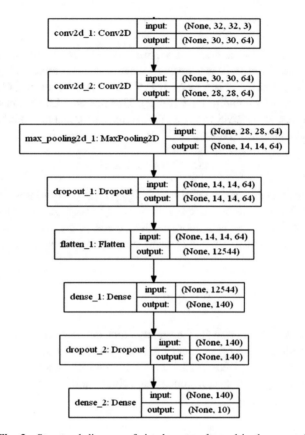

Fig. 3. Structural diagram of simple network used in the research

5   Conclusions

In contrast to our expectations, the Swish function does not overperform ReLU. In the case of accuracy Swish achieves results at best comparable with ReLU. Below are the results for training accuracy for our setup. In this simple network Swish has never correctly classified more than 70% images (Table 1). However it still clearly outperforms "s"-shaped functions like sigmoid or tanh (Fig. 4).

Table 1. Ratio of correctly classified images from validation set after full training

	sigmoid	tanh	relu	leaky_relu	swish
Final accuracy	0.6166	0.6757	0.7179	0.7295	0.6989

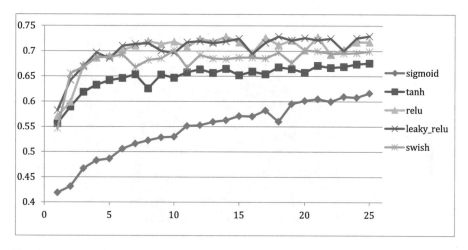

Fig. 4. Chart presenting networks performance after each of 25 epochs of training for each activation function

Regarding training time Swish performs noticeably worse. Approximately 12% longer than ReLU, however it is around 6% faster, compared to leaky ReLU based network (Fig. 5).

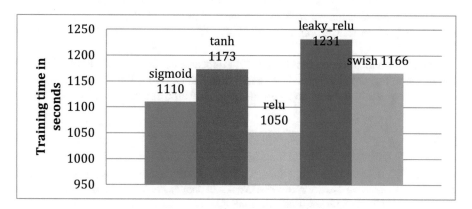

Fig. 5. Training time of each network in seconds.

The last question is how the Swish performs in runtime. Here Swish appears to 22% slower than ReLU and 5% than its Leaky ReLU variation. This is a non trivial slowdown for cases where real time performance is needed. It is even more noteworthy if the network is designed to be used in low-power IoT devices (Fig. 6).

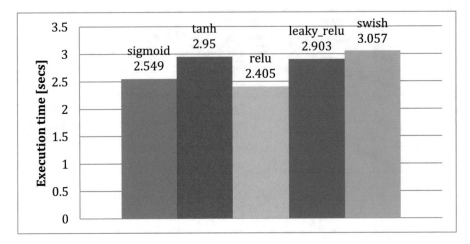

Fig. 6. Execution time of classification of 10 000 images for each network

To sum up, it appears that in trivial examples, like classification of simple images like CIFAR-10 Swish does not offer much more than ReLU or Leaky ReLU. Furthermore it is much more costly in case of computation time, both during training and runtime than classical Rectifier Linear Unit. The authors stated that Swish should behave good in small networks, but shine in much deeper ones, therefore we can decide it is not worth to engage Swish for networks with less than a dozen layers. Similar result has been also achieved by Jaiyam Sharmain in his experiment to compare Swish and ReLU [14]. In which both simple Swish and β-Swish has failed to overperform ReLU.

Taking all into consideration ReLU stays unbeaten as the first choice for researcher during preparing network model. It seems that either "dying ReLU" is not so problematic issue or still there is no more resilient activation function than ReLU.

References

1. LeCun, Y., Bottou, L., Orr, G., Müller, K.: Efficient BackProp. In: Montavon, G., Orr, G.B., Müller, K.R. (eds.) Neural Networks: Tricks of the Trade. Lecture Notes in Computer Science, vol 7700. Springer, Heidelberg (1998)
2. Hahnloser, R., Sarpeshkar, R., Mahowald, M.A., Douglas, R.J., Seung, H.S.: Digital selection and analogue amplification coexist in a cortex-inspired silicon circuit. Nature **405**, 947–951 (2000)
3. Glorot, X., Bordes, A., Bengio, Y.: Proceedings of the Fourteenth International Conference on Artificial Intelligence and Statistics, PMLR 15, pp. 315–323 (2011)
4. Tóth, L.: Phone recognition with deep sparse rectifier neural networks. In: 2013 IEEE International Conference on Acoustics, Speech and Signal Processing, Vancouver, BC, pp. 6985–6989 (2013)

5. Nair, V., Hinton, G.: Rectified linear units improve restricted boltzmann machines. In: Fürnkranz, J., Joachims, T. (eds.) Proceedings of the 27th International Conference on International Conference on Machine Learning (ICML 2010), pp. 807–814. Omnipress, USA (2010)
6. Maas, A.: Rectifier nonlinearities improve neural network acoustic models (2013)
7. Ramachandran, P., Zoph, B., Quoc, V.: Searching for activation functions, CoRR (2017)
8. Barret, Z.: Swish: a self-gated activation function (2017)
9. Mortazi, A., Bagci, U.: Automatically Designing CNN Architectures for Medical Image Segmentation: 9th International Workshop, MLMI, Held in Conjunction with MICCAI 2018, Granada, Spain, Proceedings (2018). https://doi.org/10.1007/978-3-030-00919-9_12
10. Hornik, K.: Approximation capabilities of multilayer feedforward networks. Neural Networks 4(2), 251–257 (1991). 0893-6080
11. Sharma, J.: Experiments with SWISH activation function on MNIST dataset (2016)
12. Han, J., Morag C.: The influence of the sigmoid function parameters on the speed of backpropagation learning (1995)
13. Szyc, K.: Comparison of different deep-learning methods for image classification. In: 2018 IEEE 22nd International Conference on Intelligent Engineering Systems (INES). IEEE (2018)
14. Sharma, J.: Swish in depth: a comparison of Swish & ReLU on CIFAR-10 (2017)

An Impact of Different Images Color Spaces on the Efficiency of Convolutional Neural Networks

Kamil Szyc[(✉)]

Wrocław University of Science and Technology, Wrocław, Poland
kamil.szyc@pwr.edu.pl

Abstract. A common standard while working with convolutional neural networks is using an RGB color space as an input of the network. All popular benchmarks datasets use this standard with three channels (red, green and blue) and with 8 bits per channel. We can modify images and use lower bits size per channel or use different color spaces standards like HSV, CMYK or YIQ. The main contribution of this paper is testing the influence of an image color space on convolutional neural networks. In our experiments we use DenseNet - a state of the art model and CIFAR 10 - a image benchmark dataset. Our experiments show that choosing proper color space have an impact on the final efficiency of the network and using the most popular RGB color space might not always be the best choice.

Keywords: Convolutional neural networks · Color space · RGB · HSV · CMYK · YIQ

1 Introduction

1.1 Overall

Convolutional neural networks (CNN) can solve image classification problems with very high accuracy. The typical classification problem requires choosing one class (from many others), which describes an image in the best way. Nowadays CNNs can be even better than humans [1] in tasks correlated with images. A typical CNN model is built from an input layer represented by image, hidden layers (usually convolutional [2] and pooling layers) and a fully connected layer with softmax at the end. The hidden layers determine features (from simple ones to more complex) on a single image and propagate it forward. There are more complex CNN models too [3–8]. In this paper, we wanted to test the influence of color space on the accuracy of a chosen CNN model.

Representing an image in RGB color space is a very common thing. A typical image consists of three channels red, green and blue each with 8 bits of information. This means each channel can be represented as a number from 0 (no color) to 255 (full color). In a mathematical context we can represent the image

© Springer Nature Switzerland AG 2020
W. Zamojski et al. (Eds.): DepCoS-RELCOMEX 2019, AISC 987, pp. 506–514, 2020.
https://doi.org/10.1007/978-3-030-19501-4_50

as a tensor. That tensor is a group of three (one for each channel) matrices with rows and columns equal to the image's width and height. Each value can differ from 0 to 255 and represents an individual pixel's color. In fact, that tensor is the input of each CNN.

While RGB color space is inspired by human vision system there are other types of color spaces too. For example, an HSV color space is aligned with human color-making attributes (human perception of color similarity), CMYK was created for printers and YIQ color space for TVs. All of these color spaces can be used to transform the input images without losing information about colors. A mathematical representation of this information looks different, so they can influence the efficiency of CNNs. The tensors (inputs) have a huge influence on weights in further layers in the networks. We tested all of these color space.

The normal situation is using RGB color space with 8 bits per channel. That gives $16,777,216(2^{8*3})$ color variations. It is enough for humans because the human eye can discriminate up to ten million colors [9]. Usually, we keep information about colors in 3 channels (values from 0 to 255), but we can easily convert it to one channel (value from 0 to 16,777,215). We often don't need to analyze 16 million colors to recognize objects in an image, so we can lose some information about colors and decrease the number of bits per channel. As we mentioned earlier each modification of the input of CNN can have an influence on the network's weights and consequently on it's efficiency and performance.

The main contribution of this paper is testing the influence of choosing an images color space on convolutional neural networks. We tested the different methods described above. We choose one of the state-of-the-art CNN model called DanseNet [6]. We used CIFAR-10 benchmark dataset of images, which contains 10 classes with 6000 images per class.

1.2   Color Spaces

RGB

The RGB color space is one of the most popular. Nearly all benchmark datasets with images use it (CIFAR-10, CIFAR-100, ImageNet or COCO). There are a lot of benefits in using RGB color space: almost every well-known application is compatible with it, the modern screens work using this space and it allows to make a wide range of colors.

The RGB color space is inspired by the color perception of the human eye. Our eyes can detect three light frequencies responsible for seeing red, green and blue. Depending on the activation each of these frequencies create in the eye, our brain perceives a different color. Because of that each color, which we can recognize, can be created by mixing three light beams. People typically can recognize up to ten million colors, so to achieve similar or higher amount we need 8 bits per channel - 24 bits for color. 8 bits means we can set a value from 0 to 255 for each channel, where 0 is no color and 255 is full color. We can decrease the number of bits per channel but it reduces information about the color.

In this paper we use RGB color space in the following variants: classic RGB with three channels with 8 bits per channel, RGB with only one channel and recalculated color value as $R*65536+G*256+B$, with RGB with three channels but with 2 bits (64 possible colors) and 3 bits (512 possible colors) per channel.

HSV

The HSV color space represents color as hue, saturation, and value. The hue is defined as "the degree to which a stimulus can be described as similar to or different from stimuli that are described as red, green, blue, and yellow" [10] and often is described as a pure pigment. The saturation is a "colorfulness of an area judged in proportion to its brightness" [10] and value is "brightness relative to the brightness of a similarly illuminated white" [10]. Usually, we represent HSV values from 0 to 360 for a hue and from 0 to 100 for saturation and value.

That color space is constructed in a way that is natural for humans when thinking about colors. That is the reason why it is often used by artists.

CMYK

The CMYK color space refers to the four inks used in the print industry. These colors are cyan(C), magenta(M), yellow(Y), and black(K). From historical point of view colors were obtained not by mixing the colors, but by applying color components layer by layer. The colors built according to CMYK should be viewed as a layer of colorful, light transmitting foil. We usually define each channel as a value from 0 to 100.

YIQ

The YIQ color space is used by the NTSC color TV systems. Y is luminance information - the brightness of an image (only that channel is used by black-and-white television receivers). I and Q represent the chrominance. That space is based on human color-response characteristics. Our eyes are more sensitive to changes in the orange-blue (I) range than in the purple-green range (Q). The values for Y is from 0 to 1, for I is from -0.5957 to 0.5957 and for Q from -0.5226 to 0.5226.

1.3 Convolutional Neural Networks

Convolutional neural networks (CNN) are part of deep neural networks (DNN), which are used for features detection. DNNs are most commonly known for working with images and can perform well in tasks like: classification (choosing one label for each image), classification with localization (choosing one label for each image and showing a rectangle with the object in it), object detection (similar to classification with localization, but with recognizing more than one object) and instance segmentation (shows very precisely where the objects are - outline the contours of the object). In this paper, we focused on images classification.

A neural network is made of input layer (image), hidden layers (part responsible for features generation) and final classification (choosing one label). In CNN model the input layer consists of an image represented as a tensor (a group of matrices for each channel). In hidden layers convolutional and pooling layers are used.

These layers are the most basic in this kind of DNN. The convolutional layer is used for features detection and from a mathematical point of view is just a matrix (called kernel), which we use for specific multiplication with matrix form the previous layer. A size of a kernel matrix is usually 1×1, 3×3 or 5×5 with changing values during the learning process. The pooling layers are responsible for reducing the number of parameters and computation in the network, and are also controlling overfitting. Final layers are fully connected layers normalized by function - softmax - in the end. Features are propagating through the network during the learning process and they are becoming more complex (from possibility to recognize lines up to the possibility of recognizing specific high context depending details) and later they allow to classify the image.

In CNN ReLU or Leaky ReLU activation functions are usually used - less sigmoid or tanh. SGD, Ada or Adam are most common optimizers.

There are many different types of well know CNN architecture models, which were or are the state-of-the-art models. For example VGG Net [3], Inception Network [4], ResNet [5] or DenseNet [6]. Usually, the researcher wants to increase the accuracy of models, but there are architectures where the main goal is optimizing the speed of classification or memory usage - for example, AlexNet [11], YOLO [8] or MobileNets [7].

2 Experiments

We did experiments to test the influence a color space on the efficiency of convolutional neural networks. We used CIFAR-10 images dataset and DenseNet model. By representing images in different color spaces and running the process of learning the model with the same parameters, we analyzed accuracy, speed of learning and the loss.

2.1 CIFAR 10

CIFAR-10 is benchmark dataset used very often in the images classification problems. It contains 10 classes (airplane, automobile, bird, cat, deer, dog, frog, horse, ship, and truck) with 60,000 images in total. There are 5,000 images for training and 1,000 images for testing for each class. Each image has a size of 32×32 pixels and the color in three RGB channels. The big advantage of that dataset is it's small image size. The size is enough for easy classification by humans, so researchers can test their ideas quite fast (compared to a much bigger datasets like ImageNet).

We decided not to transform the original training images, so we didn't use any image preprocessing. We know that without it, our final accuracy is not as high as in the original papers, but we wanted to focus on the influence of the color spaces.

2.2 DenseNet

DenseNet is one of the-state-of-the-art CNN architecture and an extension of ResNet. These architectures are based on an idea of making additional connections between layers. Recent works show that shorter connections between layers help CNN to make it deeper, efficient to train and more accurate. DenseNet is based on special blocks - called "dense block" - where each next convolutional layer has a short connection to all forwards layers (Fig. 1).

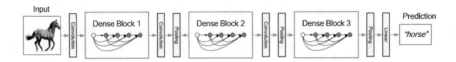

Fig. 1. A deep DenseNet with three dense blocks. Image from the original paper [6].

To configure DenseNet network we usually set a few hyperparameters. In our case the important ones are "depth", "number of dense block" and "growth rate". There are papers [6,12] which explain more carefully how to set them.

Choosing Proper Hyperparameters

We decided to use exactly the same hyperparameters for each of the models used in our experiments. We did it to show in easy way differences in efficiency. The only variable was the input data - different color spaces.

We decided to use learning hyperparameters as epochs equal to 50, a batch size equal to 32 and optimizer SGD [13] (it was used originally in DenseNet paper, so we decided to not change it to Adam) with learning rate equal to 0.1 and momentum equal to 0.9. DenseNet model was set with a depth equal to 100, growth rate equal to 12 and the number of dense blocks equal to 3.

2.3   The Course of Experiments

We prepared the images sets in the following color space: 3 channels RGB, 1 channel RGB, 3 channels RGB with 2 bits per channel, 3 channels RGB with 3 bits per channel, HSV, CMYK, and YIQ.

We used procedures described earlier to transform images from 3 RGB channels to one channel. We prepared images in that way for both sets - training and test ones. Since in different color spaces values are in different range (for example in RGB values are from 0 to 255, and in HSV from 0 to 360 or 100), we decided to normalize all values to be from 0 to 1.

Next, we used the DenseNet models with hyperparameters described earlier. Each model was the same except for input data. We ran our models using the NVIDIA GeForce GTX 1080 Ti graphics card. We used libraries Keras with TensorFlow. We created Jupiter Notebook for each color space case.

2.4   Results and Discussion

In the Tables 1 and 2 we presents our final results. We decided to measure accuracy and loss. We also present a learning process where we can see how that two

Table 1. Comparing different RGB approach

Color space	Accuracy	Loss	Learning process	Last epoch with improvement
RGB 3 channels 8 bits per channel	0.8737	0.6317		44
RGB 3 channels 2 bits per channel	0.7960	1.0566		28
RGB 3 channels 3 bits per channel	0.8186	0.9542		28
RGB 1 channel 24 bits per channel	0.8581	0.7898		34

Table 2. Comparing different color spaces

Color space	Accuracy	Loss	Learning process	Last epoch with improvement
RGB 3 channels 8 bits per channel	0.8737	0.6317		44
HSV 3 channels	0.8549	0.7285		40
CMYK 4 channels	0.8760	0.6177		49
YIQ 3 channels	0.8518	0.7339		41

values changes in time. We mark the accuracy by orange the color and loss by blue color. Axis Y is for "loss". We also mark in which (from 1 to 50) epoch was last accuracy improvement.

We achieved the best accuracy for CMYK color space, which is equal to 0.8760, however for RGB color space the result was a similar and equal to 0.8737. HSV and YIQ color systems performed worse and achieved a little more than 0.85 of accuracy.

Changing the number of bits in RGB color space had two impacts - accuracy dropped significantly - from 0.8737 to 0.7960 (for 2 bits) and 0.8186 (for 3 bits) and the speed of learning increased - the best epoch was achieved in 28 epoch (both for 2 and 3 bits) not in 44 (like for 8 bits RGB). We expected that because by reducing the number of bits we reduce information about the color too.

Converting three channel RGB to one channel RGB was a missed idea. Accuracy dropped from 0.8737 to 0.8581 and learning process stopped already in 34 epoch.

3   Conclusion

We tested different color spaces like RGB, HSV, CMYK, and YIQ and checked how they influence convolutional neural network's accuracy and loss. We also tested models without standard RGB (three 8 bits channels) inputs. We created the model with one RGB channel and another models with 2 or 3 bits per channel (losing information about color) instead of classic 8 bits per channel. To easily compare the results we normalized all values from 0 to 1 and used exactly the same model (the same hyperparameters).

The idea to convert RGB values to one channel is not good, but it was expected - considering the nature of convolutional neural networks.

The idea of changing the number of bits in RGB color space was worth considering - accuracy dropped significantly, but the speed of learning increased. The procedure of reducing the number of bits reduces also information about color, but we plan to perform a more careful test with the different number of bits in further works. It can be a good way to increase the speed of learning CNN.

The idea of changing the color space was quite good. Although the difference in the accuracy was not huge, the best result was obtained for CMYK color space - slightly better than for RGB. We are aware that we can improve those numbers by changing the model's hyperparameters. Our goal is not to convince all researchers to change their color space to CMYK. However, we think that changing the color space might be a good approach and we recommend to try it while looking for the best hyperparameters. It can be an easy way to slightly increase the final accuracy of the network.

We plan to test the influence of mixing channels from different color spaces in further works.

References

1. Wu, R., Yan, S., Shan, Y., Dang, Q., Sun, G.: Deep image: scaling up image recognition. arXiv preprint arXiv:1501.02876 (2015)
2. LeCun, Y., Boser, B., Denker, J.S., Henderson, D., Howard, R.E., Hubbard, W., Jackel, L.D.: Backpropagation applied to handwritten zip code recognition. Neural Comput. 1(4), 541–551 (1989)
3. Simonyan, K., Zisserman, A.: Very deep convolutional networks for large-scale image recognition. arXiv preprint arXiv:1409.1556 (2014)
4. Szegedy, C., Vanhoucke, V., Ioffe, S., Shlens, J., Wojna, Z.: Rethinking the inception architecture for computer vision, pp. 2818–2826 (2016)
5. Szegedy, C., Ioffe, S., Vanhoucke, V., Alemi, A.A: Inception-v4, inception-ResNet and the impact of residual connections on learning (2017)
6. Huang, G., Liu, Z., van der Maaten, L., Weinberger, K.Q.: Densely connected convolutional networks, pp. 4700–4708 (2017)
7. Howard, A.G., Zhu, M., Chen, B., Kalenichenko, D., Wang, W., Weyand, T., Andreetto, M., Adam, H.: Mobilenets: efficient convolutional neural networks for mobile vision applications. arXiv preprint arXiv:1704.04861 (2017)
8. Redmon, J., Farhadi, A.: YOLOv3: an incremental improvement. arXiv (2018)
9. Judd, D.B.: Color in business, science and industry (1952)
10. Fairchild, M.D.: Color Appearance Models. Wiley, Chichester (2013)
11. Krizhevsky, A., Sutskever, I., Hinton, G.E.: ImageNet classification with deep convolutional neural networks. In: Advances in Neural Information Processing Systems, pp. 1097–1105 (2012)
12. Pleiss, G., Chen, D., Huang, G., Li, T., van der Maaten, L., Weinberger, K.Q.: Memory-efficient implementation of densenets. arXiv preprint arXiv:1707.06990 (2017)
13. Bottou, L.: Large-scale machine learning with stochastic gradient descent, pp. 177–186 (2010)

Coordinated Resources Allocation for Dependable Scheduling in Distributed Computing

Victor Toporkov$^{(\boxtimes)}$ and Dmitry Yemelyanov

National Research University "MPEI", ul. Krasnokazarmennaya, 14,
Moscow 111250, Russia
{ToporkovVV, YemelyanovDM}@mpei.ru

Abstract. In this work, we consider heuristic algorithms for parallel jobs execution and efficient resources allocation in distributed computing environments. Existing modern job-flow execution features and realities impose many restrictions for the resources allocation procedures. Grid and many other high performance computing services operate in heterogeneous and usually geographically distributed computing environments. Emerging virtual organizations and incorporated economic scheduling models allow users and resource owners to compete for suitable allocations based on market principles and fair scheduling policies. Subject to these features a special dynamic programming scheme is proposed to select resources depending on how they fit a particular job execution duration. Based on a conservative backfilling scheduling procedure we study how different resources allocation heuristics affect integral job-flow scheduling characteristics in a dedicated simulation environment.

Keywords: Distributed computing · Grid · Economic scheduling ·
Resource management · Slot · Job · Backfilling · Optimization

1 Introduction and Related Works

Modern high-performance distributed computing systems (HPCS), including Grid, cloud and hybrid infrastructures provide access to large amounts of resources [1, 2]. These resources are typically required to execute parallel jobs submitted by HPCS users and include computing nodes, data storages, network channels, software, etc.

There are two important classes of users' parallel jobs. Bags of tasks (BoT) represent parallel applications incorporating a large number of independent or weakly connected tasks. Typical examples of BoT are parameter sweeps, Monte Carlo simulations or exhaustive search. Workflows consist of multiple tasks with control or data dependencies. Such applications may be presented as directed graphs and represent complex computational or data processing problems in many domains of science [2–4].

Most BoT and workflow applications require some assurances of quality of services (QoS) from the computing system. In order to ensure QoS requirements and constraints a coordinated allocation of suitable resources should be performed [5, 6]. Most QoS

© Springer Nature Switzerland AG 2020
W. Zamojski et al. (Eds.): DepCoS-RELCOMEX 2019, AISC 987, pp. 515–524, 2020.
https://doi.org/10.1007/978-3-030-19501-4_51

requirements are based on either time or cost constraints such as total job execution cost, deadline, response time, etc. [7–10].

Some of the most important efficiency indicators of a distributed computational environment include both system resources utilization level and users' jobs time and cost execution criteria [4, 7, 8, 11]. In distributed environments with non-dedicated resources, such as utility Grids, the computational nodes are usually partly utilized and reserved in advance by jobs of higher priority [9]. Thus, the resources available for use are represented with a set of slots - time intervals during which the individual computational nodes are capable to execute parts of independent users' parallel jobs. These slots generally have different start and finish times and a performance difference. The presence of a set of slots impedes the problem of coordinated selection of the resources necessary to execute the job-flow from computational environment users. Resource fragmentation also results in a decrease of the total computing environment utilization level [11, 12].

High-performance distributed computing systems organization and support bring certain economical expenses: purchase and installation of machinery equipment, power supplies, user support, etc. As a rule, HPCS users and service providers interact in economic terms and the resources are provided for a certain payment. In such conditions, resource management and job scheduling based on the economic models is considered as an efficient way to take into account contradictory preferences of computing participants [3, 11, 13, 14].

A metascheduler or a metabroker is considered as an intermediate link between the users, local resource management and job batch processing systems [7, 15]. It defines uniform rules of a resource sharing and consumption to improve the overall scheduling efficiency [11, 12].

The main contribution of this paper is a metaheuristic algorithm for a coordinated resources allocation for parallel jobs execution. The algorithm takes into account the system slots configuration as well as individual jobs features: size, runtime, cost, etc. When used in HPCS metaschedulers during the resources allocation step it may improve overall system utilization level by matching jobs with resources and providing better jobs placement.

The rest of the paper is organized as follows. Section 2 presents resources allocation problem in relation to job-flow scheduling algorithms and backfilling (Subsect. 2.2). Different approaches for a coordinated resources allocation are described and proposed in Sect. 3. Section 4 contains algorithms implementation details along with simulation results and analysis. Finally, Sect. 5 summarizes the paper and describes further research topics.

2 Resources Allocation for Job-Flow Scheduling

2.1 Problem Statement

In order to cover a wide range of computing systems we consider the following model for a heterogeneous resource domain.

Constituent computing nodes of a domain have different usage costs and performance levels. A space-shared resources allocation policy simulates a local queuing system (like in CloudSim [13, 14] or SimGrid [4]) and, thus, each node can process only one task at any given time. Economic scheduling model assumes that users and resource owners operate with some currency to coordinate resources allocation transactions. This model allows to regulate interaction between different organizations and to settle on fair equilibrium prices for resources usage.

Thus we consider a set R of heterogeneous computing nodes with different performance p_i and price c_i characteristics.

A node may be turned off or on by the provider, transferred to a maintenance state, reserved to perform computational jobs. Thus each node has a local utilization schedule known in advance for a considered scheduling horizon time L.

The execution cost of a single task depends on the allocated node's price and execution time, which is proportional to the node's performance level. In order to execute a parallel job one needs to allocate the specified number of simultaneously idle nodes ensuring user requirements from the resource request. The resource request specifies number n of nodes required simultaneously, their minimum applicable performance p, job's computational volume V and a maximum available resources allocation budget C. These parameters constitute a formal generalization for resource requests common among distributed computing systems and simulators [11, 13].

In heterogeneous environment, the required window length is defined based on a slot with the minimum performance. For example, if a window consists of slots with performances $p \in \{p_i, p_j\}$ and $p_i < p_j$, then we need to allocate all the slots for a time $T = \frac{V}{p_i}$. In this way V really defines a computational volume for each single job subtask. Common start and finish times ensure the possibility of inter-node communications during the whole job execution. The total cost of a window allocation is then calculated as $C_W = \sum_{i=1}^{n} T * c_i$.

2.2 Job-Flow Scheduling and Backfilling

The simplest way to schedule a job-flow execution is to use the First-Come-FirstServed (FCFS) policy. However, this approach is inefficient in terms of resources utilization. Backfilling [16] was proposed to improve system utilization.

Backfilling procedure makes use of advanced resources reservations which is an important mechanism preventing starvation of jobs requiring large number of computing nodes. Resources reservations in FCFS may create idle slots in the nodes' local schedules thus decreasing system performance. So the main idea behind backfilling is to backfill jobs into those idle slots to improve the overall system utilization. And the backfilling procedure implements this by placing smaller jobs from the back of the queue to these idle slots ahead of the priority order.

There are two common variations to backfilling - conservative and aggressive (EASY). Conservative Backfilling enforces jobs' priority fairness by making sure that jobs submitted later can't delay the start of jobs arrived earlier. EASY Backfilling aggressively backfills jobs as long as they do not delay the start of the single currently reserved job. Conservative Backfilling considers jobs in the order of their arrival and either immediately starts a job or makes an appropriate reservation upon the arrival. The jobs priority in the queue may be additionally modified in order to improve system-wide job-flow execution efficiency metrics. Under default FCFS policy the jobs are arranged by their arrival time. Other priority reordering-based policies like Shortest job First or eXpansion Factor may be used to improve overall resources utilization level [8, 9, 12].

Multiple Queues backfilling separates jobs into different queues based on metadata, such as jobs resource requirements: small, medium, large, etc. The idea behind this metaheuristic is that earlier arriving jobs and smaller-sized jobs should have higher execution priority. The number of queues and the strategy for dividing tasks among them can be set by the system administrators. Sometimes different queues may be assigned to a dedicated resource domain segments and function independently. In a single domain the metaheuristic cycles through the different queues in a round-robin fashion and may consider more jobs from the queues with smaller-sized tasks [12].

The look-ahead optimizing scheduler [9] implements dynamic programming scheme to examine all the jobs in the queue in order to maximize the current system utilization. So, instead of scanning queue for single jobs suitable for the backfilling, look-ahead scheduler attempts to find a combination of jobs that together will maximize the resources utilization.

2.3   Resources Selection Algorithms

Backfilling as well as many other job-flow scheduling algorithms in fact describe a general procedure determining high level policies for jobs prioritization and advanced resources reservations. However, the resources selection and allocation step remains sidelined since its more system specific nature. On the other hand, applying different resources allocation policies based on system or user preferences may affect scheduling results not only for individual jobs but for a whole job-flow.

In [6] we presented a Slot Subset Allocation (SSA) dynamic programming scheme for resources selection in heterogeneous computing environments based on economic principles. In a general case system nodes may be shared and reserved in advance by different users and organizations (including resource owners). So it's convenient to represent all available resources as a set of time-slots (see Sect. 2.1). SSA algorithm takes these time slots as input and performs resources selection for a specified job in accordance with the computing model and constraints described in Sect. 2.1.

Additionally SSA may perform window search optimization by a general additive criterion $Z = \sum_{i=1}^{n} c_z(s_i)$, where $c_z(s_i) = z_i$ is a target optimization characteristic value provided by a single slot s_i of window W. These criterion values z_i may represent different slot characteristics: time, cost, power, hardware and software features, etc. Thus SSA-based resources allocation is proved a flexible tool for a preference-based job-slow execution [6].

3   Coordinated Resources Allocation Heuristics

3.1   Dependable Job Placement Problem

One important aspect for a resources allocation efficiency is the resources placement in regard to an actual slots configuration. Therefore, as a practical implementation for a general z_i parameter maximization we propose to study a resources allocation placement problem. Figure 1 shows Gantt chart of 4 slots co-allocation (hollow rectangles) in a computing environment with resources pre-utilized with local and high-priority jobs (filled rectangles).

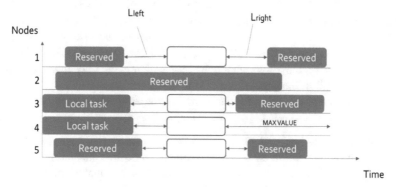

Fig. 1. Dependable window co-allocation metrics.

As can be seen from Fig. 1, even using the same computing nodes (1, 3, 4, 5) there are usually multiple window placement options with respect to the slots start time. The slots' actual placement generally may affect such job execution properties as cost, finish time, computing energy efficiency, etc.

For a quantitative placement criterion for each window slot, we can estimate times to the previous task finish time: L_{left} and to the next task start time: L_{right} (Fig. 1).

3.2   Job Placement Heuristics

However these L_{left} or L_{right} criteria alone can't improve the whole job-flow scheduling solution according to the conventional makespan or average finish time criteria. For example in [12] a special set of *breaking a tie* rules is proposed to choose between slots providing the same earliest job start time. These rules for Picking Earliest Slot for a Task (PAST) procedure may be summarized as following.

1. Minimize number of idle slots left after the window allocation: slots adjacent (succeeding or preceding) to already reserved slots have higher priority.
2. Maximize length of idle slots left after the window allocation; so the algorithm tends to left longer slots for the subsequent jobs in the queue.

With similar intentions we propose the following Coordinated Placement (CoP) heuristic rules.

1. Prioritize slots allocated on nodes with lower performance. The main idea is that when deciding between two slots providing the same window finish time it makes sense to leave higher performance slot vacant for the subsequent jobs. This breaking a tie principle is applicable for heterogeneous resources environments and do not consider slots placement configuration. However, during the preliminary simulations this heuristic alone was able to noticeably improve scheduling results so we will use it as an addition to the placement rules.
2. Prioritize slots with relatively small distances to the neighbor tasks: $L_{left\,i} \ll T$ or $L_{right\,i} \gg T$. The general idea is similar to the first rule in PAST, but CoP does not expect perfect match and defines threshold values for a satisfactory window fit.
3. Penalize slots leaving significant, but insufficient to execute a full job distances $L_{left\,i}$ or $L_{right\,i}$. For example, when $T/3 > L_{right\,i} > T/5$, the resulting slot may be to short to execute any of the subsequent jobs and will remain idle, thus, reducing the resources overall utilization.
4. On the other hand equally prioritize slots leaving sufficient compared to the job's runtime distances $L_{left\,i}$ or $L_{right\,i}$. For example with $L_{left\,i} > T$.

So the main idea behind CoP is to fill in the gaps in the resources reservation schedule by providing quite an accurate resources allocation matching jobs runtime. Unlike PAST, CoP does not expect perfect matches but makes realistic heuristic decisions to minimize general resources fragmentation.

4   Simulation Study

4.1   Implementation and Simulation Details

Based on heuristic rules described in Sect. 3.2 we implemented the following scheduling algorithms and criteria for SSA-based resources allocation.

1. Firstly we consider two conservative backfilling variations. BFstart successively implements start time minimization for each job during the resources selection step. As SSA performs criterion maximization, BFstart criterion for i-th slot has the following form: $z_i = -s_i.startTime$. By analogy BFfinish implements a more solid strategy of a finish time minimization which is different from BFstart in computing environments with heterogeneous resources. BFfinish criterion for SSA algorithm is the following: $z_i = -s_i.finishTime$.
2. PAST-like backfilling approach has a more complex criterion function which may be described with the following set of rules:

   (a) $z_i = -s_i.finishTime$; finish time is the main criterion value
   (b) $z_i = z_i - \alpha_1 * s_i.nodePerformance$; node performance amendment
   (c) if ($L_{right\,i} == 0$): $z_i = z_i + \delta_1$; PAST rule 1
   If ($L_{left\,i} == 0$): $z_i = z_i + \delta_1$; PAST rule 1
   (e) $z_i = z_i - \alpha_2 * L_{right\,i}$; PAST rule 2

3. CoP resources allocation algorithm for backfilling may be represented with the following criterion calculation:

(a) $z_i = -s_i.finishTime$; finish time is the main criterion value
(b) $z_i = z_i - \alpha_1 * s_i.nodePerformance$; node performance amendment
(c) if ($L_{right\,i} < \varepsilon_1 * T$): $z_i = z_i + \delta_1$; CoP rule 1
(d) if ($L_{left\,i} < \varepsilon_1 * T$): $z_i = z_i + \delta_1$; CoP rule 1
(e) if ($L_{right\,i} > \varepsilon_2 * T$ AND $L_{right\,i} < \varepsilon_3 * T$): $z_i = z_i - \delta_1$; CoP rule 2
(f) if ($L_{left\,i} > \varepsilon_2 * T$ AND $L_{left\,i} < \varepsilon_3 * T$): $z_i = z_i - \delta_1$; CoP rule 2
(g) if ($L_{right\,i} > T$): $z_i = z_i + \delta_3$; CoP rule 3
(h) if ($L_{left\,i} > T$): $z_i = z_i + \delta_3$; CoP rule 3

4. Finally as an additional reference solution we simulate another abstract backfilling variation BFshort which is able to reduce each job runtime for 1% during the resources allocation step. In this way each job will benefit not only from its own earlier completion time, but from earlier completion of all the preceding jobs.

The criteria for PAST and CoP contain multiple constant values defining rules behavior, namely $\alpha_1, \alpha_2, \delta_1, \delta_2, \varepsilon_1, \varepsilon_2, \varepsilon_3$. ε_i coefficients define threshold values for a satisfactory job fit in CoP approach. α_i and δ_i define each rule's effect on the criterion and are supposed to be much less compared to z_i in order to break a tie between otherwise suitable slots. However, their mutual relationship implicitly determines rules' priority which can greatly affect allocation results. Therefore there are a great number of possible α_i, δ_i and ε_i values combinations providing different PAST and CoP implementations. Based on heuristic considerations and some preliminary experiment results the values we used during the present experiment are presented in Table 1.

Table 1. PAST and CoP parameters values

Constant	α_1	α_2	δ_1	δ_2	ε_1	ε_2	ε_3
Value	0.1	0.0001	1	0.1	0.03	0.2	0.35

Because of heuristic nature of considered algorithms and their speculative parametrization (see Table 1) hereinafter by PAST [12] we will mean PAST-like approach customly implemented as an alternative to CoP.

4.2 Simulation Results

The experiment was prepared as follows using a custom distributed environment simulator [6, 11]. For our purpose, it implements a heterogeneous resource domain model: nodes have different usage costs and performance levels. A space-shared resources allocation policy simulates a local queuing system (like in CloudSim or SimGrid [13]) and, thus, each node can process only one task at any given simulation time. The execution cost of each task depends on its execution time, which is proportional to the dedicated node's performance level. The execution of a single job requires parallel execution of all its tasks. More details regarding the simulation computing model were provided in Sect. 2.1.

During each simulation experiment a new instance for the computing environment was automatically generated. Node performance level is given as a uniformly distributed random value in the interval [2, 16]. This configuration provides a sufficient resources diversity level while the difference between the highest and the lowest resource performance levels will not exceed one order. In this environment we considered job queue with 50, 100, 150 and 200 jobs accumulated at the start of the simulation. The jobs are arranged in queues by priority and no new jobs are submitted during the queue execution. Such scheduling problem statement allows to statically evaluating algorithms' efficiency in conditions with different resources utilization level. The jobs were generated with the following resources request requirements: number of simultaneously required nodes is uniformly distributed in interval $n \in [1; 8]$, computational volume $V \in [60; 1200]$ also contribute to a wide diversity in user jobs.

The results of 2000 independent simulation experiments are presented in Tables 2 and 3. Each simulation experiment includes computing environment and job queue generation, followed by a scheduling simulation independently performed using considered algorithms. The main scheduling results are then collected and contribute to the average values over all experiments.

Table 2. Simulation results: average job finish time, t.u.

Jobs number N_Q	BFstart	BFfinish	BFshort	PAST	CoP
50	318.8	302.1	298.8	300.1	298
100	579.2	555	549.2	556.1	550.7
150	836.8	805.6	796.8	809	800.6
200	1112	1072.7	1060.3	1083.3	1072.2

Table 2 contains average finish time in simulation time units (t.u.) provided by algorithms BFstart, BFfinish, BFshort, PAST and CoP for different number of jobs pre-accumulated in the queue.

As it can be seen, with a relatively small number N_Q of jobs in the queue, both CoP and PAST provide noticeable advantage by nearly 1% over a strong BFfinish variation and CoP even surpasses BFshort results. At the same time less successful BFstart approach provides almost 6% later average completion time highlighting difference between a good (BFfinish) and a regular (BFstart) possible scheduling solutions. So BFshort, CoP and PAST advantage should be evaluated against this 6% interval. Similar conclusion follows from the average makespan (i.e. the latest finish time of the queue jobs) values presented in Table 3.

However with increasing the jobs number CoP advantage over BFfinish decreases and tends to zero when $N_Q = 200$. This trend for PAST and CoP heuristics may be explained by increasing accuracy requirements for jobs placement caused with increasing N_Q number. Indeed, when considering for some intermediate job resource selection the more jobs are waiting in the queue the higher the probability that some future job will have a better fit for current resource during the backfilling procedure. In a general case all the algorithms' parameters α_i, δ_i and ε_i (more details we provided in Sect. 3.2) should be refined to correspond to the actual computing environment utilization level.

Table 3. Simulation results: average job-flow makespan, t.u.

Jobs number N_Q	BFstart	BFfinish	BFshort	PAST	CoP
50	807.8	683	675	678	673.3
100	1407	1278	1264	1272	1264
150	2003	1863	1842	1857	1844
200	2622	2474	2449	2476	2455

5 Conclusion

In this work, we address the problem of a coordinated resources allocation for parallel jobs in distributed and heterogeneous computing environments. Based on a Slots Subset Allocation resources selection algorithm we propose and implement a set of heuristic job placement criteria including PAST and CoP. The main idea behind Coordinated Placement (CoP) approach is to fill in the gaps in the resources utilization schedule by allocating resources tailored to particular jobs runtimes.

Using job placement heuristics during the resources allocation step may improve overall job-flow scheduling efficiency. Without any changes in the base algorithm and with jobs priorities and execution order preserved, the simulation study showed 1.5% improvements conservative backfilling against average job-flow finish time and makespan criteria.

Acknowledgments. This work was partially supported by the Council on Grants of the President of the Russian Federation for State Support of Young Scientists (YPhD-2979.2019.9), RFBR (grants 18-07-00456 and 18-07-00534) and by the Ministry on Education and Science of the Russian Federation (project no. 2.9606.2017/8.9).

References

1. Lee, Y.C., Wang, C., Zomaya, A.Y., Zhou, B.: Profit-driven scheduling for cloud services with data access awareness. J. Parallel Distrib. Comput. **72**(4), 591–602 (2012)
2. Bharathi, S., Chervenak, A.L., Deelman, E., Mehta, G., Su, M., Vahi, K.: Characterization of scientific workflows. In: 2008 Third Workshop on Workflows in Support of Large-Scale Science, pp. 1–10 (2008)
3. Rodriguez, M.A., Buyya, R.: Scheduling dynamic workloads in multi-tenant scientific workflow as a service platforms. Future Gener. Comput. Syst. **79**(P2), 739–750 (2018)
4. Nazarenko, A., Sukhoroslov, O.: An experimental study of workflow scheduling algorithms for heterogeneous systems. In: Malyshkin, V. (ed.) Parallel Computing Technologies, pp. 327–341. Springer International Publishing (2017)
5. Netto, M.A.S., Buyya, R.: A flexible resource co-allocation model based on advance reservations with rescheduling support. Technical report, GRIDSTR2007-17, Grid Computing and Distributed Systems Laboratory, The University of Melbourne, Australia, 9 October 2007
6. Toporkov, V., Yemelyanov, D.: Dependable slot selection algorithms for distributed computing. In: Advances in Intelligent Systems and Computing, vol. 761, pp. 482–491. Springer (2019)

7. Kurowski, K., Nabrzyski, J., Oleksiak, A., Weglarz, J.: Multicriteria aspects of grid re-source management. In: Nabrzyski, J., Schopf, J.M., Weglarz, J. (eds.) Grid Resource Management. State of the Art and Future Trends, pp. 271–293. Kluwer Academic Publishers (2003)

8. Srinivasan, S., Kettimuthu, R., Subramani, V., Sadayappan, P.: Characterization of Backfilling strategies for parallel job scheduling. In: Proceedings of the International Conference on Parallel Processing, ICPP 2002 Workshops, pp. 514–519 (2002)

9. Shmueli, E., Feitelson, D.G.: Backfilling with lookahead to optimize the packing of parallel jobs. J. Parallel Distrib. Comput. **65**(9), 1090–1107 (2005)

10. Menasc'e, D.A., Casalicchio, E.: A framework for resource allocation in grid computing. In: The 12th Annual International Symposium on Modeling, Analysis, and Simulation of Computer and Telecommunications Systems (MASCOTS 2004), Volendam, The Netherlands, pp. 259–267 (2004)

11. Toporkov, V., Toporkova, A., Tselishchev, A., Yemelyanov, D., Potekhin, P.: Heuristic strategies for preference-based scheduling in virtual organizations of utility grids. J. Ambient Intell. Humanized Comput. **6**(6), 733–740 (2015)

12. Khemka, B., Machovec, D., Blandin, C., Siegel, H.J., Hariri, S., Louri, A., Tunc, C., Fargo, F., Maciejewski, A.A.: Resource management in heterogeneous parallel computing environments with soft and hard deadlines. In: Proceedings of 11th Metaheuristics International Conference (MIC 2015) (2015)

13. Calheiros, R.N., Ranjan, R., Beloglazov, A., De Rose, C.A.F., Buyya, R.: CloudSim: a toolkit for modeling and simulation of cloud computing environments and evaluation of resource provisioning algorithms. J. Softw. Pract. Experience **41**(1), 23–50 (2011)

14. Samimi, P., Teimouri, Y., Mukhtar, M.: A combinatorial double auction resource allocation model in cloud computing. J. Inf. Sci. **357**(C), 201–216 (2016)

15. Rodero, I., Villegas, D., Bobroff, N., Liu, Y., Fong, L., Sadjadi, S.: Enabling interoperability among grid meta-schedulers. J. Grid Comput. **11**(2), 311–336 (2013)

16. Jackson, D., Snell, Q., Clement, M.: Core algorithms of the Maui scheduler. In: Revised Papers from the 7th International Workshop on Job Scheduling Strategies for Parallel Processing, JSSPP 2001, pp. 87–102 (2001)

Information Needs of Decision Makers for Risk Assessment in Road Transport

Agnieszka Tubis[(⊠)] [ID]

Faculty of Mechanical Engineering,
Wroclaw University of Science and Technology,
27 Wybrzeze Wyspianskiego Street, Wroclaw, Poland
agnieszka.tubis@pwr.edu.pl

Abstract. Transport is an area for which there are numerous studies devoted to the assessment of the occurring risk. Data collected for risk assessment depend on many factors. The demand for information to carry out the required analyses depends on the type of transport, the type of transported products, the objectives set for the tasks being performed, as well as the requirements for the implementation of the process itself. It is therefore reasonable to present the differences that exist in the information demand for risk assessment, which is carried out for the road transport process by different decision-making bodies. The purpose of this article is therefore to present the results of research on the identification of information needs of decision-makers in the risk assessment at the strategic and operational level. The information needs for strategic risk assessment were determined based on a literature review. The information needs for operational risk assessment were determined based on research conducted among managers, who plan road transport in a selected company. Finally, the performance measurement system for managers was determined based on the identified information needs.

Keywords: Risk assessment · Road transport · Logistic service

1 Introduction

The risk can be understood as *the possibility of an unfortunate occurrence* or *the potential for realisation of unwanted, negative consequences of an event* [1]. But how risk is perceived depends largely on who assesses risk, what information are available, what algorithms of methods are recognized and what acceptance criteria are used [6]. The researchers report different needs for information used in the analyses, depending on the area for which the risk assessment is prepared. For this reason, the development of information systems for risk assessment should be preceded by a process analysis. This analysis should be aimed at identifying the decision-making areas of people responsible for planning processes performance.

Transport is an area for which there are numerous studies devoted to the assessment of the occurring risk. There are many examples of publications that are aimed at identifying adverse events in various modes of transport – road [2, 10], rail [3, 12, 13], aviation [8, 9, 14] and water transport [11]. The demand for information to carry out the

© Springer Nature Switzerland AG 2020
W. Zamojski et al. (Eds.): DepCoS-RELCOMEX 2019, AISC 987, pp. 525–533, 2020.
https://doi.org/10.1007/978-3-030-19501-4_52

required analyses depends on the type of transport, the type of transported products, the objectives set for the tasks being performed, as well as the requirements for the implementation of the process itself. Detailed studies in this area have been presented in [15].

Data collected for risk assessment depend on many factors. It is therefore reasonable to present the differences that exist in the information demand for risk assessment, which is carried out for the road transport process by different decision-making bodies. The purpose of this article is therefore to present the results of research on the identification of information needs of decision-makers in the risk assessment at the strategic and operational level. Therefore, the structure of the article includes the characteristics of a strategic approach to risk assessment in road transport along with the identification of the collected data. Later, there is presented the operational approach to risk assessment together with the presentation of research results that concerned the identification of hazards in the transport process in enterprises. Finally, the range of reported indicators for the operational risk assessment in the company was proposed.

2   Strategic and Operational Approach to Risk Assessment

2.1   Strategic Approach

Risk assessment in road transport has been the subject of numerous research and scientific publications around the world for many years. However, they focus primarily on issues related to road safety. This may be proved by the fact that in the 1980's Hauer in his research defined the risk in road transport only as "*the probability of an accident*" [5]. Four years later Haight added a second element to the definition of Hauer, which he described as "*the effects of a road incident*" [4]. Such a defined risk for road transport still applies in the conducted research for this area and is dominant in scientific publications.

A literature review [15] in this area proves that publications on risk in road transport focus on analyses related to the transport system and mainly apply to the strategic risk assessment. The assessed risk is a long-term risk related to long-term decisions that are taken by institutions responsible for road safety in the analyzed area. For this reason, the strategic risk is primarily assessed in the context of driving safety aspects [16].

Basing on the last 10 years' selection of reviewed articles, the following areas of described risk analyses have been distinguished (Fig. 1).

The most publications on risk assessment refer predominantly to two areas of research – network risk assessment and hazmat risk assessment. This is mostly due to significant consequences of negative incidents, which pose a threat to human life and health and can pollute the environment. The scope of information collected for risk assessment for highlighted areas of research is shown in Fig. 2.

The risk assessed in research is related to uncertainty regarding the implementation of the strategic goal, which is to protect the life and health of road users [7]. For this reason, the scope of collected data corresponds to the information needs of decision makers at the strategic level. Based on analyses, decision-makers have the opportunity

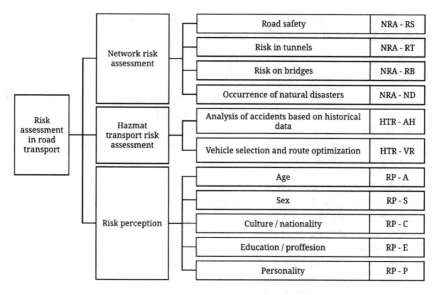

Fig. 1. Areas of risk analyses for road transport [15]

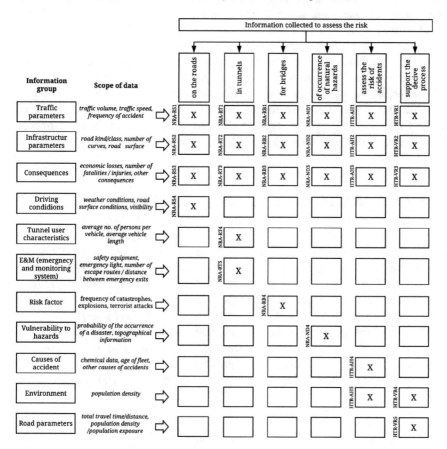

Fig. 2. The information collected for strategic risk assessment [15]

to develop and implement tools that aim to reduce the occurrence of adverse events that threaten the health and lives of road users.

2.2 Operational (Business) Approach

Specialists managing road transport in enterprises have differently defined goals for their processes. They take into account not only the safe delivery of transport, but also compliance with the logistical requirements agreed with the client in the process of planning transport service. For this reason, the risk assessment for the transport process takes on a completely different character. The risk:

– is assessed for the implementation of a specific transport order;
– applies not only to carriage, but also to activities preceding and after transport, which may affect the process as well as the transported cargo (e.g. correct load protection, timely arrival of the vehicle for loading);
– focuses on hazards that are under total or partial control of the company, i.e. those for whom the internal resources of the organization are a source of threat.

For this reason, operational risk is assessed. This risk can be defined as **the risk of occurrence of undesirable events whose occurrence disturbs the proper course of the process or limits the achievement of the adopted operational objective** [15]. The transport process in an enterprise is assessed mainly in terms of its compliance with the rule 7R (*the right product, the right quantity, the right condition, the right place, the right time, the right customer, the right cost*). Therefore, the operational objective is to handle the transport order at specified costs and with the provision of established logistic parameters of the service and product quality. The differentiation of the objective, which is defined for the process, causes decision makers to otherwise identify undesirable events and analyze the risks associated with their occurrence.

The operational risk defined in this way is the subject of further analyzes presented in this article.

3   Identification of Adverse Events in Road Transport Companies

Research carried out as part of the project "Identifying data supporting risk assessment in road transport system in the aspect of safety and meeting logistic standards" No. 2017/01/X/HS4/2017 financed as part of MINIATURA 1 program were realized among road transport companies. The study used the participant observation method as well as the unstructured (phase 1) and structured (step 2) direct interview.

As part of the first stage of the research, undesirable events were identified, which are subject to risk assessment for the process of handling transport orders. A total of 20 adverse events were identified, which were then grouped into four areas: (1) safety, (2) logistics, (3) communication, (4) costs. The outcome of the classification is shown in Fig. 3.

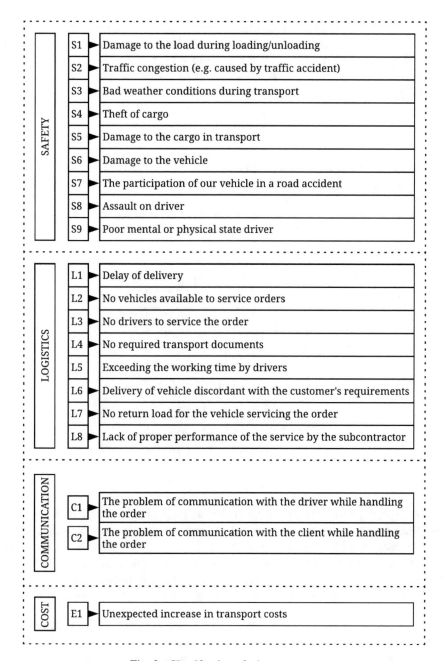

Fig. 3. Classification of adverse events

As part of stage 2, a structured interview was conducted with the managers of the selected road transport company. Managers had the task to indicate which adverse events they consider in the planning of handling transport orders. Based on the number of indications, the weight (significance) attributed to the adverse event was determined. The weight was calculated on the basis of formula 1.

$$w_i = \frac{a_i}{\sum_{i=1}^{n} a_i} \tag{1}$$

Where: a_i – number of indicating particular element – adverse event;
w_i – weight of adverse event;
n – number of adverse events.

Scales calculated for each adverse event have become the basis for prioritizing management areas. The results of the study are presented in Table 1.

Table 1. Weight and priority assigned to adverse events.

i	a_i	w_i	p_i	i	a_i	w_i	p_i
S1	6	0,070	1	L1	6	0,070	1
S2	2	0,023	5	L2	6	0,070	1
S3	2	0,023	5	L3	6	0,070	1
S4	3	0,035	4	L4	5	0,058	2
S5	6	0,070	1	L5	5	0,058	2
S6	3	0,035	4	L6	6	0,070	1
S7	3	0,035	4	L7	4	0,047	3
S8	2	0,023	5	L8	4	0,047	3
S9	2	0,023	5	C1	5	0,058	2
E1	6	0,070	1	C2	4	0,047	3

The highest priority events are events that are most often included by managers in the planning process of the transport order. These are therefore events whose occurrence frequency and/or effects are very high. Therefore, it can be assumed that managers in the surveyed enterprise assess the risk indicator for these events at a high level. According to the concept of risk management, operations managers should be targeted in the first place to reduce the risk index for the events for which priority is given to high.

Table 2 presents the list of priorities given for events qualified for each group. It should be noted that most events with the highest priority have been identified in the area of logistics. In this area, all events also received one of the three highest priorities. Events safety area received a much lower priority. Most events received priority 4 or 5.

Table 2. List of priorities given for events qualified for each group.

Group	Logistics				
Priority	1	2	3	4	5
The number of events with a given priority	4	2	2	0	0
Group	Safety				
Priority	1	2	3	4	5
The number of events with a given priority	2	0	0	3	4
Group	Communication				
Priority	1	2	3	4	5
The number of events with a given priority	0	1	1	0	0
Group	Cost				
Priority	1	2	3	4	5
The number of events with a given priority	1	0	0	0	0

4   Risk Assessment in the Selected Company Road Transport

The results obtained in the research conducted among managers who plan the transport process in enterprises allow to conclude that the data collected for risk assessment purposes, presented in Fig. 1, do not fully meet their needs. Due to the importance of logistic delivery parameters, the attention of risk managers is mainly focused on disruptions affecting the correctness of the logistics process.

A risk assessment was carried out for the process of handling transport orders in a selected road transport company to confirm these results. The extent of adverse events was limited to the events listed in Fig. 3. A linguistic approach was used to analyze the risk, in which a 10-point rating scale was used. Table 3 presents the scope of frequency assessment of the occurrence of the analyzed events and their effects.

Table 3. Interpretation of the terms used for the linguistic variable *Frequency of occurrence.*

Point range	Frequency of occurrence		Results of occurrence	
	Time	Interpretation	Result/Effect	Interpretation
1–2	Occasionally	Less often than once every 2 years	Irrelevant	A small loss of client's trust, no financial losses
3–5	Rarely	Happens once a year/once every two years	Not very significant	Costs up to 10.000 PLN
6–8	Often	Happens a few times a year	Significant	Loss of a client, costs up to 100.000 PLN
9–10	Very often	Happens at least once a month	Very significant	Loss of a market share, costs threatening with liquidity crisis

Managers of the transport department and freight forwarding department (7 managers) acted as experts. The risk (R) was calculated as the product of the frequency (F) and the effects (E) of an undesirable event occurrence. Table 4 presents the results of analysis.

Table 4. Weight and priority assigned to adverse events.

i	F_i	E_i	R_i	p_i	i	F_i	E_i	R_i	p_i
S1	5	6	30	1	L1	8	3	24	1
S2	8	2	16	5	L2	5	7	35	1
S3	8	1	8	5	L3	8	7	56	1
S4	2	8	16	4	L4	5	2	10	2
S5	4	6	24	1	L5	7	3	21	2
S6	4	7	28	4	L6	4	5	20	1
S7	4	7	28	4	L7	9	2	18	3
S8	2	7	14	5	L8	3	6	18	3
S9	5	3	15	5	C1	5	5	25	2
E1	5	5	25	1	C2	5	4	20	3

The conducted risk assessment partially confirms the results of the research described in Sect. 3. Those events, that were given high priority, received a high risk indicator also, e.g. S1, L2 and L3. However, there are also events that were given a low priority, and in the assessment they received a high risk indicator, e.g. S6 and S7. This is mainly due to the large effects that accompany their occurrence. However, due to the low frequency, managers do not include them in their planning process, which can be considered as an error. The high effects of their occurrence are the basis to consider these events as critical and to be taken into account in the planning process.

5   Summary

The traditional approach to risk assessment in road transport refers to issues of a strategic nature. The purpose of risk management in this case is primarily to protect the lives and health of road users. The objectives are defined in a different way for managers responsible for risk management in the process of handling transport loads in an enterprise. The transport is to be not only secure, but also meet the logistical requirements set by the client. For this reason, the attention of decision-makers is focused not only on information concerning traffic and infrastructure parameters or driving conditions. The scope of the required data also applies to the rate of indicators relating to the logistic parameters achieved and the company's resources used. The presented results confirm the necessity of building a database that takes into account not only the safety aspects, but also the achieved logistic results. Only such data will enable a comprehensive risk assessment to be carried out for the needs of managers, who plan a transport service process performance.

The results presented in the article are part of research on the operational risk assessment method in handling transport orders. Further research in this area will be conducted towards the development of a decision support system for managers managing transport processes in enterprises.

References

1. Aven, T.: Risk assessment and risk management: review of recent advances on their foundation. Eur. J. Oper. Res. **253**, 1–13 (2016)
2. Budzyński, M., Ryś, D., Kustra, R.: Selected problems of transport in port towns – tri-city as an example. Pol. Marit. Res. Special Issue S1 **24**(93), 16–24 (2017)
3. Friedrich, J., Restel, F.J., Wolniewicz, Ł.: Railway operation schedule evaluation with respect to the system robustness, contemporary complex systems and their dependability. In: Proceedings of the Thirteenth International Conference on Dependability and Complex Systems DepCoS-RELCOMEX, vol. 761, pp. 195–208. Springer (2019)
4. Haight, F.A.: Risk, especially risk of traffic accident. Accid. Anal. Prev. **18**(5), 359–366 (1986)
5. Hauer, E.: Traffic conflicts and exposure. Accid. Anal. Prev. **14**(5), 359–364 (1982)
6. Jamroz, K., Kadziński, A., Chruzik, K., Szymanek, A., Gucma, L., Skorupski, J.: TRANS-RISK – an integrated method for risk management in transport. J. KONBiN **1**(13), 209–220 (2010)
7. Jamroz, K.: Metoda zarządzania ryzykiem w inżynierii drogowej. Wydawnictwo Politechniki Gdańskiej, Gdańsk (2011)
8. Kierzkowski, A., Kisiel, T.: A model of check-in system management to reduce the security checkpoint variability. Simul. Model. Pract. Theory **74**, 80–98 (2017)
9. Kierzkowski, A., Kisiel, T.: Simulation model of security control system functioning: a case study of the Wroclaw airport terminal. J. Air Transp. Manag. **64**(Part B), 173–185 (2017)
10. Kustra, W., Jamroz, K., Budzyński, M.: Safety PL – a support tool for road safety impact assessment. Transp. Res. Procedia **14**, 3456–3465 (2016)
11. Montewka, J., Goerlandt, F.: Maritime transportation risk analysis: review and analysis in light of some foundational issues. Reliab. Eng. Syst. Saf. **138**, 115–134 (2015)
12. Restel, F.J., Zając, M.: Reliability model of the railway transportation system with respect to hazard state. In: International Conference on Industrial Engineering and Engineering Management (IEEM), pp. 1031–1036. IEEE (2015)
13. Restel, F.J., Wolniewicz, Ł.: Tramway reliability and safety influencing factors. Procedia Eng. **187**, 477–482 (2017)
14. Skorupski, J.: The risk of an air accident as a result of a serious incident of the hybrid type. Reliab. Eng. Syst. Saf. **140**, 37–52 (2015)
15. Tubis, A.A.: Metoda zarządzania ryzykiem operacyjnym w transporcie drogowym. Oficyna Wydawnicza Politechniki Wrocławskiej (2018)
16. Tubis, A.A.: Risk assessment in road transport – strategic and business approach. J. KONBiN **45**, 305–324 (2018)

Low-Dimensional Classification
of Text Documents

Tomasz Walkowiak[✉], Szymon Datko, and Henryk Maciejewski

Faculty of Electronics, Wrocław University of Science and Technology, Wrocław,
Poland
{tomasz.walkowiak,szymon.datko,henryk.maciejewski}@pwr.edu.pl

Abstract. In this paper we focus on overcoming a common belief
that accurate subject classification of text documents must involve high
dimensional feature vectors. We study the fastText algorithm in terms
of its ability to find and extract well distinguishable characteristics for
a text corpora. In research we compare the achieved accuracy in the
task of subject classification with various size of feature space selected.
Finally, we attempt to discover the foundation behind fastText's well
performance.

Keywords: Text mining · Feature extraction · Subject classification ·
Word embedding · fastText

1 Introduction

This work is devoted to the problem of classification of text documents in terms
of subject categories. We focus on the specific problem related to dimensionality of feature vectors required to ensure high accuracy (precision, recall) of
classification. Commonly used approaches to text classification [1,10,14] involve
representing documents by feature vectors based on the bag-of-words technique.
Components of these vectors represent (weighted) frequencies of occurrences
of words/terms in individual documents, where the number of such terms (i.e.
dimensionality of feature vectors) deemed relevant for distinguishing between
document classes ranges in the order 10^3–10^4.

Mikolov et al. [6] proposed an alternative approach based on word-embedding
techniques, where individual words are represented by high-dimensional feature
vectors trained on large text corpora, where relationships (distances) between
vector representations of words should be related to semantic similarities of words
as inferred from large corpora by observing co-occurrence of words in similar
contexts. Word embeddings can be then used to generate feature vectors for
document classification (e.g. by combining vector representations of individual
words occurring in a document - approach known as doc-to-vec). In [5] authors
proposed an efficient algorithm called *fastText* for learning word embeddings
(only on analyzed corpus) and building a classifier in the same time. Recent
studies [11,12] show that this approach seems to yield higher accuracy in the

© Springer Nature Switzerland AG 2020
W. Zamojski et al. (Eds.): DepCoS-RELCOMEX 2019, AISC 987, pp. 534–543, 2020.
https://doi.org/10.1007/978-3-030-19501-4_53

text classification as compared to the bag-of-words methods. It should be noticed that this method is totally NLP-technique-free. However, it is commonly taken for granted that successful classification should involve high-dimensional doc-to-vec feature vectors.

In this work we investigated this problem. We checked empirically, based on a Wikipedia corpus of Polish language text documents, what dimensionality of doc-to-vec feature vectors is actually required for subject classification. We observed that surprisingly low dimensional feature vectors are sufficient enough. We also attempted to analyze this phenomenon by visualizing separability of documents in very low dimensional spaces spanned by the first few doc-to-vec components.

2   Method and Data Set

2.1   fastText

The fastText algorithm is a method for text classification, recently gaining recognition for its high performance and accuracy. It was invented by Mikolov et al. [5] and utilizes the deep learning techniques to represent text with so called word embeddings. The algorithm builds a map (look-up table) with unique vectors of real numbers for each word from a training dataset (corpora) [7]. Then each text document is represented as a sum of vectors from such map, divided by a number of summed elements (so called word embeddings) [6]. Words without corresponding entry within built mapping are ignored. Due to such implementation, the order of the words in documents is also insignificant, however it is possible to consider it in the processing with modified approaches using n-grams of words. Nevertheless, a result is a highly effective, linear model.

In various research the fastText algorithm outperforms other approaches for feature extraction from text documents, including popular Bag-of-Words method [3] or Latent Dirichlet Allocation [11,12]. For more, the study [13] suggests that the size of feature vectors has minor impact on algorithm's performance. The latter phenomena was the main motivator for this paper.

The fastText algorithm originally also involves the built-in soft-max classifier [2]. However, as some studies suggest, the slightly better accuracy can be achieved when involving different classification methods, e.g. neural networks [12].

2.2   Data Set

In our study we used a corpus that consisted of almost 10000 articles originating from Polish language Wikipedia. The whole corpus was originally divided by the researchers into training set [9] (containing 6885 documents) and testing set [8] (including 2952 articles).

Both training and testing sets have documents divided into 34 subject categories (names are given in the Fig. 1). Each category contains approximately the same number of articles, so the results of experiments should not be biased into more numerous collections.

We assume that specific subject classes contain only documents that are legitimately labelled by humans and thus we expect no outliers/spurious assignments within defined categories.

2.3 Experiment Overview

The experiment was conducted in the following way. First, the fastText algorithm was involved to extract feature vectors of fixed size out of the training and testing documents sets. Then we constructed a classification model utilizing the Multi-Layer Perceptron [4] and back error propagation method in the learning process [4] on the training set. Finally, we evaluated the quality of extracted feature vectors as the achieved accuracy in the task of subject classification - the built model was used to process the vectors from testing set.

The whole study described above was conducted on various sizes of features vector. Starting from 2, up to 500, to be exact. At the end we attempted to visualize also the distribution of extracted feature vectors to discover why observed accuracy was so excellent even at low dimensions.

3 Results

In this section we discuss the results of our study. The Fig. 1 contains a legend box for most of the plots within this document. Figures 3, 4 and 5 presents the obtained accuracy measures as function of features vectors dimension. Figures 6, 7, 8 and 9 contain the visualization of low-dimensional features vectors from converted text corpora.

Fig. 1. The legend box for Figs. 3, 4, 5, 6, 7, 8 and 9.

Figure 2 presents the reached accuracy, averaged across all subject categories. What can be observed is that overall precision and recall are very similar and therefore the F1-score is as well. However, what is more important to notice is that even for the lowest analyzed dimension - 2 - we obtained the classification accuracy about 38%. Then with increasing size of feature vectors the reached accuracy increases. The highest gain is achieved for dimension 3, reaching around 65%, while for dimension 5 it is circa 78%. Finally, for dimension 10 the achieved accuracy equals approximately 82%. Further increase of features vectors dimension does not result in any significant gain of accuracy. It appears therefore as dimension sufficient enough to perform subject classification task on described dataset/corpora.

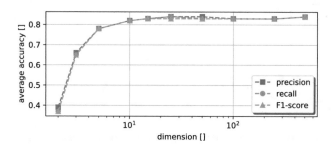

Fig. 2. Average obtained accuracy in function of feature-space dimension.

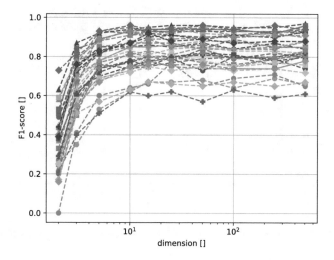

Fig. 3. Achieved F1-score (per-class), depending on size of feature vectors.

Figures 3, 4 and 5 present the detailed plots of achieved $F1-score, precision$ and $recall$, drown per each subject category (class). It can be noticed that even for the lowest researched dimension there are many categories with accuracy above 50%. The lowest accuracy in general was achieved for categories *Arabs*, *Political propaganda*, *Jews* and *Branches of law* (with initial performance for the first mentioned category around 0%).

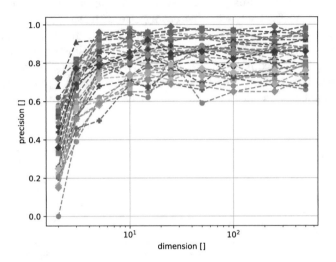

Fig. 4. Obtained performance (per-class), as a function of features space dimension.

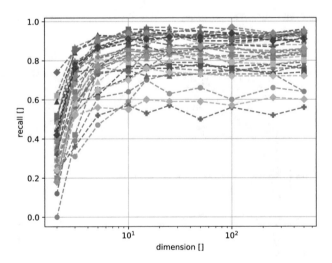

Fig. 5. Reached recall (per-class), depending on the feature vectors size.

On the other hand, there are categories like *Karnonosze mountains*, *Coints*, *Animated films* and *Computer games* that reach peaks even above 90%. Taking into account that the number of all subject classes is equal to 34, this result is at very satisfactory, high level.

Next figures present the visualization of features spaces of various dimensions. Each marker (data point) represents a single document from a corpora (dataset).

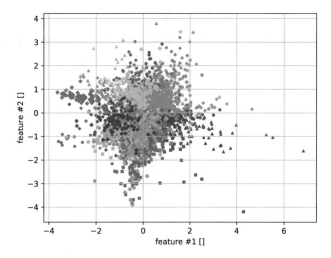

Fig. 6. Visualization of 2-dimensional features space, all categories.

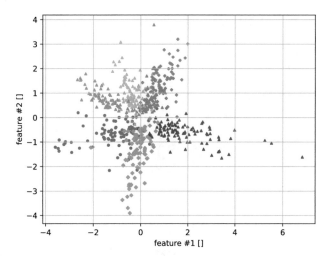

Fig. 7. Visualization of 2-dimensional features space, only selected classes.

Figures 6 and 7 show the 2-dimensional space; while the Fig. 6 appears crowded, it gives a good overall picture of how the data from text corpora are distributed in feature space using the fastText algorithm. All points are placed around the center of the coordinate system with tendency to form narrow, but long on specific axis, clusters.

To display that behaviour clearly, on the Fig. 7 there are only the markers for documents of selected categories being presented. The categories are *Drug addiction* (green diamonds), *Astronautics* (cyan triangles), *Airplanes* (orange triangles), *Albania* (blue circle), *Coins* (red triangles), *Cats* (grey triangles) and *Computer games* (grey crosses). The latter two mentioned clusters tend to overlap in this picture, so their classification accuracy is limited. However, all remaining categories form well distinguishable (separable) clusters.

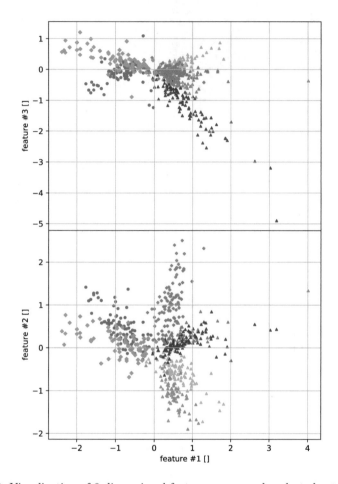

Fig. 8. Visualization of 3-dimensional features space, only selected categories.

Figure 8 presents the same analysis as on Fig. 7, with the same subject categories, but with features generated in 3-dimensional space. Upper part of a plot on the Fig. 8 displays the data in the projection between the 1st and 3rd feature, while the lower subplot displays the relation between the 1st and 2nd feature. This figure clearly explains how the fastText algorithm projects documents of given subject categories into features space.

As can be noticed, some categories (like *Drug addiction* [green diamonds] or *Coins* [red triangles]) are well distinguishable in first subplot (projection between 1st and 3rd feature) and not on the second subplot (lower one). On the other

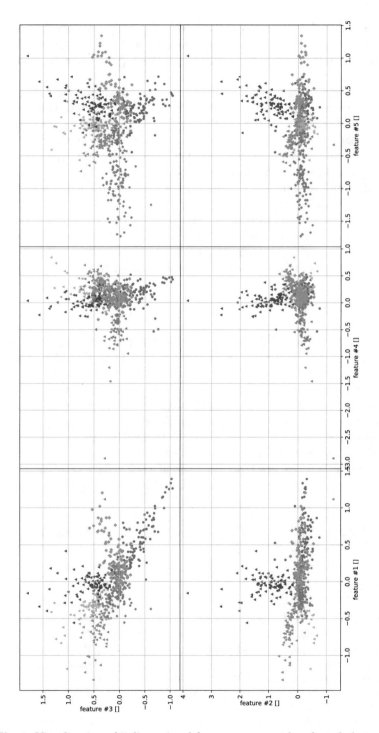

Fig. 9. Visualization of 5-dimensional features space, only selected classes.

hand, there are also categories (e.g. *Airplanes* [orange triangles] and *Computer games* [grey crosses]) that cannot be well distinguished on first subplot (upper one), while are well separable on second subplot (projection between 1st and 2nd feature).

Plots on Fig. 9 presents this phenomena even explicitly. On each projection there is at least single subject category that is well distinguishable in that subspace - for example the category *Albania* (blue circle) on subspace between 1st and 3rd feature (first subplot), the category *Computer games* (grey crosses) between 3rd and 5th feature (third subplot) and the category *Airplanes* (orange triangles) between 1st and 2nd feature (fourth subplot). The same tendency was observed in higher-dimensional spaces, however it was difficult to insert figures with clear/valuable plots of these spaces.

Last important thing to notice is related to the characteristics of data points placement in the features space - focused centrally, with narrow projections along specific axes. This should make data from fastText algorithm behaving well with angle-based techniques, such as cosine similarity measure.

4   Conclusion

In this work we evaluated the problem of dimensionality of features vectors in the task of subject classification of text documents. Utilizing the corpora containing Polish language Wikipedia articles, divided into 34 categories, we extracted the features vectors of various dimensions using the fastText algorithm. Then, we constructed a neural network model (Multi-Layer Perceptron) and examined it the task of subject classification. Comparing the achieved accuracy we prove that the relatively low sizes of features space (around 10) can be sufficient enough to obtain satisfying results. Finally, we visualized the feature space to discover the nature behind the great performance of vectors generated using fastText algorithm - its ability to produce categories as narrow, focused projections along specific axes.

Acknowledgments. This work was sponsored by National Science Centre, Poland (grant 2016/21/B/ST6/02159).

References

1. Aghila, G., et al.: A survey of naive bayes machine learning approach in text document classification. arXiv preprint arXiv:1003.1795 (2010)
2. Goodman, J.: Classes for fast maximum entropy training. In: 2001 IEEE International Conference on Acoustics, Speech, and Signal Processing, Proceedings (Cat. No. 01CH37221), vol. 1, pp. 561–564 (2001)
3. Harris, Z.: Distributional structure. Word **10**(2–3), 146–162 (1954)
4. Hastie, T.J., Tibshirani, R.J., Friedman, J.H.: The Elements of Statistical Learning: Data Mining, Inference, and Prediction. Springer Series in Statistics, 2nd edn. Springer, New York (2009). autres impressions: 2011 (corr.), 2013 (7e corr.)

5. Joulin, A., Grave, E., Bojanowski, P., Mikolov, T.: Bag of tricks for efficient text classification. In: Proceedings of the 15th Conference of the European Chapter of the Association for Computational Linguistics: Short Papers, vol. 2, pp. 427–431. Association for Computational Linguistics (2017). http://aclweb.org/anthology/E17-2068

6. Le, Q., Mikolov, T.: Distributed representations of sentences and documents. In: International Conference on Machine Learning, pp. 1188–1196 (2014)

7. Mikolov, T., Sutskever, I., Chen, K., Corrado, G.S., Dean, J.: Distributed representations of words and phrases and their compositionality. In: Advances in Neural Information Processing Systems, pp. 3111–3119 (2013)

8. Młynarczyk, K., Piasecki, M.: Wiki test - 34 categories, CLARIN-PL digital repository (2015). http://hdl.handle.net/11321/217

9. Młynarczyk, K., Piasecki, M.: Wiki train - 34 categories, CLARIN-PL digital repository (2015). http://hdl.handle.net/11321/222

10. Torkkola, K.: Discriminative features for textdocument classification. Form. Pattern Anal. Appl. $6(4)$, 301–308 (2004). https://doi.org/10.1007/s1004400301968

11. Walkowiak, T., Datko, S., Maciejewski, H.: Feature extraction in subject classification of text documents in Polish. In: Artificial Intelligence and Soft Computing - 17th International Conference, ICAISC 2018, Zakopane, Poland, 3–7 June 2018, Proceedings, Part II, pp. 445–452 (2018). https://doi.org/10.1007/978-3-319-91262-2_40

12. Walkowiak, T., Datko, S., Maciejewski, H.: Bag-of-words, bag-of-topics and word-to-vec based subject classification of text documents in polish - a comparative study. In: Zamojski, W., Mazurkiewicz, J., Sugier, J., Walkowiak, T., Kacprzyk, J. (eds.) Contemporary Complex Systems and Their Dependability, pp. 526–535. Springer, Cham (2019)

13. Walkowiak, T., Datko, S., Maciejewski, H.: Reduction of dimensionality of feature vectors in subject classification of text documents. In: Kabashkin, I., Yatskiv(Jackiva), I., Prentkovskis, O. (eds.) Reliability and Statistics in Transportation and Communication, pp. 159–167. Springer, Cham (2019)

14. Wang, L., Zhao, X.: Improved KNN classification algorithms research in text categorization. In: 2012 2nd International Conference on Consumer Electronics, Communications and Networks (CECNet), pp. 1848–1852. IEEE (2012)

Distance Measures for Clustering of Documents in a Topic Space

Tomasz Walkowiak[1(✉)] and Mateusz Gniewkowski[2]

[1] Faculty of Electronics, Wrocław University of Science and Technology,
Wrocław, Poland
tomasz.walkowiak@pwr.edu.pl
[2] Faculty of Computer Science and Management,
Wrocław University of Science and Technology, Wrocław, Poland
mateusz.gniewkowski@pwr.edu.pl

Abstract. Topic modeling is a method for discovery of topics (groups of words). In this paper we focus on clustering documents in a topic space obtained using the *MALLET* tool. We tested several different distance measures with two clustering algorithm (*spectral clustering, agglomerative hierarchical clustering*) and described those that served better (*cosine distance, correlation distance, bhattacharyya distance*) than the Euclidean metric for k-means algorithm. For evaluation purpose we used *Adjusted Mutual Information* (*AMI*) score. The need for such experiments comes from the difficulty of choosing appropriate grouping methods for the given data, which is specific in our case.

Keywords: Topic modeling · Distance · Cosine distance ·
Correlation distance · Clustering · Grouping · Clustering evaluation ·
Text analysis

1 Introduction

Topic modeling [5,22] is a method that allows us to find sets of words that best represents a given collection of documents. Those sets are called "topics". Thanks to that, thousands of files can be organized in groups and therefore easier to understand or classify. There are numerous methods and tools for topic modeling [3], but they will not be discussed here.

The main problem we focus on in this paper is how to accurately group documents with already assigned topics (those might be considered as fuzzy sets), and since measuring the distance between two samples is important part of clustering we also analyze different distance functions between two samples from our data. The need for document clusterization follows from a frequent occurrence of many interpretable topics (sets of words that describe a document) in a dataset. By increasing the given number of topics, we can find more and more subtopics (that do not necessarily remind the previous ones) and still give each of them a meaning. Grouping of documents seen in the topic space (or topics in the

© Springer Nature Switzerland AG 2020
W. Zamojski et al. (Eds.): DepCoS-RELCOMEX 2019, AISC 987, pp. 544–552, 2020.
https://doi.org/10.1007/978-3-030-19501-4_54

word space) gives a researcher, who deals with language or literature, a better view for the results of topic modelling. It also might automatically find some supersets for given topics (like a "sport" for texts about football and volleyball).

The purpose of clustering is to group together N samples described by M features (number of topics in our case), without any knowledge about the problem (labels). There are several issues with clustering like an unknown number of groups, defining a distance function or input data itself [1].

The input to the clustering algorithms is the list of N^2 pairwise dissimilarities between N points representing samples (in our case they are texts). As a result, we get an association of documents to X groups. The number of groups is often an input parameter, but some clustering algorithms are able to predict it by itself [10].

There are plenty of clustering algorithms [7,8,12] with different properties and input parameters. In our work we decided to focus on the *agglomerative hierarchical clustering* [6] because it is a common method used in document grouping that allows building hierarchy and therefore draw a dendrogram, which helps interpreting the results. For comparative reasons we also used the *spectral clustering algorithm* [17] and related to it *k-means* [12].

In this paper, we first discuss our input data format for clustering algorithms (samples and features). In Sect. 3 we focus on different distance measures. After that, in Sect. 4, we describe data sets we used for our tests. Two last sections cover the results and conclusions.

2   Topic Modelling Data

As we want to measure the similarity between documents in topics dimensions, we consider a standard output of most topic modeling methods as our input - two matrices. First of them describing the probability distribution of documents over topics, the second one topics over words (Fig. 1).

$$
\begin{array}{c}
\begin{array}{cccc} t_1 & t_2 & \cdots & t_m \end{array} \\
\begin{array}{c} d_1 \\ d_2 \\ \vdots \\ d_n \end{array}
\begin{pmatrix}
a_{11} & a_{12} & \cdots & a_{1m} \\
a_{21} & a_{22} & \cdots & a_{2m} \\
\vdots & \vdots & \ddots & \vdots \\
a_{n1} & a_{n2} & \cdots & a_{nm}
\end{pmatrix}
\end{array}
\quad
\begin{array}{c}
\begin{array}{cccc} w_1 & w_2 & \cdots & w_k \end{array} \\
\begin{array}{c} t_1 \\ t_2 \\ \vdots \\ t_m \end{array}
\begin{pmatrix}
b_{11} & b_{12} & \cdots & b_{1k} \\
b_{21} & b_{22} & \cdots & b_{2k} \\
\vdots & \vdots & \ddots & \vdots \\
b_{m1} & b_{m2} & \cdots & b_{mk}
\end{pmatrix}
\end{array}
$$

Fig. 1. Input matrices: documents (d_i) over topics (t_i) (on the left) and topics (t_i) over words (w_i)

Those matrices have some interesting hallmarks. First of all, every row represents Dirichlet distribution, which also means that they sum up to one (therefore they are basically normalized with L_1, see the picture below). Also, both of them are rather sparse, especially if ignoring all values under some threshold (which is recommended). Those characteristics are important when choosing the right distance (or similarity) measure (Fig. 2).

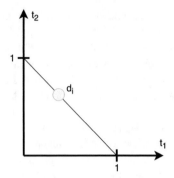

Fig. 2. 2D representation of document in topics dimensions

3 Distance Measures

One of the most crucial tasks in dimension reduction or clustering is properly selecting a "good" distance function. The goodness of the metric depends on a data set (dimension and type of numbers) and intentions of a researcher. At this point, we will discuss a few possible solutions that, according to our tests, works best (Table 1).

Table 1. Symbols description

X_n, Y_n	Two points in given dimension
x_t	t-th element of X
$\overline{X}, \overline{Y}$	Mean value of the given point
$D_{X,Y}$	Distance between x and y
D_{max}, D_{min}	Distance to the farthest/closest point from the origin
n	Dimension
$\|X\|_p$	L_p length of a vector
$\|X\|$	Euclidean length of a vector
$E[\|X\|]$	Expected value of the length of any vector
var	Variance
L_p	p-norm
θ	Angle between two vectors
$X \cdot Y$	Dot product between X and Y

3.1   Euclidean Distance and Other L_p Metrics

$$L_p = ||X||_p = (|x_1|^p + |x_2|^p + ... |x_n|^p)^{\frac{1}{p}}$$

$$D_{X,Y} = ||X - Y||$$

Plenty of works based on [4] have shown that all L_p norms fail in high dimensionality. This is because when

$$\lim_{n \to inf} var(\frac{||X_n||_p}{E[||X_n||_p]}) = 0 \quad then \quad \frac{D_{max} - D_{min}}{D_{min}} \to 0$$

The above equation shows that the difference between the farthest and closest point does not increase as fast as the minimum distance itself. Therefore all methods based on the nearest neighbour problem become incorrectly defined, especially if k is greater than three [11]. It is a useful theorem that can be used to check whether the metric is reasonable.

Moreover, in [13] authors point out, that L_2 norm is very noisy, especially with lots of irrelevant attributes, which is a case in our example due to the sparsity of input matrices. Note, that if most of the vector's elements are zeroes, then the difference may act as their sum.

What is more and also important, all L_p norms strongly rely on lengths of given vectors, which is not necessarily what we want to do. Imagine a situation where two documents (represented by vectors) are quite similar, but the given set of words (topic) is more frequent in one of them. Using L_p measurement we distinguish one sample from the other, but in linguistic meaning, they are similar. Those metrics are vulnerable to scaling.

3.2   Cosine Distance

$$D_{X,Y} = 1 - cos(\theta) = 1 - \frac{X \cdot Y}{||X||||Y||}$$

Cosine measure is widely and successfully used with document clustering problems because, by its definition, it is not vulnerable for any scaling of the given vectors and it acts better in high dimensional space. Despite its simplicity, it gives similar results to more distinguish methods (and more computationally complex) [9,18]. It is also worth to notice that the cosine similarity is much faster for sparse data because, if any position of one of the given vectors is zero, calculation are not needed.

On the other hand, there are plenty of disadvantages with the cosine distance. This is not a proper distance metric as it does not have the triangle inequality property. It is not high dimension resistant due to the fact that two randomly picked vectors on the surface of unit n-sphere have a high probability to be orthogonal but it still acts better than any L_p norm.

3.3   Correlation Distance

$$D_{X,Y} = 1 - \frac{(X - \overline{X}) \cdot (Y - \overline{Y})}{||X||||Y||}$$

Correlation Distance or *Pearson's distance* is another metric based on the cosine similarity. Its value lies between 0 and 2. Thanks to measuring the angle between mean vectors, the problem is reduced by one dimension. There might however be a problem with values lying close to the middle point (close to the point $A = (1/n, ..., 1/n)$) because, after normalization, the cosine similarity between them most likely will be equal to -1, but according to our input data (result of the topic modeling algorithm) it should not be a common issue.

3.4   Bhattacharyya Distance

$$D_{X,Y} = -ln[\sum_{t \in T} \sqrt{(x_t y_t)}]$$

Since our data is a Dirichlet distribution it seems reasonable to use some kind of a similarity measure focused on probability objects. *Bhattacharyya distance* [2] is a metric clearly based on the cosine similarity. Basically it is a cosine of two vectors built from roots of original elements. The log_e is a standard way to make it a distance measurement (as $1 - D_{cos}$).

$$X' = (\sqrt{x_1}, ..., \sqrt{x_n})$$
$$Y' = (\sqrt{y_1}, ..., \sqrt{y_n})$$
$$B_{coefficent} = \sum_{i=1}^{i=n} y_i x_i = X' \cdot Y' = cos(\theta), \quad \theta = \sphericalangle(X', Y')$$

4   Experiments

4.1   Clustering Algorithms

In order to group data, we decided to use the following algorithms:

1. *K-means* [12] algorithm is a classic method that assigns labels to the data, basing on a distance to the nearest centroid. Centroids are moved iteratively until all clusters stabilize.
2. *Agglomerative hierarchical clustering* [6] is a method that iteratively joins subgroups basing on a *linkage criterion*. In this paper, we present result for the average linkage clustering.
3. *Spectral clustering* [17] is based on the Laplacian matrix of the similarity graph and its eigenvectors. The least significant eigenvectors create new, lower dimensional space that is used with a *k-means* algorithm.

4.2   Clustering Evaluation

For evaluation purpose, we decided to use one of the information theoretic based measures - Adjusted Mutual Information (AMI). For a set of "true" labels (classes) and predicted ones (clusters) it yields an upper-bounded by 1 (the best score) value. Thanks to "correction for a chance" property the result does not increase with a number of clusters for randomly chosen labels, it should stay close to zero [19].

It is possible to compare sets with different numbers of clusters and classes. Thanks to that it should be possible to discover hidden groups in already labelled data [20].

4.3   Data Sets

We evaluated mentioned above algorithms using two collections of text documents: *Wiki* and *Press*.

The first corpus (*Wiki*) consists of articles extracted from the Polish language Wikipedia. This corpus has also good quality, however some of the class assignments may be doubtful. It is characterized by a significant number of 34 subject categories. This data set was created by merging two publicly available collections [15,16]. It includes 9, 837 articles in total, which translates into about 300 articles per class (however one class is slightly underrepresented).

The second corpus (*Press*) consists of Polish press news [21]. It is a good example of a complete, high quality and well defined data set. The texts were assigned by press agency to 5 subject categories. All the subject groups are very well separable from each other and each group contains reasonably large number of members. There are 6564 documents in total in this corpus, which gives an average of ca. 1300 documents per class.

5   Results

We performed our experiments as follows: given the documents $d_i \in D$ over topics $t_i \in T$ (the number of topics was set to 100) matrices as the result of the *MALLET* [14] for two data sets described in the previous section, we performed several tests that were evaluating the quality of clustering documents using different distance/similarity measures (like *cosine distance, correlation distance, hellinger distance, total variation distance, jensen–shannon divergence*) with different clustering algorithms (*agglomerative clustering, spectral clustering*). We have made the evaluation using *AMI* score and known classes $c_i \in C$ for every document d_i. The number of clusters to find was the same as the number of original classes. Below, we present the results for the six best combinations that were better than the *k-means* algorithm.

The results for various combinations are given in Fig. 3. We compared classic k-means algorithm with agglomerative clustering and spectral clustering algorithms. For *Wiki* dataset it can be observed that agglomerative clustering acts

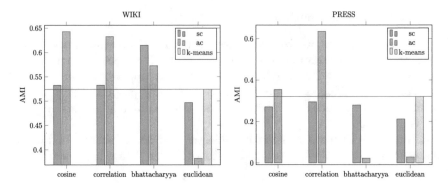

Fig. 3. AMI score for different clustering methods and distance measures (ac - agglomerative clustering, sc - spectral clustering)

better than the k-means. Notice, that the results for the cosine and correlation distances with agglomerative clustering are quite similar. It is something that we expected to see. The situation is different for the *Press* dataset. One is significantly better than the other and also any of the given solutions. As surprising it can be, it seems that using cosine measures with mean vectors is more resistant to some unique hallmarks of data distribution. In general, correlation measure with agglomerative clustering method seems like the best solution for grouping text in a topic space.

By increasing the number of clusters in the clustering algorithm and comparing the results with original labels we expected to see a global peak somewhere after the presumed number of groups. That would mean that we can suggest the researcher another input parameter (a proper amount of groups). It might help

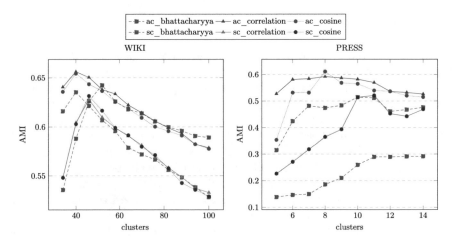

Fig. 4. AMI score for different number of clusters

discover new dependencies in a given, labelled corpus. Figure 4 shows that our expectations were correct - for the best combination of measure and clustering method (which is the agglomerative clustering with the correlation distance), the peak occurred at 8 clusters for the *Press* data set and about 40 for *Wiki* data set. The very first point of each line represents the *AMI* score for "true" number of labels.

6   Conclusion

In this paper, we discussed several options for choosing a distance/similarity measures in the context of the given clustering method. We have shown how significantly it affects the obtained results in a complex and unintuitive task which is clustering. Much of it depends on the input, and that is why it is important to classify different kinds of it and then try to find a suitable solution. After all, there is no algorithm that is appropriate for every type of data. In our work, we focused on the output of topic modelling tools (MALLET) which is a set of Dirichlet distributions for every processed document. It is worth to notice that in our dataset there should not be any vectors near to the middle point $(\frac{1}{n}, ..., \frac{1}{n})$ - this feature is probably the main reason why correlation distance works so well (every two vectors lying near to the middle point might be considered completely different although they are near to each other). This is not a problem in our case because it would mean that the algorithm did not generate topics properly. What is more, the correlation distance, is bounded from the above by two and from below by zero. It means that the interval in this measure is bigger than in cosine. In addition, we have shown that by evaluating the results of clustering for a different number of clusters it is possible to find a better fitting solution that might (or might not) be useful for the researcher.

Acknowledgements. The work was funded by the Polish Ministry of Science and Higher Education within CLARIN-PL Research Infrastructure.

References

1. Agarwal, P., Alam, M.A., Biswas, R.: Issues, challenges and tools of clustering algorithms. arXiv preprint arXiv:1110.2610 (2011)
2. Aherne, F.J., Thacker, N.A., Rockett, P.I.: The Bhattacharyya metric as an absolute similarity measure for frequency coded data. Kybernetika **34**(4), 363–368 (1998)
3. Barde, B.V., Bainwad, A.M.: An overview of topic modeling methods and tools. In: International Conference on Intelligent Computing and Control Systems (ICICCS), pp. 745–750 (2017)
4. Beyer, K., Goldstein, J., Ramakrishnan, R., Shaft, U.: When is "nearest neighbor" meaningful? In: ICDT 1999 Proceedings of the 7th International Conference on Database Theory, pp. 217–235 (1999)
5. Blei, D.M.: Probabilistic topic models. Commun. ACM **55**(4), 77–84 (2012)

6. Day, W.H.E., Edelsbrunner, H.: Efficient algorithms for agglomerative hierarchical clustering methods. J. Classif. **1**(1), 7–24 (1984)
7. Dueck, D., Frey, B.J.: Non-metric affinity propagation for unsupervised image categorization. In: 2007 IEEE 11th International Conference on Computer Vision, ICCV 2007, pp. 1–8. IEEE (2007)
8. Fung, G.: A comprehensive overview of basic clustering algorithms (2001)
9. Huang, A.: Similarity measures for text document clustering. In: Proceedings of the 6th New Zealand Computer Science Research Student Conference (2008)
10. K. Khan, S. U. Rehman, K.A.S.F., Sarasvady, S.: DBSCAN: past, present and future. In: The Fifth International Conference on the Applications of Digital Information and Web Technologies (ICADIWT 2014), pp. 232–238 (2014)
11. Keim D. A., Hinneburg A., A.C.C.: On the surprising behavior of distance metrics in high dimensional space. In: ICDT 2001 Proceedings of the 8th International Conference on Database Theory, pp. 420–434 (2001)
12. Jain, A.K.: Data clustering: 50 years beyond k-means. Pattern Recognit. Lett. **31**(8), 651–666 (2009)
13. Kriegel, H.-P., Schubert, E., Zimek, A.: A survey on unsupervised outlier detection. Stat. Anal. Data Min. **5**, 363–387 (2012)
14. Milligan Ian, Weingart Scott, G.S.: Getting started with topic modeling and mallet. Technical report (2012). http://hdl.handle.net/10012/11751
15. Młynarczyk, K., Piasecki, M.: Wiki test - 34 categories (2015). http://hdl.handle.net/11321/217. CLARIN-PL digital repository
16. Młynarczyk, K., Piasecki, M.: Wiki train - 34 categories (2015). http://hdl.handle.net/11321/222. CLARIN-PL digital repository
17. Ng, A.Y., Jordan, M.I., Weiss, Y.: On spectral clustering: analysis and an algorithm. In: Advances in neural information processing systems, pp. 849–856 (2002)
18. Subhashini, R., Kumar, V.J.: Evaluating the performance of similarity measures used in document clustering and information retrieval. In: Integrated Intelligent Computing, pp. 27–31 (2010)
19. Vinh, N.X., Epps, J., Bailey, J.: Information theoretic measures for clusterings comparison: is a correction for chance necessary? In: Proceedings of the 26th Annual International Conference on Machine Learning, ICML 2009, pp. 1073–1080. ACM, New York, NY, USA (2009). http://doi.acm.org/10.1145/1553374.1553511
20. Vinh, N.X., Epps, J., Bailey, J.: Information theoretic measures for clusterings comparison: variants, properties, normalization and correction for chance. J. Mach. Learn. Res. **11**(Oct), 2837–2854 (2010)
21. Walkowiak, T., Malak, P.: Polish texts topic classification evaluation. In: Proceedings of the 10th International Conference on Agents and Artificial Intelligence - Volume 2: ICAART, pp. 515–522. INSTICC, SciTePress (2018)
22. Wallach, H.M.: Topic modeling: Beyond bag-of-words. In: Proceedings of the 23rd International Conference on Machine Learning, ICML 2006, pp. 977–984. ACM, New York, NY, USA (2006). http://doi.acm.org/10.1145/1143844.1143967

Author Index

© Springer Nature Switzerland AG 2020
W. Zamojski et al. (Eds.): DepCoS-RELCOMEX 2019, AISC 987, pp. 553–554, 2020.
https://doi.org/10.1007/978-3-030-19501-4

Printed in the United States
By Bookmasters